MW00843594

Derivation and Computation

Cambridge Tracts in Theoretical Computer Science

Titles in the series

1. G. Chaitin *Algorithmic Information Theory*
2. L. C. Paulson *Logic and Computation*
3. M. Spivey *Understanding Z*
5. A. Ramsey *Formal Methods in Artificial Intelligence*
6. S. Vickers *Topology via Logic*
7. J.-Y. Girard, Y. Lafont & P. Taylor *Proofs and Types*
8. J. Clifford *Formal Semantics & Progmatics for Natural Language Processing*
9. M. Winslett *Updating Logical Databases*
10. K. McEvoy & J. V. Tucker (eds) *Theoretical Foundations of VLSI Design*
11. T. H. Tse *A Unifying Framework for Structured Analysis and Design Models*
12. G. Brewka *Nonmonotonic Reasoning*
14. S. G. Hoggar *Mathematics for Computer Graphics*
15. S. Dasgupta *Design Theory and Computer Science*
17. J. C. M. Baeten (ed) *Applications of Process Algebra*
18. J. C. M. Baeten & W. P. Weijland *Process Algebra*
19. M. Manzano *Extensions of First Order Logic*
21. D. A. Wolfram *The Clausal Theory of Types*
22. V. Stoltenberg-Hansen, I. Lindström & E. Griffor *Mathematical Theory of Domains*
23. E.-R. Olderog *Nets, Terms and Formulas*
26. P. D. Mosses *Action Semantics*
27. W. H. Hesselink *Programs, Recursion and Unbounded Choice*
28. P. Padawitz *Deductive and Declarative Programming*
29. P. Gärdenfors (ed) *Belief Revision*
30. M. Anthony & N. Biggs *Computational Learning Theory*
31. T. F. Melham *Higher Order Logic and Hardware Verification*
32. R. L. Carpenter *The Logic of Typed Feature Structures*
33. E. G. Manes *Predicate Transformer Semantics*
34. F. Nielson & H. R. Nielson *Two Level Functional Languages*
35. L. Feijs & H. Jonkers *Formal Specification and Design*
36. S. Mauw & G. J. Veltink (eds) *Algebraic Specification of Communication Protocols*
37. V. Stavridou *Formal Methods in Circuit Design*
38. N. Shankar *Metamathematics, Machines and Gödel's Proof*
39. J. B. Paris *The Uncertain Reasoner's Companion*
40. J. Dessel & J. Esparza *Free Choice Petri Nets*
41. J.-J. Ch. Meyer & W. van der Hoek *Epistemic Logic for AI and Computer Science*
42. J. R. Hindley *Basic Simple Type Theory*
43. A. Troelstra & H. Schwichtenberg *Basic Proof Theory*
44. J. Barwise & J. Seligman *Information Flow*
45. A. Asperti & S. Guerrini *The Optimal Implementation of Functional Programming Languages*
46. R. M. Amadio & P.-L. Curien *Domains and Lambda-Calculi*
47. W.-P. de Roever & K. Engelhardt *Data Refinement*
48. H. Kleine Büning & T. Lettman *Propositional Logic*
49. L. Novak & A. Gibbons *Hybrid Graph Theory and Network Analysis*

Derivation and Computation

Taking the Curry–Howard correspondence seriously

Harold Simmons
University of Manchester

PUBLISHED BY THE PRESS SYNDICATE OF THE UNIVERSITY OF CAMBRIDGE
The Pitt Building, Trumpington Street, Cambridge, United Kingdom

CAMBRIDGE UNIVERSITY PRESS
The Edinburgh Building, Cambridge CB2 2RU, UK http://www.cup.cam.ac.uk
40 West 20th Street, New York, NY 10011–4211, USA http://www.cup.org
10 Stamford Road, Oakleigh, Melbourne 3166, Australia
Ruiz de Alarcón 13, 28014 Madrid, Spain

First published 2000

Printed in the United Kingdom at the University Press, Cambridge

Typeset by the author in Computer Modern 10/13pt, in LaTeX 2_ε [EPC]

A catalogue record of this book is available from the British Library

Library of Congress Cataloguing in Publication data
Simmons, Harold.
Derivation and computation: taking the Curry–Howard correspondence seriously:
derivation systems, substitution algorithms, computation mechanisms / Harold Simmons.
p. cm.
Includes bibliographical references and index.
ISBN 0 521 77173 0 (hb)
1. Proof theory. 2. Lambda calculus. 3. Type theory. I. Title.
QA9.54 S55 2000
511.3 21–dc21 99-044953

ISBN 0 521 77173 0 hardback

Contents

II Solutions 213

Introduction

What is this book about? The broad area is often called Logic and Computation (within the subject of Proof Theory, within the discipline of Mathematical Logic) which is a part of Mathematics. The particular subject matter is the simply typed λ-calculus and related systems, with an application to Gödel's T. This answer gets you to the right area but doesn't say very much else. It is only a vague description of the book, and doesn't explain why this book is necessary.

Wc (I, the writer, and you, the reader) are concerned here with the two core notions

Proof   Calculation

of mathematical activity. Even though the particular expertise of a mathematician is to recognize what these are, in most mathematical texts these notions are treated informally in that there is no attempt to give a definition of them. Logic and Computation is that area of mathematics which looks at these notions in general (rather than just particular examples).

A proof is an account written in a natural language (augmented by certain technical facilities) which is used to communicate the correctness of a result from one mathematician to another. Such proofs leave out many details, and the level of detail included depends on both the transmitting and the receiving mathematician. A proof of a result given by one expert to another can look quite different from a proof of the same result given by the same expert to a bunch of undergraduates. Both the processes of constructing and reading a proof require more than a modicum of mathematical expertise.

A calculation is a transformation of input data (the problem) into output data (the answer) which is carried out according to predetermined rules. These can range from simple arithmetical calculations (for instance, one which transforms two numbers a and b given in decimal notation into the notation for the product $a \times b$) to the process which transforms the current temperature, wind speed, air pressure, ... , at a collection of locations into a forecast of tomorrow's weather.

Calculations, at least in the simpler forms, have been around much longer than proofs (or even correct arguments). The pyramid builders could certainly calculate, but it is not at all clear that they could argue correctly (rather than merely make a point by force). From Euclid onwards proofs have been an essential part, some would say the distinguishing feature, of mathematics. Nevertheless, most areas of mathematics originated in the study of certain kinds of calculations.

The two activities are related, of course. Many proofs include calculations and, indeed, a calculation can be viewed as a rudimentary proof. Each calculation is based on an algorithm and there ought to be a proof that this algorithm is correct. To analyse this interaction we use formal analogues

Derivation   Computation

of the two informal notions. Thus the book is about

Proof theory		Computation theory
regarded as the classification of the modes of correct argument	and	regarded as the classification of the discrete spaces constructed from $\mathbb{N}$

and the intimate relationship between these two topics.

The book is a journey starting from the familiar (constructive) propositional calculus, passing through the combinator and λ-calculi (both untyped and typed), and finishing with a wander round Gödel's T. Thus several topics are discussed. One of the aims is to show how these are related. One topic cannot be understood in isolation; each impinges on the other.

The book is not intended as a comprehensive account; many things are left out. On the other hand the book does not take the most direct route, and along the way several interesting side tracks that should be explored are indicated. A detailed survey of the contents is given in the Preview. Here I will suggest how you should use the book as a whole.

The material is developed in nine chapters each of which is split into several sections. The material can be read linearly, starting at page 3 and traipsing through to page 211. However, many blocks can be read out of order. Of course, some of these will refer back to earlier material, but most of this you can take on trust. Once you have some familiarity with the overall layout you can jump about the book as you want. I suggest this is how you should use the book. For instance, if you want to learn something about recursion and induction (of a kind that is not easily found in other books) you can start with Section 7.3.

Mathematics is not a spectator sport. If you want to understand it then you must do it, you can't simply watch others do it. You must do plenty of exercises as you learn a subject, which brings me to the main contribution of the book. A substantial part of the book is devoted to examples and exercises together with solutions to these exercises. These I have found useful when trying to get to grips with this material. Many of them are new, and one of my main reasons for writing this book was to collect them together in one place. Proof theory as a whole suffers from a lack of pedagogical exposition and illuminating exercises. I hope this book helps to alleviate at least the second deficiency. There is nothing worse than an exercise which you can't do and have no way of finding a solution to it. Thus I have included an almost complete set of solutions to the exercises. About half the length of the book is given over to these solutions.

Each chapter contains two numbered sequences. The first sequence is a list of items such as definitions, examples, lemmas, theorems, algorithms, ... and these are numbered consecutively throughout the chapter. The second sequence is a list of exercises split between the various sections, and numbered consecutively throughout the chapter. Thus Chapter $\heartsuit$ may contain both an item $\heartsuit$.7 (the 7th item in that chapter) and an Exercise $\heartsuit$.7 (the 7th exercise in that chapter). These numberings do not refer to the section in which the item or exercise occurs.

The solutions to the exercises are split into nine appendices (A–I) to match the nine chapters. Within each appendix the solutions are split into blocks to match the sections of the corresponding chapter. A glance at the contents pages will show what I mean.

ACKNOWLEDGEMENTS

This book arose out of a short series of lectures given at Logic Colloquium 93. I thank the organizers of that meeting for that opportunity. Martin Hyland suggested that the notes for the course should be made more widely available. After six years of rewriting this is the result.

Many people have commented on earlier drafts. All of these comments have, in some way, influenced this final version.

Except for a brief period towards the end of May 1999, David Tranah of Cambridge University Press has given encouragement, advice, and refreshment when needed. In particular, he has put up with my incompetence with LaTeX. The copy editor, Michael Behrend, did a very thorough job, and the final version is much the better for his work.

The proof trees have been produced using Paul Taylor's neat little proof tree package. In the earlier versions I also used his diagrams package, but that was not needed in this final version.

Needless to say, any mistakes, inaccuracies, bad phraseology, and persiflage are entirely my responsibility.

Finally, on a more personal note, I thank David Berstein and Joe Royle who, since they have taken charge, have made steady progress after 20 years of decline.

PREVIEW

The inspiration for this work came from the book [12]

Proofs and Types

by J-Y. Girard (in French), translated (into English) and amended by Y. Lafont and P. Taylor. This is an excellent text which anyone who is interested in proof theory should read. It is short and concise, and gives an overview of a wide selection of topics without getting bogged down in suffocating details. However, it is written with a moderately sophisticated reader in mind. Many routine details are missing and it is a little short on exercises. This work arose from the notes I made when reading that book (and related material), and the exercises I did which I found illuminating.

The original notes became public as a set of lectures

Logic and Computation

(given at Logic Colloquium 93 at the University of Keele). This title indicates the broad area covered: proof theory and related topics.

I was encouraged to expand these notes for publication, and the title

Derivation and Computation

was suggested. This indicates the two main notions used: a derivation is the formal analogue of a proof, and a computation is the formal analogue of a calculation. The book develops symbol shuffling systems (notation systems) to handle these two ideas.

This book is neither encyclopedic nor comprehensive. It is designed as a teaching tool. It doesn't always give the 'best' or fullest development of a topic, for it is often instructive to see an account that doesn't quite work or won't extend to cover later material. Several topics are not covered. I could rationalize the choice of topics, but in the end this wouldn't convince you if I have missed out your favourite.

One of its main aims is to provide a substantial collection of exercises (and solutions) in proof theory and related areas. Another aim is to illustrate that Mathematical Logic should *not* be viewed as a collection of isolated compartments (nor should it be viewed as apart *from* Mathematics, it is a part *of* Mathematics).

The 'traditional' splitting of Mathematical Logic is into

Proof Theory   Model Theory   Recursion Theory   Set Theory

to which we should now add

Category Theory

(and perhaps one or two other topics). These are not self contained units; each influences and interacts with the other. (The reason for such a splitting is entirely to do with the organization of the teaching of the subject.)

Although this book is mainly about proof theory, is contains quite a lot of recursion theory, a little bit of ordinal arithmetic, and just a hint of category theory.

Let's see what is in the book chapter by chapter.

I (the writer) assume that you (the reader) have some knowledge of basic logic. In particular, I assume that you have seen some propositional logic. Thus you have seen a propositional language, the idea of a boolean (truth-functional) valuation of the formulas, and an example of a derivation system, or propositional calculus (probably in Hilbert style). You may have seen a completeness proof. You may also have been told about a subpropositional calculus called constructive or intuitionistic logic. (You may also have seen an attempted philosophical explanation of this. If so, you should forget it.) You may have seen none of these, in which case you can use Chapter 1 to point you in the right direction.

1. Derivation Systems

This begins with a brief survey of the various styles of propositional calculi, and the principles that are used to devise and classify such calculi. Eventually the chapter concentrates on just two styles: the Hilbert style and the Natural style. These are developed and refined throughout the book. (In particular, the Gentzen style is not developed here.) Furthermore, the chapter focuses on the most important connective, $\to$, and its constructive properties. We don't look at the classical, boolean properties. This is done to match with the rest of the book where a more liberal interpretation of $\to$ is used, that of the function space constructor.

Usually a propositional calculus is presented as a system for manipulating formulas (the propositional formulas). Here we look at a more sophisticated gadget, a judgement or formula under hypothesis. The central notion is that of a derivation ∇, a tree-like collection of judgements. Each judgement has a natural reading, and the idea is that each derivation provides a complete justification of this reading of its root judgement.

We begin to develop two derivation systems H and N in the selected styles. This is done with an eye on the rest of the book, not because within the wider scheme of things these two styles are considered to have better or more important logical properties. The difference between H and N is in what they consider to be trivial judgements (judgements which do not need any justification and can be used at the leaves of a derivation), and in the way derivations can be grown (the proof rules).

Once we have understood the basics of H and N, and can produce many examples of derivations, we begin to investigate how different derivations are related, and how the two systems interact. The systems are described in a common format to make these comparisons easier. This analysis produces examples of the second persistent notion, an algorithm on derivations. The best-known example is the one underlying the Deduction Theorem (which tells us how to translate from N to H). The study of these algorithms is fundamental. Indeed, a suggested title for the book was

Algorithmic Proof Theory

but this was rejected because it could too easily be misconstrued.

As you can imagine, a detailed description of such algorithms can be a bother. This is because the data they manipulate – derivations – are themselves rather cumbersome objects. One of the novelties of this book is to show how such algorithms can be handled better using a decent notation, and this opens up pathways to other parts of Mathematical Logic. This is the

Curry–Howard correspondence

of the subtitle.

This first chapter leaves us with the problem of organizing and analysing symbol shuffling algorithms. The second chapter seems to go off in a different direction. In fact these first two chapters can be read independently, either before the other. The aim, of course, is that eventually this diverse material will be combined.

2. Computation Mechanisms

This is concerned with the basics of the untyped combinator and λ-calculi as an introduction to symbol shuffling mechanisms. This chapter is not a comprehensive survey of these subjects, but concentrates on the idea of a **reduction** (with an eye on the rest of the book). We look at two examples of this.

We introduce two kinds of terms, the combinator terms and the λ-terms. These are strings of primitive symbols built up according to certain rules. (At first sight these rules seem 'meaningless' but you must persevere for not everything is revealed at this stage.) For each kind there is a reduction relation

$$t^- \rhd\!\!\!\rhd t^+$$

on terms. As a first approximation this is intended to capture the idea that t^- 'simplifies' to t^+ by the removal of unnecessary detours. It would be useful if you have seen something of this before. You may not have, for this is not considered a central part of Mathematical Logic. If not you can read this chapter cold (but should fill in with some extra reading from elsewhere).

The important notion of this chapter is that of a **computation**

$$(\square) \quad t^- \rhd\!\!\!\rhd t^+$$

which organizes one of the two kinds of reductions. Each such computation $\square$ is a tree-like collection of instances of a reduction with the indicated instance $t^- \rhd\!\!\!\rhd t^+$ at the root. The idea is that $\square$ gives a complete account of the way its root comes about. At the leaves we find certain 1-step reductions which are considered to be so basic that they cannot be decomposed. The tree combines these to produce its root.

In the main the two systems, combinator reduction and λ-reduction, differ in what they consider to be 1-step reductions and the way reductions can be combined. At first the two different mechanisms are described separately. Then we see how each mechanism can be used to simulate the other, and so we can translate back and forth between the two kinds of terms.

By the end of the first two chapters we have seen examples of

Derivation systems   Computation mechanisms

based around the notions of

a derivation ∇ a computation $\Box$

each of which is a tree-like collection of pieces of syntax. We begin to suspect there might be a connecting structure here. The next four chapters give us three views of such a structure: the combinator aspect and the λ-aspect, and then the overall view.

In a more advanced account Chapters 3 and 4 could be subsumed under Chapter 6. The reason for not doing this here is entirely pedagogical; at first sight it is easier to take in the big picture if various important features are first highlighted and others hidden. To describe the plan it is useful to have some of the technical words available and a vague idea of what they mean. Of course, these notions are made precise as they occur in the text, so you mustn't expect to understand everything just yet. (If it was possible to explain these ideas in a few lines there wouldn't be much point in a text book on the subject, would there!)

We use a rudimentary kind of type system which, in the first instance, has two facilities, a derivation system and a computation mechanism. To describe these we first set up various syntactic categories, various kinds of strings of primitive symbols. These are

- types, which are built from variables (as place-holders) and atoms (as rigid names)
- terms, which are built from types together with identifiers (as place-holders) and constants (as rigid names)

from which we immediately obtain statements, declarations, and contexts. These are combined to give the more sophisticated notion of a judgement

$$\Gamma \vdash t : \tau$$

which consists of

- a context Γ, which is a list of declarations $x : \sigma$ where x is an identifier and σ is a type; and
- a statement $t : \tau$ with subject t (which is a term) and a predicate τ (which is a type),

where, of course, various laid down conditions must be satisfied. Such a judgement can be read as

Within the *legal* context Γ,
the *well formed* term t
inhabits the *acceptable* type τ

and a derivation system is used to provide a justification of the correctness of such a reading.

The types here are essentially the propositional formulas used in Chapter 1. The terms are enriched versions of either the combinator terms or the λ-terms of Chapter 2. The ideas of those chapters give us

$$(\nabla) \quad \Gamma \vdash t : \tau \qquad\qquad (\square) \quad t^- \rhd\!\!\!\rhd t^+$$

a derivation ∇ of a judgement $\Gamma \vdash t : \tau$, and a computation $\square$ which organizes a reduction $t^- \rhd\!\!\!\rhd t^+$.

The next three chapters fill out these ideas.

3. The Typed Combinator Calculus

Amongst other things Chapters 1 and 2 introduced the derivation system H and the computation mechanism for combinator reduction. At first these don't seem to have much to do with each other. However, a fundamental insight is that each is an enfeebled version of a wider calculus which, when seen, explains much more about the two.

This chapter develops a particular calculus **C** in the manner outlined above. All the notions there are explained for this particular calculus (and some of these apply to later calculi). As always these ideas are illustrated with plenty of examples (and exercises).

Each derivation in **C** is a collection of judgements and each judgement contains both types and terms. By deleting the terms we obtain a derivation in H. Furthermore, each H-derivation can be annotated with combinator terms to produce a **C**-derivation. This process of annotation and deletion is analysed in some detail; it is a concrete example of a Curry–Howard correspondence.

We then turn to the idea of a computation. This seems to be nothing more than a combinator computation as described in Chapter 2 with a few types scattered about. In fact there is a notion of type erasure which converts each **C**-computation into a combinator computation. This makes possible a trick on H-derivations which is difficult to see without the more sophisticated calculus **C**.

Consider a derivation and computation

$$(\nabla) \quad \Gamma \vdash t^- : \tau \qquad\qquad (\square) \quad t \rhd\!\!\!\rhd t^+$$

in **C**. Think of ∇ arising from an H-derivation by annotating with combinator terms. Then $\square$ shows how the root subject of ∇ can be reduced. What does this reduction do to the derivation?

Both ∇ and $\square$ can be described by a piece of syntax, the arboreal code of each. Using these we develop the subject reduction algorithm. This syntactically combines ∇ and $\square$ to produce a new derivation

$$(\nabla \cdot \square) \quad \Gamma \vdash t^+ : \tau$$

in the same context with the same predicate, but with a reduced subject.

We begin to see that many algorithms on derivations (as discussed in Chapter 1) can be formatted in this way. In other words combinators can be used to analyse H-derivations. We are onto something important here.

This chapter looks a bit finicky; notation and manipulation for the sake of it. However, it is merely an introduction to some important ideas which are developed much further in later chapters. The correspondence between H-derivations and combinator terms is quite straightforward (and in itself, not very exciting). It is described here in detail merely as an entry into the more intricate Curry–Howard correspondences. The second of these, often called *the* Curry–Howard correspondence, is the topic of the next chapter.

4. The Typed λ-calculus

This continues the theme of Chapter 3 by combining the derivation system N of Chapter 1 with the computation mechanism on λ-terms of Chapter 2. This produces a calculus **λ** which subsumes both N-derivations and λ-reductions. We follow the same general outline, including a discussion of annotation, deletion, erasure, and subject reduction within the present set-up. This could take quite a long time if we started from scratch. But the build-up of the previous chapters enables us to get through the material much more swiftly. However, this speed should not be taken as an indication of lesser importance. At this point we are beginning to uncover the central concepts and problems of this book.

Before we continue we need a slight interlude. One of the differences between **C** and **λ** is that λ-terms contain both free and bound identifiers whereas combinator terms contain only free identifiers. In both calculi we want to substitute for the free identifiers of a receiving term other terms (to produce a larger term). In **λ** this runs into the problem of identifier capture, so the substitution algorithm has to be designed with some care.

5. Substitution Algorithms

This analyses substitution algorithms in rather more detail than is usual. The chapter begins with a quick review of the standard algorithm and the way the various pitfalls are avoided. The usual trick is to rename the problem identifier within the receiving term. The resulting algorithm has nested recursion calls on itself. Consequently it is hard to prove that the algorithm has the desired properties. Furthermore, this version of substitution doesn't interact with derivations as we would like it to. This approach is rejected.

A neat way around the problems is not to modify the receiving term but to update the giving substitution gadget. To accommodate this we introduce a new syntactic category, the collection of replacements $\mathfrak{a}$. These are names for substitution gadgets. Updating $\mathfrak{a}$ is merely extending the piece of syntax.

With these we describe a slicker version

$$t, \mathfrak{a} \longmapsto t \cdot \mathfrak{a}$$

of a substitution algorithm. This is much easier to use, and interacts with derivations in the right way.

By this point we have mustered all the notions and techniques to begin the study of type theory in earnest. These first 100 pages can be thought of as warming-up exercises. We now come to the real thing.

6. Applied λ-calculi

This is the pivotal chapter. The preceding chapters are preparation for this one, and the following chapters build on this. The material of the preceding five chapters could be subsumed under this one. In a research monograph that would be done, but this is a teaching book, so we take things more slowly.

The chapter describes a whole family of calculi, the applied λ-calculi. These have all the facilities of **C** and **λ**, and more. So far I have suggested that you think of types as propositional formulas with '→' as implication. Now you should think of types as function spaces with '→' as the function space constructor. Terms are names of functions which live in these spaces. (In the body of the text this extended view takes place in the earlier chapters.) Any applied λ-calculus **λSig** may have primitive types, called atoms, which name concrete sets, and primitive terms, called constants, which name certain functions.

Each such calculus **λSig** has three interacting facilities. The

Derivation system   Substitution algorithm   Computation mechanism

are designed and controlled using the general methods of Chapters 1–5. There are a couple of extra features, but we needn't worry about these just here.

Although this chapter does describe these general facilities, in the main it concentrates on the more delicate matters that arise with **λSig**. We begin to look at second stage proof theory.

Unlike previous calculi, in **λSig** each derivable judgement

$$\Gamma \vdash t : \tau$$

need not have a unique derivation. This is a deliberate policy to make the calculus more flexible and avoid some silly renaming problems. But this also leads to several classification problems, two of which are particularly important.

- Type synthesis Given a pair (Γ, t), can we determine whether or not there is a type τ such that the judgement above is derivable, and if so can we describe the derivations?

- Type inhabitation Given a pair (Γ, τ), can we determine whether or not there is a term t such that the judgement above is derivable, and if so can we describe the derivations?

Answers to both these questions involve syntactic algorithms. These are more intricate than the previous algorithms, but our experience begins to pay off.

In mathematics we often obtain a result on the basis of certain hypotheses. It can happen that later we discover a different set of hypotheses. This leads to a reworking of the original material, often with better results. A miniature version of this happens in $\boldsymbol{\lambda}$**Sig**.

Suppose we have a derivation

$$(\nabla) \quad \Sigma \vdash t : \tau$$

of a judgement in a context Σ, and suppose we discover what we believe is a 'better' context Γ. In such circumstances we ought to provide an 'improved' derivation

$$(\nabla') \quad \Gamma \vdash t' : \tau$$

where t' is a suitably modified version of t. How do we do this in an algorithmic fashion?

We must start from a known comparison between Σ and Γ, some reason why Γ is 'better' than Σ. This is codified by a **mutation**

$$(\mathfrak{A}) \quad \Sigma \xrightarrow{\mathfrak{a}} \Gamma$$

a Σ-indexed battery of derivations over Γ where each declaration of Σ is 'passed through' a replacement $\mathfrak{a}$. We describe an algorithm which when supplied with ∇ and $\mathfrak{A}$ will allow $\mathfrak{A}$ to act on ∇ to return $\nabla' = \nabla \cdot \mathfrak{A}$ with $t' = t \cdot \mathfrak{a}$.

This mutation algorithm encompasses many of the informal algorithms used with λ-calculi. It is often called substitution in context, and is the derivation analogue of the substitution algorithm for terms. The trouble we took in Chapter 5 makes the work here that much easier.

The chapter concludes with a concise description of the subject reduction algorithm

$$(\nabla, \Box) \longmapsto \nabla \cdot \Box$$

which consumes a compatible derivation and computation and returns a derivation. Because of the nature of $\boldsymbol{\lambda}$**Sig**, this is more complicated than previous versions. On the other hand, we need not skirt round some details (as we did earlier) for the mutation algorithm gives us all we need.

At this juncture the development could be concluded. We have reached a natural closing point with a fairly complete description of the simplest kind of type theory, the applied λ-calculi. Of course, we don't yet have any examples beyond **C** and $\boldsymbol{\lambda}$. What we need is a good solid example which shows that all the extra effort of Chapter 6 is worth while. This is provided by the term calculus of Gödel's T, which leads us into another area of Mathematical Logic, the study of recursion and induction over the natural numbers $\mathbb{N}$.

In the initial parts of recursive function theory you learn how to describe number theoretic functions using primitive recursion. You see other forms of recursion (such as course-of-values recursion, primitive recursion with variation of parameters, etc) all of which are shown to be reducible to primitive recursion. You then see a function, a variant of Ackermann's function, which is not primitive recursive, but which is clearly recursive in some sense. If you are lucky you then see examples of more and more powerful recursive constructions (2-recursion, 3-recursion, 4-recursion, ...).

All this material is concerned with the construction of number theoretic functions (from tuples of natural numbers to natural numbers) using different kinds of recursion. Almost certainly you will see how these notions connect with the idea of (Turing) computability. You will set this against the notion of a general recursive (μ-recursive) function. Here we do not follow that path into a study of partial functions.

At a later stage you may also go back to the idea of recursion operators (which produce total functions from total functions). You may generalize these to higher order gadgets to produce more extensive classes of (total) functions. For instance, you may see a description of Ackermann's function in the form of a primitive recursion but where certain parameters are allowed to range over functions rather than natural numbers.

This higher order approach to recursive descriptions is encapsulated in a nice applied λ-calculus $\boldsymbol{\lambda G}$.

7. Multi-recursive Arithmetic

The calculus $\boldsymbol{\lambda G}$ is tailor made for describing number theoretic functions. The calculus has a single atom $\mathcal{N}$, which names $\mathbb{N}$, and all the types are generated from this. The calculus has constants to name zero, the successor function, and various iterators (the simplest kind of recursor). Thus every type and term of $\boldsymbol{\lambda G}$ has an orthodox interpretation as a number theoretic gadget. Most of these are high order, but all are rooted in $\mathbb{N}$. The derivation system of $\boldsymbol{\lambda G}$ is used to format number theoretic functions, and the computation mechanism is used to evaluate these functions.

The chapter begins with a brief description of the specifics of $\boldsymbol{\lambda G}$. This description can be brief because of the work we did in Chapter 6. Most of this chapter is concerned with various number theoretic hierarchies, and how these can be formulated in the lower reaches of $\boldsymbol{\lambda G}$. These hierarchies are used to measure the complexity of certain functions; the higher up in a hierarchy a function appears, the more complicated it is. We find that these hierarchies are intimately connected with the syntactic structure of $\boldsymbol{\lambda G}$.

I expect you know something about recursion and induction over $\mathbb{N}$. Well, the same kind of thing can be done with tuples of natural numbers. A number theoretic function may be specified by a simultaneous recursion over several of its natural number arguments. This can lead to some quite surprising constructions. This chapter centres around these multi-index recursions.

After the specifics of $\boldsymbol{\lambda G}$, there is a discussion of the various multi-index recursions and the associated multi-index inductions. Several exercises illustrate just how messy these can become.

How can we produce complicated functions? The main idea is to use an operator (a jump operator) which converts each function into a slightly more complicated function. We then iterate the use of this operator and diagonalize every so often.

When this runs out of steam we find a constructor which converts each operator into a more powerful operator, and begin to iterate the use of this constructor. As you can imagine, this can get a little out of hand. However, with a bit of care the whole process can be organized using multi-indexes. In this way a multi-index becomes a measure of complexity.

All this can be done within $\boldsymbol{\lambda G}$. Thus we can view $\boldsymbol{\lambda G}$ as a framework in which number theoretic functions can be classified using no more than the shape of the syntactic descriptions. In fact, only a small part of $\boldsymbol{\lambda G}$ is needed, so what about the rest of it?

Time for another interlude.

In Chapter 7 we go to considerable effort to understand the use of multi-indexes. To some extent all this work is unnecessary. Each multi-index is just an ordinal $\alpha < \omega^\omega$ in disguise. A direct use of ordinals leads to a cleaner version of Chapter 7, so why bother with that account? Firstly, it is nice introduction to the use of ordinals. Secondly, there is a natural jump in complexity that occurs at ω^ω, but this is not so visible without the use of multi-indexes.

8. Ordinals and Ordinal Notations

To continue with our analysis of $\boldsymbol{\lambda G}$ we need to use the ordinals $\alpha < \epsilon_0$. This chapter gives a quick survey of these ordinals. The survey is not comprehensive; you will need to fill in some details from elsewhere. One of the things we learn is the difference between an ordinal and an ordinal notation. Each notation names an ordinal, but an ordinal can have many different notations. It turns out that for use in $\boldsymbol{\lambda G}$, different notations for the same ordinal can give different results. Thus we must tread carefully.

After this short interlude we return to $\boldsymbol{\lambda G}$.

9. Higher Order Recursion

This final chapter analyses the full power of $\boldsymbol{\lambda G}$. We first observe that the material of Chapter 7 can be reformulated in a simple fashion using small ordinals in place of multi-indexes. We then extend this material using larger ordinals.

Each ordinal codifies a whole battery of iteration gadgets, one for each appropriate type. The way these gadgets combine matches closely the arithmetic of indexing ordinals. Thus the analysis of $\boldsymbol{\lambda G}$ boils down to isolating the ordinals that can be simulated within the calculus in a natural way.

This is the final mountain to climb. But, by this time we are pretty fit and it doesn't take us too long to get to the top.

And what do you think we can see from there?

This completes the development of material. We could continue further to the next natural break, but that would make the book very long. Thus we stop here.

The book contains approximately 200 exercises scattered throughout the sections. Some of these are routine, merely to illustrate the material of that section. Others indicate how the material could be extended.

Appendices A–I contain an almost complete set of solutions to the exercises. I hope these will help to reinforce your understanding of the material of the development Chapters 1–9. I have made every effort to ensure these solutions are correct (as I have with the development part of the book). But there are bound to be some mistakes, bad phrasings, or garbled explanations. I hope these don't hinder you too much.

Part I

Development and Exercises

1
DERIVATION SYSTEMS

1.1 INTRODUCTION

In the first instance propositional logic is an analysis of the behaviour of the informal connectives

not implies and or ...

as used in certain arguments. It is concerned with those statements which are either true or false, and the aim of the analysis is to describe how the connectives in such a statement influence its truth value. This is done using a formal language – the **propositional language** – built from certain atomic components – the **variables** – by combinations which mimic the use of the informal connectives. The well formed expressions in this language are called the **formulas**.

After a discussion of the semantics (i.e. the meaning) of such formulas, the next job is to analyse the notion

the formula ϕ is derivable from the batch of formulas Γ

where the derivation may depend only on the intended meaning of the connectives (but not on any perceived meaning of the variables) occurring in Γ and ϕ. This is done using a suitable notion of a **formal proof** which is defined as a certain combinatorial construction involving Γ, ϕ, and perhaps other formulas as well.

By this stage of the analysis it is clear that there are several different styles of formal proof, and these differences seem to have some significances which are not immediately clear. Thus after a while the analysis moves away from the mere existence of a formal proof to an attempt to classify proofs as objects in their own right. There may be many different derivations of the same formula from a given batch of formulas, so what is the essential content of the different derivations and how do these differences arise? To answer such questions we must study the proof styles themselves (rather than merely produce different styles of proof). This chapter describes the basis of a such an analysis.

The full propositional language has a whole battery of connectives

$$\bot\,,\ \top\,,\ \neg\,,\ \rightarrow\,,\ \wedge\,,\ \vee\,,\ \ldots$$

where

$$\bot, \top \text{ are 0-placed (i.e. constants)} \qquad \neg \text{ is 1-placed} \qquad \to, \wedge, \vee \text{ are 2-placed}$$

and the 2-placed connectives are written as infixes. In some cases we may want to use even more connectives, such as $\leftrightarrow$ (bi-implication). Any derivation system must have a sufficiently rich mechanism to handle all the connectives in its underlying language, hence the more connectives in this language, the more complicated the derivation system will be. Here we wish to get to the heart of the matter rather than give a comprehensive account. We therefore severely restrict the connectives used.

The most characteristic connective of propositional logic is '$\to$', so here we restrict our attention to the language which has this as its sole connective.

1.1 DEFINITION. There is an unlimited stock of **variables**. The two rules

Each variable is a formula. If θ, ψ are formulas then $(\theta \to \psi)$ is a formula.

generate all **propositional formulas**. If we let

$X, Y, Z, \dots$ range over the variables $\qquad \theta, \psi, \phi, \dots$ range over formulas

then

$$\phi = X \mid (\theta \to \psi)$$

is a succinct description of the set of formulas. □

The brackets in a formula are important; they are part of the syntax. Their job is to ensure that each formula is uniquely parsed. However, in practice we often omit some brackets from displayed formulas. In particular, we use the convention that

$$\theta \to \psi \to \phi \quad \text{means} \quad (\theta \to (\psi \to \phi))$$

so two pairs of brackets are omitted. This is not an arbitrary choice; it is used because of the functional interpretation of $\to$ which matches well with its logical properties.

What can we do with these formulas? Before I tell you let's look at some of the notational tricks we use throughout the book. We begin with a bookkeeping device.

Throughout we employ a uniform method of organizing syntax. Strictly speaking we use a finite, rooted, at most binary splitting tree, but we call this a tree for short.

1.2 DEFINITION. Each **tree** is grown from the singleton tree '$\bullet$' by a finite number of uses of the two rules shown to the right. Here C, L, R are trees already grown. Each tree is a structured collection of $\bullet$, the **nodes** of the tree. The topmost nodes are the **leaves**, and the unique bottommost node is the **root**. □

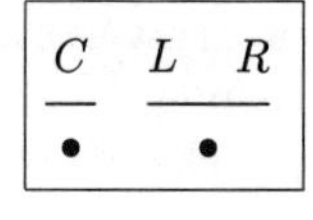

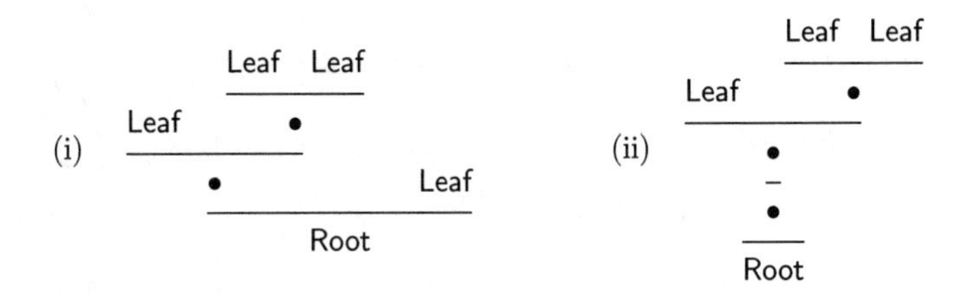

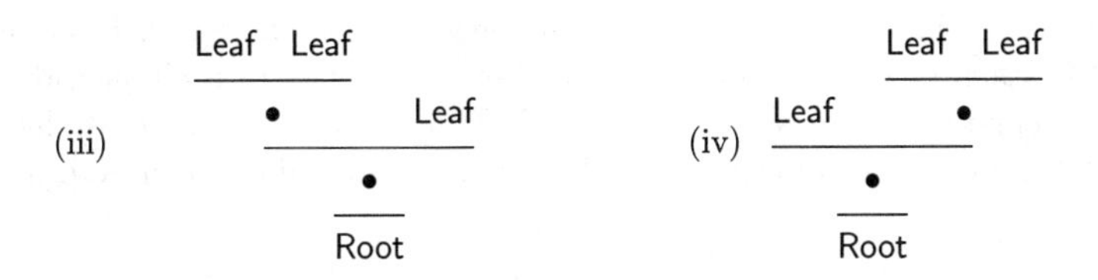

Table 1.1: Four small trees

A few examples will explain this idea.

1.3 EXAMPLE. Table 1.1 shows four small trees. Binary splitting

$$L, R \longmapsto \frac{L \quad R}{\bullet}$$

is not left–right symmetric, so trees (iii) and (iv) are different. □

We will meet several kinds of trees. In all cases each node will be filled by a piece of syntax of a certain kind. The job of the tree is to organize these pieces into a whole.

Every tree T can be flattened to a linear sequence $f(T)$ of nodes with various punctuation devices to indicate the intended construction. For instance, we may use

$$f(\bullet) = \bullet \qquad f\Big(\frac{C}{\bullet}\Big) = f(C)| \qquad f\Big(\frac{L \quad R}{\bullet}\Big) = (f(L)f(R))$$

to flatten all the trees we meet here.

1.4 EXAMPLE. The four trees of Example 1.3 flatten to

$$\text{(i)} \quad ((\bullet(\bullet\bullet))\bullet) \qquad \text{(ii)} \quad (\bullet(\bullet\bullet))|| \qquad \text{(iii)} \quad ((\bullet\bullet)\bullet)| \qquad \text{(iv)} \quad (\bullet(\bullet\bullet))|$$

respectively. Notice the difference between (iii) and (iv). □

Conversely, for each string of symbols that is uniquely parsed, its construction can be displayed by a parsing tree. It is instructive to think of formulas in this way.

1.5 EXAMPLE. For arbitrary formulas θ, ψ, ϕ consider

$$\beta = (\psi \to \phi) \to (\theta \to \psi) \to (\theta \to \phi)$$

a formula given in abbreviated form. In full this is

$$((\psi \to \phi) \to ((\theta \to \psi) \to (\theta \to \phi)))$$

and the template to the right produces its parsing tree from those of θ, ψ, ϕ. □

$$\dfrac{\dfrac{\psi \quad \phi}{\bullet} \qquad \dfrac{\dfrac{\theta \quad \psi}{\bullet} \quad \dfrac{\theta \quad \phi}{\bullet}}{\bullet}}{\bullet}$$

Each tree T has three dimensions: the height $h(T)$, the width $w(T)$, and the size $s(T)$. Each of these is a natural number. We generate each such dimension $d(T)$ by recursion on the construction of T. Thus for the base case we define $d(\bullet)$ outright (as 0 or 1, as appropriate). Then we give two rules which determine

$$d\Big(\frac{C}{\bullet}\Big) \text{ in terms of } d(C) \qquad d\Big(\frac{L \quad R}{\bullet}\Big) \text{ in terms of } d(L), d(R)$$

for the recursion steps.

1.6 DEFINITION. The rules

$$h(\bullet) = 0 \quad h\Big(\frac{C}{\bullet}\Big) = h(C) + 1 \quad h\Big(\frac{L \quad R}{\bullet}\Big) = \max(h(L), h(R)) + 1$$

$$w(\bullet) = 1 \quad w\Big(\frac{C}{\bullet}\Big) = w(C) \quad w\Big(\frac{L \quad R}{\bullet}\Big) = w(L) + w(R)$$

$$s(\bullet) = 1 \quad s\Big(\frac{C}{\bullet}\Big) = s(C) + 1 \quad s\Big(\frac{L \quad R}{\bullet}\Big) = s(L) + s(R) + 1$$

generate the height $h(T)$, the width $w(T)$, and the size $s(T)$ of a tree T. □

For instance

(i) 3, 4, 7 (ii) 4, 3, 7 (iii) 3, 3, 6 (iv) 3, 3, 6

are the dimensions of the four trees of Example 1.3.

Exercise 1.2 deals with the connections between these dimensions.

Bits of arithmetic will appear quite often. Sometimes this will be to help the analysis but in the later chapters it will be the subject of the analysis. Mostly we need not go beyond addition, multiplication, and exponentiation.

$$x, y \longmapsto y + x \qquad x, y \longmapsto y \times x \qquad x, y \longmapsto y^x$$

Here x and y are members of $\mathbb{N}$, the set of natural numbers. Every now and then more complicated arithmetic is needed. The next level of complexity is encapsulated by the stacking function

$$\beth(x, y, r) = y^{y^{\cdot^{\cdot^{\cdot^{y^x}}}}}$$

The prime $(\cdot)'$ will be used to indicate 'the successor of' several different kinds of gadgets.

For a natural number r we write r' for the natural successor $r+1$ of r.

For a set $\mathbb{S}$ we write $\mathbb{S}'$ for the set $(\mathbb{S} \longrightarrow \mathbb{S})$ of functions on $\mathbb{S}$. We think of this as the next most complicated set obtainable from $\mathbb{S}$.

For a type σ we write σ' for the type $(\sigma \rightarrow \sigma)$.

For an ordinal α we write α' for the ordinal successor of α.

There are also other, non-successor uses of the prime.

Table 1.2: Use of $(\cdot)'$ to indicate successor

---

which consumes natural numbers r, y, x and returns a stack of height r, as indicated

$$\beth(x,y,0) = x \quad \beth(x,y,1) = y^x \quad \beth(x,y,2) = y^{y^x} \quad \beth(x,y,3) = y^{y^{y^x}} \quad \cdots$$

and, in general, in the stack for $\beth(x,y,r)$ there are r occurrences of y. It is useful to set this function in a broader context.

Consider a 1-placed function

$$F : \mathbb{S} \longrightarrow \mathbb{S}$$

from some set $\mathbb{S}$ to itself. Such a function can be composed with itself, $F \circ F$, to produce a new function on $\mathbb{S}$. (Strictly speaking, this is a new function only when F is not the identity function $id_{\mathbb{S}}$ on $\mathbb{S}$.) More generally, for $r \in \mathbb{N}$ and $s \in \mathbb{S}$ we have

$$F^r s$$

the r-**fold iterate of** F **on** s. This is

$$(F \circ \cdots \circ F)(s)$$

where there are r occurrences of F. The value can be generated by recursion on r in two extreme ways.

Head	Tail
$F^0 s = s$	$F^0 s = s$
$F^{r'} s = F(F^r s)$	$F^{r'} s = F^r(F s)$

Here we write r' for $r+1$ to make the expressions more compact. This use of $(\cdot)'$ will occur throughout the book; see Table 1.2.

For a given $y \in \mathbb{N}$ consider the function $Y : \mathbb{N} \longrightarrow \mathbb{N}$ where $Y(x) = y^x$ for $x \in \mathbb{N}$. Then

$$\beth(x,y,r) = Y^r x$$

so we can generate the stacking function in at least two ways.

Exercises

1.1 Consider the formulas

$$\begin{aligned}\beta &= (\psi \to \phi) \to (\theta \to \psi) \to (\theta \to \phi)\\ \gamma &= (\theta \to \psi \to \phi) \to (\psi \to \theta \to \phi)\\ \delta &= (\theta \to \psi) \to (\psi \to \phi) \to (\theta \to \phi)\end{aligned}$$

where θ, ψ, ϕ are arbitrary. For each of these write out the parsing tree (with θ, ψ, ϕ at the leaves). You might also look at the formulas of Exercise 1.8.

1.2 Show that

$$\text{(i)}\quad h(T)+1 \le s(T) \qquad \text{(ii)}\quad s(T)+1 \le 2^{h(T)+1} \qquad \text{(iii)}\quad 2w(T) \le s(T)+1$$

holds for each tree T. Show also that for a given width the height (and hence size) of a tree can be arbitrarily large.

1.3 Let $F : \mathbb{S} \longrightarrow \mathbb{S}$ be a given function and consider the two functions H, T specified as follows.

$$\begin{array}{rclcrcl} H(0,s) &=& s & \qquad & T(0,s) &=& s\\ H(r',s) &=& F(H(r,s)) & & T(r',s) &=& T(r,F(s)) \end{array}$$

Here $r \in \mathbb{N}$ (with $r' = r+1$) and $s \in \mathbb{S}$. Devise a proof that $H = T$. You should proceed by induction over the natural number input with allowable variations of other parameters. Your proof should not depend on unjustified manipulations of syntax, but you may refer to simple properties of addition.

1.2 Generalities

The derivation systems we are concerned with here manipulate an entity

$$\Gamma \vdash \phi$$

called a **judgement**. Here ϕ is a single formula called the **predicate**, and Γ is a batch of formulas called the **context**. The symbol '$\vdash$' is a punctuation device to separate the two parts. It is commonly referred to as the **gate** (or **turnstile**). In later chapters we will meet a slightly more refined form of judgement. (There are some derivation systems which manipulate more general kinds of judgements, and some which manipulate even more complex kinds of entities. These do not occur in this book.)

Let's formally introduce these notions.

1.7 **DEFINITION**. A **context** or **hypothesis list** is a finite list

$$\Gamma = \theta_1, \theta_2, \ldots, \theta_l$$

of formulas. This context has length l, the number of components. A **judgement** or **formula under hypothesis** is a pair

$$\Gamma \vdash \phi$$

where Γ is a context and ϕ is a formula, the **predicate**. □

We wish to read the judgement as

The hypothesis Γ entails the formula ϕ

in the sense that ϕ is a 'logical consequence' of Γ. The job of a derivation system is to justify such a reading. This is done by providing a derivation

$$(\nabla) \quad \Gamma \vdash \phi$$

a tree ∇ of judgements with the particular judgement at the root. The allowable modes of construction for such a derivation are characteristic of the ambient derivation system.

The general idea is that the leaf judgements of a derivation are considered to be so basic as not to need any justification. Then, as we read the derivation from leaves to root, we see that judgements are combined in simple ways that preserve correctness. Thus the root of a derivation is correct, but not trivially so; it is justified by what sits above it.

What principles are used to construct a derivation system? In this section we give a survey of some of these methods, but before that let's look at four example derivations of the same judgement in different systems.

It is convenient to write

$$\vdash \phi \quad \text{for} \quad \emptyset \vdash \phi$$

i.e. for a judgement with an empty context.

1.8 EXAMPLE. Let

$$\beta = (\psi \to \phi) \to (\theta \to \psi) \to (\theta \to \phi)$$

for arbitrary formulas θ, ψ, ϕ. For convenience let

$$\begin{array}{lll} \rho = \theta \to \phi & \sigma = \theta \to \psi & \tau = \psi \to \phi \\ \lambda = \sigma \to \rho & \mu = \theta \to \tau & \nu = \mu \to \lambda \quad \xi = \tau \to \nu \end{array}$$

so that $\beta = \tau \to \lambda$ (after unravelling τ and λ). Also let

$$\begin{array}{lll} \sigma_1 = (\tau \to \mu \to \lambda) \to (\tau \to \mu) \to (\tau \to \lambda) & \kappa_1 = \sigma_1 \to \tau \to \sigma_1 \\ \sigma_2 = (\theta \to \psi \to \phi) \to (\theta \to \psi) \to (\theta \to \phi) & \kappa_2 = \tau \to \theta \to \tau \end{array}$$

so that $\sigma_2 = \nu$ and $\sigma_1 = \xi \to \kappa_2 \to \beta$. Finally let $\Gamma = \tau, \sigma, \theta$ to obtain a context.The four trees of Table 1.3 are derivations in systems H, N, G+, G× (to be defined later). □

Of course, I haven't yet told you what the systems H, N, G+, G× are, so you won't understand everything that is going on in these examples. That's not important just yet; concentrate on the shapes of the four trees, and the details will become clear as you read on.

Each derivation system is determined by the rules for constructing derivations. These come in two broad kinds.

H
$$\dfrac{\dfrac{\vdash \sigma_1 \qquad \dfrac{\vdash \kappa_1 \quad \vdash \sigma_2}{\vdash \xi}}{\vdash \kappa_2 \to \beta} \qquad \vdash \kappa_2}{\vdash \beta}$$

N
$$\dfrac{\dfrac{\dfrac{\dfrac{\Gamma \vdash \tau \qquad \dfrac{\Gamma \vdash \sigma \quad \Gamma \vdash \theta}{\Gamma \vdash \psi}}{\Gamma \vdash \phi}}{\tau, \sigma \vdash \rho}}{\tau \vdash \lambda}}{\vdash \beta}$$

G×
$$\dfrac{\dfrac{\dfrac{\dfrac{\dfrac{\dfrac{\dfrac{\theta \vdash \theta \qquad \dfrac{\psi \vdash \psi \quad \phi \vdash \phi}{\psi, \tau \vdash \phi}}{\theta, \sigma, \tau \vdash \phi}}{\theta, \tau, \sigma \vdash \phi}}{\tau, \theta, \sigma \vdash \phi}}{\Gamma \vdash \phi}}{\tau, \sigma \vdash \rho}}{\tau \vdash \lambda}}{\vdash \beta}$$

G+
$$\dfrac{\dfrac{\dfrac{\dfrac{\dfrac{\theta \vdash \theta \quad \theta, \psi \vdash \psi}{\theta, \sigma \vdash \psi} \qquad \theta, \sigma, \phi \vdash \phi}{\theta, \sigma, \tau \vdash \phi}}{\Gamma \vdash \phi}}{\tau, \sigma \vdash \rho}}{\tau \vdash \lambda}}{\vdash \beta}$$

Table 1.3: Four example derivations

- Leaf rules, rules governing what can appear at a leaf
- Non-leaf rules, rules governing what can appear at other nodes

In our case the non-leaf rules are either singulary or binary.

$$\frac{\mathsf{Numerator}}{\mathsf{Denominator}}\ (1\text{Rule}) \qquad \frac{\mathsf{LNumerator} \quad \mathsf{RNumerator}}{\mathsf{Denominator}}\ (2\text{Rule})$$

These should be viewed as pattern matching constructions. Thus given judgements

$$\mathsf{NJ} \qquad \mathsf{LNJ} \qquad \mathsf{RNJ}$$

which match

$$\mathsf{Numerator} \qquad \mathsf{LNumerator} \qquad \mathsf{RNumerator}$$

respectively, then

$$\frac{\mathsf{NJ}}{\mathsf{1DJ}}\ (1\text{Rule}) \qquad \frac{\mathsf{LNJ} \qquad \mathsf{RNJ}}{\mathsf{2DJ}}\ (2\text{Rule})$$

can appear as part of a derivation where $\mathsf{1DJ}$ and $\mathsf{2DJ}$ are the judgements obtained by matching the given numerators against the indicated rule.

These rules can be further subdivided.

- Leaf Axiom rules Projection rules

- Non-leaf Structural rules Rules of inference Cut rules

Also several of the rules come in two flavours, additive, $\oplus$, and multiplicative, $\otimes$. Eventually we will concentrate on just two particular systems, H and N, obtained by fixing a particular choice of rules, but before that it is instructive to survey a whole family of the more common derivation systems.

Even for the minimal language based on $\rightarrow$, there are a huge number of possible rules that can be used to construct derivations. The difference between some of these rules is quite small, and for a novice it is not at all clear why these differences should matter. The full panoply of these rules needs some careful organization; here we merely scratch the surface of this topic.

Leaf rules

There are two sorts of leaf restrictions; rules for Axioms and rules for Projections. Both come in two flavours.

Axiom
$$\begin{cases} \oplus \quad \vdash \phi & \text{where } \phi \text{ is one of a specified set of formulas} \\ \otimes \quad \Gamma \vdash \phi & \text{where } \phi \text{ is one of a specified set of formulas} \end{cases}$$

Projection
$$\begin{cases} \oplus \quad \phi \vdash \phi & \\ \otimes \quad \Gamma \vdash \phi & \text{provided } \phi \text{ appears in } \Gamma \end{cases}$$

The specified set of formulas is the set of axioms. Thus each axiom is a formula, and the whole set of these can be chosen as we please depending on what job we have in mind. Later we will fix on a particular set of axioms. Each axiom ϕ can be used to form a judgement $\Gamma \vdash \phi$ (for any legal context Γ) to produce an Axiom. The distinction between axiom and Axiom is rather pedantic but it is worth making, and we will continue to do so.

Structural rules

In a judgement $\Gamma \vdash \phi$ the context Γ is a finite *list* of formulas. But surely, in the suggested reading, the order of Γ and possible repetitions of components are not important. Can't we replace this list by a set? That is certainly true for the intended meaning of a judgement, but a derivation ∇ is supposed to provide a justification of correctness by a process of symbolic manipulation. This syntactic shuffling will need some bookkeeping, and this is easier to do using a list rather than a set. However, there are certain rules which simulate the effect of a multi-set or a set. These are the structural rules.

For our purposes there are four possible structural rules.

eXchange $\dfrac{\Gamma^l, \theta, \psi, \Gamma^r \vdash \phi}{\Gamma^l, \psi, \theta, \Gamma^r \vdash \phi}$ (X) Thinning $\dfrac{\Gamma^l, \Gamma^r \vdash \phi}{\Gamma^l, \theta, \Gamma^r \vdash \phi}$ (T)

Contraction $\dfrac{\Gamma^l, \theta, \theta, \Gamma^r \vdash \phi}{\Gamma^l, \theta, \Gamma^r \vdash \phi}$ (C) Weakening $\dfrac{\Gamma \vdash \phi}{\Gamma, \theta \vdash \phi}$ (W)

The unlimited use of X effectively converts each context into a multi-set, because the order in which a context is listed becomes irrelevant.

In the presence of X the unlimited use of C effectively converts each context into a set, because repeated components become irrelevant.

We will not allow either of these rules. Instead we show they are **admissible**: the effect of the rules can be simulated by an appropriate use of the other rules. Thus adding either of these rules does not increase the set of derivable judgements.

Sometimes in a derivation it becomes apparent that we need a hypothesis not included in the current context. Thinning allows us to insert such a new hypothesis in any convenient position. Weakening is a restricted version of T where the new hypothesis must be inserted into the gate position, immediately to the left of the gate or at the extreme right of the context.

Initially we do not allow either of these rules. We show they are admissible over the ambient derivation system. Eventually we succumb and allow Weakening as a rule.

Additive vs multiplicative

For us all formulas are generated using $\rightarrow$ as the sole connective. Thus the most common rule of inference is Modus Ponens

$$\frac{\theta \rightarrow \phi \qquad \theta}{\phi}$$

which says that if, under some circumstances, we have obtained the two formulas $\theta \rightarrow \phi$ and θ, then we can immediately obtain ϕ. When we put this rule in context we find there are two extreme versions, conveniently termed additive and multiplicative.

Additive

$$\frac{\Gamma \vdash \theta \rightarrow \phi \quad \Gamma \vdash \theta}{\Gamma \vdash \phi}\,(\mathrm{MP}\oplus)$$

Multiplicative

$$\frac{\Gamma^l \vdash \theta \rightarrow \phi \quad \Gamma^r \vdash \theta}{\Gamma^l, \Gamma^r \vdash \phi}\,(\mathrm{MP}\otimes)$$

In the additive version the two numerators must have the same context Γ and the denominator also has this context. In the multiplicative version the two numerators can have different contexts Γ^l and Γ^r, and the denominator must be the amalgam of this pair with repetitions if necessary. Even if the two numerators have the same context Γ, an application of (MP$\oplus$) may not have the same effect as an application of (MP$\otimes$)

$$\frac{\Gamma \vdash \theta \rightarrow \phi \qquad \Gamma \vdash \theta}{\Gamma, \Gamma \vdash \phi}\,(\mathrm{MP}\otimes)$$

for the latter duplicates every member of the context in the conclusion. If the hypothesis judgements have different contexts then (MP$\oplus$) cannot be applied.

There is also a 'mixed' version of MP.

$$\frac{\Gamma^l, \Gamma \vdash \theta \rightarrow \phi \qquad \Gamma, \Gamma^r \vdash \theta}{\Gamma^l, \Gamma, \Gamma^r \vdash \phi}\,(\mathrm{MP}\odot)$$

<table>
<tr><td>

$\to$	RE$\oplus$
LI$\oplus$	RI$\oplus$

</td><td>$$\frac{\Gamma \vdash \theta \to \phi \qquad \Gamma \vdash \theta}{\Gamma \vdash \phi}\,(\mathrm{RE}\oplus)$$</td></tr>
<tr><td>$$\frac{\Gamma \vdash \theta \qquad \Gamma, \psi \vdash \phi}{\Gamma, \theta \to \psi \vdash \phi}\,(\mathrm{LI}\oplus)$$</td><td>$$\frac{\Gamma, \theta \vdash \psi}{\Gamma \vdash \theta \to \psi}\,(\mathrm{RI}\oplus)$$</td></tr>
<tr><td>

$\to$	RE$\otimes$
LI$\otimes$	RI$\otimes$

</td><td>$$\frac{\Gamma^l \vdash \theta \to \phi \qquad \Gamma^r \vdash \theta}{\Gamma^l, \Gamma^r \vdash \phi}\,(\mathrm{RE}\otimes)$$</td></tr>
<tr><td>$$\frac{\Gamma^l \vdash \theta \qquad \psi, \Gamma^r \vdash \phi}{\Gamma^l, \theta \to \psi, \Gamma^r \vdash \phi}\,(\mathrm{LI}\otimes)$$</td><td>$$\frac{\Gamma, \theta \vdash \psi}{\Gamma \vdash \theta \to \psi}\,(\mathrm{RI}\otimes)$$</td></tr>
</table>

Table 1.4: The rules of inference for $\to$

Thus MP$\oplus$ and MP$\otimes$ are the extreme versions of MP$\odot$ given by $\Gamma^l = \Gamma^r = \emptyset$ and $\Gamma = \emptyset$, respectively.

Rules of inference

The majority of the rules of inference can be classified according to the following features.

- Whether it is an Introduction or an Elimination rule.
- Whether it operates on the Left or the Right of the gate.
- Whether it is Additive or Multiplicative.

Of course, for us the principal connective is $\to$, but a similar classification works for other connectives. There are some rules for which this classification is not sensible, but that does not diminish the usefulness of the classification when it does work.

For us there are six rules as shown in Table 1.4. Above there is a block of three additive rules and below a block of three multiplicative rules. At the bottom of each block we have the two Introduction rules, and at the top the Elimination rule. On the left we have the two Left rules, one in each block, and on the right the two Right rules for each block.

Notice that

$$\mathrm{MP}\oplus = \mathrm{RE}\oplus \qquad \mathrm{MP}\otimes = \mathrm{RE}\otimes$$

and the two rules RI$\oplus$, RI$\otimes$ are the same (and often go by the name of Deduction Rule). All six rules preserve correctness, but you will have to ponder a while to convince yourself of this for LI$\oplus$ and LI$\otimes$.

	Structural			Leaf		R of I		
	X	C	W	A	P	LI	RI	RE
H	No	No	No	$\otimes$	$\otimes$	No	No	$\oplus$
N	No	No	No	No	$\otimes$	No	Yes	$\oplus$
G+	Yes	Yes	No	No	$\otimes$	$\oplus$	Yes	No
G×	Yes	Yes	Yes	No	$\oplus$	$\otimes$	Yes	No

Table 1.5: The four systems H, N, G+, G×

FOUR EXAMPLE SYSTEMS

Remember the four systems H, N, G+, G× of Example 1.8. A complete description of the rules of these systems is given in Table 1.5. In the next section we begin a detailed analysis of H and N, and we continue with this throughout the book. You may have noticed that I still haven't told you the axioms for H. That will be rectified soon.

Both G+, G× are examples of **Gentzen systems**. This style of derivation is important in parts of proof theory, but does not impinge on what we do here. Thus we do not develop an analysis of this style.

CUT RULES

Earlier I included 'Cut rules' amongst the possible non-leaf rules. These rules are most often used in a Gentzen system, so I don't need to say what they are. However, you should be aware of their existence; in some situations they are an important tool.

EXERCISES

These exercises discuss some material not developed in the rest of the book.

1.4 Consider the formulas β, γ, δ of Exercise 1.1 (generated from arbitrary formulas θ, ψ, ϕ). Derive each of $\vdash \beta$, $\vdash \gamma$, $\vdash \delta$ in G+ and in G×.

1.5 Show that in the presence of suitable structural rules each of MP$\oplus$,MP$\otimes$ can be obtained from the other.

1.6 Show that the two systems G+, G× derive the same judgements. Can you describe algorithms which convert a derivation in one system into a derivation of the same judgement in the other system?

1.3 THE SYSTEMS H AND N

Throughout this book we concentrate on just two styles of derivations: the **Hilbert** style and the **Natural** style. These are not more important than other derivation

Clause	Shape	Code	H	N
Axiom	$\Gamma \vdash \zeta$	$\Gamma[\zeta]$	Yes	No
Projection	$\Gamma \vdash \theta$	$\Gamma[n]$	Yes	Yes
Introduction	$\dfrac{\overset{R}{\Gamma, \theta \vdash \phi}}{\Gamma \vdash \theta \to \phi}\,(\uparrow)$	$R\!\uparrow$	No	Yes
Elimination	$\dfrac{\overset{Q}{\Gamma \vdash \theta \to \phi} \quad \overset{P}{\Gamma \vdash \theta}}{\Gamma \vdash \phi}$	(QP)	Yes	Yes

Table 1.6: The derivation rules for H and N

styles, but they are more useful for what we do in the rest of the book. A different selection of topics might bring out different derivation styles. Furthermore, the systems H and N are perhaps the easiest for a beginner to get to grips with.

We now introduce and begin to study the first of the tree-like structures which feature throughout the book.

1.9 DEFINITION. An H- or an N-derivation

$$(\nabla) \quad \Gamma \vdash \phi$$

is a finite rooted tree of judgements grown according to the appropriate rules of Table 1.6. The particular details are set out below. □

We have two systems of derivations H and N. Both use Projection (for leaves) and Elimination (for steps). The difference between the two is that H uses certain Axioms (for leaves) and no Introduction, whereas N uses Introduction (for steps) and no Axioms.

What are the axioms used by H? In fact, there are many variants of H determined by the selection of axioms. The differences between these variants are an interesting topic (and have a significance beyond the study of propositional logic), but we do not pursue this path here. (Exercises 1.19, 1.20, and 1.21 of the next section indicate some of the questions that arise.) Here we use the simplest and most obvious set of axioms.

For arbitrary formulas θ, ψ, ϕ consider the compounds

$$\iota(\theta) = \theta \to \theta \qquad \kappa(\theta, \psi) = \psi \to \theta \to \psi \qquad \sigma(\theta, \psi, \phi) = \chi \to \xi \to \theta \to \phi$$

where we have

$$\chi = \theta \to \psi \to \phi \qquad \xi = \theta \to \psi$$

in σ. We use these as the axioms.

1.10 DEFINITION. For the system H the axioms are the judgements $\Gamma \vdash \zeta$ where ζ is one of $\iota(\theta), \kappa(\theta,\psi), \sigma(\theta,\psi,\phi)$ for arbitrary θ,ψ,ϕ. □

Some of the reasons for choosing these axioms will become clear in the next section. Other reasons will emerge as the book develops.

Derivations of the same judgement in H and N can look quite different. With

$$\beta = (\psi \to \phi) \to (\theta \to \phi) \to (\theta \to \phi)$$

we see that Example 1.8 gives derivations of $\vdash \beta$ in H and N (and in two other systems). These have quite different shapes. Perhaps at this stage you should get some practice at producing derivations. Exercises 1.7, 1.8, and 1.9 should make you sweat a bit. (It is often said that horses sweat, gentlemen perspire, and ladies glow. However, as I said in the Introduction, you cannot merely watch Mathematics. If you want to understand it then at times you must work like a horse – a reverse form of murder is a good example.)

What is the significance of column 'Code' in Table 1.6?

Our primary objects of interest are derivations, certain kinds of trees, and the way these can be manipulated. It can be quite tedious to repeatedly draw derivations, perhaps with minor changes. We need an efficient notation for such trees. The **arboreal code** of a derivation does such a job.

Each derivation can be flattened to a linear sequence of judgements with brackets used to indicate the construction of the original tree. The arboreal code is a more efficient version of this flattened sequence.

Consider first the two rules

$$\frac{\begin{matrix} R \\ \Gamma, \theta \vdash \phi \end{matrix}}{\Gamma \vdash \theta \to \phi}\,(\uparrow) \qquad \frac{\begin{matrix} Q \\ \Gamma \vdash \theta \to \phi \end{matrix} \quad \begin{matrix} P \\ \Gamma \vdash \theta \end{matrix}}{\Gamma \vdash \phi}$$

where we already have the arboreal codes P, Q, R for the numerator components. These are such that the full derivations

$$(R)\ \ \Gamma, \theta \vdash \phi \qquad (Q)\ \ \Gamma \vdash \theta \to \phi \qquad (P)\ \ \Gamma \vdash \theta$$

can be reconstructed from the codes. With these we write

$$R\uparrow \qquad (QP)$$

for the two larger derivations. Thus a use of Introduction is indicated by '$\uparrow$', and a use of Elimination is indicated by brackets.

The arboreal code for a leaf

$$\Gamma \vdash \phi$$

is a little more involved. Whatever it is this code must tell us the context Γ, the predicate ϕ, and whether the leaf is an Axiom or a Projection (and which component of Γ is projected). Let's look at these two cases.

For an Axiom using the axiom ζ,

$$(\Gamma[\zeta])\ \ \Gamma \vdash \zeta$$

the arboreal code is $\Gamma[\zeta]$. We need to give both the context Γ and the axiom ζ since neither can be inferred from the other.

For a Projection

$$(\Gamma[n]) \quad \Gamma \vdash \theta$$

the arboreal code is $\Gamma[n]$ where n is the position of the projected formula θ in the context Γ. But what is this position? It's not quite what you might expect.

Suppose, as before, the context is

$$\Gamma = \theta_1, \theta_2, \dots, \theta_{l-1}, \theta_l$$

of length l, and suppose the projected formula is θ_i where $1 \le i \le l$. The obvious way to indicate the position is to take the index i. However, eventually this has some drawbacks. It is more convenient to count the context Γ from right to left (from the gate position outwards) starting at 0. Thus in

$$\begin{array}{ccccccc} & & l-1, & l-2, & \dots, & 1 & , 0 \\ \Gamma & = & \theta_1 & , \theta_1 & , \dots, & \theta_{l-1} & , \theta_l \end{array}$$

the position of each component is indicated above that component. In particular, the component θ_i occurs in position $l-i$. In a judgement

$$\Gamma \vdash \phi$$

the hypothesis in position 0 is always the one immediately to the left of the gate. The further away from the gate a hypothesis is, the higher the position.

We think of a context Γ as a stack. This can be altered only at position 0, by either pushing on a new component and making the other components move up a position, or pulling a component from position 0 and making the other components move down a position.

You may not see all the reasons for this convention just yet, but it is the most convenient one. An illustration of this occurs in the next section.

In an arboreal code some brackets can be conventionally omitted. We let

$$RQP \quad \text{abbreviate} \quad ((RQ)P)$$

i.e. we omit brackets using the *opposite* convention to that for formulas. This is not just a way of annoying you; in the later chapters we see that it really is the best convention.

1.11 EXAMPLE. Let

$$\beta = (\psi \to \phi) \to (\theta \to \phi) \to (\theta \to \phi)$$

where θ, ψ, ϕ are arbitrary formulas. Let Σ be an arbitrary context. By modifying Example 1.8 we produce derivations of $\Sigma \vdash \beta$ in H and N. Let $\Gamma = \tau, \sigma, \theta$ and set $\Delta = \Sigma, \Gamma$. The arboreal codes for the two derivations are

$$\Sigma[\sigma_1](\Sigma[\kappa_1]\Sigma[\sigma_2])\Sigma[\kappa_2] \qquad (\Gamma[\tau]\Gamma[\sigma]\Gamma[\theta])\uparrow\uparrow\uparrow$$

respectively. □

I told you earlier to practice producing H- and N-derivations. You should now go back to these and produce the arboreal code of each.

Exercises

This collection forms the basis of a whole series of running exercises and examples.

1.7 Consider the formulas

$$\begin{aligned}\beta &= (\psi \to \phi) \to (\theta \to \psi) \to (\theta \to \phi)\\ \gamma &= (\tau \to \sigma \to \rho) \to (\sigma \to \tau \to \rho)\\ \delta &= (\theta \to \psi) \to (\psi \to \phi) \to (\theta \to \phi)\end{aligned}$$

where $\theta, \psi, \phi, \rho, \sigma, \tau$ are arbitrary. For each of these formulas π construct an H-derivation and an N-derivation of the judgement $\vdash \pi$. Give the arboreal code of each derivation you construct.

1.8 Consider the formulas

$$\begin{aligned}\epsilon &= (\psi \to \zeta \to \phi) \to (\theta \to \zeta) \to (\theta \to \psi \to \phi)\\ \omega &= (\theta \to \theta \to \phi) \to (\theta \to \phi)\\ \mu &= (\zeta \to \rho \to \theta \to \psi \to \phi) \to \zeta \to \rho \to (\theta \to \psi) \to (\theta \to \phi)\\ \nu &= (\rho \to \theta \to \psi \to \phi) \to \rho \to (\theta \to \psi) \to (\theta \to \phi)\\ \tau &= (\theta \to \psi \to \phi) \to (\zeta \to \theta) \to (\zeta \to \psi) \to (\zeta \to \phi)\end{aligned}$$

where $\theta, \psi, \phi, \rho, \zeta$ are arbitrary. For each such formula π construct a derivation of $\vdash \pi$ in H and in N. In each case give the arboreal code.

1.9 For arbitrary formulas θ, ψ let θ^{ψ} abbreviate $(\theta \to \psi) \to \psi$, i.e. θ^{ψ} is a relativized double negation of θ over ψ. Let $\theta^{\psi\phi}$ be $(\theta^{\psi})^{\phi}$, (not $\theta^{(\psi^{\phi})}$). For each of the following formulas

(i) $\omega(1) = \theta \to \theta^{\psi}$ (ii) $\omega(2) = \theta \to \theta^{\psi\phi}$

(iii) $\omega(3) = (\theta \to \psi) \to (\theta^{\phi} \to \psi^{\phi})$

(iv) $\omega(4) = \theta^{\phi} \to \theta^{\psi\phi}$ (v) $\omega(5) = \theta \to \theta^{\psi\phi}$

exhibit a derivation, with arboreal code, of $\vdash \omega(\cdot)$ in both H and N. You may use known derivations (of the $\beta, \gamma, \delta, \ldots$ formulas) to abbreviate your answers. The derivations for (ii) and (v) should be different.

1.10 Consider the three axioms $\iota = \iota(\theta)$, $\kappa = \kappa(\theta, \psi)$, $\sigma = \sigma(\theta, \psi, \phi)$ used in H (where θ, ψ, ϕ are arbitrary). Show that for each context Γ the judgements

$$\Gamma \vdash \iota \quad \Gamma \vdash \kappa \quad \Gamma \vdash \sigma$$

are derivable in N.

1.4 Some algorithms on derivations

It seems that the primary job of a derivation system is to certify the validity of formulas (under hypothesis). In other words, it seems that for a judgement $\Gamma \vdash \phi$ the principal question is whether or not there is a derivation with this judgement at the root. There may be many such derivations, but the existence of just one is sufficient to answer this principal question. If this is the only question of interest then the subject can be closed with the proof of a completeness theorem (saying that a derivation of a judgement $\Gamma \vdash \phi$ exists precisely when there is a well understood semantic relationship between Γ and ϕ). The proof of a completeness theorem can leave a nice glow of self satisfaction (at least for those who are easily satisfied).

However, we should not have such limited ambitions. It is certainly the case that a completeness result should be one thing to aim for, but it should be just one of many things. There is much more that should follow.

We have seen that the same judgement can have many different derivations even within one derivation system. Also derivations in different systems can look very different. What is it that makes two derivations different? Surely there is some information that could be extracted from the internal structure of a derivation. What is the significance of the way we translate derivations from one system to another? Why is it that when we translate a derivation from one system into another and then back again, we don't necessarily retrieve the initial derivation? Are there ways of 'improving' a derivation to make it more efficient? What do we mean by 'different derivations' of, for that matter, 'different derivation systems'?

The importance of a derivation system is not measured by the class of judgements it can derive, but by the way it derives these judgements. The 'information content' of a derivation is not just the root judgement, but the way it gets from the leaves to this root. This, of course, is a rather hazy idea, but it will achieve at least some clarity as this story unfolds. Let's make a start on this clarification.

In this section we look at several algorithms. Each such algorithm A will consume a derivation ∇ (in H or N as required by A) and possibly other input data, to return a modified derivation ∇_A (not necessarily in the same system). In each case A proceeds by recursion over ∇. The behaviour of A is specified by the base clauses and the step clauses. The base clauses say how a leaf is treated, and the step clauses describe how to pass across a use of Introduction or Elimination (as appropriate). In other words, each algorithm is a procedure for manipulating certain trees. This is where the arboreal code begins to show its worth.

The first three algorithms show that H and N are derivationally equivalent: any judgement derivable in the one is derivable in the other. However, as remarked above, the mere derivability of a judgement is not the whole story. It is the manner of its derivation where the intrigue begins.

Put differently, the measure of an algorithm is not what it achieves, but the way it achieves it. Algorithms are not equivalent because they achieve the same results; questions of efficiency must be asked.

The first algorithm is almost trivial, but it is a useful introduction to the way algorithms on derivations work. Consider the three axioms ι, κ, σ used in H (where each is a compound of other formulas). By Exercise 1.10 we know that the Axioms of H are derivable in N. This provides a basis of a translation from H to N.

1.12 ALGORITHM. *An algorithm $(\cdot)_N$ which, when supplied with an arbitrary* H-*derivation ∇, will return an* N-*derivation ∇_N with the same root.*

Construction. The result of $(\cdot)_N$ is easy to describe. Look at the leaves of the supplied H-derivation ∇. Some of these will be Axioms

$$\Gamma \vdash \iota \qquad \Gamma \vdash \kappa \qquad \Gamma \vdash \sigma$$

for axioms ι, κ, σ. We simply grow the corresponding N-derivations above these leaves.

Formally $(\cdot)_N$ proceeds by recursion over the arboreal code ∇.

$$\begin{array}{lrcl} \text{(Axiom)} & \Gamma[\zeta]_N & = & \text{explicit derivation} \\ \text{(Projection)} & \Gamma[n]_N & = & \Gamma[n] \\ \text{(Elimination)} & (QP)_N & = & Q_N P_N \end{array}$$

An easy induction shows that ∇_N is an N-derivation with the same root as ∇. □

You may think that this proof is a bit laboured. If all algorithms were as simple as this then the arboreal code wouldn't be worth the effort. Some algorithms are quite complicated, and a hint of this is seen in the next one. Eventually we want to describe a companion to Algorithm 1.12, one which converts an N-derivation into an H-derivation. For this we need a preliminary algorithm which is usually hidden within 'The Deduction Theorem'.

1.13 ALGORITHM. (Deduction) *An algorithm $(\cdot)_D$*

$$(\nabla)\ \Gamma, \theta \vdash \phi \ \longmapsto\ (\nabla_D)\ \Gamma \vdash \theta \to \phi$$

which transports the gate predicate across the gate of a supplied H-*derivation (to return an* H-*derivation).*

Construction. For convenience let $\Theta = (\Gamma, \theta)$, the root context of the supplied derivation ∇. The algorithm proceeds by recursion over ∇, but we must distinguish a Gate Projection from an Other Projection.

$$\begin{array}{lrcll} \text{(Axiom)} & \Theta[\zeta]_D & = & \Gamma[\kappa]\Gamma[\zeta] & \kappa = \zeta \to \theta \to \zeta \\ \text{(Gate Projection)} & \Theta[0]_D & = & \Gamma[\iota] & \iota = \theta \to \theta \\ \text{(Other Projection)} & \Theta[n']_D & = & \Gamma[\kappa]\Gamma[n] & \kappa = \psi \to \theta \to \psi \\ \text{(Elimination)} & (QP)_D & = & \Gamma[\sigma]Q_D P_D & \sigma = \chi \to \xi \to \theta \to \phi \end{array}$$

In the (Other Projection) clause ψ is the projected formula, in position n' in Θ but in position n in Γ. (Remember that $n' = n + 1$.)

In the (Elimination) clause ψ is the root predicate of P. Let's look at the details of this clause. We have

$$(\nabla)\quad \frac{\overset{Q}{\Theta \vdash \psi \to \phi} \quad \overset{P}{\Theta \vdash \psi}}{\Theta \vdash \phi}$$

and we can obtain derivations

$$(Q_D)\ \Gamma \vdash \chi \qquad (P_D)\ \Gamma \vdash \xi$$

by recursive calls on the algorithm. Here

$$\chi = \theta \to \psi \to \phi \qquad \xi = \theta \to \psi$$

are the two predicates. Using the axiom

$$\sigma = \chi \to \xi \to \theta \to \phi$$

we obtain the required result, as shown.

$$\dfrac{\dfrac{\Gamma \vdash \sigma \quad \overset{Q_D}{\Gamma \vdash \chi}}{\bullet} \quad \overset{P_D}{\Gamma \vdash \xi}}{\Gamma \vdash \theta \to \phi}$$

□

There are a couple of points worth noting.

The axioms ι, κ, σ of H were chosen with this algorithm in mind. Other axioms lead to different deduction algorithms. We can always try to obtain ι, κ, σ from other axioms and then use the obvious modification of Algorithm 1.13. However, for suitably chosen axioms a quite different deduction algorithm can be constructed. Exercise 1.21 illustrates this.

The algorithm treats a Gate Projection in a different way to an Other Projection. The arboreal code is such that we can tell immediately whether or not a Projection $\Gamma[n]$ is a Gate or Other by checking whether or not $n = 0$. This is one reason why we measure the position of a component of a context in this way.

In essence the algorithm $(\cdot)_D$ shows how to simulate Introduction in H. With this we can produce the companion to $(\cdot)_N$.

1.14 ALGORITHM. *An algorithm $(\cdot)_H$ which, when supplied with an arbitrary* N-*derivation ∇, will return an* H-*derivation ∇_H with the same root.*

Construction. As usual, the algorithm proceeds by recursion on ∇.

$$\begin{array}{lrcl} \text{(Projection)} & \Gamma[n]_H & = & \Gamma[n] \\ \text{(Introduction)} & (R\!\uparrow)_H & = & (R_H)_D \\ \text{(Elimination)} & (QP)_H & = & Q_H P_H \end{array}$$

The step across a use of Introduction calls on the Deduction Algorithm 1.13. □

The best way to get to know an algorithm is to use it. At this point you should try Exercises 1.11–1.16.

As the story unfolds we will meet many other algorithms. Some of these are quite complicated. To conclude this section we look at two algorithms which appear, in different guises, over and over again. At this stage we do not fill in all the details. In the later chapters we develop a more efficient approach to these algorithms.

The first algorithm is concerned with omitting a hypothesis which at a later stage we find is derivable.

1.15 ALGORITHM. (Grafting) *A pair of algorithms – one for* H *and one for* N *– which, when supplied with a pair of derivations in the parent system*

$$(R) \quad \Pi, \theta, \Delta \vdash \phi \qquad\qquad (S) \quad \Pi, \Delta \vdash \theta$$

and a nominated component θ of the R-context (as indicated), will return

$$(R * S) \quad \Pi, \Delta \vdash \phi$$

a derivation in the given system.

Construction. Informally the algorithm locates all the Projections from θ in R, replaces these by S, and then modifies R accordingly. We call this the result of grafting the side derivation S onto R at the Projections from the nominated θ.

Formally the algorithm proceeds by recursion over R. Using the given Π and Δ let $\Sigma = \Pi, \theta, \Delta$ and $\Lambda = \Pi, \Delta$. For each system we use the appropriate base and step clauses.

(Axiom)	$\Sigma[\zeta] * S$	$= \Lambda[\zeta]$
(Nominated Projection)	$\Sigma[n] * S$	$= S$
(Other Projection)	$\Sigma[r] * S$	$= \Lambda[s]$
(Introduction)	$(R\uparrow) * S$	$= (R * S\downarrow)\uparrow$
(Elimination)	$(QP) * S$	$= (Q * S)(P * S)$

In the (Other Projection) clause, r is the position of the non-nominated projected component in Σ and s is its position in Λ. Thus $s = r$ if the component is in Δ, but $s = r - 1$ if the component is in Π.

The (Introduction) clause is not needed for the H-algorithm. Thus this version of the algorithm is straightforward. For the N-algorithm the (Introduction) clause is needed. We are given a pair of derivations

$$(R) \quad \Sigma, \psi \vdash \chi \qquad\qquad (S) \quad \Pi, \Delta \vdash \theta$$

where $\phi = \psi \rightarrow \chi$, and we must construct $R\uparrow * S$. We want to make a recursion call on the algorithm but the two contexts

$$\Pi, \theta, \Delta, \psi \qquad \Pi, \Delta$$

don't match as they should. The algorithm won't accept this R and S to return '$R * S$'. We need to Weaken S to

$$(S\downarrow) \quad \Pi, \Delta, \psi \vdash \theta$$

by inserting a dummy hypothesis, as shown. At this stage we do not say how this is done. Once we have $S\downarrow$ we can form

$$(R * S\downarrow) \quad \Lambda, \psi \vdash \chi$$

and then a use of Introduction gives the required result. □

A judgement may have many different derivations (within the same system). Is it possible to say that some of these are 'better' than others? A convincing answer will need a lot more analysis than we have done so far. However, we can illustrate some aspects of 'efficient derivation'.

For the time being consider the system N. An abnormality in an N-derivation is a use of Introduction followed by an immediate use of Elimination.

$$\dfrac{\dfrac{\begin{array}{c} R \\ \Gamma, \theta \vdash \phi \end{array}}{\Gamma \vdash \sigma \to \rho}(\uparrow) \qquad \begin{array}{c} S \\ \Gamma \vdash \theta \end{array}}{\Gamma \vdash \phi}$$

An N-derivation is normal if it contains no abnormalities. Are abnormalities necessary? Informally we could guess that they are not.

> If a hypothesis θ is derivable from the preceding part Γ of the hypothesis list, then the derivation can be reorganized to omit the use of θ.

There is an informal procedure for removing abnormalities. It can be argued convincingly that this produces a 'better' derivation (for it doesn't need to hypothesize θ only to discharge it later). Let's try to make this precise.

1.16 ALGORITHM. (Normalization) *An algorithm which, when supplied with an arbitrary* N*-derivation* ∇*, will return a normal* N*-derivation* ∇^* *with the same root.*

Construction. We survey the supplied derivation ∇ to locate the abnormalities we don't like. Each abnormality, as above, can be replaced by $R * S$ using the Grafting Algorithm 1.15. This gives us a new derivation ∇' with the same root but with the offending abnormalities removed. This derivation need not be normal, for grafting can create new abnormalities. However, by repeating the process we obtain a sequence

$$\nabla, \nabla', \nabla'', \nabla''', \ldots, \nabla^{(i)}, \ldots$$

of derivations which (we hope) eventually stabilizes in a normal derivation ∇^*. □

There are quite a few holes in this algorithm. In which order should we remove the abnormalities? There is scope for several different strategies here. A more serious problem with this is that removing one abnormality can create another. In an abnormality, even if both components R and S are normal, the result $R * S$ need not be normal. In fact removing one abnormality can create *two or more* new ones. In short, on the face of it we cannot be sure that iterating this 1-step removal algorithm will eventually terminate in a normal derivation. Termination can be proved, but this is not as straightforward as it seems at first sight.

Exercise 1.17 illustrates some of the things that can happen.

These gaps can be filled in, but not here. We must wait until we have developed the appropriate techniques.

What about the system H? The notion of an abnormality doesn't make sense for H, but surely there are some inefficient derivations. At the moment it is hard to see what these inefficiencies might be. Again we must wait until we have developed the appropriate techniques before we can tackle this problem.

This selection of algorithms in this section is fairly representative of what can be done with derivations. The exercises illustrate what can happen when you use the algorithms. As you try these exercises you should get a feeling that something is missing; there is some information about derivations that is not being documented properly. That is what the Curry–Howard correspondence is about.

Exercises

1.11 Estimate the dimensions of the N-derivation ∇_N in terms of those of the parent H-derivation ∇.

1.12 Estimate the increase in height of an H-derivation a use of $(\cdot)_D$ may cause.

1.13 Consider Example 1.8 concerning the formula β. Using the abbreviations of that example consider the H- and N-derivation ∇ given to the right. Three uses of Introduction produce an N-derivation $\nabla\uparrow\uparrow\uparrow$ of $\vdash \beta$. Three uses of $(\cdot)_D$ produce an H-derivation of $\vdash \beta$. Calculate the three derivations ∇_D, ∇_{DD}, ∇_{DDD} and compare the final result with the known H-derivation of $\vdash \beta$.

$$\dfrac{\Gamma \vdash \tau \qquad \dfrac{\Gamma \vdash \sigma \qquad \Gamma \vdash \theta}{\Gamma \vdash \psi}}{\Gamma \vdash \phi}$$

1.14 Consider N-derivations of shapes

$$(QP)\uparrow \quad (QP)\uparrow\uparrow \quad (QP)\uparrow\uparrow\uparrow \quad (QP)\uparrow\uparrow\uparrow\uparrow \quad \cdots$$

for arbitrary N-derivations Q, P. Use $(\cdot)_H$ to translate these into H-derivations. The resulting shapes may refer to translations of Q and P and to the results of applying $(\cdot)_D$ to these derivations and certain leaf derivations for H.

1.15 Consider the three H-Axioms $\vdash \iota$, $\vdash \kappa$ and $\vdash \sigma$ for the usual ι, κ, σ. Translate each of these into an N-derivation, and then translate these back into H-derivations.

1.16 Show $h(\nabla_H)+1 \leq 2^{h(\nabla)+1}$ for each N-derivations ∇. Can you improve this?

1.17 Consider the formulas

$$\begin{aligned}
\beta &= (\psi \to \phi) \to (\theta \to \psi) \to (\theta \to \phi) \\
\gamma &= (\tau \to \sigma \to \rho) \to (\sigma \to \tau \to \rho) \\
\delta &= (\theta \to \psi) \to (\psi \to \phi) \to (\theta \to \phi)
\end{aligned}$$

for arbitrary formulas $\theta, \psi, \phi, \rho, \sigma, \tau$. We have seen derivations of

$$(B) \quad \vdash \beta \qquad (C) \quad \vdash \gamma \qquad (D) \quad \vdash \delta$$

in both H and N. If we now choose

$$\rho = \theta \to \phi \qquad \sigma = \theta \to \psi \qquad \tau = \psi \to \phi$$

then we have

$$\beta = \tau \to \sigma \to \rho \qquad \gamma = \beta \to \delta \qquad \delta = \sigma \to \tau \to \rho$$

so that CB provides another derivation of $\vdash \delta$.

(a) Using the two known N-derivations of B and C, normalize this new derivation CB, and compare the final derivation with the known N-derivation D.

(b) Using the two known H-derivations B and C, how does the new derivation CB compare with D?

1.18 In Section 1.2 we considered the structural rules X, C, T (and W). None of these are included as rules of H and N. Show that these rules are admissible for both these systems, i.e. the addition of these rules does not increase the class of derivable judgements. Describe algorithms which eliminate any use of these rules. How do these algorithms affect the shape of a derivation?

The exercises of Section 1.3 should have convinced you that the system H is not the most efficient we could devise. Various improvements can be made using a different selection of axioms. Most of these use the formulas $\beta, \gamma, \delta, \omega, \ldots$, and were devised from a purely logical background. The next three exercises consider a system which had its origins in functional programming.

The standard Hilbert system H is based on the three logical axioms

$$\text{(i)} \quad \iota(\theta) \qquad \text{(k)} \quad \kappa(\theta, \psi) \qquad \text{(s)} \quad \sigma(\theta, \psi, \phi)$$

for arbitrary formulas θ, ψ, ϕ. Consider the formula

$$\tau(\zeta; \lambda, \mu, \nu) = (\lambda \to \mu \to \nu) \to (\zeta \to \lambda) \to (\zeta \to \mu) \to (\zeta \to \nu)$$

for arbitrary formulas λ, μ, ν and ζ. We replace the axiom (s) by the axiom

$$\text{(t)} \quad \tau(\zeta; \lambda, \mu, \nu)$$

to obtain the Hilbert system T.

1.19 (a) Derive $\vdash \tau$ in H. Using this show how to translate T-derivations into H-derivations.

(b) Derive $\vdash \sigma$ in T. Using this show how to translate H-derivations into T-derivations.

1.20 Suppose $\lambda, \mu, \nu, \theta_1, \theta_2, \ldots, \theta_r, \ldots$ are given formulas. Set

$$\begin{array}{lll} \lambda_0 = \lambda & \mu_0 = \mu & \nu_0 = \nu \\ \lambda_{r+1} = \theta_{r+1} \to \lambda_r & \mu_{r+1} = \theta_{r+1} \to \mu_r & \nu_{r+1} = \theta_{r+1} \to \nu_r \end{array}$$

and then set

$$\xi_r = \lambda_r \to \mu_r \to \nu_r \qquad \tau_r = \tau(\theta_{r+1}; \nu_r, \mu_r, \lambda_r)$$

for each appropriate r.

(a) Assuming $\Gamma \vdash \xi_0$ is derivable in T, show that $\Gamma \vdash \xi_r$ is derivable in T. Exhibit the arboreal code and shape of the second derivation in terms of those of the first.

(b) Assuming

$$\Gamma \vdash \xi_0 \qquad \Gamma \vdash \lambda_r \qquad \Gamma \vdash \mu_r$$

are derivable in T, show that

$$\Gamma \vdash \nu_r$$

is derivable in T, and exhibit an algorithm which produces such a derivation.

1.21 (a) Describe an algorithm which, when supplied with a T-derivation with root

$$\Gamma, \theta_r, \ldots, \theta_1 \vdash \phi$$

will return a T-derivation of

$$\Gamma \vdash \theta_r \to \cdots \to \theta_1 \to \phi$$

(for an arbitrary context Γ and formulas $\theta_1, \ldots, \theta_r, \phi$). You may concentrate on the case where the concluding rule of the given derivation is a use of Elimination.

(b) Translate the N-derivations of Exercise 1.14 into T-derivations.

2
Computation Mechanisms

2.1 Introduction

In this chapter we introduce the second tree-like structure which permeates the whole book. A **computation** □ is (an instance of) a formal analogue of a kind of calculation. Suppose we have a family of terms. These are strings of certain primitive symbols put together according to given construction rules. We analyse a kind of **reduction relation** on such terms. These are re-write reductions. This chapter deals with the two basic examples which we build on in the rest of the book. Each such relation is produced in three stages, and it is useful to have a general description of this procedure.

At the first stage we produce a **1-step reduction** $\rhd$ on terms. This is defined by enumerating all instances

$$l \rhd r$$

of the relation. There may be infinitely many such instances, but in practice these fall into finitely many recognizable kinds. These instances are called the **reduction axioms** of the relation under construction.

The second stage is the heart of the construction. The idea is that if a term t^- has a subterm l which matches the left hand component of a 1-step reduction (as above), then l may be replaced in t^- by r to form a new term t^+. Let us write

$$t^- \,\dot{\rhd}\, t^+$$

to indicate that one such 1-step replacement has been carried out to transform t^- into t^+. We may take a chain of such 1-step replacements

$$t_0 \,\dot{\rhd}\, t_1 \,\dot{\rhd}\, \cdots \,\dot{\rhd}\, t_m$$

where different reduction axioms may be used at the different steps. This produces the reduction relation on terms. Thus

$$t^- \rhd\!\!\!\rhd t^+ \quad \Longleftrightarrow \quad \text{there is a chain, as above, with } t^- = t_0 \text{ and } t^+ = t_m$$

or, equivalently, $\rhd\!\!\!\rhd$ is the transitive, structural closure of $\rhd$.

Each instance $t^- \rhd\!\!\!\rhd t^+$ is witnessed by a collection of instances of $\rhd$ together with a description of how these 1-step reductions are combined. We will format this description as a tree-like structure

$$(\square)\quad t^- \mathrel{\rhd\!\!\rhd} t^+$$

of instances of $\mathrel{\rhd\!\!\rhd}$ with the instances of $\rhd$ at the leaves and the required instance of $\mathrel{\rhd\!\!\rhd}$ at the root. Such a tree is called a **computation**.

The third stage is little more than a technical convenience. By construction $\mathrel{\rhd\!\!\rhd}$ is transitive but need not be reflexive. (In fact, any term t with $t \mathrel{\rhd\!\!\rhd} t$ is a cause for concern.) There are times when a reflexive version of $\mathrel{\rhd\!\!\rhd}$ is useful, so we let $\mathrel{\rhd\!\!\rhd\!\!\rhd}$ be the reflexive closure of $\mathrel{\rhd\!\!\rhd}$.

$$t^- \mathrel{\rhd\!\!\rhd\!\!\rhd} t^+ \iff t^- \mathrel{\rhd\!\!\rhd} t^+ \text{ or } t^- = t^+$$

This will make certain properties of $\mathrel{\rhd\!\!\rhd}$ easier to discuss.

In this chapter we look at two families of terms

Combinator terms $\qquad$ λ-terms

and the appropriate reduction relations for these. These two examples should make clear the general procedure for constructing and analysing such relations. This is just the starting point for a whole collection of reduction relations considered in this book.

Exercises

2.1 This exercise introduces you to some aspects of the analysis of reduction relations which, on the whole, are not dealt with in detail in this book. The exercise is quite long, but you will find the ideas useful later.

Consider the **arithmetical terms** generated from a constant $\mathsf{0}$ by three constructions

$$\frac{t}{(\mathsf{S}t)} \qquad \frac{s \quad r}{(+sr)} \qquad \frac{s \quad r}{(\times sr)}$$

(for arbitrary terms r, s, t). There are no identifiers in these terms.

The 1-step reduction $\rhd$ has four kinds of instances.

$$+s\mathsf{0} \rhd s \quad +s(\mathsf{S}r) \rhd \mathsf{S}(+sr) \qquad \times s\mathsf{0} \rhd \mathsf{0} \quad \times s(\mathsf{S}r) \rhd +s(\times sr)$$

Here r, s are arbitrary terms and, as usual, some outermost brackets have been omitted.

The reduction relations $\dot{\rhd}$, $\mathrel{\rhd\!\!\rhd}$, $\mathrel{\rhd\!\!\rhd\!\!\rhd}$ are generated from $\rhd$ as indicated in the section.

A term t is **normal** if it has no subterm which matches the right hand side of a 1-step reduction (above). Thus t is normal if there is no term t' such that $t \mathrel{\dot{\rhd}} t'$ holds.

(a) Write down tree-like rules which generate $\dot{\rhd}$ from $\rhd$.
(b) Write down tree-like rules which generate $\mathrel{\rhd\!\!\rhd}$ from $\rhd$.
(c) Describe explicitly the class of normal terms. You should find these are in a natural correspondence with $\mathbb{N}$.

(d) Show that for each term t there is at least one normal term t^* with $t \mathrel{\rhd\!\!\rhd\!\!\rhd} t^*$. This shows that the reduction relation $\mathrel{\rhd\!\!\rhd}$ is normalizing.

(e) Using your rules from (a) show that for each divergent wedge

$$t_0 \mathrel{\dot{\rhd}} t_1 \qquad t_0 \mathrel{\dot{\rhd}} t_2$$

from a common source t_0, there is a convergent wedge

$$t_1 \mathrel{\rhd\!\!\rhd\!\!\rhd} t_3 \qquad t_2 \mathrel{\rhd\!\!\rhd\!\!\rhd} t_3$$

to a common target t_3.

(f) Can you improve (e) to obtain

$$t_1 \mathrel{\dot{\rhd}} t_3 \qquad t_2 \mathrel{\dot{\rhd}} t_3$$

as conclusion?

(g) Show that for each divergent wedge

$$t_0 \mathrel{\rhd\!\!\rhd} t_1 \qquad t_0 \mathrel{\rhd\!\!\rhd} t_2$$

from a common source t_0, there is a convergent wedge

$$t_1 \mathrel{\rhd\!\!\rhd\!\!\rhd} t_3 \qquad t_2 \mathrel{\rhd\!\!\rhd\!\!\rhd} t_3$$

to a common target t_3. This shows that the reduction relation $\mathrel{\rhd\!\!\rhd}$ is confluent.

(h) Show that for each term t there is a unique normal term t^* with $t \mathrel{\rhd\!\!\rhd\!\!\rhd} t^*$.

(i) Find a natural semantics of terms which makes parts (e, f, g, h) easier to prove.

2.2 Combinator terms

The simplest kind of terms, the combinator terms, are built from

a stock of identifiers   a stock of constants

with the punctuation symbols '(' and ')' using two rules of construction.

- Each identifier and each constant is a term.
- If q, p are terms then so is the application (qp).

It is convenient to let

$x, y, z, \ldots$	range over identifiers
Z	range over constants
$p, q, r, s, t, \ldots$	range over terms

and then

$$t = x \mid \mathsf{Z} \mid (qp)$$

is a succinct description of the set $\mathbb{C}omb$ of these terms. A term that does not contain any identifiers, i.e. is built solely from the constants, is called a **combinator**.

The idea is to select and fix a particular stock of constants, each one of which is intended to name a particular function. In this book we will fix

$$\mathsf{I} \quad \mathsf{K} \quad \mathsf{S}$$

as the three basic constants. There are other possible selections, some quite extensive. It is interesting to analyse how different selected constants affect the behaviour and efficiency of the generated system, but that is for later. Here we stick to the above three constants.

Each term has a **parsing tree**. This is a tree of terms, with the given term at the root, with identifiers and constants at the leaves, and grown to show how the root term is constructed. The way nodes are combined in the middle of the tree is shown to the right.

$$\frac{q \quad p}{(qp)}$$

2.1 EXAMPLE. When a parsing tree is used we need not fill in all the terms at the nodes. The tree to the right indicates the construction of a term Δ with subterms $\mathsf{B}, \omega, \Omega$ using all three constants but no identifiers. You should write out these terms, since all will be used again later. □

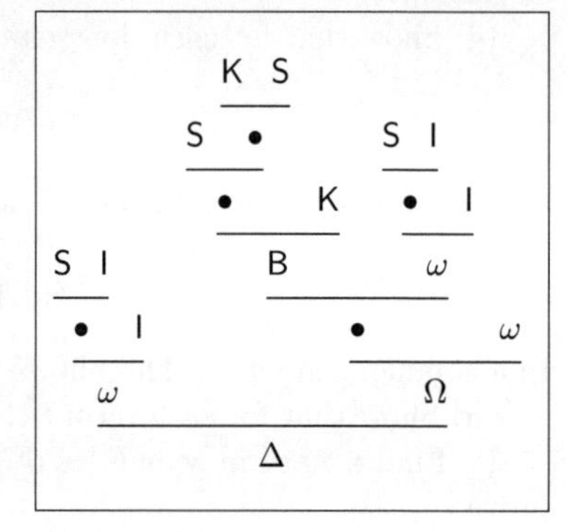

The parsing tree of a term gives the intended punctuation. However, when the term is written as a linear string in accordance with the official definition, the intended punctuation must be shown. Brackets are used to ensure that each term can be uniquely parsed. When displayed in its official version a term can contain many pairs of brackets, and this can obscure its structure. So we have some informal conventions for omitting some brackets. In particular, for terms r, s, t we write

$$tsr \quad \text{for} \quad ((ts)r)$$

(and *not* $(t(sr))$), i.e. omitted brackets collect to the left. This is opposite to the convention for formulas and types. The two conventions interact nicely; they have not been chosen at random.

The construction of combinator terms has no binding mechanism for identifiers. All identifiers occurring in a term are free (and cannot be bound). Because of this the notion of substitution is unproblematic. We write

$$t[x := s]$$

to indicate the term formed by substituting the term s for the identifier x throughout t. We will frequently meet terms

$$t(t(t(\cdots(ts)\cdots)))$$

formed by repeated application of a single term t to a term s. When there are m copies of t it is convenient to abbreviate this as $t^m s$ so that

$$t^0 s = s \qquad t^{m'} s = t(t^m s)$$

is a recursive description of this notation. This does *not* mean a term t^m applied to a term s. This will catch you out several times.

EXERCISES

2.2 For each of the combinator terms

$$\begin{array}{llll} \mathsf{B} = \mathsf{S}(\mathsf{KS})\mathsf{K} & \mathsf{C} = \mathsf{S}(\mathsf{BBS})(\mathsf{KK}) & \mathsf{D} = \mathsf{S}(\mathsf{K}(\mathsf{SB}))\mathsf{K} & \mathsf{E} = \mathsf{BBC} \\ \mathsf{W} = \mathsf{SS}(\mathsf{KI}) & \mathsf{M} = \mathsf{B}(\mathsf{BS}) & \mathsf{N} = \mathsf{S}(\mathsf{KS}) & \mathsf{T} = \mathsf{MB} \end{array}$$

write out the parsing tree.

2.3 COMBINATOR REDUCTION

Each combinator term is built up from the constants and identifiers using application. The constants are selected to name certain functions. Here the intention is that

I names the identity function

K names the left-projection function

S names a parameterized version of composition (to be described later)

but other selections will have different intended denotations. The idea is that

qp names the result of applying the denotation of q to the denotation of p

so that each term names a certain compound of the primitive functions. Strictly speaking, this is relative to an assigned interpretation for each identifier, but you can see the idea.

For the time being let us write $t \approx s$ to indicate that the two terms t, s have the same denotation. This equivalence can hold even when t and s are distinct terms. For instance

$$\mathsf{I}r \approx r \qquad \mathsf{K}sr \approx s \qquad \mathsf{S}tsr \approx (tr)(sr)$$

hold for all terms r, s, t. This is because of the intended meaning of $\mathsf{I}, \mathsf{K}, \mathsf{S}$. The third equivalence explains the phrase 'parameterized version of composition' used above.

There is a qualitative difference between the left hand member and the right hand member of these equivalences. In general we want to be able to move from the left hand term to the right hand term (but not necessarily in the other direction). We want to say

$$\mathsf{I}r \text{ reduces to } r \qquad \mathsf{K}sr \text{ reduces to } s \qquad \mathsf{S}tsr \text{ reduces to } (tr)(sr)$$

for arbitrary terms r, s, t.

2.2 EXAMPLE. Consider the two terms

$$\mathsf{KK}rst \qquad \mathsf{KK}tsr$$

for arbitrary terms r, s, t. In the first of these we can replace $\mathsf{KK}r$ by K, and in the second we can replace $\mathsf{KK}t$ by K. The two terms reduce to $\mathsf{K}st$ and $\mathsf{K}sr$ respectively, and then both of these reduce to s. Thus

$$\mathsf{KK}rst \approx \mathsf{KK}tsr$$

but neither term reduces to the other. □

We formalize this as a re-writing calculation by constructing a **reduction relation** $\rhd\!\!\!\rhd$ on terms. This is the first example of several such relations. We follow the general procedure as outlined in Section 2.1. In this section we describe the particular case for the combinator terms generated from $\mathsf{I}, \mathsf{K}, \mathsf{S}$. In later sections and chapters we refine this procedure several times.

The first stage is always to define the **1-step reductions** or **reduction axioms**.

2.3 DEFINITION. For the constants $\mathsf{I}, \mathsf{K}, \mathsf{S}$ the 1-step reduction axioms are

$$\mathsf{I}r \rhd r \qquad \mathsf{K}sr \rhd s \qquad \mathsf{S}tsr \rhd (tr)(sr)$$

for arbitrary terms r, s, t. In other words, $t^- \rhd t^+$ holds for terms t^-, t^+ if and only if t^- has one of the three left hand forms and t^+ has the companion right hand form. □

A different selection of constants will lead to a different set of 1-step reductions. (It may appear from this particular case that each selected constant Z has an associated reduction axiom of the form

$$\mathsf{Z}r_m \cdots r_1 \rhd \cdots$$

for arbitrary terms $r_1, \ldots, r_m$, with m determined by Z. This does happen quite a lot, but some reduction relations are not generated in this straightforward fashion.)

The next stage in the procedure is to extend $\rhd$ to its transitive, structural closure on terms. Informally $t^- \rhd\!\!\!\rhd t^+$ holds if we can get from t^- to t^+ in a sequence of steps where at each step a 1-step reduction $l \rhd r$ is used to replace an occurrence of l as a subterm by r. For instance

$$(\mathsf{KK}r)st \rhd\!\!\!\rhd \mathsf{K}st \rhd\!\!\!\rhd s$$

makes two calls on the reduction axiom for K.

Formally each reduction $t^- \rhd\!\!\!\rhd t^+$ is witnessed by a tree of reductions, with the given reduction at the root, with 1-step reductions at the leaves, and where the tree shows how the crucial subterms are located.

Clause	Shape	Remarks	Code
(Axiom reduction)	$\dfrac{t^- \rhd t^+}{t^- \rhd\!\!\rhd t^+}\,(0)$	1-step leaf	0
(Left application)	$\dfrac{\begin{matrix}\mathsf{q}\\ q^- \rhd\!\!\rhd q^+\end{matrix}}{q^-p \rhd\!\!\rhd q^+p}\,(\downharpoonleft)$		$\downharpoonleft\mathsf{q}$
(Right application)	$\dfrac{\begin{matrix}\mathsf{p}\\ p^- \rhd\!\!\rhd p^+\end{matrix}}{qp^- \rhd\!\!\rhd qp^+}\,(\downharpoonright)$		$\downharpoonright\mathsf{p}$
(Transitive composition)	$\dfrac{\begin{matrix}\mathsf{l} & \mathsf{r}\\ t^- \rhd\!\!\rhd t^0 & t^0 \rhd\!\!\rhd t^+\end{matrix}}{t^- \rhd\!\!\rhd t^+}\,(\circ)$		$(\mathsf{l} \circ \mathsf{r})$

Table 2.1: The computation rules for combinator terms

2.4 DEFINITION. A computation

$$(\square)\quad t^- \rhd\!\!\rhd t^+$$

is a finite rooted tree of reductions grown according to the rules of Table 2.1. Such a computation $\square$ is said to **organize** the root reduction $t^- \rhd\!\!\rhd t^+$ where t^- is the **subject** and t^+ is the **object** of the computation. The right hand column of the table shows how to generate the **arboreal code** of a computation. $\square$

As with all the tree-like structures we meet in this book, it is useful to have a notation for each computation $\square$. We take a flattened version of $\square$ to produce its arboreal code. This is generated from the constant 0 by the rules

$$\frac{\mathsf{q}}{\downharpoonleft\mathsf{q}}\,(\downharpoonleft) \qquad \frac{\mathsf{p}}{\downharpoonright\mathsf{p}}\,(\downharpoonright) \qquad \frac{\mathsf{l}\ \ \mathsf{r}}{\mathsf{l} \circ \mathsf{r}}\,(\circ)$$

as in the table. In time we will introduce conventions for omitting some brackets in a piece of code, but not just yet. The code 0 indicates the use of a 1-step reduction but doesn't tell us which. In practice we don't use a computation in isolation, and we can extract the missing information from elsewhere.

2.5 EXAMPLES. Let's look at some simple computations and a fairly complicated one.

(a) Consider $\mathsf{E} = \mathsf{SA}$ where A is an arbitrary combinator. Then we have $\mathsf{SA}sr \rhd (\mathsf{A}r)(sr)$ for all terms r, s to produce a small computation, as shown.

$$(0)\quad \frac{\mathsf{E}sr \rhd (\mathsf{A}r)(sr)}{\mathsf{E}sr \rhd\!\!\rhd (\mathsf{A}r)(sr)}\,(0)$$

$$\dfrac{\dfrac{\mathsf{G}t \rhd (\mathsf{KS}t)(\mathsf{KA}t)}{\mathsf{G}t \mathrel{\rhd\!\!\rhd} (\mathsf{KS}t)(\mathsf{KA}t)}(0) \qquad \dfrac{\dfrac{\dfrac{\mathsf{KS}t \rhd \mathsf{S}}{\mathsf{KS}t \mathrel{\rhd\!\!\rhd} \mathsf{S}}(0)}{(\mathsf{KS}t)(\mathsf{KA}t) \mathrel{\rhd\!\!\rhd} \mathsf{S}(\mathsf{KA}t)}(\downharpoonleft) \quad \dfrac{\dfrac{\mathsf{KA}t \rhd \mathsf{A}}{\mathsf{KA}t \mathrel{\rhd\!\!\rhd} \mathsf{A}}(0)}{\mathsf{S}(\mathsf{KA}t) \mathrel{\rhd\!\!\rhd} \mathsf{SA}}(\downharpoonright)}{(\mathsf{KS}t)(\mathsf{KA}t) \mathrel{\rhd\!\!\rhd} \mathsf{SA}}(\circ)}{\mathsf{G}t \mathrel{\rhd\!\!\rhd} \mathsf{SA}}(\circ)$$

$$\dfrac{\dfrac{\dfrac{\mathsf{G}t \mathrel{\rhd\!\!\rhd} \mathsf{SA}}{\mathsf{G}ts \mathrel{\rhd\!\!\rhd} \mathsf{SA}s}(\downharpoonleft)}{\mathsf{G}tsr \mathrel{\rhd\!\!\rhd} \mathsf{SA}sr}(\downharpoonleft) \qquad \dfrac{\mathsf{SA}sr \rhd (\mathsf{A}r)(sr)}{\mathsf{SA}sr \mathrel{\rhd\!\!\rhd} (\mathsf{A}r)(sr)}(0)}{\mathsf{G}tsr \mathrel{\rhd\!\!\rhd} (\mathsf{A}r)(sr)}(\circ)$$

Table 2.2: An example computation

(b) Consider the particular case $\mathsf{R} = \mathsf{SK}$ of E. For all terms r, s we have

$$(0 \circ 0) \quad \dfrac{\dfrac{\mathsf{R}sr \rhd (\mathsf{K}r)(sr)}{\mathsf{R}sr \mathrel{\rhd\!\!\rhd} (\mathsf{K}r)(sr)}(0) \quad \dfrac{(\mathsf{K}r)(sr) \rhd r}{(\mathsf{K}r)(sr) \mathrel{\rhd\!\!\rhd} r}(0)}{\mathsf{R}sr \mathrel{\rhd\!\!\rhd} r}(\circ)$$

so that R is a right-projection combinator (to match K).

(c) Let $\mathsf{J} = \mathsf{RA}$ for an arbitrary combinator A. Then, using (b), we see that

$$(0 \circ 0) \quad \mathsf{J}r \mathrel{\rhd\!\!\rhd} r$$

for all terms r. (This shows that an identity combinator can be constructed out of K and S, and so we could omit I from our selection of constants. However, at this stage there are good reasons for including I.)

(d) Let $\mathsf{G} = \mathsf{S}(\mathsf{KS})(\mathsf{KA})$ for an arbitrary combinator A. Then the tree of Table 2.2 shows that for all r, s, t

$$\square = (\downharpoonleft\downharpoonleft(0 \circ (\downharpoonleft 0 \circ \downharpoonright 0))) \circ 0 \quad \text{organizes} \quad (\square) \quad \mathsf{G}tsr \mathrel{\rhd\!\!\rhd} (\mathsf{A}r)(sr)$$

hence the combinator E of (a) can be modified to deal with dummy arguments. □

The last example, (d), illustrates some facets of computations. The various linear branches of this computation can be put together in different ways to achieve the same effect. Thus

$$\square = (\downharpoonleft\downharpoonleft 0 \circ \downharpoonleft\downharpoonleft\downharpoonleft 0) \circ (\downharpoonleft\downharpoonleft\downharpoonright 0 \circ 0)$$

also organizes the same root reduction. You may like to fill in the details of this tree. These differences are concerned with **evaluation strategies**, a subject not dealt with here.

Computations can be quite large. In practice (unless we need the details) we often give a condensed, linear, version. Thus, using the G of (d), for each r, s, t

$$\mathsf{G}tsr = \mathsf{S}(\mathsf{KS})(\mathsf{KA})tsr \mathrel{\rhd\!\!\rhd} (\mathsf{KS}t)(\mathsf{KA}t)sr \mathrel{\rhd\!\!\rhd} \mathsf{SA}sr \mathrel{\rhd\!\!\rhd} (\mathsf{A}s)(\mathsf{A}r)$$

is an abbreviated version of the full computation.

The third stage in the construction of a reduction relation is to move to the reflexive closure $\rhd\!\!\rhd\!\!\rhd$ of $\rhd\!\!\rhd$, i.e. to the reflexive, transitive, structural closure of the 1-step relation $\rhd$. In this particular case the relation $\rhd\!\!\rhd$ is irreflexive, so can be retrieved from $\rhd\!\!\rhd\!\!\rhd$

$$t^- \rhd\!\!\rhd t^+ \iff t^- \rhd\!\!\rhd\!\!\rhd t^+ \text{ and } t^- \neq t^+$$

and the difference between $\rhd\!\!\rhd$ and $\rhd\!\!\rhd\!\!\rhd$ is little more than a technical convenience. Almost always we use $\rhd\!\!\rhd$. The reason for $\rhd\!\!\rhd\!\!\rhd$ will become clear in Section 2.7.

EXERCISES

2.3 Consider the eight combinators $\mathsf{B}, \mathsf{C}, \dots, \mathsf{T}$ of Exercise 2.2. Show that the following hold for all terms a, b, c, x, y, z.

$$\begin{array}{llll}
\text{(i)} & \mathsf{B}z \rhd\!\!\rhd \mathsf{S}(\mathsf{K}z) & & \mathsf{B}zyx \rhd\!\!\rhd z(yx) \\
\text{(ii)} & \mathsf{C}z \rhd\!\!\rhd \mathsf{B}(\mathsf{S}z)\mathsf{K} & & \mathsf{C}zyx \rhd\!\!\rhd zxy \\
\text{(iii)} & \mathsf{D}zy \rhd\!\!\rhd \mathsf{B}yz & & \mathsf{D}zyx \rhd\!\!\rhd y(zx) \\
\text{(iv)} & \mathsf{E}a \rhd\!\!\rhd \mathsf{B}(\mathsf{C}a) & & \mathsf{E}azyx \rhd\!\!\rhd (ax)(zy) \\
\text{(v)} & \mathsf{W}y \rhd\!\!\rhd \mathsf{S}y\mathsf{I} & & \mathsf{W}yx \rhd\!\!\rhd yxx \\
\text{(vi)} & \mathsf{M}cba \rhd\!\!\rhd \mathsf{S}(cba) & & \mathsf{M}abcyx \rhd\!\!\rhd (cbax)(yx) \\
\text{(vii)} & \mathsf{N}az \rhd\!\!\rhd \mathsf{S}(az) & & \mathsf{N}azyx \rhd\!\!\rhd (azx)(yx) \\
\text{(viii)} & \mathsf{T}az \rhd\!\!\rhd \mathsf{S}(\mathsf{B}az) & & \mathsf{T}azyx \rhd\!\!\rhd a(zx)(yx)
\end{array}$$

Organize each reduction as a computation and give the arboreal code. Indicate how alternative computations can arise. How do these relate to the H-derivations of Exercises 1.7 and 1.8?

2.4 Let

$$\mathsf{Y} = \mathsf{XX} \text{ where } \mathsf{X} = \mathsf{B}(\mathsf{SI})\mathsf{A} \qquad \mathsf{Z} = \mathsf{SXX} \text{ where } \mathsf{X} = \mathsf{CBA}$$

where $\mathsf{A} = \mathsf{SII}$ and B, C are as in Exercise 2.3. These terms Y, Z are, respectively, the Turing fixed point combinator and the Church fixed point combinator.

(a) Show that $\mathsf{Y} \rhd\!\!\rhd \mathsf{SIY}$ and hence $\mathsf{Y}f \rhd\!\!\rhd f(\mathsf{Y}f)$ for each term f.

(b) Show that for each term f there are terms g, h with $\mathsf{Z}f \rhd\!\!\rhd g \rhd\!\!\rhd fg$ and $\mathsf{Z}f \rhd\!\!\rhd h$, $f(\mathsf{Z}f) \rhd\!\!\rhd h$.

Often we need a combinator which behaves in a predetermined way. The next few exercises show how such a combinator can be constructed.

2.5 Let $\mathsf{L} = \mathsf{S}(\mathsf{KK})$. By using compounds of the form $\mathsf{K}^i(\mathsf{L}^j\mathsf{I})$ for $i, j \in \mathbb{N}$, show that for each pair of integers k, n with $1 \leq k \leq n$ and combinator Z, there are combinators $\binom{n}{k}$ and $\binom{n}{\mathsf{Z}}$ satisfying

$$\binom{n}{k}x_1 \cdots x_n \rhd\!\!\rhd x_k \qquad \binom{n}{\mathsf{Z}}x_1 \cdots x_n \rhd\!\!\rhd \mathsf{Z}$$

for all $x_1, \dots, x_n$.

2.6 Show that each combinator built up using only K and I is a projection combinator, i.e. behaves like $\binom{n}{k}$ for some $1 \le k \le n$. Can every projection combinator be built from K and I?

2.7 Using the Turner combinator $\mathsf{T} = \mathsf{B}(\mathsf{BS})\mathsf{B}$ show that

$$\mathsf{T}^n\mathsf{I}vux_1\cdots x_n \rhd\!\!\!\rhd (vx_1\cdots x_n)(ux_1\cdots x_n)$$

holds for all $u, v, x_1, \ldots, x_n$.

2.8 (a) Let X be an expression built up from a given list $x_1, \ldots, x_n$ of identifiers and certain other combinators. We wish to produce a combinator A satisfying

$$\mathsf{A}x_1\cdots x_n \rhd\!\!\!\rhd X \quad \text{i.e.} \quad \mathsf{A}a_1\cdots a_n \rhd\!\!\!\rhd X[x_1 := a_1, \ldots, x_n := a_n]$$

for all terms $a_1, \ldots, a_n$. By analysing the structure of X, show how to produce A.

(b) Exhibit combinators A satisfying

(i) $\mathsf{A}xyz \rhd\!\!\!\rhd zxy$ (ii) $\mathsf{A}xyz \rhd\!\!\!\rhd zyx$ (iii) $\mathsf{A}xyz \rhd\!\!\!\rhd \mathsf{S}(\mathsf{K}x)(\mathsf{S}yz)$

for all x, y, z.

The next four exercises show how the behaviour of a numeric function can be embedded in a computation. This theme is developed throughout the book.

2.9 Let $\overline{\mathrm{Suc}} = \mathsf{SB}$ and $\overline{0} = \mathsf{KI}$ and set $\overline{m} = (\mathsf{SB})^m\overline{0}$ for each $m \in \mathbb{N}$. Show that

$$\overline{m}yx \rhd\!\!\!\rhd y^m x \qquad \overline{\mathrm{Suc}}\,\overline{m} \rhd\!\!\!\rhd \overline{m+1}$$

hold for all $m \in \mathbb{N}$ and terms y, x.

2.10 Let $\overline{\mathrm{Add}} = \mathsf{TB}$, $\overline{\mathrm{Mlt}} = \mathsf{B}$, and $\overline{\mathrm{Exp}} = \mathsf{I}$ using the standard combinators B, T as in Exercise 2.2. Show that

$$\overline{\mathrm{Add}}\,\overline{n}\,\overline{m}yx \rhd\!\!\!\rhd y^{n+m}x \qquad \overline{\mathrm{Mlt}}\,\overline{n}\,\overline{m}yx \rhd\!\!\!\rhd y^{n\times m}x \qquad \overline{\mathrm{Exp}}\,\overline{n}\,\overline{m}yx \rhd\!\!\!\rhd y^{n^m}x$$

for all $m, n \in \mathbb{N}$ and terms y, x.

2.11 Let Swp and Jmp be any combinators such that

$$\mathsf{Swp}wvu \rhd\!\!\!\rhd uwv \qquad \mathsf{Jmp}xwvu \rhd\!\!\!\rhd u(wv)x$$

for all terms u, v, w, x. Let

$$\mathsf{Add} = \mathsf{Swp}(\mathsf{SB}) \quad \mathsf{Mlt} = \mathsf{Jmp}\overline{0}\mathsf{Add} \quad \mathsf{Exp} = \mathsf{Jmp}\overline{1}\mathsf{Mlt} \quad \mathsf{Bth} = \mathsf{CJmpExp}$$

using the numerals $\overline{0}, \overline{1}$ of Exercise 2.9 and the standard combinator C. Reduce

$$\mathsf{Add}\,\overline{n}\,\overline{m} \qquad \mathsf{Mlt}\,\overline{n}\,\overline{m} \qquad \mathsf{Exp}\,\overline{n}\,\overline{m} \qquad \mathsf{Bth}\,\overline{k}\,\overline{n}\,\overline{m}$$

to normal form (for $n, m, k \in \mathbb{N}$).

2.12 Let $\mathsf{Grz} = \mathsf{Swp}(\mathsf{Jmp}\overline{1})$ using the combinators Swp and Jmp of Exercise 2.11 and the numeral $\overline{1}$. Suppose the function $F : \mathbb{N} \longrightarrow \mathbb{N} \longrightarrow \mathbb{N}$ is represented by a combinator term F in the sense that $\mathsf{F}\overline{n}\,\overline{m} \rhd\!\!\!\rhd \overline{Fnm}$ for all $n, m \in \mathbb{N}$. Show that the term $\mathsf{Grz}\,\mathsf{F}\,\overline{i}\,\overline{n}\,\overline{m}$ reduces to a normal form for all $i, n, m \in \mathbb{N}$. Estimate the size of the associated computation.

2.4 λ-TERMS

You may think the development of Sections 2.2 and 2.3 is a bit neurotic; a lot of fuss to achieve very little. So it is, but it is meant as an introduction to the ideas involved, and when we deal with more complicated terms this preparation begins to pay dividends. In particular, this is the case when we have a binding construct which converts free occurrences of identifiers into non-free ones.

The λ-**terms** are built from

a stock of **identifiers**

with the abstraction symbol 'λ' and three punctuation symbols '.', '(', and ')' using three rules of construction.

- Each identifier is a term.
- If q, p are terms then so is the **application** (qp).
- If y is an identifier and r is a term, then the **abstraction** $(\lambda y\,.\,r)$ is a term.

It is convenient to let

$x, y, z, \ldots$	range over identifiers
$p, q, r, s, t, \ldots$	range over terms

and then

$$t = x \mid (qp) \mid (\lambda y\,.\,r)$$

is a succinct description of the set of terms. Let $\mathbb{L}amb$ be the set of such λ-terms.

Each λ-term has a **parsing tree** with the term at the root and certain identifiers at the leaves. The shape of the tree indicates the construction of the term. Thus

$$\frac{q \quad p}{(qp)} \qquad \frac{r}{(\lambda y\,.\,r)}\,(y)$$

are the construction rules. The abstracted identifier annotates the abstraction rule.

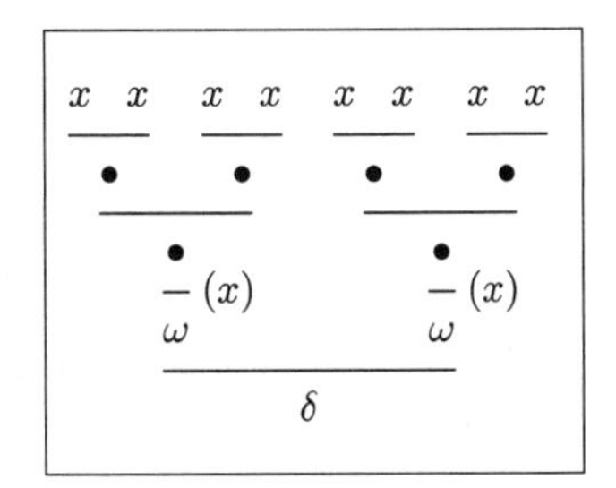

2.6 **EXAMPLE**. The tree to the right indicates the construction of a term δ with a subterm ω which occurs twice in δ. □

The parsing tree of a term indicates the intended punctuation. When terms are written as linear strings, as in the official definition, we must have explicit punctuation. This is the job of the brackets. In practice we do not display all brackets in a term. There are several conventions for omitting brackets. As with combinator terms, we write

$$tsr \quad \text{for} \quad ((ts)r)$$

for arbitrary terms r, s, t. As can be seen in the example

$$(\lambda x_1 . (\lambda x_2 . (\cdots . (\lambda x_n . s) \cdots)))$$

iterated abstractions can contain a lot of brackets. We omit some of these and write

$$\lambda x_1 . \lambda x_2 . \cdots . \lambda x_n . s \quad \text{or even} \quad \lambda x_1 , x_2 , \cdots , x_n . s$$

for this term.

The compound $\lambda x . r$ indicates a binding of x, i.e. all free occurrences of x in r (if any) become bound. This means that λ-terms have both free and bound occurrences of identifiers (perhaps of the same identifier), and we should be aware of the problems this can cause. For instance, the term

$$(\lambda x . ((\lambda x . ((\lambda x . x)x))x))x$$

is not what it seems at first sight.

The **support** ∂t or the **set of free occurring identifiers** of the term t is obtained by recursion over t.

$$\partial x = \{x\} \quad \partial(qp) = \partial q \cup \partial p \quad \partial(\lambda y . r) = \partial r - \{y\}$$

Even this does not tell us the full story, but it will do for the time being.

A λ-term t with no free occurring identifiers, i.e. with $\partial t = \emptyset$, is called a **λ-combinator**. We often refer to these as combinators, but they should not be confused with the combinators of the combinator calculus (although they do a similar job).

Two terms t_1, t_2 are **alphabetic variants** if they differ only in the choice of bound variables. Thus the two parsing trees have the same shape and differ only at the leaves with identifiers which become bound at some node in the construction. For most purposes alphabetic variants are interchangeable, but in some circumstances these minor differences can become important. For the time being we will mostly ignore the differences between alphabetic variants.

This brings us the problem of substitution.

For each pair of terms s, t and identifier x we write

$$t[x := s] \quad \text{to indicate} \quad \begin{array}{l}\text{the result of substituting } s \text{ for all}\\ \text{free occurrences of } x \text{ throughout } t\end{array}$$

as we did with combinator terms. Unlike these previous substitutions, this one cannot be performed by mere syntactic replacement; we must be careful to prevent the unwanted capture of identifiers.

The intention is that all freely occurring identifiers of s remain free after the substitution into t. In particular,

$$\partial(t[x := s]) = (\partial t - \{x\}) \cup \partial s$$

is the intended relationship between the various sets of free identifiers.

As an illustration of what can go wrong consider the two terms

$$t = \lambda y . yx \qquad t' = \lambda z . zx \qquad \text{with} \qquad \partial t = \partial t' = \{x\}$$

and where, for most purposes, the two terms t and t' are interchangeable. In particular, since t and t' are alphabetic variants, we expect that

$$t[x := s] \qquad t'[x := s]$$

are also alphabetic variants. What are these two terms

$$(\lambda y \,.\, yx)[x := s] \qquad (\lambda z \,.\, zx)[x := s]$$

for an arbitrary term s? A simple syntactic replacement gives

$$(\lambda y \,.\, ys) \qquad (\lambda z \,.\, zs)$$

which, for most cases, are alphabetic variants. But look at what happens when s is y (or has a free occurrence of y). We get

$$(\lambda y \,.\, yy) \qquad (\lambda z \,.\, zy)$$

which are not alphabetic variants (for the first one is a combinator and the second has a free occurrence of y). The second is a correct substitution but the first is incorrect.

The problem is that in

$$(\lambda \overset{1}{y}.\overset{1}{y}x)[x := \overset{2}{y}]$$

there are two distinct uses of y (as indicated) and these become confused by the syntactic manipulations. In particular, the abstraction '$\lambda\overset{1}{y}$' mistakenly binds the identifier $\overset{2}{y}$, which it shouldn't do since $\overset{1}{y}$ and $\overset{2}{y}$ are doing two quite unrelated jobs.

This is known as **identifier capture**; any free occurrence of y in s (the substituted term) is unintentionally caught in the scope of the internal abstraction λy in t (the receiving term). The standard way to avoid this is to change the overactive identifier y in t before the syntactic replacement.

In the example above, to calculate $t[x := s]$ we change t to t' and then perform the syntactic replacement to get $(\lambda z \,.\, zs)$ as the correct result. Rather, this is the correct result provided z does not occur free in s. If z does occur free in s then we must rename y and z by some identifier which does not occur free in s (and so prevent the capture of any freely occurring identifiers of s).

There are other unintended results if a substitution is carried out incorrectly. We will guard against these by suitable changes of abstracting identifiers, but this must be done with some care.

2.7 **DEFINITION**. For all terms s, t and identifiers x the term

$$t[x := s]$$

is generated by recursion over the structure of t using the following clauses.

$$\begin{array}{rclrcl} x[x := s] & = & s & (qp)[x := s] & = & (q[x := s])(p[x := s]) \\ y[x := s] & = & y & (\lambda y \,.\, r)[x := s] & = & \lambda z \,.\, (r'[x := s]) \end{array}$$

In the bottom left clause the identifier y is distinct from x. In the crucial bottom right clause we set

$$r' = r[y := v]$$

where v is a 'safe' identifier which avoids unwanted capture. □

For the time being we do not explain what is 'safe'. In Chapter 5 we look in detail at substitution algorithms and then we discuss this problem. If we wish we may set

$$\mathcal{U}\text{nsafe} = (\partial r - \{y\}) \cup \partial s \cup \{x\}$$

and take $v \notin \mathcal{U}\text{nsafe}$ with $v = y$ if possible. In practice we usually take a fresh identifier.

EXERCISES

2.13 Consider the term

$$(\lambda x \,.\, ((\lambda x \,.\, ((\lambda x \,.\, x)x))x))x$$

(as in the section). Write out a complete construction and parsing of this term, and indicate the free occurring identifiers of all subterms in the construction. Give an alphabetic variant of this term with the different bindings indicated by different identifiers.

2.14 Sticking strictly to the letter of the law calculate $(\lambda x \,.\, x)[x := x]$.

2.15 For the simple λ-term $t = (\lambda y \,.\, yx)$ calculate $t[x := s]$ where s is each of x, y, xy, xyz, t. You may assume that x, y, z are distinct.

2.16 Calculate $t[x := s]$ where $t = (\lambda y \,.\, ((\lambda z \,.\, zx)y))$ and where the identifiers x, y, z may or may not be distinct. You should set out all the gory details and indicate how various equalities amongst the identifiers affect the calculation.

2.5 λ-REDUCTION

What should we make of the term $(\lambda y \,.\, r)s$ and similar compounds? The intention is that $(\lambda y \,.\, r)$ describes the behaviour of a function where the value at y is given by r. Thus the value of the function at s can be found by substituting s for y in r. In short we should expect a relationship between the term $(\lambda y \,.\, r)s$ and the term $r[y := s]$ and there ought to be a standard mechanism which converts the first term to the second.

2.8 **DEFINITION.** A **reducible expression** or **redex** is a compound term

$$(\lambda y \,.\, r)s \qquad r[y := s]$$

for terms r, s and an identifier y as shown on the left. The term on the right is the **immediate reduct** of the redex. □

We want a procedure which replaces redex subterms of a term by 'equivalent terms'. Thus we generate a reduction relation $\rhd\!\!\!\rhd$ on terms to achieve redex removals. As usual, we start with the 1-step reductions.

Clause	Shape	Remarks	Code
(Redex reduction)	$\dfrac{t^- \rhd t^+}{t^- \rhd\!\!\!\rhd t^+}(1)$	$t^- = (\lambda y\,.\,r)s$ $t^+ = r[y := s]$	1
(Left application)	$\dfrac{\overset{\mathsf{q}}{q^- \rhd\!\!\!\rhd q^+}}{q^-p \rhd\!\!\!\rhd q^+p}(\downharpoonleft)$		$\downharpoonleft\mathsf{q}$
(Right application)	$\dfrac{\overset{\mathsf{p}}{p^- \rhd\!\!\!\rhd p^+}}{qp^- \rhd\!\!\!\rhd qp^+}(\downharpoonright)$		$\downharpoonright\mathsf{p}$
(Abstraction)	$\dfrac{\overset{\mathsf{r}}{r^- \rhd\!\!\!\rhd r^+}}{t^- \rhd\!\!\!\rhd t^+}(\uparrow)$	$t^- = (\lambda y\,.\,r^-)$ $t^+ = (\lambda y\,.\,r^+)$	$\uparrow\mathsf{r}$
(Transitive composition)	$\dfrac{\overset{\mathsf{l}}{t^- \rhd\!\!\!\rhd t^0} \quad \overset{\mathsf{r}}{t^0 \rhd\!\!\!\rhd t^+}}{t^- \rhd\!\!\!\rhd t^+}(\circ)$		$(\mathsf{l} \circ \mathsf{r})$

Table 2.3: The computation rules for λ-terms

2.9 DEFINITION. For the λ-terms the 1-step reductions are

$$(\lambda y\,.\,r)s \rhd r[y := s]$$

for terms r, s and an identifier y, i.e. each reduction axiom is given by a redex and its immediate reduct. □

With this we form $\rhd\!\!\!\rhd$ by taking the transitive, structural closure. Instances of this relation are witnessed by certain trees.

2.10 DEFINITION. A computation

$$(\square)\ \ t^- \rhd\!\!\!\rhd t^+$$

is a finite rooted tree of reductions grown according to the rules of Table 2.3. Such a computation □ is said to **organize** the root reduction $t^- \rhd\!\!\!\rhd t^+$ where t^- is the **subject** and t^+ is the **object** of the computation. The right hand column of the table shows how to generate the **arboreal code** of a computation. □

The arboreal code for a computation is generated from the constant 1 by the rules

$$\frac{\mathsf{q}}{\downharpoonleft\mathsf{q}}(\downharpoonleft) \qquad \frac{\mathsf{p}}{\downharpoonright\mathsf{p}}(\downharpoonright) \qquad \frac{\mathsf{r}}{\uparrow\mathsf{r}}(\uparrow) \qquad \frac{\mathsf{l} \quad \mathsf{r}}{\mathsf{l} \circ \mathsf{r}}(\circ)$$

$$\dfrac{\dfrac{\dfrac{\dfrac{\overline{2}s \rhd \lambda x\,.\,s^2x}{\overline{2}s \mathrel{\rhd\!\!\!\rhd} (\cdot)}(1)}{\overline{2}sx \mathrel{\rhd\!\!\!\rhd} (\cdot)x}(\rfloor) \quad \dfrac{(\cdot)x \rhd s^2x}{(\cdot)x \mathrel{\rhd\!\!\!\rhd} s^2x}(1)}{\overline{2}sx \mathrel{\rhd\!\!\!\rhd} s^2x}(\circ) \quad \dfrac{\dfrac{\dfrac{sx \rhd y^3x}{sx \mathrel{\rhd\!\!\!\rhd} y^3x}(1)}{s^2x \mathrel{\rhd\!\!\!\rhd} s(\cdot)}(\lfloor) \quad \dfrac{s(y^3x) \rhd y^3(y^3x)}{s(\cdot) \mathrel{\rhd\!\!\!\rhd} y^6x}(1)}{s^2x \mathrel{\rhd\!\!\!\rhd} y^6x}(\circ)}{\overline{2}sx \mathrel{\rhd\!\!\!\rhd} y^6x}(\circ)$$

$$\dfrac{\dfrac{t\overline{2} \rhd (\cdot)}{t\overline{2} \mathrel{\rhd\!\!\!\rhd} (\cdot)}(1) \quad \dfrac{\dfrac{\overline{2}sx \mathrel{\rhd\!\!\!\rhd} y^6x}{\lambda x\,.\,\overline{2}sx \mathrel{\rhd\!\!\!\rhd} \lambda x\,.\,y^6x}(\uparrow)}{\lambda y, x\,.\,\overline{2}sx \mathrel{\rhd\!\!\!\rhd} \overline{6}}(\uparrow)}{t\overline{2} \mathrel{\rhd\!\!\!\rhd} \overline{6}}(\circ)$$

Table 2.4: An example computation

as in the table. The code **1** indicates a use of redex removal but, in isolation, doesn't give us all the components. Computations are rarely used in isolation and the missing information can be found elsewhere.

More precisely, the code **1** indicates a call on the substitution algorithm. The primary job of a computation

$$(\square)\quad t^- \mathrel{\rhd\!\!\!\rhd} t^+$$

is to unravel the subject t^- to locate the redex subterms which are to be removed. These occur at the leaves of (the left hand end of) $\square$. For each such leaf the redex $(\lambda y\,.\,r)s$ is decomposed and then the triple

$$r \quad y \quad s$$

is passed to the substitution algorithm which returns $r[y := s]$ as the object of that leaf. The tree $\square$ reconstructs these reducts to produce the object of the computation.

This kind of reduction is often called β-**reduction**, and we will use that name here.

Before the next example we recall the abbreviation '$t^m s$' for the m-fold application of a term t to a term s. This was introduced in Section 2.2 for combinator terms, but we can also use it with λ-terms.

2.11 EXAMPLE. Using the informal iteration notation '$y^m x$' let

$$t = \lambda u, y, x\,.\,usx \quad s = \lambda x\,.\,y^3x \quad \overline{2} = \lambda y, x\,.\,y^2x \quad \overline{6} = \lambda y, x\,.\,y^6x$$

and consider the compound $t\overline{2}$. The computation given in Table 2.4 produces

$$(\square)\quad t\overline{2} \mathrel{\rhd\!\!\!\rhd} \overline{6} \quad \text{where} \quad \square = \mathbf{1} \circ \uparrow\uparrow((\rfloor\mathbf{1} \circ \mathbf{1}) \circ (\lfloor\mathbf{1} \circ \mathbf{1}))$$

and hints how arithmetic becomes involved with reduction. $\square$

As usual $\mathrel{\rhd\!\!\!\rhd\!\!\!\rhd}$ is the reflexive closure of $\mathrel{\rhd\!\!\!\rhd}$. This will be used in Section 2.7.

EXERCISES

2.17 Using the terms $t = \lambda z.zx$, $s = \lambda y.yz$, $r = \lambda x.xy$, reduce each of

$$tsr \quad t(sr) \quad trs \quad t(rs) \qquad str \quad s(tr) \quad srt \quad s(rt) \qquad rts \quad r(ts) \quad rst \quad r(st)$$

as much as possible.

2.18 Consider the term $t = (\lambda x\,.\,((\lambda x\,.\,((\lambda x\,.\,x)x))x))x$ (as in Exercise 2.13). Show that $t \gg x$ via at least six different computations.

2.19 This is an exercise on fixed-point λ-combinators.

(a) Let $\mathsf{Y} = \lambda y.\Omega\Omega$ where $\Omega = \lambda x.y(xx)$. Show that for each term f there are terms F, G such that $\mathsf{Y}f \gg F \gg fF$, and G and fG reduces to a common term.

(b) Consider the following terms.

(i) $\Omega\Omega$	where $\Omega = \lambda x, y.y(xxy)$
(ii) $\Omega\Omega\Omega$	where $\Omega = \lambda x_1, x_2, y.y(x_i x_j x_k y)$ where $i, j, k \in \{1, 2\}$
(iii) $\Omega\Omega\cdots\Omega$ ($n+1$ terms)	where $\Omega = \lambda x_1, x_2, \ldots, x_n, y.y(Xy)$ where $X = x_{i(1)}x_{i(2)}\cdots x_{i(n+1)}$ where $i(1), i(2), \ldots, i(n+1) \in \{1, 2, \ldots, n\}$
(iv) $\Omega\Omega\cdots\Omega$ (25 terms)	where $\Omega = \lambda a, b, c, \ldots, y.y(Xy)$ where $X \equiv \mathit{thisisafixpointcombinator}$

For each such term Y show that for each term f the reduction $\mathsf{Y}f \gg f(\mathsf{Y}f)$ holds.

2.20 (a) Let $\overline{m} = \lambda y, x\,.\,y^m x$ for each $m \in \mathbb{N}$ and consider the four terms to the right. By reducing the terms $\overline{\mathrm{Suc}}\,\overline{m}$, $\overline{\mathrm{Add}}\,\overline{n}\,\overline{m}$, $\overline{\mathrm{Mlt}}\,\overline{n}\,\overline{m}$, and $\overline{\mathrm{Exp}}\,\overline{n}\,\overline{m}$ as much as is possible, indicate how a substantial proportion of arithmetic can be simulated in the λ-calculus.

$$\begin{aligned}
\overline{\mathrm{Suc}} &= \lambda u, y, x\,.\,y(uyx)\\
\overline{\mathrm{Add}} &= \lambda v, u, y, x\,.\,(uy)(vyx)\\
\overline{\mathrm{Mlt}} &= \lambda v, u, y, x\,.\,u(vy)x\\
\overline{\mathrm{Exp}} &= \lambda v, u, y, x\,.\,uvyx
\end{aligned}$$

(b) Find a term $\overline{\mathrm{Bth}}$ satisfying

$$\overline{\mathrm{Bth}}\,\overline{k}\,\overline{n}\,\overline{m} \gg \overline{\beth(k, n, m)}$$

for all $m, n, k \in \mathbb{N}$ (where $\beth$ is the stacking function).

2.6 INTERTRANSLATABILITY

Both the set $\mathbb{C}omb$ of combinator terms and the set $\mathbb{L}amb$ of λ-terms have their own internal structure (given by the term forming operations) and both carry a reduction relation $\gg$. It seems intuitively clear that anything that can be done with combinator terms can be done with λ-terms. This can be made precise.

The translation

$$\begin{array}{c}
\mathbb{C}omb \longrightarrow \mathbb{L}amb\\
t \longmapsto t_\lambda
\end{array}$$

is defined by recursion on combinator terms using the clauses

$$x_\lambda = x \qquad \mathsf{Z}_\lambda = \text{explicit } \lambda\text{-term} \qquad (qp)_\lambda = q_\lambda p_\lambda$$

for some suitably chosen 'explicit λ-term'. In our case these are

$$\mathsf{I}_\lambda = \lambda x.x \qquad \mathsf{K}_\lambda = \lambda y, x.y \qquad \mathsf{S}_\lambda = \lambda z, y, x.(zx)(yx)$$

for the three constants. This translation $(\cdot)_\lambda$ respects the common structure of $\mathbb{C}omb$ and $\mathbb{L}amb$ (the identifiers and application). It also respects the reduction relations.

2.12 LEMMA. *For all $s, t \in \mathbb{C}omb$ the implication*

$$t \rhd\!\!\!\rhd s \Rightarrow t_\lambda \rhd\!\!\!\rhd s_\lambda$$

holds.

Sketch proof. By induction on the computation for $t \rhd\!\!\!\rhd s$. □

This implication is by no means an equivalence.

2.13 EXAMPLE. A simple calculation shows that $(\mathsf{SK})_\lambda \rhd\!\!\!\rhd \lambda y\,.\,\mathsf{I}_\lambda$ and hence $(\mathsf{SK}t)_\lambda \rhd\!\!\!\rhd \mathsf{I}_\lambda$ for any term t. However, when t is an identifier, the term $\mathsf{SK}t$ will not reduce. □

With this we are getting into the realms of computational efficiency, which is not something we wish to pursue here.

What about a converse translation? For this we need an observation.

2.14 LEMMA. *For each identifier x and $t \in \mathbb{C}omb$, there is an x-free combinator term $[x]t$ such that $([x]t)x \rhd\!\!\!\rhd t$ holds.*

Sketch proof. (There are several different ways to construct $[x]t$. The important differences between these are concerned with computational efficiency. Here we use the one with the simplest definition.)

The term $[x]t$ is constructed by recursion on t using the clauses

$$[x]x = \mathsf{I} \qquad [x]y = \mathsf{K}y \qquad [x]\mathsf{Z} = \mathsf{KZ} \qquad [x](qp) = \mathsf{S}([x]q)([x]p)$$

where, in the second clause, $y \not\equiv x$. □

Given this result the translation

$$\begin{array}{rcl} \mathbb{L}amb & \longrightarrow & \mathbb{C}omb \\ t & \longmapsto & \langle t \rangle \end{array}$$

is defined by recursion on t using the clauses

$$\langle x \rangle = x \qquad \langle qp \rangle = \langle q \rangle \langle p \rangle \qquad \langle \lambda y.r \rangle = [y]\langle r \rangle$$

for the usual λ-terms.

It can be shown that

$$\langle t \rangle_\lambda \rhd\!\!\!\rhd t$$

holds for all λ-terms t. The other composite is not so pleasant.

2.15 EXAMPLE. A few calculations give

$$\langle \mathsf{I}_\lambda \rangle = \mathsf{I} \qquad \langle \mathsf{K}_\lambda \rangle = \mathsf{S}(\mathsf{KK})\mathsf{I} \qquad \langle \mathsf{S}_\lambda \rangle = \mathsf{SS}(\mathsf{KS})(\mathsf{S}(\mathsf{S}(\mathsf{KS})(\mathsf{S}(\mathsf{KK})(\mathsf{KS})))\cdots)$$

where the last term contains 79 atomic combinators. □

There are many translations between $\mathbb{C}omb$ and $\mathbb{L}amb$, all with good and bad aspects.

Exercises

2.21 Suppose $\mathbb{C}omb$ is built up using a constant Z for which $\rhd$ has a defining property

$$\mathsf{Z}x_1 \cdots x_n \rhd X$$

where X is an explicit Z-free term. How should the explicit λ-term Z_λ be defined?

2.22 Compute the terms

$$[x](xy) \quad [x](yx) \quad [y][x](xy) \quad [y][x](yx) \quad [y][x]((xy)(yx))$$

using the translation of this section.

2.23 Show that

$$([x]t)s \mathrel{\rhd\!\!\!\rhd} t[x := s] \qquad \langle(\lambda x.t)s\rangle e \mathrel{\rhd\!\!\!\rhd} \langle t \rangle[x := \langle s \rangle]$$

for all identifiers x, and combinator terms t, s on the left, and λ-terms on the right.

2.24 Show that $\langle t \rangle_\lambda \mathrel{\rhd\!\!\!\rhd} t$ holds for all $t \in \mathbb{L}amb$.

2.25 We have seen that for the two translations $(\cdot)_\lambda$ and $\langle\cdot\rangle$ given in the section, it is not the case that $\langle t \rangle_\lambda = t$ holds. This can be achieved by a simple modification to the definition of $\langle\cdot\rangle$. Add the clause

$$[x](tx) = t \quad \text{if } t \text{ is } x\text{-free}$$

with the obligation that this is used whenever possible. Show that with this modification the identity does hold. Are the other translation results still valid?

2.7 Confluence and normalization

Each reduction relation $\mathrel{\rhd\!\!\!\rhd}$ is generated from a 1-step reduction relation $\rhd$ by taking the transitive, structural closure. To witness a reduction $t^- \mathrel{\rhd\!\!\!\rhd} t^+$ we locate a subterm l of t^- which matches the left hand side of an instance $l \rhd r$ of $\rhd$, and replace l by r to form t^+. We then repeat this manoeuvre as often as is necessary, perhaps using different instances of $\rhd$ at each step. What happens if there is more than one subterm l of t^- which matches the right hand side of an instance of $\rhd$, perhaps different instances? Does it matter which such subterm we replace?

2.16 DEFINITION. A reduction relation is **confluent** if for each divergent wedge

$$t^- \mathrel{\rhd\!\!\rhd\!\!\rhd} t_1 \qquad t^- \mathrel{\rhd\!\!\rhd\!\!\rhd} t_2$$

from a common source t^-, there is a convergent wedge

$$t_1 \mathrel{\rhd\!\!\rhd\!\!\rhd} t^+ \qquad t_2 \mathrel{\rhd\!\!\rhd\!\!\rhd} t^+$$

to a common target t^+. □

(This definition is one of the primary reason for introducing the reflexive closure $\mathrel{\rhd\!\!\rhd\!\!\rhd}$ of $\mathrel{\rhd\!\!\rhd}$. In the definition each of the instances of $\mathrel{\rhd\!\!\rhd\!\!\rhd}$ may be an equality. It is possible to rephrase this condition entirely in terms of $\mathrel{\rhd\!\!\rhd}$, but it becomes a bit messy.)

Although we won't prove it, all the reduction relations we meet in this book are confluent. This means that if for a term t^- there are alternative 1-step reductions which could be used, then it doesn't matter which we choose. The eventual effect of the other can be achieved from any later stage. There may be good reasons for choosing one alternative over the another, but these are concerned with evaluation strategies. We don't deal with that topic in this book.

Suppose we wish to reduce a term until it can't be reduced any more. Confluence ensures that we can perform the 1-step reductions in any order we wish. The eventual irreducible term will be the same no matter which order we take. At least it will be if there is any eventual term.

2.17 EXAMPLES. ($\mathbb{C}omb$) Consider the terms

$$\mathsf{B} = \mathsf{S}(\mathsf{KS})\mathsf{K} \qquad \omega = \mathsf{SII} \qquad \Omega = \mathsf{B}\omega\omega \qquad \Delta = \omega\Omega.$$

produced in Example 2.1. Simple calculations give

$$\mathsf{B}tsr \mathrel{\rhd\!\!\rhd} t(sr) \qquad \omega r \mathrel{\rhd\!\!\rhd} rr$$

for all terms r, s, t. Thus

$$\Delta \mathrel{\rhd\!\!\rhd} \Omega\Omega = \mathsf{B}\omega\omega\Omega \mathrel{\rhd\!\!\rhd} \omega(\omega\Omega) = \omega\Delta \mathrel{\rhd\!\!\rhd} \Delta\Delta$$

and hence

$$\Delta \mathrel{\rhd\!\!\rhd} \Delta\Delta \mathrel{\rhd\!\!\rhd} (\Delta\Delta)(\Delta\Delta) \mathrel{\rhd\!\!\rhd} ((\Delta\Delta)(\Delta\Delta))((\Delta\Delta)(\Delta\Delta)) \mathrel{\rhd\!\!\rhd} \ldots$$

so that Δ may be reduced for ever.

($\mathbb{L}amb$) Consider the terms

$$\omega = (\lambda x\,.\,(xx)(xx)) \qquad \delta = \omega\omega$$

produced in Example 2.6. We have

$$\omega t \mathrel{\rhd\!\!\rhd} (tt)(tt)$$

for all terms t. Thus

$$\delta = \omega\omega \mathrel{\rhd\!\!\rhd} (\omega\omega)(\omega\omega) = \delta\delta$$

and hence

$$\delta \mathrel{\rhd\!\!\rhd} \delta\delta \mathrel{\rhd\!\!\rhd} (\delta\delta)(\delta\delta) \mathrel{\rhd\!\!\rhd} ((\delta\delta)(\delta\delta))((\delta\delta)(\delta\delta)) \mathrel{\rhd\!\!\rhd} \cdots$$

so that δ may be reduced for ever. □

These examples show that for the combinator terms and λ-terms of this chapter, the reduction relation has some unpleasant properties. This is a flaw in the construction of these terms which will be corrected in later chapters. To anticipate that we introduce a couple of notions.

2.18 DEFINITION. Let $\bowtie\!\!\!\triangleright$ be a reduction relation on terms generated from a 1-step reduction relation $\triangleright$.

(a) A term t is **normal** (relative to $\triangleright$) if it has no subterm l which matches the left hand side of any instance of $\triangleright$.

(b) The relation $\bowtie\!\!\!\triangleright$ is **normalizing** if for each term t there is a normal term t^* with $t \bowtie\!\!\!\triangleright t^*$. □

Examples 2.17 show that the two reduction relations of this chapter are not normalizing. However, although we won't prove this, all the remaining reduction relations that we meet in this book will be normalizing.

Exercises

2.26 Prove an analogue of Exercise 2.1(e) for both $\mathbb{C}omb$ and $\mathbb{L}amb$. In other word show that for each divergent wedge

$$t_0 \mathrel{\dot{\triangleright}} t_1 \qquad t_0 \mathrel{\dot{\triangleright}} t_2$$

from a common source t_0, there is a convergent wedge

$$t_1 \bowtie\!\!\!\triangleright t_3 \qquad t_2 \bowtie\!\!\!\triangleright t_3$$

to a common target t_3. Here $\dot{\triangleright}$ indicates an instance of the reduction $\bowtie\!\!\!\triangleright$ in which transitive composition is not used.

You need not give all the details, but you should locate the special features which ensure this confluence. Do the two cases $\mathbb{C}omb$ and $\mathbb{L}amb$ separately.

3
THE TYPED COMBINATOR CALCULUS

3.1 INTRODUCTION

Why should the material of Chapters 1 and 2 appear in the same book, and so close together? Because the two chapters describe the two sides of the same coin; we cannot fully understand the one topic without understanding the other.

In Chapter 1 we introduced the idea of a derivation ∇ and a derivation system. We concentrated on two systems H and N, and we saw how to translate between these.

In Chapter 2 we introduced the idea of a computation $\square$ and a computation mechanism. We concentrated on two such mechanisms, one using combinator terms and one using λ-terms. Again we saw how to translate between these.

In this chapter we describe a calculus **C** which is an amalgam of the two families of H-derivations and combinator reductions. This is the first and simplest calculus we will see. The later chapters describe several more calculi, some not so simple.

In general a calculus (of the kind studied in this book) has three facilities.

Derivation system   Substitution algorithm   Computation mechanism

Chapters 1 and 2 describe the rudiments of the first and last of these. For the time being substitution will be done informally, but later we will be forced into a formal description.

Each calculus has various syntactic categories including

Type   Term   Judgement   Derivation   Computation

where each such syntactic object is built from certain primitives in a predetermined way. We have seen some of these constructions already. Now things start to get a little more complicated, and each new calculus will introduce a new level of complexity.

Let's begin the description of the calculus **C**.

3.1 **DEFINITION.** (**Types**) There is an unlimited stock of variables. The types are then generated as follows.

- Each variable is a type.

- If ρ, σ are types then $(\sigma \to \rho)$ is a type.

(Terms) There is an unlimited stock of identifiers and a given stock of constants. The terms are then generated as follows.

- Each constant is a term.
- Each identifier is a term.
- If p, q are terms then (qp) is a term.

We write $\mathbb{C}omb_\bullet$ for the set of such terms. □

These types are nothing more than the formulas of Chapter 1. We carry over the notations and conventions from there. Thus $X, Y, Z, \ldots$ are typical variables, and $\tau \to \sigma \to \rho$ abbreviates $(\tau \to (\sigma \to \rho))$ etc. We may assign to each variable X a set $[\![X]\!]$ as its interpreted value. This assignment lifts to produce a value $[\![\tau]\!]$ for each type where

$$[\![\sigma \to \rho]\!] = [\![\sigma]\!] \longrightarrow [\![\rho]\!]$$

using the function space construction on the right hand side. If each set $[\![X]\!]$ is either empty or a singleton, then each set $[\![\tau]\!]$ is empty or a singleton, and we obtain the classical truth value semantics of formulas (as types). However, we are concerned with the more general function space semantics of types. Much of what we do is motivated by this.

The terms of Definition 3.1 seem to be just the combinator terms of Chapter 2. So they are, but we will soon impose some extra sophistication. We carry over much of the notations and conventions of Chapter 2. Thus $x, y, z, \ldots$ are typical identifiers, and

$$tsr \quad \text{abbreviates} \quad ((ts)r)$$

etc. Strictly speaking Definition 3.1 gives the **raw terms**. Eventually we will extract a subclass of **well formed terms**, but for the most part we don't labour this distinction.

To complete this definition we need to say what the constants are. This is where the extra sophistication appears. In Chapter 2 we used three constants $\mathsf{I}, \mathsf{K}, \mathsf{S}$. Here these become three families

$$\mathsf{I}(\theta) \quad \mathsf{K}(\theta, \psi) \quad \mathsf{S}(\theta, \psi, \phi)$$

indexed by types θ, ψ, ϕ, as shown. Thus for each such triple θ, ψ, ϕ there is one constant $\mathsf{S}(\theta, \psi, \phi)$, for each pair θ, ψ there is one constant $\mathsf{K}(\theta, \psi)$, and for each type θ there is one constant $\mathsf{I}(\theta)$.

More generally we might want to base a calculus on a family of constants

$$\mathsf{Z}(\theta_1, \ldots, \theta_m)$$

indexed by an m-tuple $\theta_1, \ldots, \theta_m$ of types. Similarly we might construct a family of compound terms $\mathsf{Z}(\theta_1, \ldots, \theta_m)$ indexed by an m-tuple of types. In these circumstances we often write

$$\mathsf{Z}_\bullet \quad \text{for} \quad \mathsf{Z}(\theta_1, \ldots, \theta_m)$$

when it is convenient not to display the indexing types. Thus the terms of the calculus **C** are generated from $\mathsf{I}_\bullet, \mathsf{K}_\bullet, \mathsf{S}_\bullet$. The blob$_\bullet$ on $\mathbb{C}omb_\bullet$ should remind you of the type indexing.

EXERCISES

3.1 Consider a function

$$f : R \longrightarrow S \longrightarrow T$$

of compound type. This will require two arguments, r from R and s from S, to produce an eventual value in T. Write out an evaluation tree which illustrates that the two bracketing conventions (for terms and types) are well matched.

3.2 DERIVATION

The first of the important facilities of **C**, the derivation system, is very like the system of H-derivations of Chapter 1. Each **C**-derivation is an H-derivation augmented with certain **C**-terms. We need some preliminary notions.

3.2 DEFINITION. (a) A **statement** is a pair $t : \tau$ where the **subject** t is a term and the **predicate** τ is a type.

(b) A **declaration** is a statement $x : \sigma$ where the subject x is an identifier.

(c) A **context** is a list

$$\Gamma = x_1 : \sigma_1, \dots, x_l : \sigma_l$$

of declarations. The context Γ is **legal** if the identifiers $x_1, \dots, x_l$ are distinct.

(d) A **judgement**

$$\Gamma \vdash t : \tau$$

is a statement $t : \tau$ in context Γ. □

The colon ':' in a statement and the gate '$\vdash$' in a judgement are punctuation devices. They have no intrinsic meaning. A context may be empty, in which case it is legal (vacuously).

We wish to read a judgement

$\Gamma \vdash t : \tau$ as
Within the *legal* context Γ,
the *well formed* term t
inhabits the *acceptable* type τ

where the legality, well-formedness, and acceptability impose certain restrictions on the components and the judgement as a whole. We know what a legal context is, and we will see that all types are acceptable. The well formed terms are not so easy to extract, which is why we need the derivation system.

To justify the reading of a judgement we provide a derivation

$$(\nabla) \quad \Gamma \vdash t : \tau$$

of that judgement. This is a finite rooted tree of judgements grown according to certain rules which track the rules for H-derivations. Some of the leaves of ∇ are justified by appealing to certain predetermined axioms.

Clause	Shape	Proviso	Code
Axiom	$\Gamma \vdash \mathsf{Z} : \zeta$	Γ legal and $\mathsf{Z} : \zeta$ is a housing axiom	$\Gamma[\mathsf{Z}]$
Projection	$\Gamma \vdash x : \sigma$	Γ legal and $x : \sigma$ occurs in Γ	$\Gamma[x]$
Elimination	$\dfrac{\overset{Q}{\Gamma \vdash q : \pi \to \tau} \quad \overset{P}{\Gamma \vdash p : \pi}}{\Gamma \vdash qp : \tau}$		(QP)

Table 3.1: The derivation rules for C

3.3 REQUIREMENT. Each constant Z of the calculus comes furnished with a housing type ζ. The statement $\mathsf{Z} : \zeta$ is the **housing axiom** of Z. □

Here we use the three families

$$\mathsf{I}(\theta) \quad \mathsf{K}(\theta, \psi) \quad \mathsf{S}(\theta, \psi, \phi)$$

of constants (for arbitrary types θ, ψ, ϕ). These have housing axioms

$$\mathsf{I}(\theta) : \iota(\theta) \quad \mathsf{K}(\theta, \psi) : \kappa(\theta, \psi) \quad \mathsf{S}(\theta, \psi, \phi) : \sigma(\theta, \psi, \phi)$$

where

$$\iota(\theta) = (\theta \to \theta) \quad \kappa(\theta, \psi) = (\psi \to \theta \to \psi) \quad \sigma(\theta, \psi, \phi) = \chi \to \xi \to (\theta \to \phi)$$

where

$$\chi = \theta \to \psi \to \phi \qquad \xi = \theta \to \psi$$

(for arbitrary types θ, ψ, ϕ). These compound types ι, κ, σ are just the axioms of the system of H-derivations of Chapter 1.

3.4 DEFINITION. A **derivation**

$$(\nabla) \quad \Gamma \vdash t : \tau$$

is a finite rooted tree of judgements grown according to the rules of Table 3.1.

There are two base clauses, the Leaf rules (Axiom and Projection), each of which is restricted by a proviso; and a recursion clause (Elimination). The fourth column of the table gives the arboreal code for each rule. □

You should compare this definition with 1.9. There is more going on here. Notice that the arboreal code $\Gamma[x]$ for a Projection refers to the identifier x of the declaration $x : \sigma$ rather than its position. This use of an identifier as a label makes life a little easier. In the same way the code $\Gamma[\mathsf{Z}]$ for an Axiom refers to the constant rather than the type. This allows the same type to be labelled by more than one constant.

3.5 EXAMPLE. For arbitrary θ, ψ, ϕ consider the following type.

$$\beta = (\psi \to \phi) \to (\theta \to \psi) \to (\theta \to \phi)$$

We know that, as a formula, β is H-derivable. We now show that, as a type, it is inhabited by a typed combinator. We modify Solution 1.7(β). As there let

$$\rho = \theta \to \phi \qquad \sigma = \theta \to \psi \qquad \tau = \psi \to \phi \qquad \nu = \sigma \to \rho$$

so that $\beta = \tau \to \sigma \to \rho$ is the target formula. Also let

$$\begin{array}{lll} \kappa_2 = \tau \to \theta \to \tau & \alpha = \theta \to \tau & \sigma_2 = \alpha \to \sigma \to \rho \\ \kappa_1 = \sigma_2 \to \tau \to \sigma_2 & \mu = \tau \to \sigma_2 & \sigma_1 = \mu \to \kappa_2 \to \beta \end{array}$$

to produce four axioms $\kappa_1, \kappa_2, \sigma_1, \sigma_2$. From the expanded versions of these we have four housing axioms

$$\begin{array}{ll} \mathsf{K}_1 = \mathsf{K}(\tau, \sigma_2) & \mathsf{K}_1 : \kappa_1 \\ \mathsf{K}_2 = \mathsf{K}(\theta, \tau) & \mathsf{K}_2 : \kappa_2 \\ \mathsf{S}_1 = \mathsf{S}(\tau, \alpha, \nu) & \mathsf{S}_1 : \sigma_1 \\ \mathsf{S}_2 = \mathsf{S}(\theta, \psi, \phi) & \mathsf{S}_2 : \sigma_2 \end{array}$$

$$\dfrac{\dfrac{\vdash \mathsf{S}_1 : \sigma_1 \qquad \dfrac{\vdash \mathsf{K}_1 : \kappa_1 \quad \vdash \mathsf{S}_2 : \sigma_2}{\vdash \mathsf{K}_1\mathsf{S}_2 : \mu}}{\vdash \mathsf{S}_1(\mathsf{K}_1\mathsf{S}_2) : \kappa \to \beta} \qquad \vdash \mathsf{K}_2 : \kappa_2}{\vdash \mathsf{B}_\bullet : \beta}$$

and hence the indicated derivation shows that

$$\mathsf{B}_\bullet = \mathsf{B}(\theta, \psi, \phi) = \mathsf{S}_1(\mathsf{K}_1\mathsf{S}_2)\mathsf{K}_2$$

inhabits β in the empty context. This $\mathsf{B}_\bullet$ is a typed version of the untyped combinator B introduced in Exercise 2.4. □

It is time for you to construct some derivations yourself.

EXERCISES

3.2 In Exercises 1.7 and 1.8 you were asked to provide H-derivations for the standard formulas $\beta, \gamma, \delta, \epsilon, \omega, \mu, \nu, \tau$ built up from formulas $\theta, \psi, \phi, \zeta, \rho, \sigma, \tau$. In Exercise 2.4 you were asked to verify certain properties of untyped combinators B, C, D, E, W, M, N, T. For each of these formulas π and corresponding combinator P, show there is a typed version $\mathsf{P}_\bullet$ of P and a derivation of $\vdash \mathsf{P}_\bullet : \pi$ which codifies the original H-derivation.

3.3 Consider the types $\omega(i)$ of Exercise 1.9 (for arbitrary θ, ψ, ϕ). Exhibit inhabitants of these types (in the empty context). The inhabitants for (ii) and (v) should be different.

3.4 Show that if the judgement $\Gamma \vdash t : \tau$ is derivable, then Γ is legal and each identifier occurring in t is declared in Γ.

3.5 For contexts Γ, Σ let

$$\Gamma \sqsubseteq \Sigma \quad \text{mean} \quad \Sigma \text{ is legal and each declaration in } \Gamma \text{ also appears in } \Sigma$$

i.e. Σ is obtained from Γ by a sequence of Weakenings and eXchanges. Describe an algorithm which, when supplied with

$$(\nabla) \quad \Gamma \vdash t : \tau \qquad \Gamma \sqsubseteq \Sigma$$

(a derivation and a modified context), will return a derivation

$$(\nabla^{+}) \quad \Sigma \vdash t : \tau$$

with the same subject. How are the shapes of the two derivations related?

This shows that Weakening and eXchange are admissible rules. Can you modify the algorithm so that it also handles Contraction?

3.3 Annotation and deletion

The system of **C**-derivations looks very like the system of **H**-derivations described in Chapter 1. In fact, a **C**-derivation is nothing more than an **H**-derivation annotated with combinator terms, where these terms keep track of the construction of the derivation. This is the first and simplest example of a Curry–Howard correspondence. In this section we look at the details of this simple case.

As you read this description you may begin to think it is a lot of fuss over a rather shallow observation. You will be right. This instance of a CH correspondence isn't very deep and, on its own, doesn't need an elaborate discussion. However, later we look at more complicated calculi each with a CH correspondence. These need more careful handling. The analysis of this chapter is a preparation for these later examples.

Before we begin let's look at the result which supports everything else we do.

3.6 **THEOREM.** *For each context Γ and term t there is at most one derivation*

$$(\nabla) \quad \Gamma \vdash t : \tau$$

(for arbitrary types τ).

This shows that the pair (Γ, t) uniquely determines τ and ∇. Thus it ought to be possible to use (Γ, t) as a notation for ∇. In fact, if you compare the constructions of the arboreal code ∇ and the term t, you will see there is a tight match. This is the essence of the CH correspondence. Of course, there may be some pairs (Γ, t) which do not code any derivation; for instance, if Γ is illegal or t contains an identifier not declared in Γ.

To justify Theorem 3.6 we convert the proof into an algorithm.

3.7 **ALGORITHM.** (**Generation**) *An algorithm which, when supplied with a context Γ and a term t, will decide whether or not there is a derivation*

$$(\nabla) \quad \Gamma \vdash t : \tau$$

(for some type τ), and if so will return such a derivation.

Construction. The algorithm proceeds by recursion on the input t. There are two base cases and one recursion step.

(Axiom) When t is a constant Z, there is a derivation ∇ if and only if Γ is legal, in which case $\Gamma[\mathsf{Z}]$ is an example. The housing axiom $\mathsf{Z} : \zeta$ of Z gives the predicate $\tau = \zeta$.

(Projection) When t is an identifier x, there is a derivation ∇ if and only if Γ is legal and x is declared in Γ, in which case $\Gamma[x]$ is the only example. The declaration $x : \sigma$ of x in Γ gives the predicate $\tau = \sigma$.

(Elimination) When t is a compound qp, there is a derivation ∇ if and only if there are derivations

$$(Q) \quad \Gamma \vdash q : \pi \to \tau \qquad (P) \quad \Gamma \vdash p : \pi$$

with compatible predicates, as indicated. In this case QP is such a derivation.

To justify this algorithm note that (Elimination) always produces a compound subject. □

To describe the CH correspondence we need to be careful with the terminology. There are two kinds of contexts involved. For an H-derivation a context is a list

$$\sigma_1, \sigma_2, \dots, \sigma_l$$

of formulas/types thought of as the current hypotheses. For the remainder of this section we speak of a context (hypothesis list) to emphasize this. For a C-derivation a context is a list

$$\Gamma = x_1 : \sigma_1, x_2 : \sigma_2, \dots, x_l : \sigma_l$$

of declarations, a list of types each labelled with an identifier. For the remainder of this section we speak of a context (declaration list) to emphasize this.

Each context (declaration list) Γ, as above, can be converted into a context (hypothesis list)

$$\Gamma^\delta = \sigma_1, \sigma_2, \dots, \sigma_l$$

by deleting the labels. This process is lifted up to derivations.

3.8 ALGORITHM. (Deletion) *A translation algorithm*

$$(\nabla) \quad \Gamma \vdash t : \tau \quad \longmapsto \quad (\nabla^\delta) \quad \Gamma^\delta \vdash \tau$$

which, when supplied with a C*-derivation, will return an* H*-derivation in the deleted context with the same predicate.*

Construction. The algorithm proceeds by recursion on the supplied derivation ∇. There are two base cases and one recursion step.

(Axiom)	$\Gamma[\mathsf{Z}]^\delta = \Gamma^\delta[\zeta]$	$\mathsf{Z} : \zeta$ is the housing axiom of Z
(Projection)	$\Gamma[x]^\delta = \Gamma^\delta[n]$	x labels the nth position of Γ
(Elimination)	$(QP)^\delta = Q^\delta P^\delta$	

(In the second clause remember that the declarations of Γ are counted from right to left beginning with the gate declaration in position 0. Thus the position is not the same as the index used above.)

It is routine to check that $(\cdot)^\delta$ sends a **C**-derivation into an H-derivation of the required form. □

Each **C**-derivation translates into an H-derivation. We wish to show that every H-derivation arises in this way from an essentially unique **C**-derivation. There are two problems, both of which indicate the need for extra data.

The supplied context (hypothesis list) Γ is merely a list $\sigma_1, \ldots, \sigma_l$ of formulas/types. From this we must produce a (legal) context (declaration list)

$$x_1 : \sigma_1, \ldots, x_l : \sigma_l$$

using identifiers $x_1, \ldots, x_l$. These labels cannot be extracted from the given H-derivation; they must be supplied as part of the input. Thus we assume the H-derivation is given in the form

$$(\nabla) \quad \Gamma^\delta \vdash \tau$$

where Γ is a context (declaration list) with known labels.

Suppose the supplied H-derivation is a leaf with an axiom ζ as predicate. This must convert into a leaf

$$\Gamma \vdash \mathsf{Z} : \zeta$$

for some constant Z. Where does Z come from? It cannot be found within the system of H-derivation; it must be part of the environment in which the translation is carried out. Thus the axiom ζ which helps to generate H-derivations must match a housing axiom $\mathsf{Z} : \zeta$ which helps to generate **C**-derivations.

3.9 ALGORITHM. (Annotation) *An algorithm*

$$(\nabla) \quad \Gamma^\delta \vdash \tau \quad \longmapsto \quad ((\Gamma, \nabla)^\alpha) \quad \Gamma \vdash t : \tau$$

which, when supplied with a context (declaration list) Γ and an H*-derivation in the deleted context, will return a* **C***-derivation in the given context with the same predicate.*

Construction. The algorithm proceeds by recursion on the supplied derivation ∇. There are two base cases and one recursion step.

(Axiom)	$(\Gamma, \Gamma^\delta[\zeta])^\alpha = \Gamma[\mathsf{Z}]$	$\mathsf{Z} : \zeta$ is the housing axiom of Z
(Projection)	$(\Gamma, \Gamma^\delta[n])^\alpha = \Gamma[x]$	x labels the nth position of Γ
(Elimination)	$(\Gamma, QP)^\alpha = (\Gamma, Q)^\alpha(\Gamma, P)^\alpha$	

It can be checked that the result of ${}^\alpha(\cdot)$ is a **C**-derivation of the required form. □

These two algorithms clarify the notion of a CH correspondence.

3.10 THEOREM. *Consider a context (declaration list) Γ.*
For each H*-derivation ∇ over Γ^{δ}, we have $(\Gamma, \nabla)^{\alpha\delta} = \nabla$.*
For each **C***-derivation ∇ over Γ, we have $(\Gamma, \nabla^{\delta})^{\alpha} = \nabla$.*
The two algorithms $(\cdot)^{\alpha}$ and $(\cdot)^{\delta}$ are inverse companions.

By Theorem 3.6 each **C**-derivation

$$(\nabla) \quad \Gamma \vdash t : \tau$$

is uniquely determined by the root pair (Γ, t). In fact, this pair is an uncoupled version of the arboreal code for ∇. By Theorem 3.10 the **C**-derivation ∇ is uniquely determined by Γ and the H-derivation

$$(\nabla^{\delta}) \quad \Gamma^{\delta} \vdash \tau$$

and every H-derivation arises in this way. Thus, for a given context (declaration list) there is a matching between

combinator terms t H-derivations ∇

(although not all combinator terms can occur). Imprecisely we say that combinator terms and H-derivations are 'the same thing'.

This correspondence gives us a neat way of manipulating H-derivations. To illustrate this consider $(\cdot)_D$, the Deduction Algorithm 1.13.

Suppose we start from a **C**-derivation

$$(\nabla) \quad (\Gamma, x : \theta) \vdash t : \tau$$

with $x : \theta$ as the gate declaration. By deletion we have an H-derivation

$$(\nabla^{\delta}) \quad (\Gamma^{\delta}, \theta) \vdash \tau \quad \text{which converts into} \quad (\nabla^{\delta}{}_D) \quad \Gamma^{\delta} \vdash \theta \to \tau$$

using the deduction algorithm. We know this must arise from a **C**-derivation

$$(\nabla^{d}) \quad \Gamma \vdash (-) : \theta \to \tau$$

for some term $(-)$. Let us write $[x : \theta]t$ for this term. (It is plausible that $(-)$ depends only on $x : \theta$ and t.) Thus there is an algorithm $\nabla \longmapsto \nabla^{d}$ on **C**-derivations which converts the given ∇ into a new derivation

$$(\nabla^{d}) \quad \Gamma \vdash [x : \theta]t : (\theta \to \tau) \quad \text{with} \quad \nabla^{d\delta} = \nabla^{\delta}{}_D$$

i.e. the result of the original deduction algorithm can be obtained from the constructed term $[x : \theta]t$.

Exercises

3.6 In each of the following you are given a typed combinator $\mathsf{A}_\bullet$ with the indexing types obscured. In each case find necessary and sufficient conditions on these indexing types for $(\emptyset, \mathsf{A})$ to be the code of a derivation. Write down the housing type α with $\emptyset \vdash \mathsf{A}_\bullet : \alpha$ and indicate the shape of the derivation of this judgement.

$$\text{(a)}\ \mathsf{C}_\bullet\mathsf{S}_\bullet\mathsf{I}_\bullet \qquad \text{(b)}\ \mathsf{S}_2(\mathsf{K}_\bullet\mathsf{S}_1) \qquad \text{(c)}\ S^2 0 \ \text{ where } S = \mathsf{S}_\bullet\mathsf{B}_\bullet,\ 0 = \mathsf{K}_\bullet\mathsf{I}_\bullet$$

3.7 Prove Theorem 3.6.

3.8 Prove Theorem 3.10.

3.9 Describe the algorithm $\nabla \longmapsto \nabla^d$ on C-derivations which simulates the deduction algorithm on H-derivations.

3.10 Describe an algorithm which, when supplied with a pair of derivations

$$(R)\ \ \Pi, x : \sigma, \Delta \vdash r : \rho \qquad (S)\ \ \Pi, \Delta \vdash s : \sigma$$

and a nominated component $x : \sigma$ of the R-context, will return a derivation

$$(R * S)\ \ \Pi, \Delta \vdash r * s : \tau$$

where $r * s = r[x := s]$.

3.4 Computation

There is a reduction relation $\bowtie$ on the set $\mathbb{C}omb_\bullet$ of typed combinator terms. As with the untyped version, this is generated from a 1-step relation $\rhd$.

3.11 DEFINITION. For the constants $\mathsf{I}_\bullet, \mathsf{K}_\bullet, \mathsf{S}_\bullet$, the 1-step reduction axioms are

$$\mathsf{I}_\bullet r \bowtie r \qquad \mathsf{K}_\bullet sr \bowtie s \qquad \mathsf{S}_\bullet tsr \bowtie (tr)(sr)$$

for terms r, s, t. □

These are the 1-step reductions we use here. In general, for a constant $\mathsf{Z}_\bullet$, the reduction axiom need *not* be of the form

$$\mathsf{Z}_\bullet r_m \cdots r_1 \rhd \cdots$$

for terms $r_1, \ldots, r_m$. Some of the reduction axioms may involve several constants, and some constants may appear in more than one reduction axiom.

Once we have the 1-step reductions we generate $\bowtie$ as the transitive, structural closure of $\rhd$. We use a refined version of Definition 2.4.

Clause	Shape	Remarks	Code
(Axiom reduction)	$\dfrac{t^- \rhd t^+}{t^- \rhd\!\!\rhd t^+}\,(0)$	1-step leaf	0
(Left application)	$\dfrac{\overset{\mathsf{q}}{q^- \rhd\!\!\rhd q^+}}{q^-p \rhd\!\!\rhd q^+p}\,(\downharpoonleft)$		$\downharpoonleft\mathsf{q}$
(Right application)	$\dfrac{\overset{\mathsf{p}}{p^- \rhd\!\!\rhd p^+}}{qp^- \rhd\!\!\rhd qp^+}\,(\downharpoonright)$		$\downharpoonright\mathsf{p}$
(Transitive composition)	$\dfrac{\overset{\mathsf{l}}{t^- \rhd\!\!\rhd t^0} \quad \overset{\mathsf{r}}{t^0 \rhd\!\!\rhd t^+}}{t^- \rhd\!\!\rhd t^+}\,(\circ)$		$(\mathsf{l} \circ \mathsf{r})$

Table 3.2: The computation rules for typed combinator terms

3.12 DEFINITION. A computation

$$(\square) \quad t^- \rhd\!\!\rhd t^+$$

is a finite rooted tree of reductions grown according to the rules of Table 3.2. Such a computation $\square$ is said to **organize** the root reduction $t^- \rhd\!\!\rhd t^+$ where t^- is the **subject** and t^+ is the **object** of the computation. The right hand column of the Table shows how the arboreal code for a computation is generated. $\square$

Tables 3.2 and 2.1 seem to be exactly the same. The hidden difference is the terms used. Here the terms are typed, but there they are untyped. However, notice that the typing information does not appear in the arboreal code for a computation.

3.13 EXAMPLE. Consider the typed combinator

$$\mathsf{B}_\bullet = \mathsf{B}(\theta, \psi, \phi) = \mathsf{S}_1(\mathsf{K}_1\mathsf{S}_2)\mathsf{K}_2$$

of Example 3.5 (where θ, ψ, ϕ are arbitrary types). Then the computation

$$\dfrac{\dfrac{\dfrac{\dfrac{\dfrac{\mathsf{B}_\bullet z \rhd (\mathsf{K}_1\mathsf{S}_2 z)(\mathsf{K}_2 z)}{\mathsf{B}_\bullet z \rhd\!\!\rhd (\mathsf{K}_1\mathsf{S}_2 z)(\mathsf{K}_2 z)}\,(0) \quad \dfrac{\dfrac{\mathsf{K}_1\mathsf{S}_2 z \rhd \mathsf{S}_2}{\mathsf{K}_1\mathsf{S}_2 z \rhd\!\!\rhd \mathsf{S}_2}\,(0)}{(\mathsf{K}_1\mathsf{S}_2 z)(\mathsf{K}_2 z) \rhd\!\!\rhd \mathsf{S}_2(\mathsf{K}_2 z)}\,(\downharpoonleft)}{\mathsf{B}_\bullet z \rhd\!\!\rhd \mathsf{S}_2(\mathsf{K}_2 z)}\,(\circ)}{\mathsf{B}_\bullet zy \rhd\!\!\rhd \mathsf{S}_2(\mathsf{K}_2 z)y}\,(\circ)}{\mathsf{B}_\bullet zyx \rhd\!\!\rhd \mathsf{S}_2(\mathsf{K}_2 z)yx}\,(\circ) \quad \dfrac{\dfrac{\mathsf{S}_2(\mathsf{K}_2 z)yx \rhd (\mathsf{K}_2 zx)(zy)}{\mathsf{S}_2(\mathsf{K}_2 z)yx \rhd\!\!\rhd (\mathsf{K}_2 zx)(zy)}\,(0) \quad \dfrac{\dfrac{\mathsf{K}_2 zx \rhd z}{\mathsf{K}_2 zx \rhd\!\!\rhd z}\,(0)}{(\mathsf{K}_2 zx)(yx) \rhd\!\!\rhd z(yx)}\,(\downharpoonleft)}{\mathsf{S}_2(\mathsf{K}_2 z)yx \rhd\!\!\rhd z(yx)}\,(\circ)}{\mathsf{B}_\bullet zyx \rhd\!\!\rhd z(yx)}\,(\circ)$$

is a typed version of an untyped computation given in Solution 2.3. As there with $\mathsf{a} = 0 \circ \downharpoonleft 0$ we see that $\mathsf{b} = \downharpoonleft\downharpoonleft\mathsf{a} \circ \mathsf{a}$ is the arboreal code for this computation (where we have now written a for b'). There are other ways to organize this reduction. $\square$

Definition 3.12 is word for word the same as Definition 2.4. The fact that the terms are typed doesn't seem to matter. We can erase the types.

3.14 DEFINITION. The **type erasing** operation from typed terms to untyped terms

$$\mathbb{C}omb_{\bullet} \xrightarrow{(\cdot)^{\epsilon}} \mathbb{C}omb$$

is generated by recursion using

$$x^{\epsilon} = x \qquad \mathsf{Z}_{\bullet}^{\epsilon} = \mathsf{Z} \qquad (qp)^{\epsilon} = q^{\epsilon}p^{\epsilon}$$

for each identifier x, constant $\mathsf{Z}_{\bullet}$, and terms q, p. $\square$

What role is played by the types in a computation? For each typed 1-step reduction $t^- \rhd t^+$ we may erase the types of the terms to produce two untyped terms $t^{-\epsilon}, t^{+\epsilon}$. Observe that $t^{-\epsilon} \rhd t^{+\epsilon}$ is an untyped 1-step reduction. In fact, we can erase the types from each node of a typed computation.

3.15 THEOREM. *Type erasure*

$$(\square) \;\; t^- \bowtie t^+ \quad \longmapsto \quad (\square^{\epsilon}) \;\; t^{-\epsilon} \bowtie t^{+\epsilon}$$

converts each typed computation $\square$ *into an untyped computation* $\square^{\epsilon}$ *where these have the same arboreal code. Furthermore, given an untyped computation*

$$(\square^*) \;\; t^{-\epsilon} \bowtie t^*$$

where the subject $t^{-\epsilon}$ *is the erasure of a typed term* t^-*, there is a unique typed computation* $\square$ *(as above) with* $\square^{\epsilon} = \square^*$ *(and hence* $t^* = t^{+\epsilon}$*).*

Proof. The remarks above observe that type erasure converts each typed 1-step reduction into an untyped 1-step reduction. With these the full algorithm $\square \longmapsto \square^{\epsilon}$ is now immediate.

Conversely, given an untyped computation

$$(\square^*) \;\; t^{-\epsilon} \bowtie t^*$$

how do we produce the unique typed computation $\square$ with $\square^{\epsilon} = \square^*$? We proceed by recursion over $\square^*$.

There are three base cases $\square^* = 0$ where

$$t^{-\epsilon} = \mathsf{I}r^! \;\; t^* = r^! \qquad t^{-\epsilon} = \mathsf{K}s^!r^! \;\; t^* = s^! \qquad t^{-\epsilon} = \mathsf{S}t^!s^!r^! \;\; t^* = (t^!r^!)(s^!r^!)$$

respectively. Here $\mathsf{I}, \mathsf{K}, \mathsf{S}$ are untyped combinators and $r^!, s^!, t^!$ are untyped terms which are not assumed to be erased typed terms. Let's look at the S-case (which is the most complicated).

From the shape of $t^{-\epsilon}$ and the way erasure works we must have

$$t^{-} = \mathsf{S}_{\bullet}tsr$$

for a unique typed combinator $\mathsf{S}_{\bullet}$ and unique typed terms t, s, r. Since

$$t^{-\epsilon} = (\mathsf{S}_{\bullet}tsr)^{\epsilon} = \mathsf{S}_{\bullet}^{\epsilon}t^{\epsilon}s^{\epsilon}r^{\epsilon} = \mathsf{S}t^{\epsilon}s^{\epsilon}r^{\epsilon}$$

we see that

$$t^{\epsilon} = t^{!} \qquad s^{\epsilon} = s^{!} \qquad r^{\epsilon} = r^{!}$$

hold. Let $t^{+} = (tr)(sr)$, so that $t^{+\epsilon} = t^{*}$. There is a unique 1-step computation with t^{-} as subject, namely

$$(\square) \;\; t^{-} \rhd\!\!\!\rhd t^{+}$$

and this satisfies $\square^{\epsilon} = \square^{*}$.

The induction steps across the construction rules follow in a similar way. □

This proof makes use of the special form of the reduction axioms here. A similar proof works from most other axioms, but the base cases must be checked afresh.

Exercises

3.11 Consider the term $\mathsf{C}_{\bullet}\mathsf{S}_{\bullet}\mathsf{I}_{\bullet}$ used in Exercise 3.6(a). Organize a computation of the reduction $\mathsf{C}_{\bullet}\mathsf{S}_{\bullet}\mathsf{I}_{\bullet} \rhd\!\!\!\rhd \mathsf{S}_{*}\mathsf{S}_{\bullet}(\mathsf{K}_{*}\mathsf{I}_{\bullet})$ for certain constants $\mathsf{K}_{*}, \mathsf{S}_{*}$.

3.12 Consider the combinators $\mathsf{W}(i)$ you obtained in Exercise 3.3. For each case reduce the untyped version $\mathsf{W}(i)^{\epsilon}$ as much as possible and give, in arboreal code, a computation $\mathsf{w}(i)$ which organizes this reduction.

3.13 Consider the untyped fixed point combinators Y, Z given in Exercise 2.4. Show there are no derivations

$$\vdash \mathsf{Y}_{\bullet} : \eta \qquad \vdash \mathsf{Z}_{\bullet} : \zeta$$

for typed combinators $\mathsf{Y}_{\bullet}, \mathsf{Z}_{\bullet}$ with $\mathsf{Y}_{\bullet}^{\epsilon} = \mathsf{Y}$ and $\mathsf{Z}_{\bullet}^{\epsilon} = \mathsf{Z}$.

3.14 Complete the proof of Theorem 3.15, i.e. give the induction steps.

3.5 Subject reduction

The calculus $\mathbf{C}$ has two important facilities

the derivation system   the computation mechanism

and a mediating facility, the substitution algorithm. Here this algorithm is unproblematic so we haven't spent too much time on it. Later calculi will need a more detailed discussion. This section looks at the first example of the main interaction between a derivation system and a computation mechanism.

Consider a pair

$$(\nabla)\quad \Gamma \vdash t^- : \tau \qquad (\Box)\quad t^- \rhd\!\!\!\rhd t^+$$

which are compatible in the sense that they have the same subject t^-. The derivation ∇ shows that t^- is well formed, but what about the object t^+ of $\Box$? Since $\Box$ makes no explicit reference to the typing discipline, couldn't t^+ be ill formed? It is an important property of **C** (and of all the later calculi) that subject reduction is satisfied.

3.16 EXAMPLE. Consider the types

$$\begin{array}{ll} \beta = (\psi \to \phi) \to (\theta \to \psi) \to (\theta \to \phi) & \rho = \theta \to \phi \\ \gamma = (\tau \to \sigma \to \rho) \to (\sigma \to \tau \to \rho) & \sigma = \theta \to \psi \\ \delta = (\theta \to \psi) \to (\psi \to \phi) \to (\theta \to \phi) & \tau = \psi \to \phi \end{array}$$

for arbitrary types θ, ψ, ϕ. We know that $\vdash \mathsf{B}_\bullet : \beta$, $\vdash \mathsf{C}_\bullet : \gamma$ and $\vdash \mathsf{D}_\bullet : \delta$ are derivable. In fact we can take

$$\mathsf{B}_\bullet = \mathsf{S}_1(\mathsf{K}_1\mathsf{S}_2)\mathsf{K}_2 \quad \mathsf{C}_\bullet = \mathsf{S}_4\mathsf{P}_\bullet\mathsf{Q}_\bullet \quad \mathsf{D}_\bullet = \mathsf{S}_5(\mathsf{K}_5(\mathsf{S}_6\mathsf{B}_3))\mathsf{K}_6 \quad \mathsf{P}_\bullet = \mathsf{B}_1\mathsf{B}_2\mathsf{S}_3 \quad \mathsf{Q}_\bullet = \mathsf{K}_3\mathsf{K}_4$$

for appropriate constants $\mathsf{K}_1, \dots, \mathsf{K}_6, \mathsf{S}_1, \dots, \mathsf{S}_6$ (and, of course, where $\mathsf{B}_1, \mathsf{B}_2, \mathsf{B}_3$ are built from other constants). We have

$$\beta = \tau \to \sigma \to \rho \qquad \gamma = \beta \to \delta \qquad \delta = \sigma \to \tau \to \rho$$

so that $\vdash \mathsf{C}_\bullet\mathsf{B}_\bullet : \delta$ and $\vdash \mathsf{D}_\bullet : \delta$ give two inhabitants of δ. These two derivations have shapes

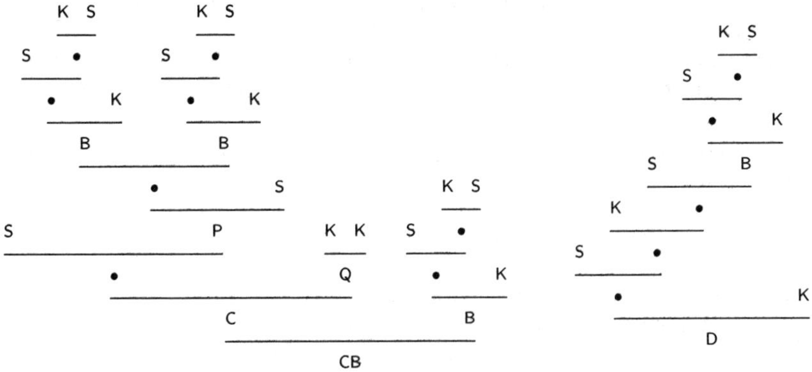

```
        K S         K S
        ---                                                K S
   S     •     S     •                                     ---
   -------     -------                                S     •
      •   K       •   K                               -------
      -----       -----                                  •   K
        B           B                                    -----
        ---------------                             S      B
              •       S        K S                  --------
              ---------        ---              K      •
 S                P     K K  S  •               ---------
 ------------------     ---  ----            S     •
        •               Q     •   K          -------
        -----------------     -----            •           K
                C               B              -------------
                -----------------                    D
                        CB
```

where all the labels have been erased. Is there any connection between these derivations?

For each term t we have

$$\mathsf{B}_\bullet t \rhd (\mathsf{K}_1\mathsf{S}_2 W)(\mathsf{K}_2 W) \rhd\!\!\!\rhd \mathsf{S}_2(\mathsf{K}_2 W) \qquad \mathsf{P}_\bullet \rhd\!\!\!\rhd \mathsf{S}_{12}(\mathsf{K}_{12}\mathsf{B}_2)\mathsf{S}_3$$

where S_{12} and K_{12} are components of B_1. Then

$$\mathsf{P}_\bullet t \rhd\!\!\!\rhd (\mathsf{K}_{12}\mathsf{B}_2 W)(\mathsf{S}_3 W) \rhd\!\!\!\rhd \mathsf{B}_2(\mathsf{S}_3 W) \rhd\!\!\!\rhd \mathsf{S}_{22}(\mathsf{K}_{22}(\mathsf{S}_3 W)) \qquad \mathsf{Q}_\bullet W = \mathsf{K}_3\mathsf{K}_4 W \rhd\!\!\!\rhd \mathsf{K}_4$$

∇	Shape	$\mathsf{0}$	$\nabla \bullet \mathsf{0}$
$\Gamma[\mathsf{I}]R$	$\dfrac{\Gamma \vdash \mathsf{I}_\bullet : \iota \quad \overset{R}{\Gamma \vdash r : \theta}}{\Gamma \vdash \mathsf{I}_\bullet r : \theta}$	$\dfrac{\mathsf{I}_\bullet r \rhd r}{\mathsf{I}_\bullet r \rhd\!\!\rhd r}$	R
$\Gamma[\mathsf{K}]SR$	$\dfrac{\dfrac{\Gamma \vdash \mathsf{K}_\bullet : \kappa \quad \overset{S}{\Gamma \vdash s : \psi}}{\bullet} \quad \overset{R}{\Gamma \vdash r : \theta}}{\Gamma \vdash \mathsf{K}_\bullet sr : \psi}$	$\dfrac{\mathsf{K}_\bullet sr \rhd s}{\mathsf{I}_\bullet sr \rhd\!\!\rhd s}$	S
$\Gamma[\mathsf{S}]TSR$	$\dfrac{\dfrac{\dfrac{\Gamma \vdash \mathsf{S}_\bullet : \sigma \quad \overset{T}{\Gamma \vdash t : \chi}}{\bullet} \quad \overset{S}{\Gamma \vdash s : \xi}}{\bullet} \quad \overset{R}{\Gamma \vdash r : \theta}}{\Gamma \vdash \mathsf{S}_\bullet tsr : \phi}$	$\dfrac{\mathsf{S}_\bullet tsr \rhd (tr)(sr)}{\mathsf{S}_\bullet tsr \rhd\!\!\rhd (tr)(sr)}$	$(TR)(SR)$

Table 3.3: The recipes for $\nabla \bullet \mathsf{0}$

where S_{22} and K_{22} are components of B_2. Putting these together we have

$$\mathsf{C}_\bullet\mathsf{B}_\bullet = \mathsf{S}_4\mathsf{P}_\bullet\mathsf{Q}_\bullet\mathsf{B}_\bullet \rhd\!\!\rhd (\mathsf{P}_\bullet\mathsf{B}_\bullet)(\mathsf{Q}_\bullet\mathsf{B}_\bullet) \rhd\!\!\rhd \mathsf{S}_{22}(\mathsf{K}_{22}(\mathsf{S}_3\mathsf{B}_\bullet))\mathsf{K}_4$$

which has the same general shape as $\mathsf{D}_\bullet$. In fact, we can arrange that

$$\mathsf{S}_5 = \mathsf{S}_{22} \quad \mathsf{K}_5 = \mathsf{K}_{22} \quad \mathsf{S}_6 = \mathsf{S}_3 \quad \mathsf{B}_3 = \mathsf{B}_\bullet \quad \mathsf{K}_6 = \mathsf{K}_4$$

and then $\mathsf{C}_\bullet\mathsf{B}_\bullet \rhd\!\!\rhd \mathsf{D}_\bullet$. This reduction can be organized into a computation which will act on the left hand derivation to produce the right hand one. □

We describe an algorithm which, when supplied with a compatible pair $\nabla, \square$ (as above), will return a derivation

$$(\nabla \cdot \square) \quad \Gamma \vdash t^+ : \tau$$

and hence show that the object t^+ is well formed.

Before we discuss the general algorithm we look at the case where $\square$ is a 1-step computation. We do this for the particular cases of the constants $\mathsf{I}_\bullet, \mathsf{K}_\bullet, \mathsf{S}_\bullet$. More general reduction axioms may not be so easy to handle.

3.17 ALGORITHM. (Leaf recipes) *An algorithm*

$$(\nabla) \quad \Gamma \vdash t^- : \tau \quad \longmapsto \quad (\nabla \bullet \mathsf{0}) \quad \Gamma \vdash t^+ : \tau$$

which, when supplied with a derivation where the root subject is the subject of a reduction axiom, will return a derivation of the corresponding object in the same context.

Construction. There are three cases as given in Table 3.3. Thus we have

$$(\Gamma[\mathsf{I}_\bullet]R) \bullet \mathsf{0} = R \qquad (\Gamma[\mathsf{K}_\bullet]SR) \bullet \mathsf{0} = S \qquad (\Gamma[\mathsf{S}_\bullet]TSR) \bullet \mathsf{0} = (TR)(SR)$$

for the three cases. Only the third of these needs some justification. The types involved are generated from arbitrary θ, ψ, ϕ in the usual way. Thus

$$\chi = \theta \to \psi \to \phi \qquad \xi = \theta \to \psi \qquad \sigma = \chi \to \xi \to \theta \to \phi$$

where σ is the type of $\mathsf{S}_\bullet$. With these we see that

$$\dfrac{\dfrac{\overset{T}{\Gamma \vdash t : \chi} \quad \overset{R}{\Gamma \vdash r : \theta}}{\Gamma \vdash tr : (\psi \to \phi)} \qquad \dfrac{\overset{S}{\Gamma \vdash s : \xi} \quad \overset{R}{\Gamma \vdash r : \theta}}{\Gamma \vdash sr : \psi}}{\Gamma \vdash (tr)(sr) : \phi}$$

justifies the compound $(TR)(SR)$. ☐

These leaf recipes feed the general algorithm.

3.18 ALGORITHM. (Subject reduction) *An algorithm which, when supplied with*

$$(\nabla) \ \ \Gamma \vdash t^- : \tau \qquad\qquad (\square) \ \ t^- \rhd\!\!\!\rhd t^+$$

a compatible derivation and computation, will return a derivation

$$(\nabla \cdot \square) \ \ \Gamma \vdash t^+ : \tau$$

the result of the action *of* $\square$ *on* ∇.

Construction. The algorithm proceeds by recursion over $\square$ with variation of the parameter ∇. For each compatible pair $(\nabla, \square)$ there is precisely one action step that can be taken, and this step is determined by $\square$.

(Axiom)	$\nabla \cdot \mathsf{0} = \nabla \bullet \mathsf{0}$	(Trans)	$\nabla \cdot (\mathsf{l} \circ \mathsf{r}) = (\nabla \cdot \mathsf{l}) \cdot \mathsf{r}$
(LAppl)	$(QP) \cdot \rfloor\mathsf{q} = (Q \cdot \mathsf{q})P$	(RAppl)	$(QP) \cdot \lfloor\mathsf{p} = Q(P \cdot \mathsf{p})$

At its base the algorithm calls on the leaf recipes given in Table 3.3. For the three recursion steps (LA,RA,T) we need to check that the algorithm returns a derivation of the required form. The arguments for the two applications are similar and easy. Only the transitive step needs a little work.

The algorithm is supplied with

$$(\nabla) \ \ \Gamma \vdash t^- : \tau \qquad\qquad (\square) \ \ \dfrac{\overset{\mathsf{l}}{t^- \rhd\!\!\!\rhd t^0} \quad \overset{\mathsf{r}}{t^0 \rhd\!\!\!\rhd t^+}}{t^- \rhd\!\!\!\rhd t^+}$$

for some intermediate term t^0. By recursion the pair (∇, l) produces

$$(\nabla \cdot \mathsf{l}) \ \ \Gamma \vdash t^0 : \tau$$

which is compatible with r, and then a second recursion gives

$$((\nabla \cdot \mathsf{l}) \cdot \mathsf{r}) \quad \Gamma \vdash t^+ : \tau$$

which is the required form of $\nabla \cdot (\mathsf{l} \circ \mathsf{r})$.

The termination of this algorithm follows by a straight forward induction over the computation. □

As observed in Theorem 3.15, the types in a computation have no importance. In the same way they have no role to play in a subject reduction.

Suppose we have a derivation

$$(\nabla) \quad \Gamma \vdash t^- : \tau$$

with a typed subject t^-. Suppose we now erase the types of t^- to produce an untyped term $t^{-\epsilon}$. Suppose also we produce an untyped computation

$$(\square^*) \quad t^{-\epsilon} \rhd\!\!\!\rhd t^*$$

to an untyped object t^*. How do we know that t^* makes sense in Γ? By Theorem 3.15 there is a typed computation

$$(\square) \quad t^- \rhd\!\!\!\rhd t^+$$

with $\square^\epsilon = \square^*$ and $t^{+\epsilon} = t^*$, and then the subject reduction algorithm produces

$$(\nabla \cdot \square) \quad \Gamma \vdash t^+ : \tau$$

to show that we may insert types into t^* in a coherent fashion. In fact, the two computations $\square$ and $\square^\epsilon$ have exactly the same arboreal code, so we may determine $\nabla \cdot \square$ directly from the given codes of ∇ and $\square^*$.

Why should the material of Chapters 1 and 2 appear in the same book? One reason is that combinator reduction can be used to simplify H-derivations.

Suppose we have an H-derivation

$$(\nabla) \quad \Gamma \vdash \tau$$

given by its arboreal code ∇. By annotating this we produce a **C**-derivation

$$(\nabla) \quad \Gamma \vdash t : \tau$$

for some subject t. As indicated, this new derivation has essentially the same arboreal code. The only difference is that identifiers are used in place of position indexes. Erasure gives us an untyped term t^ϵ. This has no trace of the types in the original H-derivation. By analysing t^ϵ we may produce a reduction

$$(\square) \quad t^\epsilon \rhd\!\!\!\rhd t^*$$

to an untyped term t^*. But now a use of subject reduction followed by deletion produces a new H-derivation

$$(\nabla \cdot \square) \quad \Gamma \vdash \tau$$

which in some sense is simpler than ∇. If we arrange that the object t^* is a normal term, then we say the resulting derivation $\nabla \cdot \square$ is normal.

The idea of a computation acting on a derivation with a resulting subject reduction is important, and will be developed throughout the book. We will return to subject reduction algorithms several times, each time with a little more complexity. Each time you should make sure you understand this simpler case before you move on.

EXERCISES

3.15 For arbitrary types θ, ψ, ϕ set

$$\rho = \theta \to \phi \qquad \sigma = \theta \to \psi \qquad \tau = \psi \to \phi \qquad \beta = \tau \to \sigma \to \rho$$

in the usual way. Also set $\Gamma = z : \tau, y : \sigma, x : \theta$ to produce a context.

(a) Write down derivations

$$(B) \quad \Gamma \vdash \mathsf{B}_\bullet : \beta \qquad\qquad (\nabla) \quad \Gamma \vdash \mathsf{B}_\bullet zyx : \phi$$

for a suitable typed combinator $\mathsf{B}_\bullet$.

(b) With $\mathsf{B} = \mathsf{B}_\bullet^\epsilon$, write down an untyped computation

$$(\square) \quad \mathsf{B}zyx \rhd\!\!\!\rhd z(yx)$$

where z, y, x are taken from Γ.

(c) Calculate $\nabla \cdot \square$, the action of $\square$ on ∇. You should set out the details of the workings of the algorithm.

3.16 Consider the derivations

$$(B) \quad \vdash \mathsf{B}_\bullet : \beta \qquad (C) \quad \vdash \mathsf{C}_\bullet : \gamma \qquad (D) \quad \vdash \mathsf{D}_\bullet : \delta$$

and $\nabla = CB$ of Example 3.16. There it is suggested there is a computation

$$(\square) \quad \mathsf{C}_\bullet \mathsf{B}_\bullet \rhd\!\!\!\rhd \mathsf{D}_\bullet$$

such that $\nabla \cdot \square = D$.

(a) Write down the derivation B, C, D, ∇ in the informal notation.

(b) Write down a computation $\square$.

(c) Verify $\nabla \cdot \square = D$.

3.17 Consider the term $\mathsf{C}_\bullet \mathsf{S}_\bullet \mathsf{I}_\bullet$ used in Exercise 3.6(a). There you should have proved that, when well disciplined, this term inhabits the coercion type

$$\omega = (\theta \to \theta \to \phi) \to (\theta \to \phi)$$

(for some indexing types θ, ϕ). How does this term relate to the term $\mathsf{W}(\theta, \phi)$ of Exercise 3.2(ω)?

3.18 For arbitrary types $\theta, \psi, \phi, \rho, \zeta$ consider the compound types

$$\begin{aligned}
\mu &= (\zeta \to \rho \to \theta \to \psi \to \phi) \to \zeta \to \rho \to (\theta \to \psi) \to (\theta \to \phi) \\
\nu &= (\rho \to \theta \to \psi \to \phi) \to \rho \to (\theta \to \psi) \to (\theta \to \phi) \\
\tau &= (\theta \to \psi \to \phi) \to (\zeta \to \theta) \to (\zeta \to \psi) \to (\zeta \to \phi)
\end{aligned}$$

as used in Exercise 3.2. Produce normal inhabitants of these types. How do these inhabitants relate to previously produced inhabitants?

3.19 For each $1 \le i \le 5$ consider the derivation $\nabla(i)$ and computation $\square(i)$ you obtained in Exercises 3.3 and 3.12. Use these to find a normal inhabitant of the type $\omega(i)$. Are your inhabitants of $\omega(1)$ and $\omega(5)$ different?

3.20 Consider the following four pairs of arboreal codes for computations

(i)	$\mathsf{l} \circ (\mathsf{m} \circ \mathsf{r})$	$(\mathsf{l} \circ \mathsf{m}) \circ \mathsf{r}$	(ii)	$\rfloor(\mathsf{l} \circ \mathsf{r})$	$\rfloor\mathsf{l} \circ \rfloor\mathsf{r}$
(iii)	$\lfloor(\mathsf{l} \circ \mathsf{r})$	$\lfloor\mathsf{l} \circ \lfloor\mathsf{r}$	(iv)	$\rfloor\mathsf{l} \circ \lfloor\mathsf{r}$	$\lfloor\mathsf{r} \circ \rfloor\mathsf{l}$

where in each case the components $\mathsf{l}, \mathsf{m}, \mathsf{r}$ are assumed to have the appropriate compatibility. Can you find a useful 'equivalence' between the members of each pair?

4
THE TYPED λ-CALCULUS

4.1 INTRODUCTION

In Chapter 1 we introduced two derivation systems **H** and **N**. In Chapter 2 we introduced two computation mechanisms using combinator terms and λ-terms. In Chapter 3 we amalgamated the **H**-system with combinator reduction to produce a calculus **C**. In this chapter we amalgamate the **N**-system with λ-reduction to produce a calculus $\boldsymbol{\lambda}$, the **simply typed λ-calculus.**

As with **C** the calculus $\boldsymbol{\lambda}$ has three facilities.

Derivation system   Substitution algorithm   Computation mechanism

Here the substitution algorithm is more important because λ-terms contain free and bound identifiers. However, much of what we do with $\boldsymbol{\lambda}$ is a replica of the material for **C**. This enables us to develop $\boldsymbol{\lambda}$ quite quickly, for it isn't necessary to repeat all the details, and we can concentrate on the new aspects.

4.1 DEFINITION. (**Types**) There is an unlimited stock of variables. The types are then generated as follows.

- Each variable is a type.
- If ρ, σ are types then $(\sigma \to \rho)$ is a type.

(**Terms**) There is an unlimited stock of identifiers. The terms are then generated as follows.

- Each identifier is a term.
- If p, q are terms then (qp) is a term.
- If r is a term, σ is a type, and y is an identifier, then $(\lambda y : \sigma \,.\, r)$ is a term.

We write $\mathbb{L}amb_{\bullet}$ for the set of such terms. □

These types are just the same as in Chapter 1 (where they are called formulas) and Chapter 3. We carry over the notations and conventions from there.

The terms of Definition 4.1 are very like those of Section 2.4. The difference is that here an abstraction

$$\lambda y : \sigma \,.\, r$$

contains a housing type σ for the bound identifier y. This restricts the range of variation of y. We carry over the notations and conventions for λ-terms.

Exercises

4.1 Consider the term

$$t = (\lambda x \,.\, ((\lambda x \,.\, ((\lambda x \,.\, x)x))x))x$$

as in Exercise 2.13. Show there is a typed version of this where different uses of x inhabit different types. Is this typed version well formed?

4.2 Derivation

Strictly speaking Definition 4.1 gives the **raw terms**. We want the **well formed terms**. To do this we take the usual route.

4.2 DEFINITION. (a) A **statement** is a pair $t : \tau$ where the **subject** t is a term and the **predicate** τ is a type.

(b) A **declaration** is a statement $x : \sigma$ where the subject x is an identifier.

(c) A **context** is a list

$$\Gamma = x_1 : \sigma_1, \dots, x_l : \sigma_l$$

of declarations. The context Γ is **legal** if the identifiers $x_1, \dots, x_l$ are distinct.

(d) A **judgement**

$$\Gamma \vdash t : \tau$$

is a statement $t : \tau$ in context Γ. □

This definition is word for word the same as Definition 3.2. However, a term is now a λ-term (as in Definition 4.1) not a combinator term. We wish to read a judgement

$\Gamma \vdash t : \tau$ as Within the *legal* context Γ, the *well formed* term t inhabits the *acceptable* type τ

where the legality, well-formedness, and acceptability need to be justified. For us all types are acceptable, and the legality of a context is easy to detect. The well-formedness of a term is more complicated.

4.3 DEFINITION. A **derivation**

$$(\nabla) \quad \Gamma \vdash t : \tau$$

is a finite rooted tree of judgements grown according to the rules of Table 4.1.

There is one base clause (Projection), which is restricted by a proviso; and two recursion clauses (Introduction, Elimination). The fourth column of the table gives the arboreal code for each rule. □

Clause	Shape	Proviso	Code
Projection	$\Gamma \vdash x : \sigma$	Γ legal and $x : \sigma$ occurs in Γ	$\Gamma[x]$
Introduction	$\dfrac{\overset{R}{\Gamma, y : \sigma \vdash r : \rho}}{\Gamma \vdash t : \tau}\,(\uparrow)$	$t = \lambda y : \sigma . r$ $\tau = \sigma \to \rho$	$R\!\uparrow$
Elimination	$\dfrac{\overset{Q}{\Gamma \vdash q : \pi \to \tau} \quad \overset{P}{\Gamma \vdash p : \pi}}{\Gamma \vdash qp : \tau}$		(QP)

Table 4.1: The derivation rules for $\boldsymbol{\lambda}$

You should compare this with the definition of an N-derivation. There is more going on here. Notice that the arboreal code $\Gamma[x]$ for a Projection uses the identifier x of a declaration rather than its position. This use of identifiers as labels makes some things a little easier (but also causes some problems).

4.4 EXAMPLES. Consider the types

$$\iota = \theta \to \theta \qquad \kappa = \psi \to \theta \to \psi \qquad \sigma = \chi \to \xi \to \theta \to \phi$$

where

$$\chi = \theta \to \psi \to \phi \qquad \xi = \theta \to \psi$$

and θ, ψ, ϕ are arbitrary. These, of course, are the axioms used in the system H and the housing types of the constants $\mathsf{I}(\theta), \mathsf{K}(\theta, \psi), \mathsf{S}(\theta, \psi, \phi)$ of the calculus **C**. We have three N-derivations

$$\dfrac{\theta \vdash \theta}{\vdash \iota} \qquad \dfrac{\dfrac{\psi, \theta \vdash \psi}{\psi \vdash \theta \to \psi}}{\vdash \kappa} \qquad \dfrac{\dfrac{\dfrac{\dfrac{\dfrac{\chi,\xi,\theta \vdash \chi \quad \chi,\xi,\theta \vdash \theta}{\chi,\xi,\theta \vdash \psi \to \phi} \quad \dfrac{\chi,\xi,\theta \vdash \xi \quad \chi,\xi,\theta \vdash \theta}{\chi,\xi,\theta \vdash \psi}}{\chi,\xi,\theta \vdash \phi}}{\chi,\xi \vdash \theta \to \phi}}{\chi \vdash \xi \to \theta \to \phi}}{\vdash \sigma}$$

which form the basis of a translation of H into N.

Consider the contexts

$$\Delta = x : \theta \qquad \Pi = y : \psi, x : \theta \qquad \Sigma = z : \chi, y : \psi, x : \theta$$

where the identifiers x, y, z are distinct (so the contexts are legal). Then

$$
\dfrac{\Delta \vdash x : \theta}{\vdash \mathsf{I}(\theta) : \iota}
\qquad
\dfrac{\dfrac{\Pi \vdash y : \psi}{y : \psi \vdash - : \theta \to \psi}}{\vdash \mathsf{K}(\theta, \psi) : \kappa}
\qquad
\dfrac{\dfrac{\dfrac{\dfrac{\Sigma \vdash z : \chi \quad \Sigma \vdash x : \theta}{\Sigma \vdash zx : \psi \to \phi} \quad \dfrac{\Sigma \vdash y : \xi \quad \Sigma \vdash x : \theta}{\Sigma \vdash yx : \psi}}{\Sigma \vdash (zx)(yx) : \phi}}{- \vdash - : \theta \to \phi}}{\dfrac{z : \chi \vdash - : \xi \to \theta \to \phi}{\vdash \mathsf{S}(\theta, \psi, \phi) : \sigma}}
$$

are $\boldsymbol{\lambda}$-derivations with root subjects

$$
\mathsf{I}(\bullet) = \lambda x : \theta \, . \, x \qquad \mathsf{K}(\bullet) = \lambda y : \psi, x : \theta \, . \, y \qquad \mathsf{S}(\bullet) = \lambda z : \chi, y : \psi, x : \theta \, . \, (zx)(yx)
$$

respectively. Some components have been omitted from the derivations. You should fill these in.

Reading the trees from leaves to root we obtain the arboreal codes

$$
I(\theta) = \Delta[x]\uparrow \quad K(\theta, \psi) = \Pi[x]\uparrow\uparrow \quad S(\theta, \psi, \phi) = ((\Sigma[z]\Sigma[x])(\Sigma[y]\Sigma[x]))\uparrow\uparrow\uparrow
$$

for these derivations. Observe the close correspondence between the arboreal code and the root subject of each derivation. □

At this point you should learn to grow $\boldsymbol{\lambda}$-derivations. The practice you had with N-derivations will help a lot.

Exercises

4.2 Consider the three standard types β, γ, δ (where β, δ are built from arbitrary types θ, ψ, ϕ, and γ is built from arbitrary types ρ, σ, τ). Produce $\boldsymbol{\lambda}$-derivations

$$
\vdash \mathsf{B} : \beta \qquad \vdash \mathsf{C} : \gamma \qquad \vdash \mathsf{D} : \delta
$$

for suitable λ-terms B, C, D. In each case you should write down the arboreal code of the derivation you produce.

4.3 Consider the types $\epsilon, \omega, \mu, \nu, \tau$ used in Exercise 1.8. (These are built from arbitrary types $\theta, \psi, \phi, \zeta, \rho$.) For each such type π Solution 1.8 provides two N-derivations of $\vdash \pi$. One of these, the left hand one, relies heavily on appropriate H-derivations; and the other, the right hand one, makes better use of the special facilities of N. For each such π produce $\boldsymbol{\lambda}$-derivations

$$
(L) \quad \vdash \mathsf{L} : \pi \qquad\qquad (R) \quad \vdash \mathsf{R} : \pi
$$

which codify these two N-derivations. You should write down the arboreal code for each derivation. Can you see any relationship between L and R?

4.4 For a term t the support ∂t is the set of identifiers occurring free in t. For a context Γ the support $\partial\Gamma$ is the set of identifiers declared in Γ. Write down a recursive construction for ∂t. Show that if the judgement $\Gamma \vdash t : \tau$ is derivable, then Γ is legal and $\partial t \subseteq \partial\Gamma$.

4.5 There is an obvious translation

$$\mathbb{C}omb_{\bullet} \xrightarrow{(\cdot)_\lambda} \mathbb{L}amb_{\bullet}$$

from typed combinator terms to typed λ-terms given by

$$x_\lambda = x \qquad (qp)_\lambda = q_\lambda p_\lambda \qquad \mathsf{Z}_{\bullet\lambda} = \text{explicit term}$$

using the explicit terms of Example 4.4. Show how this extends to a translation

$$(\nabla)\ \Gamma \vdash t : \tau \quad \longmapsto \quad (\nabla_\lambda)\ \Gamma \vdash t_\lambda : \tau$$

from **C**-derivations to **λ**-derivations.

4.3 Annotation and deletion

The system of **λ**-derivations is nothing more than the system of N-derivations with each derivation annotated by λ-terms, where these terms keep track of the construction of the derivation. This is the best-known example of a Curry–Howard correspondence, and is sometimes referred to as *the* Curry–Howard correspondence. In this section we look at the details of this case.

In this discussion we are helped by the corresponding material for **C** given in Section 3.3. Many details here are the same, so we may concentrate on those aspects special to **λ**.

4.5 THEOREM. *For each context* Γ *and term* t *there is at most one derivation*

$$(\nabla)\ \Gamma \vdash t : \tau$$

(for arbitrary types τ*).*

This shows that the pair (Γ, t) uniquely determines τ and ∇. In fact, the arboreal code for ∇ is nothing more than a fused version of the pair (Γ, t).

To justify Theorem 4.5 we convert the proof into an algorithm.

4.6 ALGORITHM. (Generation) *An algorithm which, when supplied with a context* Γ *and a term* t*, will decide whether or not there is a derivation*

$$(\nabla)\ \Gamma \vdash t : \tau$$

(for some type τ*), and if so will return such a derivation.*

Construction. The algorithm proceeds by recursion on the input t. There is one base case and two recursion steps. The clue is to ask what the root rule of a derivation ∇ can be. It is determined by the shape of t.

(Projection) When t is an identifier x, there is a derivation ∇ if and only if Γ is legal and x is declared in Γ, in which case $\Gamma[x]$ is the only example. The declaration $x : \sigma$ of x in Γ gives the predicate $\tau = \sigma$.

(Introduction) When t is an abstraction $(\lambda y : \sigma . r)$, there is a derivation ∇ if and only if there is a derivation

$$(R) \quad \Gamma, y : \sigma \vdash r : \rho$$

for some type ρ. Given R we may set $\tau = \sigma \to \rho$ and $\nabla = R\uparrow$ to produce the required output.

(Elimination) When t is a compound qp, there is a derivation ∇ if and only if there are derivations

$$(Q) \quad \Gamma \vdash q : \pi \to \tau \qquad (P) \quad \Gamma \vdash p : \pi$$

with compatible predicates, as indicated. In this case QP is such a derivation. □

To describe the CH correspondence for **λ** we carry over the notation and terminology for **C**. Thus given a context (declaration list)

$$\Gamma = x_1 : \sigma_1, x_2 : \sigma_2, \ldots, x_l : \sigma_l$$

we may delete the labelling identifiers x_i to obtain

$$\Gamma^\delta = \sigma_1, \sigma_2, \ldots, \sigma_l$$

a context (hypothesis list). Every context (hypothesis list) can be annotated to produce a legal context (declaration list) whose deletion is the original context (hypothesis list). We lift this deletion process to derivations.

4.7 ALGORITHM. (Deletion) *An algorithm*

$$(\nabla) \quad \Gamma \vdash t : \tau \quad \longmapsto \quad (\nabla^\delta) \quad \Gamma^\delta \vdash \tau$$

which, when supplied with a **λ**-*derivation, will return an* N-*derivation in the deleted context with the same predicate.*

Construction. The algorithm proceeds by recursion on the supplied derivation ∇. There is one base case and two recursion steps.

(Projection)	$\Gamma[x]^\delta = \Gamma^\delta[n]$	x labels the nth position of Γ
(Introduction)	$R\uparrow^\delta = R^\delta\uparrow$	
(Elimination)	$(QP)^\delta = Q^\delta P^\delta$	

(In the (Projection) clause remember that the declarations of Γ are counted from right to left beginning with the gate declaration in position 0. Thus the position is not the same as the index as used above.)

It is easy to check that $(\cdot)^\delta$ sends a **λ**-derivation into an N-derivation of the required form. □

The companion annotation algorithm which attempts to insert labelling terms is not so straight forward. As with an H-derivation, if we start from an N-derivation we can not expect to conjure up the 'intended' context labels out of thin air. We must start from a derivation

$$(\nabla) \ \ \Gamma^\delta \vdash \tau$$

where the context (declaration list) is the deletion Γ^δ of a context (declaration list) with known labels. We can then attempt to construct a labelling subject term t and produce a derivation of

$$\Gamma \vdash t : \tau$$

whose deletion is ∇. But with $\boldsymbol{\lambda}$, in the middle of ∇ we may need a context that is longer than Γ. Where do these extra labels come from? They are the required bound identifiers of t. Again, these identifiers can not be conjured up out of nothing.

We need a selection policy which, given a context (declaration list) Γ and a type σ, will choose a fresh label y so that the context (declaration list)

$$\Gamma, y : \sigma$$

is legal. It doesn't matter too much what the selection policy is, and the choice of y may depend on many other factors not mentioned here. But with a selection policy the annotation algorithm becomes deterministic.

4.8 ALGORITHM. (Annotation) *Given a selection policy, an algorithm*

$$(\nabla) \ \ \Gamma^\delta \vdash \tau \quad \longmapsto \quad ((\Gamma, \nabla)^\alpha) \ \ \Gamma \vdash t : \tau$$

which, when supplied with a context (declaration list) Γ *and an* N*-derivation in the deleted context, will return a* $\boldsymbol{\lambda}$*-derivation in the given context with the same predicate.*

Construction. The algorithm proceeds by recursion on the supplied derivation ∇. There is one base case and two recursion steps.

(Projection)	$(\Gamma, \Gamma^\delta[n])^\alpha = \Gamma[x]$	x labels the nth position of Γ
(Introduction)	$(\Gamma, R\uparrow)^\alpha = (\Sigma, R)^\alpha\uparrow$	
(Elimination)	$(\Gamma, QP)^\alpha = (\Gamma, Q)^\alpha(\Gamma, P)^\alpha$	

For the (Introduction) clause we are given an N-derivation

$$(R) \ \ \Gamma^\delta, \sigma \vdash \rho$$

in an extension by a type σ of Γ^δ. Using the selection policy we find a label y so that

$$\Sigma = \Gamma, y : \sigma$$

is legal. By recursion we obtain a $\boldsymbol{\lambda}$-derivation

$$((\Sigma, R)^\alpha) \ \ \Sigma \vdash r : \rho$$

for some term r. An Introduction now gives the required $\boldsymbol{\lambda}$-derivation in Γ. □

These two algorithms and the need for a selection policy clarify the notion of a CH correspondence.

4.9 THEOREM. *Consider a context (declaration list) Γ.*

For each N*-derivation ∇ over Γ^δ, we have $(\Gamma, \nabla)^{\alpha\delta} = \nabla$.*

For each $\boldsymbol{\lambda}$*-derivation ∇ over Γ, we have $(\Gamma, \nabla^\delta)^\alpha = \nabla$ provided the labels are chosen appropriately.*

Relative to an appropriate choice of labels, the two algorithms $(\cdot)^\alpha$ and $(\cdot)^\delta$ are inverse companions. The deletion of an annotation is always the original N*-derivation, but the annotation of a deletion is merely an alphabetic variant of the original* H*-derivation.*

Proof. As in Solution 3.8 each part is proved by induction over ∇. The first part is straight forward. For the second part the induction step across a use of I should be looked at.

Suppose Γ is a given context (declaration list) and $\nabla = R\uparrow$ is a $\boldsymbol{\lambda}$-derivation over Γ. Thus we have $\boldsymbol{\lambda}$-derivations

$$(R) \quad \Gamma, y : \sigma \vdash r : \rho \qquad (\nabla) \quad \Gamma \vdash t : \tau$$

where $t = (\lambda y : \sigma \,.\, r)$ and $\tau = (\sigma \to \rho)$. By deletion we produce N-derivations

$$(R^\delta) \quad \Gamma, \sigma \vdash \rho \qquad (\nabla^\delta) \quad \Gamma \vdash \tau$$

with $\nabla^\delta = R^\delta\uparrow$. Now

$$(\Gamma, \nabla^\delta)^\alpha = (\Gamma, R^\delta\uparrow)^\alpha = (\Sigma, R^\delta)^\alpha\uparrow$$

where $\Sigma = (\Gamma, z : \sigma)$ for a selected identifier z. Clearly we need to select $z = y$, and then

$$(\Sigma, R^\delta)^\alpha = R$$

follows by the induction hypothesis, to give

$$(\Gamma, \nabla^\delta)^\alpha = R\uparrow = \nabla$$

as required. □

Is there any selection policy for which deletion followed by annotation always returns the original? No; for any selection policy this deletion followed by annotation produces the 'standard' version of the original derivation. But there are always derivations with a 'non-standard' use of identifiers.

The need to rename identifiers is a persistent problem in λ-calculus. It is usually solved by either

- agreeing to work 'up to alphabetic variants', i.e. by handwaving, or
- enforcing a rigid selection policy so that only 'standard' derivations occur, i.e. by rejecting most λ-terms

neither of which is what we want to do in practice. Here, for the time being, we take a more pragmatic approach; we ignore the problem. We return to this topic in Chapter 5.

Theorem 4.5 shows that each derivation

$$(\nabla) \quad \Gamma \vdash t : \tau$$

is uniquely determined by the root pair (Γ, t), and Algorithm 4.6 indicates how to construct ∇ from the pair. The arboreal code of ∇ is a concise description of this construction. More generally, given any pair (Γ, t) we can determine whether or not it is the root pair of some derivation ∇, and if so we can produce the predicate τ. In practice this means that certain components can be omitted from a derivation, for they can be reinstated using the remaining material.

Exercises

4.6 Consider the following terms which depend on unknown types θ, ψ, ϕ.

$$\begin{array}{lll} \overline{S} = & \lambda u : \phi, y : \psi, x : \theta \,.\, y(uyx) & \overline{0} = \lambda y : \psi, x : \theta \,.\, x \\ \overline{A} = & \lambda v : \phi, u : \phi, y : \psi, x : \theta \,.\, (uy)(vyx) & \overline{1} = \lambda y : \psi, x : \theta \,.\, yx \\ \overline{M} = & \lambda v : \phi, u : \phi, y : \psi, x : \theta \,.\, u(vy)x & \overline{2} = \lambda y : \psi, x : \theta \,.\, y^2 x \end{array}$$

For each of the compounds t

$$\text{(i)} \quad t = \overline{S}(\overline{S}\,\overline{0}) \qquad \text{(ii)} \quad t = \overline{A}\,\overline{1} \qquad \text{(iii)} \quad t = \overline{M}\,\overline{2}$$

find necessary and sufficient conditions on these types θ, ψ, ϕ so that $\vdash t : \tau$ is derivable for some type τ. Relate τ to θ, ψ, ϕ.

4.4 Substitution

As with the untyped λ-calculus of Section 2.4, substitution for $\boldsymbol{\lambda}$ is not entirely straightforward. We need to be aware of the threat of unwanted identifier capture. Luckily much of what we did in Section 2.4 can be transferred to $\boldsymbol{\lambda}$.

4.10 DEFINITION. For all terms s, t and identifiers x the term

$$t[x := s]$$

is generated by recursion over the structure of t using the following clauses.

$$\begin{array}{rclrcl} x[x := s] & = & s & (qp)[x := s] & = & (q[x := s])(p[x := s]) \\ y[x := s] & = & y & (\lambda y : \sigma \,.\, r)[x := s] & = & \lambda v : \sigma \,.\, (r'[x := s]) \end{array}$$

In the bottom left clause the identifier y is distinct from x. In the crucial bottom right clause we set

$$r' = r[y := v]$$

where v is a 'safe' identifier which avoids unwanted capture. □

This definition is almost word for word the same as Definition 2.7. The only difference is the appearance of a housing type in the crucial clause passing across an abstraction. However, that type does not seem to effect the substitution too much. As in Section 2.2 we do not explain what a 'safe' identifier is; we postpone that until Chapter 5.

As well as substitution into terms, with **λ** we need to substitute into a derivation. To avoid confusion we call this **grafting** for reasons that will become clear. (It is sometimes called **substitution in context**.)

4.11 ALGORITHM. (Grafting) *An algorithm which, when supplied with a pair of derivations*

$$(R)\ \ \Pi, x : \sigma, \Delta \vdash r : \rho \qquad (S)\ \ \Pi, \Delta \vdash s : \sigma$$

and a nominated component $x : \sigma$ of the R-context (as indicated), will return a derivation

$$(R * S)\ \ \Pi, \Delta \vdash r[x := s] : \rho$$

the result of **grafting** *S onto R at the nominated leaves.*

Construction. The algorithm proceeds by recursion over R. There are two base cases and two recursion steps. For convenience let $\Sigma = \Pi, x : \sigma, \Delta$.

	R	$R * S$
(Nominated)	$\Sigma[x]$	S
(Projection)	$\Sigma[y]$	$(\Pi, \Delta)[y]$
(Introduction)	$P\uparrow$	$(P' * S')\uparrow$
(Elimination)	QP	$(Q * S)(P * S)$

In the (Projection) clause the identifier y is not the same as the identifier x. Thus y is declared in (Π, Δ), and then $R * S$ is the corresponding Projection. The (Introduction) clause will be explained in a moment.

Informally this algorithm locates those leaves of R which project from the nominated identifier x. Further branches are grown from those leaves to form replicas of S.

What does the (Introduction) step do? We have a derivation

$$(P)\ \ \Sigma, y : \xi \vdash p : \pi$$

where

$$R = P\uparrow \qquad r = (\lambda y : \xi \,.\, p) \qquad \rho = \xi \to \pi$$

for appropriate components y, ξ, p, π. Recall that

$$r[x := s] = \lambda v : \xi \,.\, (p'[x := s]) \quad \text{where} \quad p' = p[y := v]$$

and v is 'safe'. With such a v we modify P and S to produce

$$(P')\ \ \Sigma, v : \xi \vdash p' : \pi \qquad (S')\ \ \Pi, \Delta, v : \xi \vdash s : \sigma$$

where P' is obtained from P by replacing each use of y by v, and S' is obtained from S by weakening the context. The choice of v must make this possible.

With this pair P', S' a recursive call on the algorithm gives

$$(P' * S') \quad \Pi, \Delta, v : \xi \vdash p'[x := s] : \pi$$

and then a use of I produces

$$P\!\uparrow * S = (P' * S')\!\uparrow$$

as required. □

To make this algorithm watertight we really should say how the modifications P' of P and S' of S are produced. What can that renaming identifier be? Since there is a derivation P we know that y is not declared in Σ (otherwise the context $(\Sigma, y : \xi)$ is not legal). Thus, from S we know that y is not free in s (otherwise it would be declared in (Π, Δ)). Thus there is no danger of identifier capture, and we may leave $v = y$. Thus $P' = P$ and S' is a minor variant of S.

This looks very nice, but it is wrong.

Consider the case

$$(P) \quad x : \sigma, y : \xi \vdash y : \xi \qquad (S) \quad \vdash s : \sigma$$

where $\Pi = \Delta = \emptyset$ and P is a Projection. The previous argument suggests that we take

$$(S') \quad y : \xi \vdash s : \sigma$$

and then $P' * S' = S'$. But what if $s = \lambda y : \xi \,.\, q$ with $\sigma = \xi \to \chi$ for some q and χ? Then

$$y : \xi \vdash s : \sigma$$

is not derivable. (Try it.) We can not leave $v = y$ for this obstructs the wanted S'.

The renaming of y to v must avoid identifier capture *and* it must allow the weakening of S to S'. In fact, to give a watertight immersion algorithm we ought to design a more powerful algorithm which anticipates the problems before it meets them. We don't do this just yet, but leave it until Section 6.4. You will find that Exercises 4.7, 4.8, 4.9 illustrate some of the problems to be overcome. Algorithm 4.11 still works but it is not entirely mechanistic; it requires a bit of intelligence from the user.

Exercises

4.7 Describe an algorithm which, when supplied with a derivation

$$(\nabla) \quad \Pi, x : \sigma, \Delta \vdash t : \tau$$

and a nominated declaration, as shown, with $x \notin \partial t$, will return a derivation

$$(\nabla^-) \quad \Pi, \Delta \vdash t : \tau$$

with the nominated declaration omitted (and no change to t).

4.8 Describe an algorithm which, when supplied with a derivation

$$(\nabla)\quad \Pi, x : \sigma, \Delta \vdash t : \tau$$

a nominated declaration, as shown, and an identifier u which does not appear anywhere in ∇ (either free or bound), will return a derivation

$$(\nabla')\quad \Pi, u : \sigma, \Delta \vdash t' : \tau$$

where t' is obtained from t by replacing each occurrence of x by u.

4.9 Describe an algorithm which, when supplied with a derivation

$$(\nabla)\quad \Gamma \vdash t : \tau$$

and a declaration $u : \sigma$ where the identifier u does not appear anywhere in ∇ (either free or bound), will return a derivation

$$(\nabla^{+})\quad \Gamma, u : \sigma \vdash t : \tau$$

in the lengthened context.

4.10 Exercise 4.5 produced a translation from **C**-derivations to **λ**-derivations. Describe a companion translation which converts each **λ**-derivation into a **C**-derivation.

4.5 Computation

There is a reduction relation $\rhd\!\!\!\rhd$ on the set $\mathbb{L}amb_{\bullet}$ of typed λ-terms. This is almost the same as that on the set $\mathbb{L}amb$ of untyped λ-terms, and is generated in the standard way.

4.12 DEFINITION. (a) A **reducible expression** or **redex** is a compound term

$$(\lambda y : \sigma \,.\, r)s \qquad r[y := s]$$

for terms r, s and an identifier y as shown on the left. The term on the right is the **immediate reduct** of the redex.

(b) For **λ** the 1-step **reduction axioms** are

$$t^{-} \rhd t^{+}$$

where t^{-} is a redex and t^{+} is the immediate reduct. □

We generate the relation $\rhd\!\!\!\rhd$ as the transitive, structural closure of the 1-step relation $\rhd$.

4.13 DEFINITION. A **computation**

$$(\square)\quad t^{-} \rhd\!\!\!\rhd t^{+}$$

is a finite rooted tree of reductions grown according to the rules of Table 4.2. Such a computation $\square$ is said to **organize** the root reduction $t^{-} \rhd\!\!\!\rhd t^{+}$ where t^{-} is the **subject** and t^{+} is the **object** of the computation. The right hand column of the table shows how to generate the arboreal code for a computation. □

Clause	Shape	Remarks	Code
(Redex reduction)	$\dfrac{t^- \rhd t^+}{t^- \mathrel{\rhd\!\!\!\rhd} t^+}\,(1)$	$t^- = (\lambda y : \sigma \,.\, r)s$ $t^+ = r[y := s]$	1
(Left application)	$\dfrac{\overset{\mathsf{q}}{q^- \mathrel{\rhd\!\!\!\rhd} q^+}}{q^- p \mathrel{\rhd\!\!\!\rhd} q^+ p}\,(\downharpoonleft)$		$\downharpoonleft\mathsf{q}$
(Right application)	$\dfrac{\overset{\mathsf{p}}{p^- \mathrel{\rhd\!\!\!\rhd} p^+}}{qp^- \mathrel{\rhd\!\!\!\rhd} qp^+}\,(\downharpoonright)$		$\downharpoonright\mathsf{p}$
(Abstraction)	$\dfrac{\overset{\mathsf{r}}{r^- \mathrel{\rhd\!\!\!\rhd} r^+}}{t^- \mathrel{\rhd\!\!\!\rhd} t^+}\,(\uparrow)$	$t^- = (\lambda y : \sigma \,.\, r^-)$ $t^+ = (\lambda y : \sigma \,.\, r^+)$	$\uparrow\mathsf{r}$
(Transitive composition)	$\dfrac{\overset{\mathsf{l}}{t^- \mathrel{\rhd\!\!\!\rhd} t^0} \quad \overset{\mathsf{r}}{t^0 \mathrel{\rhd\!\!\!\rhd} t^+}}{t^- \mathrel{\rhd\!\!\!\rhd} t^+}\,(\circ)$		$(\mathsf{l} \circ \mathsf{r})$

Table 4.2: The computation rules for $\boldsymbol{\lambda}$

Table 4.2 seems to be the same as Table 2.3. Here the terms are typed, but there they are untyped. However, these types don't seem to do very much. You should also compare Table 4.2 with Table 3.2. There is a bit more going on here. We have Redex reduction in place of Axiom reduction, and a new rule of Abstraction.

The types in a computation don't do very much. In fact, we can ignore them.

4.14 DEFINITION. The **type erasing** operation from typed terms to untyped terms

$$\mathbb{L}amb_{\bullet} \xrightarrow{(\cdot)^{\epsilon}} \mathbb{L}amb$$

is generated by recursion by

$$x^{\epsilon} = x \quad (\lambda y : \sigma \,.\, r)^{\epsilon} = (\lambda y \,.\, r^{\epsilon}) \quad (qp)^{\epsilon} = q^{\epsilon} p^{\epsilon}$$

for identifiers x, y, type σ, and terms q, p, r. □

It can be shown that type erasure passes through substitution.

$$(r[x := s])^{\epsilon} = r^{\epsilon}[x := s^{\epsilon}]$$

Thus, if we erase the types from a typed 1-step reduction then we get an untyped 1-step reduction. More generally, type erasure passes through computations.

4.15 THEOREM. *Type erasure*

$$(\square)\ t^- \rhd\!\!\!\rhd t^+ \quad \longmapsto \quad (\square^\epsilon)\ t^{-\epsilon} \rhd\!\!\!\rhd t^{+\epsilon}$$

converts each typed computation $\square$ *into an untyped computation* $\square^\epsilon$ *where these have the same arboreal code. Furthermore, given an untyped computation*

$$(\square^*)\ t^{-\epsilon} \rhd\!\!\!\rhd t^*$$

where the subject $t^{-\epsilon}$ *is the erasure of a typed term* t^-*, there is a unique typed computation* $\square$ *(as above) with* $\square^\epsilon = \square^*$ *(and hence* $t^* = t^{+\epsilon}$*).*

This is similar to the corresponding result, Theorem 3.15, for combinator terms.

Exercises

4.11 Exercise 4.3 produced a pair of derivations

$$(L) \quad \vdash \mathsf{L} : \pi \qquad (R) \quad \vdash \mathsf{R} : \pi$$

for each type π taken from $\epsilon, \omega, \mu, \nu, \tau$ (as defined in Exercise 1.8). For each such pair describe two or more computations which organize a reduction $\mathsf{L} \rhd\!\!\!\rhd \mathsf{R}$.

4.12 Show that $(t[x := s])^\epsilon = t^\epsilon[x := s^\epsilon]$ holds for all terms t, s and identifiers x.

4.13 Prove Theorem 4.15.

4.6 Subject reduction

The calculus $\boldsymbol{\lambda}$ has two important facilities

the derivation system the computation mechanism

and a mediating facility, the substitution algorithm. We have seen that for $\boldsymbol{\lambda}$ this algorithm is not entirely satisfactory. Later, in Chapter 5, we will take another look at substitution. Here we look at the direct interaction between derivation and computation.

We wish to produce an algorithm which, when supplied with a compatible pair

$$(\nabla)\ \Gamma \vdash t^- : \tau \qquad (\square)\ t^- \rhd\!\!\!\rhd t^+$$

with a common subject t^-, will return a derivation

$$(\nabla \cdot \square)\ \Gamma \vdash t^+ : \tau$$

which shows that the object t^+ of $\square$ is well formed in the context Γ of ∇. We wish to calculate $\nabla \cdot \square$ by a syntactic manipulation of the arboreal code for ∇ and $\square$. The algorithm will proceed by recursion over $\square$, so we will need something to start from. We look first at the case where $\square$ is a leaf computation, a redex removal.

4.16 ALGORITHM. (Leaf recipe) *An algorithm which, when supplied with*

$$(\nabla)\quad \Gamma \vdash t^- : \tau \qquad t^- = (\lambda x : \sigma \,.\, r)s$$

a derivation with a redex root subject, will return a derivation

$$(\nabla \bullet \mathbf{1})\quad \Gamma \vdash t^+ : \tau \qquad t^+ = r[x := s]$$

with the corresponding immediate reduct as subject (in the same context).

Construction. We know that ∇ unravels as $\nabla = (R\uparrow)S$ where

$$(R)\quad \Gamma, x : \sigma \vdash r : \tau \qquad (S)\quad \Gamma \vdash s : \sigma$$

are the two components. Using the Grafting Algorithm 4.11 we set

$$\nabla \bullet \mathbf{1} = R * S$$

for the required result. □

This leaf recipe feeds the general algorithm.

4.17 ALGORITHM. (Subject reduction) *An algorithm which, when supplied with*

$$(\nabla)\quad \Gamma \vdash t^- : \tau \qquad (\Box)\quad t^- \gg t^+$$

a compatible derivation and computation, will return a derivation

$$(\nabla \cdot \Box)\quad \Gamma \vdash t^+ : \tau$$

the result of the action *of* $\Box$ *on* ∇.

Construction. The algorithm proceeds by recursion over $\Box$ with variation of the parameter ∇. For each compatible pair $(\nabla, \Box)$ there is precisely one action step that can be taken, and this step is determined by $\Box$.

(Proj) $\nabla \cdot \mathbf{1} = \nabla \bullet \mathbf{1}$ (Trans) $\nabla \cdot (\mathsf{l} \circ \mathsf{r}) = (\nabla \cdot \mathsf{l}) \cdot \mathsf{r}$

(LAppl) $(QP) \cdot \rfloor\mathsf{q} = (Q \cdot \mathsf{q})P$ (RAppl) $(QP) \cdot \lfloor\mathsf{p} = Q(P \cdot \mathsf{p})$

(Abstr) $R\uparrow \cdot \uparrow\mathsf{r} = (R \cdot \mathsf{r})\uparrow$

At its base the algorithm calls on the leaf recipe of Algorithm 4.16. We should check that the algorithm returns a derivation of the required form, and does terminate. These proofs are routine. □

Superficially this reduction algorithm looks very like the corresponding Algorithm 3.18 for **C** with an extra clause to handle abstractions. In fact, the extra clause makes the algorithm considerably more complicated. There is a constant need to check for clashing identifiers and rename where necessary. You should do several examples to make sure you understand exactly what is involved.

EXERCISES

4.14 Exercises 4.3 and 4.11 provide several pairs of derivations and a computation

$$(L)\quad \vdash \mathsf{L} : \pi \qquad (R)\quad \vdash \mathsf{R} : \pi \qquad (\Box)\quad \mathsf{L} \gg \mathsf{R}$$

relating the two subject terms. For these L, R and $\Box$, show that $L \cdot \Box = R$ holds.

5
Substitution Algorithms

5.1 Introduction

Until now the idea of substituting in a term for the free occurrences of an identifier has been used only in an informal manner. In any particular instance we can see what we want, can recognize the potential problems, and we can use our common sense to achieve the desired effect. However, substitution is a crucial component of many of the fundamental algorithms that we are concerned with, so the notion can't be left on this informal level. The time has come for a fuller discussion of this notion and a more precise description of a suitable algorithm.

So far for each pair t, s of terms and identifier x we have written

$$t[x := s] \quad \text{for} \quad t \text{ with each free occurrence of } x \text{ replaced by } s$$

with a suitable renaming of bound identifiers in t to avoid the unwanted capture of free identifiers of s. We have even given an algorithm (of sorts) which calculates $t[x := s]$. In this chapter we begin by analysing some of the defects of this informal algorithm and then produce a better algorithm which does this job. This new algorithm will have wider applicability.

In what way do we want to extend the use of substitution?

The first extension is relatively minor, but has an eye on future developments. We want a single algorithm which applies to λ-terms, combinator terms, and more general terms to be used later. In short we want to allow constants to appear in terms. Thus we assume given a stock of constants k. The raw terms are then generated as follows.

- Each constant k is a term.
- Each identifier is a term.
- If q, p are terms, then so is (qp).
- If r is a term, σ is a type, and y is an identifier, then $(\lambda y : \sigma \,.\, r)$ is a term.

The terms which contain no constants are just the λ-terms of Chapter 4. By allowing combinators as constants we generate the combinator terms and more as well. Later we will use other kinds of constants, but for substitution there is no syntactic difference between these and combinators. In use these terms will be immersed in a typing discipline, but that need not concern us here. In fact, it is important that substitution can be performed without the hindrance of any surrounding regime; substitution is an entirely syntactic operation.

The **support** ∂t of a term is the set of identifiers occurring free in t. The clauses

$$\partial\mathsf{k} = \emptyset \quad \partial x = \{x\} \quad \partial(qp) = \partial(q) \cup \partial(p) \quad \partial(\lambda y : \sigma \,.\, r) = \partial r - \{y\}$$

generate this finite set of identifiers.

So far we have considered substitution only for one identifier at a time. We now want to perform simultaneous substitution for several identifiers at once. This needs a more general kind of substitution operator. Let Idf and Trm be, respectively, the set of all identifiers and all terms. We consider certain assignments

$$\begin{array}{c} \mathsf{Idf} \xrightarrow{\alpha} \mathsf{Trm} \\ x \longmapsto x\alpha \end{array}$$

where the argument 'x' of the function 'α' is written on the left to produce the value '$x\alpha$'. Let $\partial\alpha$ be the set of identifiers moved by α. Thus

$$x \in \partial\alpha \Longleftrightarrow x\alpha \neq x$$

and, as the notation suggests, we call $\partial\alpha$ the support of α. A **substitution operator** is such an assignment α with finite support.

Given a list $x_1, x_2, \ldots, x_m$ of distinct identifiers together with a matching list $s_1, s_2, \ldots, s_m$ of (not necessarily distinct) terms there is a substitution operator α defined by

$$x_1\alpha = s_1 \quad x_2\alpha = s_2 \quad \cdots \quad x_m\alpha = s_m$$

with $y\alpha = y$ for $y \notin \{x_1, \ldots, x_m\}$. This moves no more than $x_1, x_2, \ldots, x_m$. It is convenient to write

$$[x_1 := s_1,\ x_2 := s_2,\ \ldots,\ x_m := s_m]$$

for this substitution operator. Every substitution operator can be described in this way. However, you should remember the corresponding substitution will be *simultaneous* so the order of the components $x_i := s_i$ is irrelevant. The case $m = 1$ gives the case of a simple operator $[x := s]$ we have used already.

For each term t and substitution operator α we write

$$t \cdot \alpha \quad \text{for} \quad \begin{array}{l} t \text{ with each free occurrence of an identifier} \\ x \text{ replaced by } x\alpha \text{ in a simultaneous fashion} \end{array}$$

with a suitable renaming of bound identifiers in t to avoid the unwanted capture of free identifiers. Thus the clauses

$$\begin{array}{rclcrcl} \mathsf{k} \cdot \alpha & = & \mathsf{k} & \quad & (qp) \cdot \alpha & = & (q \cdot \alpha)(p \cdot \alpha) \\ x \cdot \alpha & = & x\alpha & & (\lambda y : \sigma \,.\, r) \cdot \alpha & = & (\lambda v : \sigma \,.\, (r' \cdot \alpha)) \\ & & & & \text{where } r' & = & r \cdot [y := v] \text{ and } v \text{ is 'safe'} \end{array}$$

provide a recursive way to calculate $t \cdot \alpha$. In particular, when $\alpha = [x := s]$ and there are no constants, this is just the substitution algorithm used in Chapter 4.

There are two problems with this. The first is that to calculate $(\lambda y : \sigma . r) \cdot \alpha$ requires two recursion calls on the algorithm, one to calculate $r' = r \cdot [y := v]$ and one to calculate $r' \cdot \alpha$. Admittedly the first call is only a renaming, but if there are several abstractions in a term t, then the full calculation of $t \cdot \alpha$ will be quite tortuous. Algorithms which involve nested recursion calls are not a good idea; they can be very hard to analyse.

The second problem is the lack of a definition of 'safe'. Before we can analyse this algorithm we must correct this defect.

5.1 DEFINITION. For each term t and substitution operator α let

$$v \in M(t, \alpha) \iff \text{There is some } w \in \partial t \cap \partial\alpha \text{ with } v \in \partial(w\alpha)$$

and let

$$U(t, \alpha) = \partial t \cup \partial\alpha \cup M(t, \alpha)$$

to produce two finite sets of identifiers. □

Notice that ∂t and $\partial\alpha$ are finite, so there are only finitely many $w \in \partial t \cap \partial\alpha$, and for each such w the support $\partial(w\alpha)$ is finite. Hence $M(t, \alpha)$ and $U(t, \alpha)$ are finite. You should convince yourself that for the case $t = (\lambda y : \sigma . r)$ the set of unsafe renaming identifiers is exactly $U(t, \alpha)$. Here are some remarks to help.

- If v is free in t, then v is free in r, so

  $$r[y := v]$$

  will coalesce the distinct identifiers y and v in r. (These are distinct since y is not free in t.)

- If v is moved by α, then the effect of

  $$(r[y := v])\alpha$$

  will be to move y (to $v\alpha$). But y is bound in t, so can only be renamed in $t\alpha$ (not moved to an arbitrary term).

- Suppose w is free in t and v is free in $w\alpha (\neq w)$. The term $t \cdot \alpha$ ought to have a subterm $w\alpha$ where w occurs in t, and so have v as a free identifier. But $(\lambda v : \sigma . -)$ would bind this occurrence.

The following example demonstrates these points. We will use it later to illustrate the new algorithm.

5.2 EXAMPLE. Consider $t = \lambda y : \sigma . yxw$ and $\alpha = [w := z]$ where w, x, y, z are four distinct identifiers. Hitting t with α ought to produce

$$t \cdot \alpha = \lambda v : \sigma . vxz$$

for some identifier v different from x and z. In fact the algorithm gives

$$t \cdot \alpha = \lambda v : \sigma . (((yxw) \cdot [y := v]) \cdot \alpha)$$

for some suitably chosen v. Let's check that

- the proviso does give an acceptable result
- ignoring any part of the proviso leads to an unacceptable result

and hence the proviso is just what we want (at least in this case).

Note first that

$$\partial t = \{w, x\} \qquad \partial\alpha = \{w\} \qquad M(t, \alpha) = \partial(w\alpha) = \{z\}$$

so that

$$U(t, \alpha) = \{w, x, z\}$$

and $v \notin U(t, \alpha)$ does lead to an acceptable result.

But what does

$$\mathit{body} = ((yxw) \cdot [y := v]) \cdot [w := z]$$

become if we ignore these restrictions? Look at the three cases $v = w, v = x, v = z$ in turn.

$$\begin{array}{ll} (v = w) & \mathit{body} = (vxv) \cdot [v := z] = zxz \\ (v = x) & \mathit{body} = (vvw) \cdot [w := z] = vvz \\ (v = z) & \mathit{body} = (vxw) \cdot [w := v] = vxv \end{array}$$

Thus all three cases lead to unacceptable results, and hence the whole of the proviso seems necessary.

You may like to ponder this point. An acceptable result of the substitution algorithm ought to be

$$t\alpha = \lambda v : \sigma \,.\, vxz$$

where $v \notin \{x, z\}$. In particular, $v = w$ should be acceptable. However, the proviso specifically excludes $v = w$, and we have just seen that this choice does lead to an unacceptable result! □

At this point you should try using the algorithm yourself.

Exercises

5.1 Consider the term and substitution operator

$$t = \lambda y : \sigma \,.\, z(\lambda x : \rho \,.\, yx)x \qquad \alpha = [x := y, y := z, z := x]$$

where x, y, z are distinct identifiers. Calculate $t \cdot \alpha$, describing how the algorithm proceeds. At each renaming the chosen identifier should be the first available one from z, y, x, w, v, u.

5.2 Repeat Exercise 5.1 with

$$t = \lambda z : \tau \,.\, (\lambda y : \sigma \,.\, zy)(\lambda x : \rho \,.\, yx) \qquad \alpha = [y := x, x := y]$$

as the term and substitution operator.

5.2 Formal replacements

Eventually we will *not* use the substitution algorithm described in Section 5.1. We will use a neater algorithm with unnested recursion calls which produces the required result in a smoother fashion. To describe this we need a change of perspective.

So far we have thought of a substitution operator

$$\mathsf{Idf} \xrightarrow{\alpha} \mathsf{Trm}$$

as a set theoretic object, a function of indicated type with finite support. However, it is neater to work with names for such gadgets where each operator will have many different names. We thus introduce a syntactic category of **replacements** $\mathfrak{a}, \mathfrak{b}, \mathfrak{c}, \ldots$, each of which names a substitution operator. Each such replacement is generated from a base replacement $\mathfrak{i}$ (the canonical name for the identity substitution operator) using a sequence of **updates**

$$[y \mapsto s]$$

determined by an identifier y and a term s.

5.3 DEFINITION. Each **replacement** is generated recursively by

- $\mathfrak{i}$ is a replacement
- if $\mathfrak{a}$ is a replacement, then so is $'\mathfrak{a} = [y \mapsto s]\mathfrak{a}$

where, in the recursion clause, $[y \mapsto s]$ is an update. □

For each such replacement $\mathfrak{a}$ and identifier x we obtain a term $x\mathfrak{a}$ by

$$x\mathfrak{i} = x \qquad x'\mathfrak{a} = \begin{cases} s & \text{if } x = y \\ x\mathfrak{a} & \text{if } x \neq y \end{cases}$$

where $'\mathfrak{a} = [y \mapsto s]\mathfrak{a}$ in the recursion clause. The assignment

$$\begin{array}{c} \mathsf{Idf} \longrightarrow \mathsf{Trm} \\ x \longmapsto x\mathfrak{a} \end{array}$$

is a substitution operator, the operator **named** by $\mathfrak{a}$. The support $\partial\mathfrak{a}$ of a replacement $\mathfrak{a}$ is generated by

$$\partial\mathfrak{i} = \emptyset \qquad \partial'\mathfrak{a} = \begin{cases} \partial\mathfrak{a} \cup \{y\} & \text{if } y \neq s \\ \partial\mathfrak{a} - \{y\} & \text{if } y = s \end{cases}$$

where $'\mathfrak{a} = [y \mapsto s]\mathfrak{a}$ in the recursion clause.

Perhaps now the intention is clear. Each replacement $\mathfrak{a}$ has the form

$$[y_1 \mapsto s_1][y_2 \mapsto s_2] \cdots [y_n \mapsto s_n]\mathfrak{i}$$

where $y_1, y_2, \dots, y_n$ is a list of *not necessarily distinct* identifiers and $s_1, s_2, \dots, s_n$ is a matching list of terms. To determine $x\mathfrak{a}$ read this description from left to right to locate the first (leftmost) component $[y_i \mapsto s_i]$ with $x = y_i$. We then set $x\mathfrak{a} = s_i$. If there is no such component, then $x\mathfrak{a} = x$. If there are two or more components

$$[x \mapsto s_1] \qquad [x \mapsto s_2] \qquad \cdots$$

then the leftmost one overrides the others. In particular, the value of

$$[y \mapsto y]\mathfrak{a}$$

at y is y even if there is an update $[y \mapsto s]$ in $\mathfrak{a}$. This is the reason for the 'if $y = s$' clause in the construction of $\partial\mathfrak{a}$.

The use of just one update produces the canonical name for the simplest kind of substitution operator

$$[y \mapsto s]\mathfrak{i} \quad \text{names} \quad [y := s]$$

(unless $s = y$, in which case we have a non-canonical name of the identity substitution operator).

For the later analysis it is convenient to set up a labelled transition structure on identifiers and an associated battery of diamond operations on sets of identifiers.

5.4 DEFINITION. (a) For each replacement $\mathfrak{a}$ and identifiers v, w let

$$w \xleftarrow{\mathfrak{a}} v \qquad \text{mean} \qquad v \in \partial(w\mathfrak{a})$$

to set up a labelled transition relation on identifiers.

(b) For each replacement $\mathfrak{a}$ let $\cdot^{\langle\mathfrak{a}\rangle}$ be the operation on sets $\mathcal{W}$ of identifiers given by

$$v \in \mathcal{W}^{\langle\mathfrak{a}\rangle} \iff \text{There is some } w \in \mathcal{W} \text{ with } w \xleftarrow{\mathfrak{a}} v$$

(for identifiers v). □

Strictly speaking since

$$\mathcal{W}^{\langle\mathfrak{a}\rangle} = \bigcup\{\partial(w\mathfrak{a}) \mid w \in \mathcal{W}\}$$

we can do without the transition relation $\xleftarrow{\mathfrak{a}}$ but it does help to explain what is going on. It is routine to check that

$$\emptyset^{\langle\mathfrak{a}\rangle} = \emptyset \qquad \{x\}^{\langle\mathfrak{a}\rangle} = \partial(x\mathfrak{a}) \qquad \left(\bigcup \mathbb{W}\right)^{\langle\mathfrak{a}\rangle} = \bigcup\{\mathcal{W}^{\langle\mathfrak{a}\rangle} \mid \mathcal{W} \in \mathbb{W}\}$$

for all families $\mathbb{W}$ of sets of identifiers. Note also that $\mathcal{W}^{\langle\mathfrak{a}\rangle}$ is finite whenever $\mathcal{W}$ is finite.

As well as these general properties of the diamond operations $\cdot^{\langle\mathfrak{a}\rangle}$ we also need a more particular property.

5.5 LEMMA. *For each set $\mathcal{W}$ of identifiers and replacement $\mathfrak{a}$, both*

$$\mathcal{W}^{\langle\mathfrak{a}\rangle} = (\mathcal{W} - \partial\mathfrak{a}) \cup (\mathcal{W} \cap \partial\mathfrak{a})^{\langle\mathfrak{a}\rangle} \qquad \mathcal{W}^{\langle\mathfrak{a}\rangle} \cup \partial\mathfrak{a} = (\mathcal{W} \cup \partial\mathfrak{a}) \cup (\mathcal{W} \cap \partial\mathfrak{a})^{\langle\mathfrak{a}\rangle}$$

hold.

Proof. For each identifier v we have

$$\begin{aligned} v \in \mathcal{W}^{\langle\mathfrak{a}\rangle} &\iff (\exists w \in \mathcal{W})[v \in \partial(w\mathfrak{a})] \\ &\iff \left\{\begin{array}{l} (\exists w \in \mathcal{W} \cap \partial\mathfrak{a})[v \in \partial(w\mathfrak{a})] \\ \text{or} \\ (\exists w \in \mathcal{W} - \partial\mathfrak{a})[v = w] \end{array}\right\} \iff v \in (\mathcal{W} \cap \partial\mathfrak{a})^{\langle\mathfrak{a}\rangle} \cup (\mathcal{W} - \partial\mathfrak{a}) \end{aligned}$$

which proves the first identity.

Using this we have

$$\mathcal{W}^{\langle\mathfrak{a}\rangle} \cup \partial\mathfrak{a} = (\mathcal{W} - \partial\mathfrak{a}) \cup \partial\mathfrak{a} \cup (\mathcal{W} \cap \partial\mathfrak{a})^{\langle\mathfrak{a}\rangle} = \mathcal{W} \cup \partial\mathfrak{a} \cup (\mathcal{W} \cap \partial\mathfrak{a})^{\langle\mathfrak{a}\rangle}$$

which is the second identity. □

Suppose now we have a replacement $\mathfrak{a}$ which names a substitution operator α, and consider the sets

$$M(t, \mathfrak{a}) = M(t, \alpha) \qquad U(t, \mathfrak{a}) = U(t, \alpha)$$

attached to $\mathfrak{a}$ and a term t by Definition 5.1. Using the transition relation $\overset{\mathfrak{a}}{\longleftarrow}$ we have

$$M(t, \mathfrak{a}) = (\partial t \cap \partial\mathfrak{a})^{\langle\mathfrak{a}\rangle}$$

and hence

$$U(t, \mathfrak{a}) = \partial t \cup \partial\mathfrak{a} \cup (\partial t \cap \partial\mathfrak{a})^{\langle\mathfrak{a}\rangle} = (\partial t)^{\langle\mathfrak{a}\rangle} \cup \partial\mathfrak{a}$$

(using Lemma 5.5 with $\mathcal{W} = \partial t$). This is quite a neat way of describing the unsafe set. We will use a similar restriction in the new algorithm but with an extra twist. Furthermore, the two disjuncts play quite separate roles in the algorithm.

We conclude this section with a result for later use.

5.6 LEMMA. *Let $\mathcal{R}, \mathcal{T}$ be sets of identifiers and let $\mathfrak{a}$ be a replacement with*

$$\mathcal{T} = \mathcal{R} - \{y\} \qquad {}'\mathfrak{a} = [y \mapsto v]\mathfrak{a}$$

for identifiers y, v with $v \notin \mathcal{T}^{\langle\mathfrak{a}\rangle}$. Then both

$$\mathcal{R}^{\langle '\mathfrak{a}\rangle} = \begin{cases} \mathcal{T}^{\langle\mathfrak{a}\rangle} \cup \{v\} & \text{if } y \in \mathcal{R} \\ \mathcal{T}^{\langle\mathfrak{a}\rangle} & \text{if } y \notin \mathcal{R} \end{cases} \qquad \mathcal{R}^{\langle '\mathfrak{a}\rangle} - \{v\} = \mathcal{T}^{\langle\mathfrak{a}\rangle}$$

hold.

Proof. Recall that

$$w\,'\mathfrak{a} = \begin{cases} v & \text{if } w = y \\ w\mathfrak{a} & \text{if } w \neq y \end{cases}$$

so that, for each identifier z, we have

$$z \in \mathcal{R}^{\langle '\mathfrak{a} \rangle} \iff (\exists w \in \mathcal{R})[z \in \partial(w\,'\mathfrak{a})] \iff \begin{cases} y \in \mathcal{R} \text{ and } z = v \\ \text{or} \\ (\exists w \in \mathcal{R} - \{y\} = \mathcal{T})[z \in \partial(w\mathfrak{a})] \end{cases} \iff \begin{cases} y \in \mathcal{R} \text{ and } z = v \\ \text{or} \\ z \in \mathcal{T}^{\langle \mathfrak{a} \rangle} \end{cases}$$

which gives the left hand identity.

With this we have

$$\mathcal{R}^{\langle '\mathfrak{a} \rangle} - \{v\} = \mathcal{T}^{\langle \mathfrak{a} \rangle} - \{v\} = \mathcal{T}^{\langle \mathfrak{a} \rangle}$$

(since $v \notin \mathcal{T}^{\langle \mathfrak{a} \rangle}$) to give the right hand identity. □

The above description of $U(t, \mathfrak{a})$ indicates how the $\mathfrak{a}$-modified support $(\partial t)^{\langle \mathfrak{a} \rangle}$ of t will be useful. The two lemmas of this section help us to calculate such sets.

EXERCISES

5.3 Calculate

$$\partial([z \mapsto y][x \mapsto w]\mathsf{i})$$

where w, x, y, z are not necessarily distinct identifiers. Consider the effect of the possible equalities amongst these identifiers.

5.4 Show that for a replacement $\mathfrak{a}$ the diamond operation $\cdot^{\langle \mathfrak{a} \rangle}$ is monotone, but need be neither inflationary, deflationary, nor idempotent.

5.3 THE GENERIC ALGORITHM

We can immediately give the basics of the new algorithm.

5.7 ALGORITHM. (Generic substitution) *An algorithm*

$$t, \mathfrak{a} \longmapsto t \cdot \mathfrak{a}$$

which, when supplied with a term t *and a replacement* $\mathfrak{a}$*, will return* $t \cdot \mathfrak{a}$*, the* action *of* $\mathfrak{a}$ *on* t.

Construction. This algorithm proceeds by recursion over t with variation of $\mathfrak{a}$. Using the standard notation for terms we have

$$\begin{aligned} \mathsf{k} \cdot \mathfrak{a} &= \mathsf{k} & (qp) \cdot \mathfrak{a} &= (q \cdot \mathfrak{a})(p \cdot \mathfrak{a}) \\ x \cdot \mathfrak{a} &= x\mathfrak{a} & (\lambda y : \sigma \,.\, r) \cdot \mathfrak{a} &= (\lambda v : \sigma \,.\, (r \cdot {}'\mathfrak{a})) \\ & & \text{where } {}'\mathfrak{a} &= [y \mapsto v]\mathfrak{a} \end{aligned}$$

where in the fourth clause the renaming identifier v must be 'safe'. □

The crucial difference between the old algorithm and the new algorithm is the abstraction clause. Previously we renamed the bound identifier, converted r into r', and then we hit r' with the substitution operator. Here we merely change the bound identifier and then update the replacement. In a calculation of $t \cdot \mathfrak{a}$ the replacement $\mathfrak{a}$ will move deeper and deeper into t being updated as it goes. Eventually we are left to calculate $x \cdot {}^*\mathfrak{a}$ for several identifiers x, and several updated versions ${}^*\mathfrak{a}$ of $\mathfrak{a}$. This last step is done by unravelling ${}^*\mathfrak{a}$.

The algorithm is described as generic for it still doesn't say what a 'safe' identifier is. We now address that problem. Consider the crucial case

$$t = \lambda y : \sigma \,.\, r$$

with a replacement $\mathfrak{a}$. If we get our choice right then the free identifiers of $t \cdot \mathfrak{a}$ ought to be those identifiers v which are free in some $w\mathfrak{a}$ for some identifier w free in t. These are the unsafe identifiers (for we don't want any of these to become bound). Thus

$$v \text{ is unsafe} \iff (\exists w \in \partial t)[v \in \partial(w\mathfrak{a})] \iff v \in (\partial t)^{\langle \mathfrak{a} \rangle}$$

so we must keep away from $(\partial t)^{\langle \mathfrak{a} \rangle}$. It may turn out that certain other identifiers are unsafe also, but in all cases we must at least exclude $(\partial t)^{\langle \mathfrak{a} \rangle}$. Let us refer to this as **taking the minimum precautions**.

We now return to Example 5.2.

5.8 EXAMPLE. Consider

$$t = \lambda y : \sigma \,.\, yxw \qquad \mathfrak{a} = [w \mapsto x]\mathfrak{i}$$

where w, x, y, z are four distinct identifiers. Observe that $\mathfrak{a}$ names $\alpha = [w := z]$, and we have seen that the old algorithm gives

$$t \cdot \alpha = \lambda v : \sigma \,.\, vxz$$

provided $v \notin \{w, x, z\}$. However, intuitively $v = w$ should give an acceptable result.

We have

$$(\partial t)^{\langle \mathfrak{a} \rangle} = \{w, x\}^{\langle \mathfrak{a} \rangle} = \{w\}^{\langle \mathfrak{a} \rangle} \cup \{x\}^{\langle \mathfrak{a} \rangle} = \{z, x\}$$

(since $w\mathfrak{a} = z$, $x\mathfrak{a} = x$). With $v \notin \{z, x\}$ let $'\mathfrak{a} = [y \mapsto v]\mathfrak{a}$, so that

$$y\,'\mathfrak{a} = v \qquad x\,'\mathfrak{a} = x\mathfrak{a} = x \qquad w\,'\mathfrak{a} = w\mathfrak{a} = z$$

since w, x, y are distinct. Thus, taking the minimum precautions, we have

$$t \cdot \mathfrak{a} = \lambda v : \sigma \,.\, ((yxw) \cdot {}'\mathfrak{a}) = \lambda v : \sigma \,.\, ((y\,'\mathfrak{a})(x\,'\mathfrak{a})(w\,'\mathfrak{a})) = \lambda v : \sigma \,.\, vxz$$

and we get an acceptable result even when $v = w$. □

With the generic algorithm there are fewer things to worry about. This is because the algorithm doesn't have nested recursion calls, and updates of replacements neutralize some of the potential problems.

In many circumstances the minimum precautions are the only precautions needed.

5.9 LEMMA. *For each term t and replacement $\mathfrak{a}$ we have*

$$\partial(t \cdot \mathfrak{a}) = (\partial t)^{\langle \mathfrak{a} \rangle}$$

provided at each crucial step in the calculation we take at least the minimum precautions.

Proof. This is proved by induction over the structure of t with variation of the parameter $\mathfrak{a}$. Only the induction step across an abstraction is not immediate.

For this step consider $t = (\lambda y : \sigma . r)$ and let $\mathcal{R} = \partial r$ and $\mathcal{T} = \partial t$, so that $\mathcal{T} = \mathcal{R} - \{y\}$. We have

$$t \cdot \mathfrak{a} = (\lambda v : \sigma . (r \cdot {'\mathfrak{a}}))$$

where ${'\mathfrak{a}} = [y \mapsto v]\mathfrak{a}$ with $v \notin \mathcal{T}^{\langle \mathfrak{a} \rangle}$ (since we are taking at least the minimum precautions). By Lemma 5.6 we have

$$\mathcal{R}^{\langle '\mathfrak{a} \rangle} - \{v\} = \mathcal{T}^{\langle \mathfrak{a} \rangle}$$

and so, using the induction hypothesis, we have

$$\partial(t \cdot \mathfrak{a}) = \partial(r \cdot {'\mathfrak{a}}) - \{v\} = \mathcal{R}^{\langle '\mathfrak{a} \rangle} - \{v\} = \mathcal{T}^{\langle \mathfrak{a} \rangle} = (\partial \tau)^{\langle \mathfrak{a} \rangle}$$

as required. □

This result verifies the internal coherence of the generic algorithm (provided we take at least the minimum precautions). It also shows that we can calculate the support of the modified term without going through the substitution.

Exercises

5.5 Consider the term and substitution operator

$$t = \lambda y : \sigma . z(\lambda x : \rho . yx)x \qquad \alpha = [x := y, y := z, z := x]$$

used in Exercise 5.1. Write down a replacement $\mathfrak{a}$ which names α, and calculate $t \cdot \mathfrak{a}$ taking no more than the minimum precautions. Describe how the algorithm proceeds. At each renaming the chosen identifier should be the first available one from z, y, x, w, v, u.

5.6 Repeat Exercise 5.5 using the term an operator of Exercise 5.2.

5.7 For an arbitrary replacement $\mathfrak{a}$ and identifiers y, v consider the updated replacement ${'\mathfrak{a}} = [y \mapsto v]\mathfrak{a}$. Describe the transition relation $\overset{'\mathfrak{a}}{\longleftarrow}$ of the modified replacement in terms of $\overset{\mathfrak{a}}{\longleftarrow}$.

5.8 Complete the details of the proof of Lemma 5.9.

5.4 THE MECHANISTIC ALGORITHM

With the generic algorithm it is possible to take no more than the minimum precautions when choosing a renaming identifier. However, there are other problems which such a choice should anticipate.

For instance, suppose we have a judgement $\Gamma \vdash t : \tau$ and we wish to calculate $t \cdot \mathfrak{a}$. Let $\partial\Gamma$ be the support of Γ, the set of identifiers declared in Γ. We know that $\partial t \subseteq \partial\Gamma$ but $\partial\Gamma$ may be strictly bigger. The minimum precautions would avoid $(\partial t)^{\langle\mathfrak{a}\rangle}$ when choosing a renaming identifier, but it seems reasonable to avoid the larger set $(\partial\Gamma)^{\langle\mathfrak{a}\rangle}$, especially if we want to put $t \cdot \mathfrak{a}$ back in context.

For another example suppose we have two replacements $\mathfrak{a}$ and $\mathfrak{b}$ and we wish to calculate both

$$t \cdot \mathfrak{a} \qquad t \cdot \mathfrak{b}$$

for some common term t, and then perhaps combine or compare the results in some way. The separate calculation would avoid

$$(\partial t)^{\langle\mathfrak{a}\rangle} \qquad (\partial t)^{\langle\mathfrak{b}\rangle}$$

respectively. However, it seems sensible to do the two calculations in parallel, in which case we should avoid

$$(\partial t)^{\langle\mathfrak{a}\rangle} \cup (\partial t)^{\langle\mathfrak{b}\rangle}$$

to make a common choice of renaming identifiers where possible.

There are other problems as well.

5.10 EXAMPLE. Consider

$$t = \lambda y : \sigma \,.\, \lambda x : \rho \,.\, r$$

where $x \neq y$ and r is arbitrary. For convenience let

$$s = \lambda x : \rho \,.\, r \quad \text{so that} \quad t = \lambda y : \sigma \,.\, s$$

and let

$$\mathcal{T} = \partial t \qquad \mathcal{S} = \partial s \qquad \mathcal{R} = \partial r \quad \text{so that} \quad \mathcal{T} = \mathcal{S} - \{y\} \qquad \mathcal{S} = \mathcal{R} - \{x\}$$

are the relevant connections. We ought to get

$$t \cdot \mathfrak{i} = \lambda v : \sigma \,.\, \lambda u : \rho \,.\, r''$$

for some suitable body r'' and *distinct* identifiers u, v (chosen in some way). Let's see what can happen.

The algorithm gives

$$t \cdot \mathfrak{i} = \lambda v : \sigma \,.\, (s \cdot {}'\mathfrak{i}) = \lambda v : \sigma \,.\, \lambda u : \rho \,.\, (r \cdot {}''\mathfrak{i})$$

where

$${}'\mathfrak{i} = [y \mapsto v]\mathfrak{i} \qquad {}''\mathfrak{i} = [x \mapsto u]\,{}'\mathfrak{i}$$

where the identifiers v, u are suitable safe. How are these chosen?

The algorithm requires

$$v \notin \mathcal{T}^{\langle i \rangle} = \mathcal{T} \qquad u \notin \mathcal{S}^{\langle i \rangle} \quad \text{where} \quad \mathcal{S}^{\langle ' i \rangle} = \begin{cases} \mathcal{T}^{\langle i \rangle} \cup \{v\} & \text{if } y \in \mathcal{S} \\ \mathcal{T}^{\langle i \rangle} & \text{if } y \notin \mathcal{S} \end{cases}$$

by Lemma 5.6 with $\mathcal{R} = \mathcal{S}$. Thus if $y \notin \mathcal{S} = \partial s$ then $\mathcal{S}^{\langle ' i \rangle} = \mathcal{T}$.

When this is the case, i.e. when $(\lambda y : \sigma . -)$ is a vacuous abstraction in t, the minimum precautions could give $u = v$ to produce

$$t \cdot i = \lambda v : \sigma . \lambda v : \rho . r''$$

for a suitable body. This is definitely not what we want; the two bound identifiers should be distinct. □

To avoid these and similar problems we assume that at each instance of a use of the algorithm there is a global finite set $\mathcal{U}$ of **untouchable** identifiers which must be avoided when selecting a renaming identifier. As the algorithm proceeds this set $\mathcal{U}$ will change accordingly.

Let Fin be the set of finite sets $\mathcal{W}$ of identifiers. A **renaming function** is a function

$$\mathsf{Fin} \xrightarrow{\ fresh\ } \mathsf{Idf}$$

which assigns to each $\mathcal{W} \in \mathsf{Fin}$ an identifier $fresh(\mathcal{W}) \notin \mathcal{W}$. With this we can give the full substitution algorithm

5.11 ALGORITHM. (**Mechanistic substitution**) *An algorithm which, when supplied with a term t, a replacement $\mathfrak{a}$, and a finite set $\mathcal{U}$ of untouchables, will call on a renaming function fresh to return a term $t \cdot \mathfrak{a}$ and a finite set $\mathcal{U}(t, \mathfrak{a})$ of untouchables to be carried forward.*

Construction. The algorithm proceeds by recursion on t.

$$\begin{array}{rclrcl} \mathsf{k} \cdot \mathfrak{a} & = & \mathsf{k} & \mathcal{U}(\mathsf{k}, \mathfrak{a}) & = & \mathcal{U} \\ x \cdot \mathfrak{a} & = & x\mathfrak{a} & \mathcal{U}(x, \mathfrak{a}) & = & \mathcal{U} \\ (qp) \cdot \mathfrak{a} & = & (q \cdot \mathfrak{a})(p \cdot \mathfrak{a}) & \mathcal{U}(t, \mathfrak{a}) & = & \mathcal{U} \\ (\lambda y : \sigma . r) \cdot \mathfrak{a} & = & (\lambda v : \sigma . (r \cdot {}'\mathfrak{a})) & \mathcal{U}(t, \mathfrak{a}) & = & \mathcal{U} \cup \{v\} \end{array}$$

where, in the fourth clause, ${}'\mathfrak{a} = [y \mapsto v]\mathfrak{a}$ and

$$v = fresh((\partial t)^{\langle \mathfrak{a} \rangle} \cup \mathcal{U})$$

is the renaming identifier. □

Let's see how this refinement avoids the problem of Example 5.10.

5.12 EXAMPLE. Consider an iterated abstraction

$$t = \lambda x_1 : \rho_1, \ldots, x_m : \rho_m . s$$

and suppose $\mathcal{U}$ is the current set of untouchables. Given a replacement $\mathfrak{a}$ generate

$$\begin{array}{l} \mathcal{U}_0, \mathcal{U}_1, \mathcal{U}_2, \mathcal{U}_3, \quad \dots \\ \quad u_1,\ u_2,\ u_3, \quad \dots \end{array}$$

by

$$\mathcal{U}_0 = (\partial t)^{\langle \mathfrak{a} \rangle} \cup \mathcal{U} \qquad u_{i+1} = \mathit{fresh}(\mathcal{U}_i) \qquad \mathcal{U}_{i+1} = \mathcal{U}_i \cup \{u_{i+1}\}$$

(for $i = 0, 1, 2, \dots$). Observe that this sequence of identifiers can be generated from ∂t without decomposing t. Let

$$^{*}\mathfrak{a} = [x_m \mapsto u_m] \cdots [x_1 \mapsto u_1]\mathfrak{a}$$

using the generated sequence of identifiers. Then

$$t \cdot \mathfrak{a} = \lambda u_1 : \rho_1, \dots, u_m : \rho_m \,.\, (s \cdot {}^{*}\mathfrak{a})$$

is the result of the algorithm. The proof of this is left as an Exercise. □

From now on we will use this mechanistic algorithm. Of course, we won't specify any particular renaming function, but we do assume that renaming identifiers are chosen in a uniform way.

What happens when we calculate $t \cdot \mathfrak{i}$ using this algorithm? We find that $t \cdot \mathfrak{i} = t$ need not hold. Each time we pass across a binding of an identifier y we need an update $[y \mapsto v]$ where the renaming identifier v is determined, precisely, by the environment and the term. (Informally we can often leave y unchanged, but the precise algorithm may not allow this even when there is no danger of identifier capture.)

Let us say that a **renaming replacement** is one of the form

$$[y_m \mapsto v_m] \cdots [y_1 \mapsto v_1]\mathfrak{i}$$

where $y_1 \dots, y_m, v_1, \dots, v_m$ are identifiers, i.e. a replacement where each update (overridden or not) merely changes the name of an identifier.

As we calculate $t \cdot \mathfrak{i}$ we build up one, or several, renaming replacements. Eventually these are used to change the bound identifiers in the term.

Let us write

$$t_1 \approx t_2$$

to indicate that the two terms t_1, t_2 are alphabetic variants (i.e. differ only in the choice of bound identifiers). We find that

$$t \cdot \mathfrak{i} \approx t \qquad t_1 \approx t_2 \Rightarrow t_1 \cdot \mathfrak{i} = t_1 \cdot \mathfrak{i}$$

and hence

$$t_1 \approx t_2 \Longleftrightarrow t_1 \cdot \mathfrak{i} = t_1 \cdot \mathfrak{i}$$

holds. Thus we can think of $t \cdot \mathfrak{i}$ as a **standard form** of t.

How does this algorithm compare with the usual algorithm as given in Section 5.1? The crucial clause there has more local restrictions.

5.13 THEOREM. *For each substitution operator α named by a replacement $\mathfrak{a}$, and each term t, the result $t \cdot \alpha$ of the usual algorithm can be calculated as $t \cdot \mathfrak{a}$ provided we add $\partial\mathfrak{a}$ to the current set of untouchables.*

This new algorithm is much more flexible, especially when it takes into account the needs of the global environment. Furthermore, with this algorithm it is easier to verify the required properties of substitution.

Exercises

5.9 Using the term t of Exercises 5.1 and 5.5, and the replacement $\mathfrak{a}$ of Solution 5.5, calculate $t \cdot \mathfrak{a}$ using the mechanistic algorithm with $\mathcal{U} = \{z\}$ as the current set of untouchables. As before, each renaming identifier should be the first available one from z, y, x, w, v, u.

5.10 Repeat Exercise 5.9 using the term t of Exercises 5.2 and 5.6, and the replacement $\mathfrak{a}$ of Solution 5.5.

5.11 The term

$$t = \lambda x : \tau \,.\, ((\lambda y : \tau \,.\, (y(\lambda x : \sigma \,.\, ((\lambda y : \rho \,.\, xy)z))))x)$$

is rather hard to read because of the overuse of 'x' and 'y'. By calculating $t \cdot \mathfrak{i}$ put t into a standard form. You should assume that $\mathcal{U} = \partial t$ is the current set of untouchables, and any fresh identifiers are chosen from $z, y, x, w, v, u, \ldots$ in this order.

5.12 Complete the details of Example 5.12.

5.13 Recalling the definition of $\approx$, show that

$$t \cdot \mathfrak{i} \approx t \qquad t_1 \approx t_2 \Rightarrow t_1 \cdot \mathfrak{i} = t_1 \cdot \mathfrak{i}$$

holds for all terms t, t_1, t_2.

5.5 Some properties of substitution

There are several intuitive properties that any form of substitution should have. Perhaps the most basic is that it should only be support sensitive (more precisely, it should be non-support insensitive). The result $t \cdot \mathfrak{a}$ of a substitution should depend only on the behaviour of $\mathfrak{a}$ on the identifiers occurring free in t.

5.14 LEMMA. *For each pair $\mathfrak{a}, \mathfrak{b}$ of replacements and term t, if $x \cdot \mathfrak{a} = x \cdot \mathfrak{b}$ for all $x \in \partial t$ then $t \cdot \mathfrak{a} = t \cdot \mathfrak{b}$ holds.*

Proof. This is proved by induction over the structure of t with allowable variations in the replacements. Only the passage across an abstraction $t = (\lambda y : \sigma . r)$ is not immediate.

For this step let $\mathcal{R} = \partial r$ and $\mathcal{T} = \mathcal{R} - \{y\}$ so that $\mathcal{T} = \partial t$. Since $\mathfrak{a}$ and $\mathfrak{b}$ agree on $\mathcal{T}$ we have $\mathcal{T}^{\langle \mathfrak{a} \rangle} = \mathcal{T}^{\langle \mathfrak{b} \rangle}$ so that the results $t \cdot \mathfrak{a}$ and $t \cdot \mathfrak{b}$ have a common set of variables unsafe for renaming y. Thus

$$t \cdot \mathfrak{a} = (\lambda v : \sigma . (r \cdot {}'\mathfrak{a})) \qquad t \cdot \mathfrak{b} = (\lambda v : \sigma . (r \cdot {}'\mathfrak{b}))$$

for some $v \notin \mathcal{T}^{\langle \mathfrak{a} \rangle} = \mathcal{T}^{\langle \mathfrak{b} \rangle}$. Both ${}'\mathfrak{a}$ and ${}'\mathfrak{b}$ are obtained using the update $[y \mapsto v]$. We show that ${}'\mathfrak{a}$ and ${}'\mathfrak{b}$ agree on $\mathcal{R}$ and hence, by the induction hypothesis, we have $r \cdot {}'\mathfrak{a} = r \cdot {}'\mathfrak{b}$, to give the required result.

Trivially $y \cdot {}'\mathfrak{a} = v = y \cdot {}'\mathfrak{b}$ holds. For any other $x \in \mathcal{R}$ we have $x \in \mathcal{T}$ and then

$$x \cdot {}'\mathfrak{a} = x \cdot \mathfrak{a} = x \cdot \mathfrak{b} = x \cdot {}'\mathfrak{b}$$

to complete the proof. □

A more general requirement is that an iterated substitution should be equivalent (in some sense) to a single substitution. The data for the single substitution should be obtainable from the data of the two component substitutions. In fact, all of the usual standard properties of substitutions are variations on this theme, and are often quite tricky to prove. The approach here makes the proofs much simpler.

Given a pair $\mathfrak{b}, \mathfrak{a}$ of replacements and a term t, we can perform the first substitution $t \cdot \mathfrak{b}$ followed by the second substitution $(t \cdot \mathfrak{b}) \cdot \mathfrak{a}$ to obtain a translated version of t. This ought to be the result of hitting t with a single replacement. We show this is obtained in the expected way.

5.15 DEFINITION. The **sequential composite**

$$\mathfrak{b} ; \mathfrak{a}$$

of a pair of replacements $\mathfrak{b}, \mathfrak{a}$ is constructed by recursion on $\mathfrak{b}$.

$$\mathfrak{i} ; \mathfrak{a} = \mathfrak{a} \qquad ([y \mapsto s]\mathfrak{b}) ; \mathfrak{a} = [y \mapsto s \cdot \mathfrak{a}](\mathfrak{b} ; \mathfrak{a})$$

Notice how in the recursion clause the updating component is modified by the passive parameter $\mathfrak{a}$. □

This puts a binary operation $\cdot ; \cdot$ on replacements. By construction $\mathfrak{i}$ is left neutral, $\mathfrak{i} ; \mathfrak{a} = \mathfrak{a}$, but it is *not* right neutral, i.e. $\mathfrak{b} ; \mathfrak{i} = \mathfrak{b}$ need not hold. For instance

$$([x \mapsto s]\mathfrak{i}) ; \mathfrak{i} = [x \mapsto s \cdot \mathfrak{i}](\mathfrak{i} ; \mathfrak{i}) = [x \mapsto s \cdot \mathfrak{i}]\mathfrak{i}$$

so that $\cdot ; \mathfrak{i}$ has a standardizing effect on replacements (just as $\cdot \mathfrak{i}$ has a standardizing effect on terms).

The good news is that $\cdot ; \cdot$ can be associative. This follows from the next result.

5.16 THEOREM. *For all terms t and all replacements $\mathfrak{b}$ and $\mathfrak{a}$,*

$$t \cdot (\mathfrak{b} \,;\mathfrak{a}) = (t \cdot \mathfrak{b}) \cdot \mathfrak{a}$$

provided the two calculations are done in parallel with appropriate extra precautions.

Proof. We proceed by induction over t with variation of $\mathfrak{b}$ and $\mathfrak{a}$. The base case

$$x \cdot (\mathfrak{b} \,;\mathfrak{a}) = (x \cdot \mathfrak{b}) \cdot \mathfrak{a}$$

requires a simple induction over $\mathfrak{b}$. This is left as an exercise.

The base case for a constant is trivial, and the induction step across an application is immediate. Only the passage across an abstraction

$$t = \lambda y : \sigma \,.\, r$$

requires some work.

For this crucial step let $\mathcal{R} = \partial r$ and $\mathcal{T} = \partial t = \mathcal{R} - \{y\}$, and suppose $\mathcal{U}$ is the common set of untouchables at the start of both calculations. Then

$$t \cdot (\mathfrak{b} \,;\mathfrak{a}) = \lambda u : \sigma \,.\, (r \cdot {}'(\mathfrak{b} \,;\mathfrak{a})) \quad \text{where } {}'(\mathfrak{b} \,;\mathfrak{a}) = [y \mapsto u](\mathfrak{b} \,;\mathfrak{a}) \text{ and } u \notin \mathcal{T}^{\langle \mathfrak{b} ;\mathfrak{a} \rangle} \cup \mathcal{U}$$

and u becomes untouchable. Similarly

$$t \cdot \mathfrak{b} = \lambda w : \sigma \,.\, (r \cdot {}'\mathfrak{b}) \quad \text{where } {}'\mathfrak{b} = [y \mapsto w]\mathfrak{b} \text{ and } w \notin \mathcal{T}^{\langle \mathfrak{b} \rangle} \cup \mathcal{U}$$

and w becomes untouchable, to give

$$(t \cdot \mathfrak{b}) \cdot \mathfrak{a} = \lambda v : \sigma \,.\, ((r \cdot {}'\mathfrak{b}) \cdot {}'\mathfrak{a}) \quad \text{where } {}'\mathfrak{a} = [w \mapsto v]\mathfrak{a} \text{ and } v \notin \partial(t \cdot \mathfrak{b})^{\langle \mathfrak{b} \rangle} \cup \mathcal{U} \cup \{w\}$$

and v becomes untouchable. Using Lemma 5.9 and a simple argument we have

$$\partial(t \cdot \mathfrak{b})^{\langle \mathfrak{a} \rangle} = (\mathcal{T}^{\langle \mathfrak{b} \rangle})^{\langle \mathfrak{a} \rangle} = \mathcal{T}^{\langle \mathfrak{b} ;\mathfrak{a} \rangle}$$

so that

$$u \notin \mathcal{T}^{\langle \mathfrak{b} ;\mathfrak{a} \rangle} \cup \mathcal{U} \qquad v \notin \mathcal{T}^{\langle \mathfrak{b} ;\mathfrak{a} \rangle} \cup \mathcal{U} \cup \{w\}$$

are the restrictions on the choice of u, v. In some circumstances we might choose $u = w \neq v$, which we don't want. We specifically exclude $u = w$ so that

$$u \notin \mathcal{W} \qquad v \notin \mathcal{W}$$

for a common set $\mathcal{W}$. This ensures that $u = v$ so that, using the induction hypothesis on r, we have

$$\begin{aligned} t \cdot (\mathfrak{b} \,;\mathfrak{a}) &= \lambda v : \sigma \,.\, (r \cdot {}'(\mathfrak{b} \,;\mathfrak{a})) \\ (t \cdot \mathfrak{b}) \cdot \mathfrak{a} &= \lambda v : \sigma \,.\, (r \cdot ({}'\mathfrak{b} \,; {}'\mathfrak{a})) \end{aligned}$$

where

$${}'(\mathfrak{b} \,;\mathfrak{a}) = [y \mapsto v](\mathfrak{b} \,;\mathfrak{a}) \qquad {}'\mathfrak{b} = [y \mapsto w]\mathfrak{b} \qquad {}'\mathfrak{a} = [w \mapsto v]\mathfrak{a}$$

are the component replacements. It suffices to show

$$r \cdot {}'(\mathfrak{b} \,;\mathfrak{a}) = r \cdot ({}'\mathfrak{b} \,; {}'\mathfrak{a})$$

and for this we use Lemma 5.14, i.e. we show that $'(\mathfrak{b}\,;\mathfrak{a})$ and $'\mathfrak{b}\,;\,'\mathfrak{a}$ agree on $\partial r = \mathcal{R}$.

For the identifier y we have

$$y \cdot '(\mathfrak{b}\,;\mathfrak{a}) = v \qquad y \cdot ('\mathfrak{b}\,;\,'\mathfrak{a}) = y \cdot [y \mapsto w \cdot '\mathfrak{a}](\mathfrak{b}\,;\,'\mathfrak{a}) = w \cdot '\mathfrak{a} = v$$

as required. For any other $x \in \mathcal{R}$, i.e. $x \in \mathcal{T}$, the base case gives

$$\begin{array}{rclcl} x \cdot '(\mathfrak{b}\,;\mathfrak{a}) & = & x \cdot (\mathfrak{b}\,;\mathfrak{a}) & = & (x \cdot \mathfrak{b}) \cdot \mathfrak{a} \\ x \cdot ('\mathfrak{b}\,;\,'\mathfrak{a}) & = & (x \cdot '\mathfrak{b}) \cdot '\mathfrak{a} & = & (x \cdot \mathfrak{b}) \cdot '\mathfrak{a} \end{array}$$

since $y \neq x$. Thus it suffices to show that $'\mathfrak{a}$ and $\mathfrak{a}$ agree on $\partial(x \cdot \mathfrak{b})$. Only the identifier w can cause a difference. But if $w \in \partial(x \cdot \mathfrak{b})$ then $x \overset{\mathfrak{b}}{\longleftarrow} w$ which, since $x \in \mathcal{T}$, gives $w \in \mathcal{T}^{\langle \mathfrak{b} \rangle}$, and this is not so (by the choice of w). Thus we have the required result. □

I said earlier that the transition relations $\overset{\bullet}{\longleftarrow}$ are not necessary but do help to smooth some of the calculations. For instance, it can be shown that

$$w \overset{\mathfrak{b}\,;\mathfrak{a}}{\longleftarrow} u \Longleftrightarrow (\exists v)[w \overset{\mathfrak{b}}{\longleftarrow} v \overset{\mathfrak{a}}{\longleftarrow} u]$$

holds for all replacements $\mathfrak{b}, \mathfrak{a}$ and identifiers w, u. This is essentially the observation

$$(\mathcal{T}^{\langle \mathfrak{b} \rangle})^{\langle \mathfrak{a} \rangle} = \mathcal{T}^{\langle \mathfrak{b}\,;\mathfrak{a} \rangle}$$

used in the proof of the previous lemma.

Exercises

5.14 Show that

$$x \cdot (\mathfrak{b}\,;\mathfrak{a}) = (x \cdot \mathfrak{b}) \cdot \mathfrak{a}$$

holds for all replacements $\mathfrak{a}, \mathfrak{b}$ and identifiers x.

5.15 Show that

$$w \overset{\mathfrak{b}\,;\mathfrak{a}}{\longleftarrow} u \Longleftrightarrow (\exists v)[w \overset{\mathfrak{b}}{\longleftarrow} v \overset{\mathfrak{a}}{\longleftarrow} u] \qquad (\mathcal{T}^{\langle \mathfrak{b} \rangle})^{\langle \mathfrak{a} \rangle} = \mathcal{T}^{\langle \mathfrak{b}\,;\mathfrak{a} \rangle}$$

hold for all replacements $\mathfrak{b}, \mathfrak{a}$ and identifiers w, u and sets $\mathcal{T}$ of identifiers.

5.16 In what sense is the sequential composition of replacements associative?

5.17 We say a term t is standard if $t \cdot \mathfrak{i} = t$, and a replacement $\mathfrak{a}$ is standard if $\mathfrak{a}\,;\mathfrak{i} = \mathfrak{a}$.

(a) Show that a replacement $\mathfrak{a}$ is standard if and only if for each of its component updates $[y \mapsto s]$ the term s is standard.

(b) Show that if $\mathfrak{a}$ is standard then $t \cdot \mathfrak{a}$ is standard for all terms t.

(c) Show the converse of (b).

5.18 The previous exercise introduced the notion of a standard term (and that of a standard replacement). How stable is the notion, i.e. to what extent does it depend on the current environment?

5.19 This exercise gives a version of the crucial interchange property for substitution.

Suppose $\mathfrak{a}$ is standard. Show that for all terms t, s and each identifier y

$$(t \cdot [y \mapsto s]\mathfrak{i}) \cdot \mathfrak{a} = (t \cdot {}'\mathfrak{a}) \cdot [v \mapsto s \cdot \mathfrak{a}]\mathfrak{i}$$

where ${}'\mathfrak{a} = [y \mapsto v]\mathfrak{a}$, provided $v \notin (\partial t - \{y\})^{\langle \mathfrak{a} \rangle}$.

6
Applied λ-calculi

6.1 Introduction

In this chapter we describe a whole family of calculi – the applied λ-calculi – which subsume **C** and **λ**, and have certain new facilities to make them more useful. Each such calculus **λSig** is determined by its signature. This may be empty, in which case we obtain the ground calculus **λ∅**, a cleaner version of **λ**. In general

- we allow atomic types
- we allow constant terms
- we add to the derivation system a structural rule Weakening

and these extensions have several consequence, some unexpected.

An applied λ-calculus **λSig** is determined by two packets of information.

(Specifics) Three families of conditions and restrictions. These are characteristic of **λSig**, so changing any part of them changes the calculus. These Specifics must adhere to certain general requirements which will be explained as we proceed.

(Generalities) The global constructions used in all applied λ-calculi. Roughly speaking these are the constructs of **λ** modified to handle the extra features.

These two packets are intertwined. We can't give all the Specifics without knowing some of the Generalities, and these in turn depend on some of the Specifics. A self contained account would develop these in tandem, moving from one to the other as necessary. However, our experience with **λ** will help us proceed at a faster pace.

The first extension is an enrichment of the language; there may be more types and more terms. Both of these are generated in the same way as before, but there are extra primitives as supplied by the Specifics.

(Language Specifics) There are two kinds of language primitives (specific to **λSig**).

- A family **A** of atoms $\mathcal{A}$. This (with the appropriate Generalities) allows us to generate the types.

- A family **K** of constants k. This (with the appropriate Generalities) allows us to generate the terms.

These atoms $\mathcal{A}$ and constants k will appear in later Specifics.

(Language Generalities) These merely tell us how types, terms, statements, declarations, contexts, and judgements are generated.

As with **C** and **λ**, we have to hand stocks of variables and identifiers. We allow the stock of variables to be empty (i.e. there are no variables) or unlimited, but the stock of identifiers is always unlimited.

6.1 DEFINITION. (Types) The types of **λSig** are generated recursively as follows.

- Each atom $\mathcal{A}$ is a type.
- Each variable is a type.
- If ρ, σ are types, then $(\sigma \to \rho)$ is a type.

We often refer to $(\sigma \to \rho)$ as an 'implication' or a 'function space'. A molecular type is one built up without the use of variables, i.e. solely from atoms.

(Terms) The terms of **λSig** are generated recursively as follows.

- Each constant k is a term.
- Each identifier is a term.
- If p, q are terms, then (qp) is a term.
- If r is a term, σ is a type, and y is an identifier, then $(\lambda y : \sigma \, . \, r)$ is a term.

We often refer to (qp) as an application, and to $(\lambda y : \sigma \, . \, r)$ as an abstraction. □

Notice that when there are no variables available, all types are molecular. As before, we will omit certain brackets to improve readability. We also condense sequences of abstractions.

The initial stages of the development of an applied λ-calculus produce six syntactic categories.

Type Term Statement Declaration Context Judgement

The first two, defined above, are an amalgam of the two notions used earlier. The remaining four

$$t : \tau \qquad x : \sigma \qquad \Gamma \qquad \Gamma \vdash t : \tau$$

are defined in exactly the same way. This is just syntax; the facilities are more interesting. Each calculus **λSig** has three distinct facilities

Derivation system Substitution algorithm Computation mechanism

together with the interactions between these. The substitution algorithm is described in Chapter 5. In the remainder of this chapter we describe the other two facilities and the various interactions.

Exercises

6.1 Suppose the calculus **λSig** has two atoms $\mathcal{N}$ and $\mathcal{B}$, where $\mathcal{N}$ is a name for the set of natural numbers and $\mathcal{B}$ is a name for the set of boolean values. What constants should **λSig** have?

6.2 Derivation

The derivation system of **λSig** is determined entirely by the derivation Specifics and Generalities.

(Derivation Specifics) Each constant k has an allocated type κ. The statement $\mathsf{k} : \kappa$ is called the housing axiom of k. Thus each constant is the subject of exactly one housing axiom.

(Derivation Generalities) These tell us how derivations are generated. The remainder of this section is devoted to these.

We have seen already examples of housing axioms. We may add to any applied λ-calculus any of the standard typed combinators Z, each such Z has an associated type ζ, and then $\mathsf{Z} : \zeta$ is its housing axiom. Here we are concerned with a more radical kind of enrichment: one which extends the range of applicability of a calculus.

To illustrate these facilities we will use a simple applied λ-calculus **λAdd**. This has an atom $\mathcal{N}$ and no variables (so each type is molecular). There are four constants with housing axioms

$$\mathsf{0} : \mathcal{N} \quad \mathsf{S} : \mathcal{N}' \quad \mathsf{H} : \mathcal{N}^+ \quad \mathsf{T} : \mathcal{N}^+$$

where $\mathcal{N}' = \mathcal{N} \to \mathcal{N}$ and $\mathcal{N}^+ = \mathcal{N} \to \mathcal{N}'$. The idea is that $\mathcal{N}$ names the set $\mathbb{N}$ of natural numbers, $\mathsf{0}$ names the number zero, S names the successor function, and H, T are two different names for addition. (The difference is used later to illustrate a point.)

The derivation system of **λSig** builds on that of **λ** with two extra facilities: housing axioms are allowed at leaves, and Weakening is allowed at any node of a derivation. This second addition means that the root judgement of a derivation no longer determines that derivation (for, in some cases, the same judgement can be derived in several different ways). To describe derivations we need an efficient notation, and this is where the arboreal code, suitably extended, comes into its own. All this information is gathered together into the following definition.

6.2 DEFINITION. A derivation

$$(\nabla) \quad \Gamma \vdash t : \tau$$

is a finite rooted tree of judgements grown according to the rules of Table 6.1.

There are two base clauses, the Leaf rules (Axiom and Projection) each of which is restricted by a proviso; and three recursion clauses, a Structural rule (Weakening) which is also restricted by a proviso, and two Construction rules (Introduction and Elimination). The fourth column gives the arboreal code for that rule. □

Clause	Shape	Proviso	Code
Axiom	$\Gamma \vdash \mathsf{k} : \kappa$	Γ legal and $\mathsf{k} : \kappa$ an axiom statement	$\Gamma[\mathsf{k}]$
Projection	$\Gamma \vdash x : \sigma$	Γ legal and $x : \sigma$ occurs in Γ	$\Gamma[x]$
Weakening	$\dfrac{\overset{S}{\Gamma \vdash t : \tau}}{\Gamma, x : \sigma \vdash t : \tau}\ (\downarrow)$	$\Gamma, x : \sigma$ legal	$(S, x, \sigma)\downarrow$
Introduction	$\dfrac{\overset{R}{\Gamma, y : \sigma \vdash r : \rho}}{\Gamma \vdash t : \tau}\ (\uparrow)$	$t = \lambda y : \sigma . r$ $\tau = \sigma \to \rho$	$R\uparrow$
Elimination	$\dfrac{\overset{Q}{\Gamma \vdash q : \pi \to \tau} \quad \overset{P}{\Gamma \vdash p : \pi}}{\Gamma \vdash qp : \tau}$		(QP)

Table 6.1: The derivation rules for **λSig**

The rules P, I, E are just as before. The rule A is the natural extension of the way we use combinators. The new rule W is a restricted version of Thinning, for the new declaration can be inserted only at the gate.

The two rules W and I change the context. If the current context is Γ then

$$(\downarrow)\ \ \text{W lengthens } \Gamma \text{ to } \Gamma^{\downarrow} \qquad (\uparrow)\ \ \text{I shortens } \Gamma \text{ to } \Gamma^{\uparrow}$$

where in both cases the change is by one declaration. The arboreal code $(S, x, \sigma)\downarrow$ for a Weakening can become a little cumbersome. In practice we often write it as $S\downarrow$, especially when the new declaration can be inferred from the environment. We omit some brackets to improve readability. Thus we write

$$RQP \quad \text{for} \quad (R(QP))$$

as with λ-terms.

The explicit use of Weakening gives many more derivations, and leads to some unexpected behaviour. Here is an example which will be useful later.

6.3 EXAMPLE. With $\tau = \sigma \to \rho$ where σ, ρ are arbitrary types, consider the contexts

$$\Xi = z : \tau \qquad \Sigma = z : \tau, y : \sigma \qquad \Gamma = z : \tau, y : \sigma, v : \sigma$$

where z, y, v are distinct (to ensure that Σ, Γ are legal). With the simple derivations

$$Z = \Xi[z] \qquad Y = \Sigma[y] \qquad V = \Gamma[v] \qquad Z^* = Z\downarrow\downarrow$$

(where Z, Y, V are Projections and Z^* is a projection followed by two Weakenings) let

$$\nabla = ((Z\downarrow)Y)\uparrow \qquad \nabla^\eta = ((\nabla\downarrow)Y)\uparrow \qquad \nabla^* = ((Z^*V)\uparrow Y)\uparrow$$

where the Weakenings in ∇ and ∇^η extend Ξ to Σ. Let

$$t = (\lambda y : \sigma \,.\, zy) \qquad t^\eta = (\lambda y : \sigma \,.\, ty) \qquad t^* = (\lambda y : \sigma \,.\, (\lambda v : \sigma \,.\, zv)y)$$

(so that t^η and t^* are alphabetic variants). We have

$$(\nabla)\ \ \Xi \vdash t : \tau \qquad (\nabla^\eta)\ \ \Xi \vdash t^\eta : \tau \qquad (\nabla^*)\ \ \Xi \vdash t^* : \tau$$

with shapes

$$\dfrac{\dfrac{Z}{\bullet}(\downarrow)\quad Y}{\dfrac{\bullet}{\nabla}(\uparrow)} \qquad \dfrac{\dfrac{\dfrac{\dfrac{\dfrac{Z}{\bullet}(\downarrow)\quad Y}{\bullet}}{\nabla}(\uparrow)}{\bullet}(\downarrow)\quad Y}{\dfrac{\bullet}{\nabla^\eta}(\uparrow)} \qquad \dfrac{\dfrac{\dfrac{\dfrac{\dfrac{Z}{\bullet}(\downarrow)}{Z^*}(\downarrow)\quad V}{\bullet}}{\bullet}(\uparrow)\quad Y}{\dfrac{\bullet}{\nabla^*}(\uparrow)}$$

respectively. Notice that $\nabla^\eta \neq \nabla^*$. □

Because most judgements have several different derivations there is a question of classifying these derivations. We won't address that here except for the simplest kind of derivation.

An **extraction** (from a context Γ) is a derivation of

$$\Gamma \vdash \mathsf{k} : \kappa \quad \text{or} \quad \Gamma \vdash x : \sigma$$

where $\mathsf{k} : \kappa$ is an axiom statement and $x : \sigma$ is a declaration of Γ. Every extraction from Γ has the form

$$\Gamma^{\Uparrow}[\cdot]\Downarrow$$

where $\Gamma^{\Uparrow}$ is a shortened version of Γ and the Weakenings $\Downarrow$ build up Γ from $\Gamma^{\Uparrow}$. When this sequence $\Downarrow$ is empty, so $\Gamma^{\Uparrow} = \Gamma$, we have a leaf (an Axiom or a Projection). When $\Downarrow$ is as long as possible, so $\Gamma^{\Uparrow}$ is as short as possible, we have a **standard** extraction

(k) If the subject is a constant k then $\Gamma^{\Uparrow} = \emptyset$

(x) If the subject is an identifier x then x is the gate identifier of $\Gamma^{\Uparrow}$

and the final Weakenings $\Downarrow$ build up Γ from $\Gamma^{\Uparrow}$.

We write

$$\Gamma[\mathsf{k}] \quad \Gamma(\mathsf{k}) \qquad \Gamma[x] \quad \Gamma(x)$$

for, respectively, the leaf extraction and the standard extraction from Γ. Every other extraction sits between these two extremes.

By design we can check whether or not an attempted derivation truly is a derivation. We read the attempt from leaves to root and at each node we check that one of the five rules has been used correctly. Thus the problem of **derivation checking** is comparatively straight forward. What about the problem of **derivation generation**, i.e. the problem of generating a derivation satisfying certain properties? We are going to look at two forms of this problem.

> (**Type synthesis**) Given a pair (Γ, t) – a context Γ and a term t – can we determine whether or not there is a type τ such that $\Gamma \vdash t : \tau$ is derivable, and enumerate all possible examples of such derivations?

> (**Type inhabitation**) Given a pair (Γ, τ) – a context Γ and a type τ – can we determine whether or not there is a term t such that $\Gamma \vdash t : \tau$ is derivable, and enumerate all possible examples of such derivations?

We discuss both of these in this chapter. The first is discussed immediately in the next section, but the second has to wait until Section 6.6.

EXERCISES

6.2 Consider the terms $\mathsf{B}, \mathsf{C}, \mathsf{D}$ from Solution 4.2. Thus

$$\begin{array}{lccc} \mathsf{B} = \lambda z : \tau, y : \sigma, x : \theta \,.\, z(yx) & & \rho = \theta \to \phi & \beta = \tau \to \sigma \to \rho \\ \mathsf{C} = \lambda w : \beta, y : \sigma, z : \tau \,.\, wzy & \text{where} & \sigma = \theta \to \psi & \gamma = \beta \to \delta \\ \mathsf{D} = \lambda y : \sigma, z : \tau, x : \theta \,.\, z(yx) & & \tau = \psi \to \phi & \delta = \sigma \to \tau \to \rho \end{array}$$

for arbitrary θ, ψ, ϕ. Consider the possible derivations $\vdash \mathsf{B} : \beta$, $\vdash \mathsf{C} : \gamma$, $\vdash \mathsf{D} : \delta$ in $\boldsymbol{\lambda\emptyset}$. Using the contexts

$$\Pi = z : \tau, y : \sigma, x : \theta \qquad \Gamma = w : \beta, y : \sigma, z : \tau \qquad \Delta = y : \sigma, z : \tau, x : \theta$$

for each term write down the arboreal codes of six different derivations which conclude with three Introductions $(\cdots)\uparrow\uparrow\uparrow$.

6.3 Consider the terms

$$r = \lambda z : \sigma \,.\, z \qquad s = \lambda z : \sigma \,.\, rz \qquad t = \lambda z : \sigma \,.\, sz$$

where there are three separate uses of the identifier z. Write down the arboreal codes of derivations T of $\vdash t : \sigma'$ indicating the different positions in which Weakening may be used. What happens if the different bindings in t are indicated by different identifiers?

6.4 Review all the derivations you have produced so far in $\boldsymbol{\lambda}$ (and in $\mathbf{C}$). For each of these consider how uses of Weakening could change these, perhaps to produce 'better' derivations. In each case write out the arboreal code of the new derivation.

6.3 TYPE SYNTHESIS

The input data for a type synthesis problem is a pair (Γ, t) where Γ is a context and t is a term. A solution to such a problem is a derivation

$$(\nabla) \quad \Gamma \vdash t : \tau$$

for some type τ, the type **synthesized** by that solution. We will find that for each pair (Γ, t) there is at most one type τ that can be synthesized, and there are only finitely many synthesizing derivations. This uniqueness is called the **unicity of types**.

What can the root rule of a derivation be? Either it is a Weakening, or it is not, in which case we say the derivation is **principal**. The root rule of a principal derivation is determined entirely by the root subject. Each term t has an **associated principal rule** $R(t)$ and a **rank** $|t|$ as follows.

t	$R(t)$	$\lvert t\rvert$
k	A	0
x	P	0
$\lambda y : \sigma . r$	I	$\lvert r\rvert + 1$
qp	E	$\max(\lvert q\rvert, \lvert p\rvert) + 1$

The rank $|t|$ will be useful later.

Each derivation ∇ (as above) has the form

$$\nabla = \nabla(t)\Downarrow \qquad (\nabla(t)) \quad \Gamma^{\Uparrow} \vdash t : \tau$$

where $\nabla(t)$ is principal, $\Gamma^{\Uparrow}$ is an initial part of Γ, and $\Downarrow$ is a sequence of Weakenings which extend $\Gamma^{\Uparrow}$ to Γ. This gives us some information useful when searching for a synthesized type. But it also indicates that such a search must follow several paths. A type synthesis algorithm must be organized with some thought.

6.4 ALGORITHM. (Type synthesis) *An algorithm which, when supplied with a synthesis problem (Γ, t), i.e. a context Γ and a term t, will return (a tree-like tableau of) all derivations*

$$(\nabla) \quad \Gamma \vdash t : \tau$$

for all possible predicates τ.

Construction. The algorithm proceeds by recursion over the pair (Γ, t) where both components Γ and t may vary. At each recursion step the current problem (Γ, t) is replaced by one, two, or three problems (Σ, s), each of which must be tested. The whole search space is a ternary splitting tree. Every solution to the original problem occurs in this tree.

At each stage there is a current problem (Γ, t). The algorithm first determines the legality of Γ. If Γ is illegal, then there is no derivation in Γ and that search branch is aborted. If Γ is legal, then there may be alternatives. If Γ is non-empty, we omit the last (rightmost) declaration to obtain $\Gamma^{\uparrow}$. Any solution to $(\Gamma^{\uparrow}, t)$ gives a solution to (Γ, t) by a use of W. In all cases with Γ legal, there is a path to follow as determined by t. There are two ground cases where the current subject is a constant or an identifier. Both of these may terminate in success (✓) with a returned leaf, or failure (⫛).

Here is the procedure in detail.

Determine the legality of Γ.

- If Γ is illegal then abort in failure, (⫛).
- If Γ is legal then both

  (l) follow the path indicated by t (r) test $(\Gamma^{\uparrow}, t)$ (if $\Gamma \neq \emptyset$)

  where either of these may return a success.

The paths determined by t are as follows. (Note that these are accessed only when Γ is legal.)

$(t = \mathsf{k})$ Use the housing axiom to return $\Gamma[\mathsf{k}]$.

$(t = x)$ Determine whether or not x is declared in Γ.

(⫛) If x is not declared then abort (in failure).

(✓) If x is declared then return $\Gamma[x]$.

The returned predicate (if any) is given by the declaration in Γ.

$(t = \lambda y : \sigma . r)$ Test $((\Gamma, y : \sigma), r)$ and, for all possible solutions R, return $R\!\uparrow$.

$(t = qp)$ Test both (Γ, q) and (Γ, p) and require both to succeed with compatible types $\pi \to \tau$, τ (for some π, τ). Return all derivations QP for all possible solutions Q of (Γ, t) and P of (Γ, p) with compatible types.

Starting with an input problem (Γ, t) this procedure will generate a tree of subproblems (Σ, s). If and when it closes each branch of the tree will either

(✓) succeed and return a leaf derivation, or

(⫛) fail (because of an illegal context or an undeclared identifier)

and then we may discount the failed branches. The successful part of the tree contains enough information to construct all possible solutions to the original problem (and all subproblems generated on the way).

Note that at this stage we have *not* proved that each branch does close. This will be done later. □

We can annotate the generated search tree so that all solutions can be read off. A problem (Γ, qp) with $\Gamma \neq \emptyset$ splits into three problems

$$\frac{(\Gamma, q) \text{ and } (\Gamma, p) \quad \text{or} \quad (\Gamma^{\uparrow}, qp)}{(\Gamma, qp)}$$

to give us two chances of success. However, if $(\Gamma^{\uparrow}, qp)$ succeeds then both (Γ, q) and (Γ, p) will succeed. This leads to a classification of all possible solutions together with various search strategies, but we won't pursue this question.

Here is an example to illustrate these ideas.

6.5 EXAMPLE. Consider the synthesis problem (Γ, t) where $\Gamma = x : \theta, y : \psi$ and $t = \lambda z : \phi \,.\, zx$ where x and y are distinct. You can see immediately that $\Gamma \vdash t : \tau$ is derivable only if $\phi = \theta \to \xi$ (to allow zx) and $\tau = \phi \to \xi$ for some type ξ. But can you write down all possible derivations? Also, what happens if $z = x$ or $z = y$?

For convenience let

$$\Theta = x : \theta \qquad \Gamma = x : \theta, y : \psi \qquad \Pi = x : \theta, y : \psi, z : \phi \qquad \Lambda = x : \theta, z : \phi$$

so that Π is legal only if $z \notin \{x, y\}$, and Λ is legal only if $z \neq x$.

The given problem

$$(\bot) \quad (\Gamma, t)$$

splits into two problems

$$(\uparrow) \quad (\Pi, zx) \qquad (\downarrow) \quad (\Theta, t)$$

either of which will lead to a solution (or solutions) of $(\bot)$. Notice that path $(\uparrow)$ is continued only if Π is legal, but we must always pursue path $(\downarrow)$.

Following each path we generate several subproblems.

$$\begin{array}{lllllllll} \text{From } (\uparrow) & (l\uparrow) & (\Pi, z) & \text{and} & (r\uparrow) & (\Pi, x) & \text{or} & (\downarrow\uparrow) & (\Gamma, zx) \ \pitchfork \\ \text{From } (\downarrow) & (\uparrow\downarrow) & (\Lambda, zx) & \text{or} & (\downarrow\downarrow) & (\emptyset, t) \ \pitchfork & & & \end{array}$$

We can see that $(\downarrow\downarrow)$ is doomed to failure (for x is free in t and not declared in $\emptyset$). The algorithm will eventually detect this, but we need not follow that path. Similarly, $(\downarrow\uparrow)$ can succeed only if z is declared in Γ, i.e. $z \in \{x, y\}$. But then Π is illegal, so the algorithm will never get that far. Again we can abort that path.

Continuing to unravel the non-aborted paths we obtain new nodes

$$\begin{array}{llllllllll} \text{From } (l\uparrow) & (\cdot l\uparrow) & \Pi[z] & \checkmark & \text{or} & (\downarrow l\uparrow) & (\Gamma, z) \ \pitchfork & & & \\ \text{From } (r\uparrow) & (\cdot r\uparrow) & \Pi[x] & \checkmark & \text{or} & (\downarrow r\uparrow) & (\Gamma, x) & & & \\ \text{From } (\uparrow\downarrow) & (l\uparrow\downarrow) & (\Lambda, z) & & \text{and} & (r\uparrow\downarrow) & (\Lambda, x) & \text{or} & (\downarrow\uparrow\downarrow) & (\Theta, zx) \ \pitchfork \end{array}$$

two of which, $(\downarrow l\uparrow,\ \downarrow\uparrow\downarrow)$, are doomed to failure, and two of which, $(\cdot l\uparrow,\ \cdot r\uparrow)$, have terminated in success.

There are still three to pursue

$$\begin{array}{llllll} \text{From } (\downarrow r\uparrow) & (\cdot\downarrow r\uparrow) & \Gamma[x] \ \checkmark & \text{or} & (\downarrow\downarrow r\uparrow) & (\Theta, x) \\ \text{From } (l\uparrow\downarrow) & (\cdot l\uparrow\downarrow) & \Lambda[z] \ \checkmark & \text{or} & (\downarrow l\uparrow\downarrow) & (\Theta, z) \ \pitchfork \\ \text{From } (r\uparrow\downarrow) & (\cdot r\uparrow\downarrow) & \Lambda[x] \ \checkmark & \text{or} & (\downarrow r\uparrow\downarrow) & (\Theta, x) \end{array}$$

and these bring three terminations in success, one eventual failure, and two open problems (which are, in fact, the same problem).

Finally we obtain

$$\begin{array}{llllll} \text{From } (\downarrow\downarrow r\uparrow) & (\cdot\downarrow\downarrow r\uparrow) & \Theta[x] & \checkmark & \text{or} & (\downarrow\downarrow\downarrow r\uparrow) & (\emptyset, x) & \pitchfork \\ \text{From } (\downarrow r\uparrow\downarrow) & (\cdot\downarrow r\uparrow\downarrow) & \Theta[x] & \checkmark & \text{or} & (\downarrow\downarrow r\uparrow\uparrow) & (\emptyset, x) & \pitchfork \end{array}$$

and we see the whole search eventually terminates.

The algorithm generates the search tree given in Table 6.2. (As you can see the search space of even a simple synthesis problem can be quite large.) in this example some of the branches have terminated successfully in a Projection, and others have been abandoned (because we can see they are doomed to failure). When we eliminate these failures we obtain the following.

$$\dfrac{\dfrac{\dfrac{\begin{array}{c}\Pi[z]\\ \cdot l\uparrow\end{array}}{l\uparrow}\quad\text{and}\quad\dfrac{\begin{array}{c}\Pi[x]\\ \cdot r\uparrow\end{array}\ \text{or}\ \dfrac{\begin{array}{c}\Gamma[x]\\ \cdot\downarrow r\downarrow\end{array}\ \text{or}\ \dfrac{\begin{array}{c}\Theta[x]\\ \cdot\downarrow\downarrow r\uparrow\end{array}}{\downarrow\downarrow r\uparrow}}{\downarrow r\uparrow}}{r\uparrow}}{\uparrow}\quad\text{or}\quad\dfrac{\dfrac{\dfrac{\begin{array}{c}\Lambda[z]\\ \cdot l\uparrow\downarrow\end{array}}{l\uparrow\downarrow}\quad\text{and}\quad\dfrac{\begin{array}{c}\Lambda[x]\\ \cdot r\uparrow\downarrow\end{array}\ \text{or}\ \dfrac{\begin{array}{c}\Theta[x]\\ \cdot\downarrow r\uparrow\downarrow\end{array}}{\downarrow r\uparrow\downarrow}}{r\uparrow\downarrow}}{\uparrow\downarrow}}{\downarrow}}{\bot}$$

This tree gives five different derivations in total. You should write down all five. Two of these work when $z = y$ (provided $\psi = \phi$). But none work when $z = x$ (because the term zz is not possible). □

There are two related questions to be answered. How can we prove each branch of a search space closes off, and more generally how can we prove the whole search algorithm does terminate? How can we prove unicity of types? At each recursion step the current problem (Γ, t) can produce several other problems (Σ, s), but it is not clear why these are simpler than (Γ, t). How can we get round this?

We attach to each problem (Γ, t) a **cost** which is a pair

$$|\Gamma, t| = (l(\Gamma), |t|)$$

of natural numbers where $l(\Gamma)$ is the length of Γ and $|t|$ is the rank of t (defined earlier). Look how this cost changes as we pass across a recursion step.

$$\text{(W)}\ \ \frac{(l,a)}{(l+1,a)} \qquad \text{(I)}\ \ \frac{(l+1,a)}{(l,a+1)} \qquad \text{(E)}\ \ \frac{(l,b) \quad (l,a)}{(l,c)}$$

where $c = \max(b+1, a+1)$

Here the denominator indicates the complexity of the given problem, and the numerator indicates the complexity of the problem or problems to be solved after applying the indicated unravelling.

We need a well founded transitive relation $\sqsubset$ on pairs $(l, a) \in \mathbb{N}^2$ where

$$(w)\ \ (l,a) \sqsubset (l+1,a) \qquad (i)\ \ (l+1,a) \sqsubset (l,a+1) \qquad (e)\ \ b < c \Rightarrow (l,b) \sqsubset (l,c)$$

hold for all $a, b, c, l \in \mathbb{N}$. In fact, if we can achieve (w, i) then we have

$$(l,a) \sqsubset (l+1,a) \sqsubset (l,a+1)$$

which ensures (e). Once we have such a relation we can proceed by induction. The details are left to the exercises.

$$
\dfrac{
\dfrac{
\dfrac{\Pi[z]\cdot l{\uparrow} \text{ or } \dfrac{\pitchfork}{{\downarrow}l{\uparrow}}}{l{\uparrow}}
\text{ and }
\dfrac{\Pi[x]\cdot r{\uparrow} \text{ or } \dfrac{\Gamma[x]\cdot{\downarrow}r{\uparrow} \text{ or } \dfrac{\Theta[x]\cdot{\downarrow}{\downarrow}r{\uparrow} \text{ or } \dfrac{\pitchfork}{{\downarrow}{\downarrow}{\downarrow}r{\uparrow}}}{{\downarrow}{\downarrow}r{\uparrow}}}{{\downarrow}r{\uparrow}}}{r{\uparrow}}
\text{ or } \dfrac{\pitchfork}{{\downarrow}{\uparrow}}
}{\uparrow}
\text{ or }
\dfrac{
\dfrac{
\dfrac{\Lambda[z]\cdot l{\uparrow}{\downarrow} \text{ or } \dfrac{\pitchfork}{{\downarrow}l{\uparrow}{\downarrow}}}{l{\uparrow}{\downarrow}}
\text{ and }
\dfrac{\Lambda[x]\cdot r{\uparrow}{\downarrow} \text{ or } \dfrac{\Theta[x]\cdot{\downarrow}r{\uparrow}{\downarrow} \text{ or } \dfrac{\pitchfork}{{\downarrow}{\downarrow}r{\uparrow}{\downarrow}}}{{\downarrow}r{\uparrow}{\downarrow}}}{r{\uparrow}{\downarrow}}
\text{ or } \dfrac{\pitchfork}{{\downarrow}{\uparrow}{\downarrow}}
}{{\uparrow}{\downarrow}}
\text{ or } \dfrac{\pitchfork}{{\downarrow}{\downarrow}}
}{\downarrow}
}{\bot}
$$

Table 6.2: An example search tree

Exercises

6.5 Write down all five solutions to the synthesis problem of Example 6.3. Which of these are still solutions when $z = y$?

6.6 Consider the three terms $\mathsf{B}, \mathsf{C}, \mathsf{D}$ as given in Exercise 6.2. Synthesize the types for each of $(\emptyset, \mathsf{B})$, $(\emptyset, \mathsf{C})$, and $(\emptyset, \mathsf{D})$ and hence show that Solution 6.2 does give all possible derivations. In each case draw the successful part of the search tree to show how all six derivations arise.

6.7 Consider the three terms r, s, t of Exercise 6.3. Use the synthesis algorithm to show there is just one derivation T. What happens if the different bindings in t are indicated by different identifiers?

6.8 (a) Show how to convert each pair (l, a) of natural numbers into a single natural number $|l, a|$ so that

$$(l, a) \sqsubset (m, b) \Longleftrightarrow |l, a| < |m, b|$$

defines a relation $\sqsubset$ with the required well founded properties (w, i).

(b) Show that for each derivation ∇ of $\Gamma \vdash t : \tau$ the comparisons

$$|t| \leq h(\nabla) \leq |l(\Gamma), |t||$$

hold.

(c) Show that each synthesis problem has only finitely many solutions.

6.9 Show that for each synthesis problem (Γ, t) there is at most one type τ such that $\Gamma \vdash t : \tau$ is derivable.

6.10 The synthesis algorithm is designed to return all solutions to a supplied synthesis problem (Γ, t). Rework the algorithm so that it decides whether or not there is a solution, and returns an example of such a solution when there is one. Can you devise a notion of 'canonical' solution and modify the algorithm so that it returns only this solution (if it exists)?

6.4 Mutation

Suppose we have a derivation

$$(\nabla) \quad \Sigma \vdash t : \tau$$

obtained earlier or given to us by some algorithm, and suppose we wish to modify the root judgement.

For instance, we may want to modify the root context Σ.

(T) Perhaps we want to insert into Σ more declarations (for later use).

(X) Perhaps we would prefer to list Σ in a different order.

(C) Perhaps we now realize that two declarations $x : \sigma$ and $y : \sigma$ in Σ should be coalesced into $z : \sigma$ with a common label.

We have seen that in the unannotated systems such structural manipulations

Thinning   eXchange   Contraction

are admissible (i.e. can be added to the system without enlarging the set of derivable judgements). What about the corresponding situation for **λSig**? For such a modification we know what the new root context Γ should be, and we require a new derivation

$$(\nabla^+) \quad \Gamma \vdash t^+ : \tau$$

where the new root subject t^+ is no more than a minor modification of t. The new rule W gives us a certain amount of thinning, but does it give us all we want? The usual proof of admissibility of T and X (which drives the use of these rules up towards the leaves) produces $t^+ = t \cdot \mathfrak{i}$, an alphabetic variant of t.

In a different direction we may have a replacement $\mathfrak{a}$ and require a new derivation ∇^+ (as above) where $t^+ = t \cdot \mathfrak{a}$ and the new root context Γ is somehow related to Σ. Each old declaration $x : \sigma$ will produce a statement $s : \sigma$ where $s = x \cdot \mathfrak{a}$, and this must be derivable in Γ. But s need not be an identifier, so the choice of Γ is not clear.

We want an algorithm $\nabla \longmapsto \nabla^+$ which **mutates** a source derivation ∇ into a target context ∇^+. This should do for derivations what the substitution algorithm does for terms. Accordingly we need the analogue of a replacement.

6.6 DEFINITION. The data for a **mutation**

$$(\mathfrak{A}) \quad \Sigma \xrightarrow{\mathfrak{a}} \Gamma$$

has several components.

- A **source** context Σ and a **target** context Γ, both of which are legal.
- A replacement $\mathfrak{a}$, called the **shaft** of the mutation.
- A **nominated** derivation

$$(x \cdot \mathfrak{A}) \quad \Gamma \vdash s : \sigma$$

  for each component $x : \sigma$ of Σ where $s = x \cdot \mathfrak{a}$.

Observe the required restrictions on this data. Both the source and target contexts must be legal, and the nominated derivations must cohere with the shaft. □

In practice we use such a mutation $\mathfrak{A}$ only in the presence of a derivation $\Sigma \vdash t : \tau$ over source context Σ. This ensures that Σ is legal. If Σ is non-empty then there is at least one nominated derivation $\Gamma \vdash s : \sigma$ and this ensures that Γ is legal. When $\Sigma = \emptyset$ we impose this condition on Γ.

Each source declaration $x : \sigma$ produces a target derivation $\Gamma \vdash s : \sigma$ where $s = x \cdot \mathfrak{a}$ but this may not be uniquely determined by Γ and s. For this reason we must nominate the target derivation $x \cdot \mathfrak{A}$ to be used. If we change this nomination then we change the mutation. The reason for the notation '$x \cdot \mathfrak{A}$' will become clear shortly.

6.7 EXAMPLES. (T, X) Let Σ, Γ be a pair of legal contexts where Γ is obtained from Σ by several insertions and swaps. There are several mutations

$$(\mathfrak{A}) \quad \Sigma \xrightarrow{\mathfrak{i}} \Gamma$$

with $\mathfrak{i}$, the base replacement, as shaft. Each nominated derivation

$$(x \cdot \mathfrak{A}) \quad \Gamma \vdash x : \sigma$$

must be an extraction, but we can choose these to fit in with other possible requirements.

(C) Consider legal contexts

$$\Sigma = \Pi, x : \zeta, y : \zeta, \Delta \qquad \Gamma = \Pi, z : \zeta, \Delta$$

(so z may be one of x or y, or be fresh). Let $\mathfrak{a}$ name the simultaneous substitution

$$[x := z, y := z]$$

(e.g. $\mathfrak{a} = [x \mapsto z][y \mapsto z]\mathfrak{i}$). There are several possible mutations

$$(\mathfrak{A}) \quad \Sigma \xrightarrow{\mathfrak{a}} \Gamma$$

where both $x \cdot \mathfrak{A}$ and $y \cdot \mathfrak{A}$ are extractions $\Gamma \vdash z : \zeta$ (and not necessarily the same one), and each other nominated derivation is an extraction of an identifier from Π or Δ. (Remember that the legality of Σ ensures that $x \neq y$.)

(S) Consider a context $\Sigma = \Pi, x : \sigma, \Delta$ and a simple replacement $\mathfrak{a} = [x \mapsto s]\mathfrak{i}$. Let Γ be any legal context where each declaration of Π or Δ appears in Γ and suppose S is a derivation of $\Gamma \vdash s : \sigma$. Then we have a mutation

$$(\mathfrak{A}) \quad \Sigma \xrightarrow{\mathfrak{a}} \Gamma$$

where $x \cdot \mathfrak{A}$ is S and each other nominated derivation is an extraction. $\square$

Recall that we have the supports

$$\partial t \text{ of a term } t \qquad \partial \mathfrak{a} \text{ of a replacement } \mathfrak{a} \qquad \partial \Gamma \text{ of a context } \Gamma$$

each of which is a finite set of identifiers. We know that $\partial t \subseteq \partial \Gamma$ holds for each derivable judgement $\Gamma \vdash t : \tau$, and $(\partial t)^{\langle \mathfrak{a} \rangle} = \partial (t \cdot \mathfrak{a})$ is a consequence of the substitution algorithm. Mutations have a similar property since $(\partial \Sigma)^{\langle \mathfrak{a} \rangle} \subseteq \partial \Gamma$ for each mutation $\mathfrak{A}$ (as above).

We are going to describe an algorithm

$$(\nabla, \mathfrak{A}) \longmapsto \nabla \cdot \mathfrak{A}$$

which converts a compatible derivation ∇ and mutation $\mathfrak{A}$ into a derivation $\nabla \cdot \mathfrak{A}$. This will be the analogue of the substitution algorithm and will proceed by recursion over ∇ with variation of $\mathfrak{A}$. In particular, there are two important modifications of $\mathfrak{A}$ that we will need.

6.8 DEFINITION. (Restriction) Consider a mutation

$$(\mathfrak{A}) \quad \Sigma \xrightarrow{\mathfrak{a}} \Gamma$$

where $\Sigma = \Pi, y : \sigma$, i.e. $\Pi = \Sigma^{\uparrow}$. The restriction

$$(\uparrow\mathfrak{A}) \quad \Sigma^{\uparrow} \xrightarrow{\mathfrak{a}} \Gamma$$

has the indicated source, shaft, and target, with nominated derivations

$$x \cdot \uparrow\mathfrak{A} = x \cdot \mathfrak{A}$$

for all identifiers x declared in $\Sigma^{\uparrow} = \Pi$.

(Lengthening) Consider a mutation

$$(\mathfrak{A}) \quad \Sigma \xrightarrow{\mathfrak{a}} \Gamma$$

and a pair

$$\Sigma^{\downarrow} = \Sigma, y : \sigma \qquad \Gamma^{\downarrow} = \Gamma, v : \sigma$$

of legal extensions. Let $'\mathfrak{a} = [y \mapsto v]\mathfrak{a}$. The lengthening

$$('\mathfrak{A}) \quad \Sigma^{\downarrow} \xrightarrow{'\mathfrak{a}} \Gamma^{\downarrow}$$

has nominated derivations

$$x \cdot {'\mathfrak{A}} = (x \cdot \mathfrak{A}, v, \sigma)\downarrow \qquad y \cdot {'\mathfrak{A}} = \Gamma^{\downarrow}[v]$$

for all identifiers x declared in Σ. □

In practice when we use a lengthening we are given $\Sigma^{\downarrow} = \Sigma, y : \sigma$ and we can choose the identifier v as we see fit. We must have $v \notin \partial\Gamma$, but it may be convenient to restrict the choice even further.

6.9 ALGORITHM. (Mutation) *An algorithm which, when supplied with*

$$(\nabla) \quad \Sigma \vdash t : \tau \qquad (\mathfrak{A}) \quad \Sigma \xrightarrow{\mathfrak{a}} \Gamma$$

a derivation and a mutation of indicated context compatibility, will return a derivation, the action *of* $\mathfrak{A}$ *on* ∇,

$$(\nabla \cdot \mathfrak{A}) \quad \Gamma \vdash t^{+} : \tau$$

where $t^{+} = t \cdot \mathfrak{a}$, *and this must be calculated using* $\partial\Gamma$ *as part of the untouchables.*

Construction. The algorithm proceeds by recursion over the derivation ∇ with variation of the mutation $\mathfrak{A}$. There are five clauses, two base clauses (A,P) and three recursion clauses (W, I, E). We state these first using the arboreal code, and then fill in the necessary details.

$$\begin{array}{llll}
\text{(A)} & \Sigma[\mathsf{k}]\cdot\mathfrak{A} = \Gamma(\mathsf{k}) & \text{(W)} & S\downarrow\cdot\mathfrak{A} = S\cdot\uparrow\mathfrak{A} \\
\text{(P)} & \Sigma[x]\cdot\mathfrak{A} = x\cdot\mathfrak{A} & \text{(I)} & R\uparrow\cdot\mathfrak{A} = (R\cdot{}'\mathfrak{A})\uparrow \\
 & & \text{(E)} & (QP)\cdot\mathfrak{A} = (Q\cdot\mathfrak{A})(P\cdot\mathfrak{A})
\end{array}$$

We must check that this algorithm does return a derivation

$$(\nabla\cdot\mathfrak{A})\quad \Gamma\vdash t^+ : \tau$$

of the required form with $t^+ = t\cdot\mathfrak{a}$. There are also one or two other points to clear up.

(A) When $t = \mathsf{k}$ we have $t^+ = t\cdot\mathfrak{a} = \mathsf{k}$, so the A-case does return a derivation of the required form. Notice that this derivation is always standard.

(P) When $t = x$ where $x : \sigma$ is in Σ, we have a nominated derivation

$$(x\cdot\mathfrak{A})\quad \Gamma\vdash s : \sigma \qquad \text{where } s = x\cdot\mathfrak{a}$$

and hence $t^+ = t\cdot\mathfrak{a} = x\cdot\mathfrak{a} = s$. This case always returns a nominated derivation, so a different nomination will change the effect of the mutation.

(W) Here we have an input derivation

$$(\nabla = S\downarrow)\quad \cfrac{\overset{S}{\Pi\vdash t:\tau}}{\Sigma\vdash t:\tau}$$

where $\Sigma = \Pi, y : \sigma$ for some declaration $y : \sigma$. By Definition 6.8 we have

$$(\uparrow\mathfrak{A})\quad \Pi\xrightarrow{\ \mathfrak{a}\ }\Gamma$$

and then, by recursion, we obtain a derivation

$$(S\cdot\uparrow\mathfrak{A})\quad \Gamma\vdash t^+ : \tau \qquad \text{where } t^+ = t\cdot\mathfrak{a}$$

and so we may take $\nabla\cdot\mathfrak{A} = S\downarrow\cdot\mathfrak{A} = S\cdot\uparrow\mathfrak{A}$ for the required result.

(I) In this case we have an input derivation

$$(\nabla = R\uparrow)\quad \cfrac{\overset{R}{\Sigma, y:\sigma\vdash r:\rho}}{\Sigma\vdash t:\tau}$$

where $t = (\lambda y : \sigma\,.\,r)$ and $\tau = \sigma\to\rho$. The substitution algorithm gives $t\cdot\mathfrak{a} = \lambda v : \sigma\,.\,r^+$ where ${}'\mathfrak{a} = [y\mapsto v]\mathfrak{a}$, $r^+ = r\cdot{}'\mathfrak{a}$ for some suitable identifier v outside a known finite set. We may impose $v\notin\partial\Gamma$ as a further restriction, so the context $\Gamma, v : \sigma$ is legal. Let

$$({}'\mathfrak{A})\quad \Sigma, y:\sigma\xrightarrow{\ {}'\mathfrak{a}\ }\Gamma, v:\sigma$$

be the lengthening as given by Definition 6.8. By recursion we have a derivation

$$(R\cdot{}'\mathfrak{A})\quad \Gamma, v:\sigma\vdash r^+ : \rho$$

where $r^+ = r \cdot {'\mathfrak{a}}$. An Introduction gives

$$((R \cdot {'\mathfrak{A}})\uparrow) \quad \Gamma \vdash t^+ : \tau$$

where

$$t^+ = \lambda v : \sigma \,.\, r^+ = \lambda v : \sigma \,.\, (r \cdot {'\mathfrak{a}}) = (\lambda y : \sigma \,.\, r) \cdot \mathfrak{a} = t \cdot \mathfrak{a}$$

and we may take $\nabla \cdot \mathfrak{A} = R\uparrow \cdot \mathfrak{A} = (R \cdot {'\mathfrak{A}})\uparrow$ for the required result.

(E) This case is straight forward. □

The important clause is the step across a use of Introduction. This requires a lengthening of the mutation. A longish example will illustrate some of the subtleties involved.

6.10 EXAMPLE. Recall the contexts Ξ, Σ, Γ of Example 6.3 and the associated derivations $\nabla, \nabla^\eta, \nabla^*$. Consider

$$(\mathfrak{A}) \quad \Xi \xrightarrow{\mathfrak{a}} \Xi \quad \text{where } \mathfrak{a} = \mathfrak{i} \text{ and } z \cdot \mathfrak{A} = Z$$

to produce what looks like a rather trivial mutation. The lengthening

$$({'\mathfrak{A}}) \quad \Sigma \xrightarrow{'\mathfrak{a}} \Sigma \quad \text{where } {'\mathfrak{a}} = [y \mapsto y]\mathfrak{i} \text{ and } z \cdot {'\mathfrak{A}} = Z\downarrow,\ y \cdot {'\mathfrak{A}} = Y$$

will be needed in our calculations. With these we have

$$\begin{aligned} \nabla \cdot \mathfrak{A} &= ((Z\downarrow)Y)\uparrow \cdot \mathfrak{A} \\ &= ((Z\downarrow)Y \cdot {'\mathfrak{A}})\uparrow \\ &= ((Z\downarrow \cdot {'\mathfrak{A}})(Y \cdot {'\mathfrak{A}}))\uparrow \\ &= ((Z \cdot \uparrow{'\mathfrak{A}})(Y \cdot {'\mathfrak{A}}))\uparrow \\ &= ((z \cdot \uparrow{'\mathfrak{A}})(y \cdot {'\mathfrak{A}}))\uparrow \quad = \quad \nabla \end{aligned}$$

since $z \cdot \uparrow{'\mathfrak{A}} = z \cdot {'\mathfrak{A}} = Z\downarrow$ and $y \cdot {'\mathfrak{A}} = Y$ to show that $\mathfrak{A}$ leaves ∇ unchanged.

We have

$$(\uparrow{'\mathfrak{A}}) \quad \Xi \xrightarrow{'\mathfrak{a}} \Sigma$$

and the second lengthening

$$({'\!\uparrow}{'\mathfrak{A}}) \quad \Sigma \xrightarrow{''\mathfrak{a}} \Gamma \quad \text{where } {''\mathfrak{a}} = [y \mapsto v]\,{'\mathfrak{a}}$$

will be needed in our calculations. With this we have

$$\nabla^\eta \cdot \mathfrak{A} = ((\nabla\downarrow)Y)\uparrow \cdot \mathfrak{A} = ((\nabla \cdot \uparrow{'\mathfrak{A}})Y)\uparrow$$

where

$$\nabla \cdot \uparrow{'\mathfrak{A}} = ((Z \cdot \uparrow{'\!\uparrow}{'\mathfrak{A}})(Y \cdot {'\!\uparrow}{'\mathfrak{A}}))\uparrow = ((z \cdot \uparrow{'\!\uparrow}{'\mathfrak{A}})(y \cdot {'\!\uparrow}{'\mathfrak{A}}))\uparrow$$

where

$$z \cdot \uparrow{'\!\uparrow}{'\mathfrak{A}} = z \cdot {'\!\uparrow}{'\mathfrak{A}} = (z \cdot \uparrow{'\mathfrak{A}})\downarrow = (z \cdot {'\mathfrak{A}})\downarrow = Z\downarrow\downarrow = Z^* \qquad y \cdot {'\!\uparrow}{'\mathfrak{A}} = \Gamma[v] = V$$

to give

$$\nabla \cdot \uparrow'\mathfrak{A} = (Z^*V)\uparrow \qquad \nabla^\eta \cdot \mathfrak{A} = (((Z^*V)\uparrow)Y)\uparrow = \nabla^*$$

and hence $\nabla^\eta \cdot \mathfrak{A} \neq \nabla^\eta$.

It can be shown that $\nabla^* \cdot \mathfrak{A} = \nabla^*$.

Next consider

$$(\mathfrak{B}) \quad \Xi \xrightarrow{\mathfrak{b}} \Xi \quad \text{where } \mathfrak{b} = [z \mapsto t]\mathfrak{i},\ z \cdot \mathfrak{B} = \nabla$$

to produce a less trivial mutation. The lengthening

$$('\mathfrak{B}) \quad \Sigma \xrightarrow{'\mathfrak{b}} \Sigma \quad \text{where } '\mathfrak{b} = [y \mapsto y]\mathfrak{b},\ z \cdot {}'\mathfrak{B} = \nabla\downarrow,\ y \cdot {}'\mathfrak{B} = Y$$

will be needed in our calculations. With these we have

$$\begin{aligned} \nabla \cdot \mathfrak{B} &= ((Z\downarrow)Y)\uparrow \cdot \mathfrak{B} \\ &= ((Z\downarrow)Y \cdot {}'\mathfrak{B})\uparrow \\ &= ((Z\downarrow \cdot {}'\mathfrak{B})(Y \cdot {}'\mathfrak{B}))\uparrow \\ &= ((Z \cdot \uparrow'\mathfrak{B})(Y \cdot {}'\mathfrak{B}))\uparrow \\ &= ((z \cdot \uparrow'\mathfrak{B})(y \cdot {}'\mathfrak{B}))\uparrow \\ &= ((\nabla\downarrow)Y)\uparrow \qquad = \nabla^\eta \end{aligned}$$

to show that $\mathfrak{B}$ transforms ∇ into ∇^η.

Putting these together we have

$$(\nabla \cdot \mathfrak{B}) \cdot \mathfrak{A} = \nabla^\eta \cdot \mathfrak{A} = \nabla^*$$

i.e. in two steps ∇ can be mutated into ∇^*. □

When we discuss the action of a computation on a derivation we will need a redex removal algorithm which, when supplied with a derivation with a redex root subject, will reduce that redex and reconstruct an appropriate derivation around the reduct. Before we can describe such an algorithm we need some notational and terminological preparation.

Each redex has the form $t^- = ts$ where $t = (\lambda y : \sigma \,.\, r)$ where we will call

r the **applicator term** $\qquad$ s the **applicant term**

of t^-. We call

$$\mathfrak{s} = [y \mapsto s]\mathfrak{i}$$

the **applicant replacement** of t^-, so eventually we want to relate t^- to $t^+ = r \cdot \mathfrak{s}$ (via a reduction).

A **redex derivation** has a root

$$\Gamma \vdash t^- : \rho$$

with a redex subject t^- (as above). We are interested in principal redex derivations, i.e. those of the form $\nabla = TS$ where

$$(T) \quad \Gamma \vdash t : \sigma \to \rho \qquad (S) \quad \Gamma \vdash s : \sigma$$

are the components. Note that T need not be principal. (Nor need S, but that is not important.) We may unravel T to produce $T = R{\uparrow}{\Downarrow}$ where

$$(R)\quad \Sigma \vdash r : \rho \qquad \Sigma = \Gamma^{\Uparrow}, y : \sigma \qquad \Gamma^{\Uparrow} \text{ is an initial part of } \Gamma$$

and the final sequence of Weakenings in T builds Γ from $\Gamma^{\Uparrow}$. Continuing with the terminology above we call

R the applicator derivation $\qquad$ S the applicant derivation

of ∇. We need the analogue of the applicant replacement to hit a derivation with.

6.11 DEFINITION. Given a principal redex derivation with decomposition

$$\nabla = (R{\uparrow}{\Downarrow})S$$

(as above) the applicant mutation

$$(\mathfrak{S})\quad \Sigma \xrightarrow{\mathfrak{s}} \Gamma$$

has the associated applicant replacement $\mathfrak{s}$ as shaft with nominated derivations

$$z \cdot \mathfrak{S} = \Gamma(z) \qquad y \cdot \mathfrak{S} = S$$

(for all identifiers z declared in $\Gamma^{\Uparrow}$). Notice that, apart from $y \cdot \mathfrak{S}$, all the nominated derivations are standard extractions. □

With this we can introduce the interface between the subject reduction algorithm (to be described in Section 6.7) and the mutation algorithm (already described this section). We introduce a notation '$- \bullet \mathbf{1}$' which indicates a call on the redex removal algorithm.

6.12 ALGORITHM. (Redex removal) *An algorithm which, when supplied with a principal redex derivation*

$$(\nabla)\quad \Gamma \vdash t^{-} : \rho$$

(as above), will return a derivation

$$(\nabla \bullet \mathbf{1})\quad \Gamma \vdash t^{+} : \tau$$

where $t^{+} = r \cdot \mathfrak{s}$.

Construction. This is almost trivial. We decompose the supplied derivation $\nabla = (R{\uparrow}{\Downarrow})S$ (as above) to produce the applicator derivation R and the applicant mutation $\mathfrak{S}$ with shaft $\mathfrak{s}$, and set

$$\nabla \bullet \mathbf{1} = R \cdot \mathfrak{S}$$

as the output derivation. □

Of course, this algorithm is not quite as simple as it looks. To determine $R \cdot \mathfrak{S}$ will require repeated calls on the mutation algorithm, so the whole calculation could be quite involved.

There are cases when we come across a redex $t^- = (\lambda x : \sigma . r)x$ which should reduce to r (or rather, $r \cdot \mathfrak{i}$). The above algorithm with $\mathfrak{s} = [x \mapsto x]\mathfrak{i}$ will produce this. Notice that this names the same concrete replacement as $\mathfrak{i}$, and there is a temptation to use $\mathfrak{i}$ in place of $\mathfrak{s}$. You will find it less confusing if you stick to the official format.

We have seen that in general a derivable judgement can have many different derivations, some which seem more useful than others. With experience we can select from amongst all possible derivations those which are canonical in some sense. The synthesis algorithm helps with this. The mutation algorithm can be used to produce a different canonical derivation.

6.13 DEFINITION. For each legal context Γ the standardizing mutation on Γ

$$(\mathfrak{I}_\Gamma) \quad \Gamma \xrightarrow{\mathfrak{i}} \Gamma$$

has shaft $\mathfrak{i}$ and standard extractions

$$x \cdot \mathfrak{I}_\Gamma = \Gamma(x)$$

for the nominated derivations (for each identifier x declared in Γ). □

Given a derivation ∇ of $\Gamma \vdash t : \tau$ we may think of

$$(\nabla \cdot \mathfrak{I}_\Gamma) \quad \Gamma \vdash (t \cdot \mathfrak{i}) : \tau$$

as the standard solution of (Γ, t). (Of course, the result $\nabla \cdot \mathfrak{I}_\Gamma$ depends on the selection policy used to determine any identifier renaming.) What kind of derivation does $\mathfrak{I}_\Gamma$ produce?

6.14 DEFINITION. A derivation ∇ is (fully) standard if it can be obtained from standard extractions by the use of I and E only, i.e. if the only uses of W in ∇ are to produce standard extractions. □

It can be checked that $\nabla \cdot \mathfrak{I}_\Gamma$ is (fully) standard. In fact, there is an algorithm which when supplied with (Γ, t) will return the standard solution (assuming there is at least one solution). One such algorithm is to use the synthesis algorithm to produce any solution ∇ and then hit ∇ with $\mathfrak{I}_\Gamma$. However, there is a much faster algorithm. This is dealt with in the exercises.

Exercises

6.11 Show that for each mutation

$$(\mathfrak{A})\ \Sigma \xrightarrow{\mathfrak{a}} \Gamma$$

the inclusion $(\partial\Sigma)^{(\mathfrak{a})} \subseteq \partial\Gamma$ holds.

6.12 Let

$$(\mathfrak{A})\ \Sigma \xrightarrow{\mathfrak{a}} \Gamma$$

be an arbitrary mutation. Show that for each declaration $x : \sigma$ of the source context Σ and extraction

$$(\nabla)\ \Sigma \vdash x : \sigma$$

the derivation $\nabla \cdot \mathfrak{A}$ depends only on x (and not the particular extraction used).

6.13 Consider the context

$$\Gamma = z : \sigma'', y : \sigma', x : \sigma$$

(where σ is arbitrary).

(i) Write down all solutions ∇ to (Γ, zyx).

(ii) Write down all mutations $\mathfrak{A}$ of the form $\Gamma \xrightarrow{\mathfrak{i}} \Gamma$.

(iii) For each such ∇ and $\mathfrak{A}$ calculate $\nabla \cdot \mathfrak{A}$.

(iv) How does the output $\nabla \cdot \mathfrak{A}$ depend on the input $(\nabla, \mathfrak{A})$?

6.14 For an arbitrary mutation

$$(\mathfrak{A})\ \Sigma \xrightarrow{\mathfrak{a}} \Gamma$$

describe, in general terms, all the mutations that can be built from $\mathfrak{A}$ by a sequence of no more than three restrictions and lengthenings. You may assume that Σ contains at least three declarations (so that $\Sigma^{\uparrow\uparrow\uparrow}$ makes sense); and the first, second, and third lengthening of any context are by

$$u : \theta \quad v : \psi \quad w : \phi$$

respectively where u, v, w are distinct and fresh to both Σ and Γ. How many such mutations are generated in this way?

6.15 The base mutation

$$(\mathfrak{I})\ \emptyset \xrightarrow{\mathfrak{i}} \emptyset$$

requires no nominated derivations. Consider the derivations R, S, T generated in the solution to Exercise 6.3. What is the effect of hitting each of these with $\mathfrak{I}$?

6.16 Consider the five derivations

$$(\nabla)\quad \Gamma \vdash t : \tau$$

generated in Example 6.5 and enumerated in Exercise 6.5. Observe that there are just two different mutations

$$(\mathfrak{A})\quad \Gamma \xrightarrow{\mathfrak{i}} \Gamma$$

where the nominated derivations are extractions. Write down these two mutations $\mathfrak{A}$. Hit each of the five derivations ∇ with both mutations $\mathfrak{A}$. Show that the derivation $\nabla \cdot \mathfrak{A}$ depends only on $\mathfrak{A}$.

6.17 Using the terms $\mathsf{B}, \mathsf{C}, \mathsf{D}$ of Exercises 6.2 and 6.6 let

$$\mathsf{D}^* := \lambda y : \sigma, z : \tau \,.\, \mathsf{B}zy \quad \text{so that} \quad \vdash \mathsf{D}^* : \delta$$

where the types $\rho, \sigma, \dots, \delta$ are generated from arbitrary types θ, ψ, ϕ in the usual way.

For derivations

$$(C)\quad \vdash \mathsf{C} : \gamma \qquad (B)\quad \vdash \mathsf{B} : \beta$$

go through a calculation of $CB \bullet \mathbf{1}$ to produce a derivation D^*. The construction of the various mutations and the choice of renaming identifiers should be explained.

Once you have produced D^* you should hit it with the base mutation $(\mathfrak{I})$ where again the details of the calculation should be displayed. Indicate the shape of the final derivation.

You should do the calculation for the different versions of C and B to illustrate the different effects.

6.18 We say a replacement $\mathfrak{a}$ is **renaming** if all of its updates have the form $[y \mapsto v]$ for identifiers y, v. In particular, for each identifier x the terms $x\mathfrak{a}$ is an identifier. (This is the analogue of a renaming replacement introduced in Section 5.4.) We say a mutation $\mathfrak{A}$ is **renaming** if it is carried by a renaming replacement *and* each of its nominated derivations, $x \cdot \mathfrak{A} = \Gamma(x\mathfrak{a})$, is a standard extraction. In particular, each standardizing mutation $\mathfrak{I}_\Gamma$ is renaming. Show that for each renaming mutation $\mathfrak{A}$ and compatible derivation ∇, the derivation $\nabla \cdot \mathfrak{A}$ is (fully) standard.

6.19 Extend the result of Exercise 6.18 by showing that if each nominated derivation of a mutation $\mathfrak{A}$ is (fully) standard then $\nabla \cdot \mathfrak{A}$ is (fully) standard for all compatible derivations ∇.

6.20 Given a pair of mutations

$$(\mathfrak{B})\quad \Pi \xrightarrow{\mathfrak{b}} \Sigma \qquad (\mathfrak{A})\quad \Sigma \xrightarrow{\mathfrak{a}} \Gamma$$

we may attempt to define a composite mutation

$$(\mathfrak{B}\,;\mathfrak{A})\quad \Pi \xrightarrow{\mathfrak{b}\,;\,\mathfrak{a}} \Gamma$$

with shaft $\mathfrak{b}\,;\mathfrak{a}$ by taking

$$x \cdot (\mathfrak{B}\,;\mathfrak{A}) = (x \cdot \mathfrak{B}) \cdot \mathfrak{A}$$

as nominated derivations for each identifier x declared in Π.

(a) Show that this does define a mutation.

(b) Show that

$$(\nabla \cdot \mathfrak{B}) \cdot \mathfrak{A} = \nabla \cdot (\mathfrak{B}\,;\mathfrak{A})$$

for all derivations ∇ over the context Π.

(c) Show that this composition of mutations is associative.

6.21 Here is an algorithm which, when supplied with

a context Γ a term t a renaming replacement $\mathfrak{a}$

will either fail, or succeed and return

a (fully) standard derivation $\nabla[\Gamma, t, \mathfrak{a}]$

which solves $(\Gamma, t \cdot \mathfrak{a})$. The algorithm proceeds by recursion over t with variation of the parameters Γ and $\mathfrak{a}$.

t	$\nabla[\Gamma, t, \mathfrak{a}]$	Proviso
k	$\Gamma(\mathsf{k})$	Γ is legal
x	$\Gamma(x)$	Γ is legal and $x \cdot \mathfrak{a}$ is declared in Γ
qp	$\nabla[\Gamma, q, \mathfrak{a}]\nabla[\Gamma, p, \mathfrak{a}]$	
$\lambda y : \sigma . r$	$\nabla[\Gamma^{\downarrow}, r, {'\mathfrak{a}}]\uparrow$	$v = \mathit{fresh}(\partial\Gamma)$, $\Gamma^{\downarrow} = \Gamma, v : \sigma$, ${'\mathfrak{a}} = [y \mapsto v]\mathfrak{a}$

The algorithm fails when it reaches a base case ($t = \mathsf{k}$ or $t = x$) where the proviso isn't satisfied. Observe that for the recursion steps ($t = qp$ and $t = \lambda y : \sigma . r$) if the smaller derivations exist then so does the larger one.

Show that for each compatible pair

$$(\nabla)\ \ \Sigma \vdash t : \tau \qquad (\mathfrak{A})\ \ \Sigma \xrightarrow{\mathfrak{a}} \Gamma$$

where $\mathfrak{A}$ is renaming, the algorithm succeeds and $\nabla \cdot \mathfrak{A} = \nabla[\Gamma, t, \mathfrak{a}]$ holds.

6.5 Computation

As with $\boldsymbol{\lambda}$ and $\mathbf{C}$, an applied λ-calculus $\boldsymbol{\lambda}\mathbf{Sig}$ has a reduction relation $t^- \rhd\!\!\!\!\rhd\, t^+$ on terms. We call t^- the **subject** and t^+ the **object** of the reduction. A computation is a tree of instances of such reductions designed to isolate at the leaves the 1-step reductions involved in the root reduction. Each redex gives such a 1-step reduction, but now certain other reductions are allowed.

(Computation Specifics) There is a collection of reduction axioms

$$(0)\quad t^- \rhd t^+ \quad \text{with associated recipe} \quad \nabla \longmapsto \nabla \bullet 0$$

each of which converts

$$(\nabla)\quad \Gamma \vdash t^- : \tau \quad \text{into} \quad (\nabla \bullet 0)\quad \Gamma \vdash t^+ : \tau$$

where ∇ is a principal derivation with a compatible root subject.

(Computation Generalities) By the discussion leading up to Definition 6.11, each redex $t^- = (\lambda y : \sigma . r)s$ provides a replacement $\mathfrak{s} = [y \mapsto s]\mathrm{i}$ and a redex removal

$$(1)\quad t^- \rhd t^+ \quad \text{with associated recipe} \quad \nabla \longmapsto \nabla \bullet 1$$

where $t^+ = r \cdot \mathfrak{s}$ and the recipe converts

$$(\nabla)\quad \Gamma \vdash t^- : \tau \quad \text{into} \quad (\nabla \bullet 1)\quad \Gamma \vdash t^+ : \tau$$

where ∇ is a principal derivation.

Of course, the need for a recipe $-\bullet 0$ for each reduction axiom puts a considerable restriction on these. Let's use the simple calculus $\boldsymbol{\lambda}\mathbf{Add}$ to illustrate this.

6.15 EXAMPLE. The calculus $\boldsymbol{\lambda}\mathbf{Add}$ has two housing axioms $\mathsf{H} : \mathcal{N}^+$ and $\mathsf{T} : \mathcal{N}^+$ where the intention is that both the constants H, T name the addition operator. Each has two associated reduction axioms

$$\begin{array}{llll} \mathsf{H}v\mathsf{0} & \rhd v & \qquad \mathsf{T}v\mathsf{0} & \rhd v \\ \mathsf{H}v(\mathsf{S}u) & \rhd \mathsf{S}(\mathsf{H}vu) & \qquad \mathsf{T}v(\mathsf{S}u) & \rhd \mathsf{T}(\mathsf{S}v)u \end{array}$$

for arbitrary terms v, u (where $\mathsf{0} : \mathcal{N}$ and $\mathsf{S} : \mathcal{N}'$ are the other two housing axioms). These two pairs of reduction axioms correspond to the head and tail recursive specifications of addition. All four axioms must have an associated recipe which shows how a derivation can pass across the axiom.

Consider the bottom right axiom. We need a recipe which converts

$$(\nabla)\quad \Gamma \vdash \mathsf{T}v(\mathsf{S}u) : \mathcal{N} \quad \text{into} \quad (\nabla^+)\quad \Gamma \vdash \mathsf{T}(\mathsf{S}v)u : \mathcal{N}$$

for arbitrary Γ, v, u. Notice how the source and target derivation subjects are, respectively, the subject and object of the reduction axiom.

Unravelling ∇ produces component derivations

$$(T)\ \Gamma^l \vdash \mathsf{T} : \mathcal{N}^+ \quad (V)\ \Gamma^l \vdash v : \mathcal{N} \quad (S)\ \Gamma^r \vdash \mathsf{S} : \mathcal{N}' \quad (U)\ \Gamma^r \vdash u : \mathcal{N}$$

where Γ^l, Γ^r are shortenings of Γ and

$$\nabla = ((TV)\Downarrow)((SU)\Downarrow)$$

for suitable Weakenings $\Downarrow$ (not necessarily the same). Let

$$T^+ = T\Downarrow \quad V^+ = V\Downarrow \quad S^+ = S\Downarrow \quad U^+ = U\Downarrow$$

where these Weakenings build the root contexts up to Γ. Then

$$\nabla^+ = T^+(S^+V^+)U^+$$

will do. There are also other possibilities. □

Clause	Shape	Remarks	Code
(Axiom reduction)	$\dfrac{t^- \rhd t^+}{t^- \mathrel{\rhd\!\!\!\rhd} t^+}\,(0)$	From Specifics	0
(Redex reduction)	$\dfrac{t^- \rhd t^+}{t^- \mathrel{\rhd\!\!\!\rhd} t^+}\,(1)$	$t^- = (\lambda y : \sigma . r)s$ $t^+ = r \cdot \mathfrak{s}$ $\mathfrak{s} = [y \mapsto s]\mathfrak{i}$	1
(Left application)	$\dfrac{\overset{\mathsf{q}}{q^- \mathrel{\rhd\!\!\!\rhd} q^+}}{q^- p \mathrel{\rhd\!\!\!\rhd} q^+ p}\,(\downharpoonleft)$		$\downharpoonleft\mathsf{q}$
(Right application)	$\dfrac{\overset{\mathsf{p}}{p^- \mathrel{\rhd\!\!\!\rhd} p^+}}{qp^- \mathrel{\rhd\!\!\!\rhd} qp^+}\,(\downharpoonright)$		$\downharpoonright\mathsf{p}$
(Abstraction)	$\dfrac{\overset{\mathsf{r}}{r^- \mathrel{\rhd\!\!\!\rhd} r^+}}{t^- \mathrel{\rhd\!\!\!\rhd} t^+}\,(\uparrow)$	$t^- = (\lambda y : \sigma . r^-)$ $t^+ = (\lambda y : \sigma . r^+)$	$\uparrow\mathsf{r}$
(Transitive composition)	$\dfrac{\overset{\mathsf{l}}{t^- \mathrel{\rhd\!\!\!\rhd} t^0} \quad \overset{\mathsf{r}}{t^0 \mathrel{\rhd\!\!\!\rhd} t^+}}{t^- \mathrel{\rhd\!\!\!\rhd} t^+}\,(\circ)$		$\mathsf{l} \circ \mathsf{r}$

Table 6.3: The computation rules for $\boldsymbol{\lambda}$**Sig**

Computations for an applied λ-calculus combine features of those for **C** and $\boldsymbol{\lambda}$.

6.16 DEFINITION. A computation

$$(\square) \quad t^- \mathrel{\rhd\!\!\!\rhd} t^+$$

is a finite rooted tree grown according to the rules of Table 6.3. The fourth column of that table gives the arboreal code each rule. A computation is said to organize its root reduction. $\square$

In calculi such as $\boldsymbol{\lambda}$**Add** computations can be used to do arithmetic.

6.17 EXAMPLE. In $\boldsymbol{\lambda}$**Add**, for each $m \in \mathbb{N}$ we set $\ulcorner m \urcorner = \mathsf{S}^m 0$ to produce the numeral $\ulcorner m \urcorner$ whose job is to name m. We use the reduction axioms to generate computations $\square(H, m)$ and $\square(T, m)$ by

$$\begin{aligned} \square(H,0) &= \mathsf{0} & \qquad \square(T,0) &= \mathsf{0} \\ \square(H,m') &= \mathsf{0} \circ \downharpoonright\square(H,m) & \qquad \square(T,m') &= \mathsf{0} \circ \square(T,m) \end{aligned}$$

and then check

$$(\square(H,m)) \quad \mathsf{H}\ulcorner n \urcorner \ulcorner m \urcorner \mathrel{\rhd\!\!\!\rhd} \ulcorner n+m \urcorner \qquad (\square(T,m)) \quad \mathsf{T}\ulcorner n \urcorner \ulcorner m \urcorner \mathrel{\rhd\!\!\!\rhd} \ulcorner n+m \urcorner$$

for all $n, m \in \mathbb{N}$. For both computations the base case, $m = 0$, is immediate. For the steps $m \mapsto m'$ we have

$$(\square(H,m')) \qquad \dfrac{\dfrac{\mathsf{H}\ulcorner n\urcorner\ulcorner m'\urcorner \rhd \mathsf{S}(\mathsf{H}\ulcorner n\urcorner\ulcorner m\urcorner)}{\mathsf{H}\ulcorner n\urcorner\ulcorner m'\urcorner \rhd\!\!\rhd \mathsf{S}(\mathsf{H}\ulcorner n\urcorner\ulcorner m\urcorner)}\,(0) \qquad \dfrac{\dfrac{\square(H,m)}{\mathsf{H}\ulcorner n\urcorner\ulcorner m\urcorner \rhd\!\!\rhd \ulcorner n+m\urcorner}}{\mathsf{S}(\mathsf{H}\ulcorner n\urcorner\ulcorner m\urcorner) \rhd\!\!\rhd \mathsf{S}\ulcorner n+m\urcorner}\,(\downharpoonleft)}{\mathsf{H}\ulcorner n\urcorner\ulcorner m\urcorner \rhd\!\!\rhd \ulcorner n+m'\urcorner}\,(\circ)$$

$$(\square(T,m')) \qquad \dfrac{\dfrac{\mathsf{T}\ulcorner n\urcorner\ulcorner m'\urcorner \rhd \mathsf{T}(\mathsf{S}\ulcorner n\urcorner)\ulcorner m\urcorner}{\mathsf{T}\ulcorner n\urcorner\ulcorner m'\urcorner \rhd\!\!\rhd \mathsf{T}\ulcorner n'\urcorner\ulcorner m\urcorner)}\,(0) \qquad \dfrac{\square(T,m)}{\mathsf{T}\ulcorner n'\urcorner\ulcorner m\urcorner \rhd\!\!\rhd \ulcorner n'+m\urcorner}}{\mathsf{T}\ulcorner n\urcorner\ulcorner m'\urcorner \rhd\!\!\rhd \ulcorner n+m'\urcorner}\,(\circ)$$

as required. You should observe the differences between these two families of computations. (If you can't see this, write out the trees for $m = 3$). □

As is usual, the reduction relation $\rhd\!\!\rhd$ is irreflexive so, by definition, a term t is normal if there is no term t' with $t \rhd\!\!\rhd t'$. As before, a normal term can not contain a redex subterm (for otherwise a use of 1 could effect a reduction). However, the lack of redex subterms does not imply normality. This is because of the reduction axioms. For instance, in **λAdd** the term H0(S0) contains no identifiers at all (and therefore has no redex subterms) but it is not normal since

$$(0 \circ 0) \quad \mathsf{H0(S0)} \rhd\!\!\rhd \mathsf{S0}$$

holds. This means that the notion of 'normal term' is more delicate than before.

Exercises

6.22 Suppose we decide to have in **λSig** the combinators $\mathsf{I}(\bullet), \mathsf{K}(\bullet), \mathsf{S}(\bullet)$ with the standard reduction properties. Describe the Computation Specifics for these.

6.23 Consider the three terms r, s, t of Exercise 6.3, and observe that $t \rhd\!\!\rhd r$. Write down what appear to be two different computations which organize this reduction.

6.24 Using the terms B, C, D of Exercises 6.2, 6.6, 6.17, organize a computation of CB $\rhd\!\!\rhd$ D and indicate which replacements are needed.

6.25 Consider the computations $\square(H,m)$ and $\square(T,m)$ generated in Example 6.17. Calculate the height and width of these as a functions of m.

6.6 Type inhabitation

Superficially the type inhabitation problem is similar to the type synthesis problem discussed in Section 6.3. Here we are given a pair (the problem data) (Γ, τ) where Γ is a (legal) context and τ is a type. We are required to determine whether or not there is a derivation $\Gamma \vdash t : \tau$ for some term t, and classify all possible such derivations (for varying t). It is clearly possible to cobble together an algorithm which will enumerate all such derivations ∇, but this is not very satisfactory. In this generality nothing very sensible can be said about the inhabitation problem. We need to impose further restrictions which help to structure the problem.

Let us restrict the problem to a search for and classification of normal derivations, i.e. derivations where the root subject is normal. This is a quite severe restriction, but it is worthwhile, for it enables us to make considerable progress.

We need some terminology.

6.18 DEFINITION. (a) For a legal context Γ the terms anchored in Γ are generated recursively as follows.

- Each constant is anchored at depth 0 in Γ.
- Each identifier declared in Γ is anchored at depth 0 in Γ.
- If q is anchored at depth m in Γ, and p is normal, and there are derivations

$$\Gamma \vdash q : \pi \to \tau \qquad \Gamma \vdash p : \pi$$

(for some types π, τ), then qp is anchored at depth m' in Γ.

(b) For a context Γ and a type τ, a term t is Γ-anchored at τ if t is anchored at some depth in Γ and $\Gamma \vdash t : \tau$ is derivable.

(c) For a context Γ and a type τ, a term t is Γ-docked at τ if t has the shape $t = \lambda y : \sigma \, . \, r$ where r is normal and

$$\Gamma^{-}, y : \sigma \vdash r : \rho$$

is derivable for some initial part Γ^{-} of Γ and $\tau = \sigma \to \rho$. □

The Γ-anchored and Γ-docked terms will feature quite strongly in any solution to a normal inhabitation problem. It can be seen that if t is Γ-anchored at τ then

$$t = hp_1 \cdots p_m$$

where h is a constant or an identifier declared in Γ. We call this the head of t. Furthermore, we find there are derivations

$$\Gamma \vdash h : \pi_1 \to \cdots \to \pi_m \to \tau \qquad \Gamma \vdash p_i : \pi_i \quad (1 \leq i \leq m)$$

with $p_1, \dots, p_m$ normal, and then t is anchored at depth m in Γ.

A Γ-anchored term need not be normal. In **λAdd** the term H00 is anchored at depth 2 in any (legal) context, but it is not normal (since H00 $\rhd$ 0). There may be more intricate ways in which a Γ-anchored term fails to be normal, and any solution to the normal inhabitation problem must find a way of sifting out these terms.

We say a derivation is normal if its root subject is normal. This, of course, does not restrict the use of Weakening in the derivation.

6.19 LEMMA. (Type inhabitation) *For each normal derivation*

$$(\nabla) \quad \Gamma \vdash t : \tau$$

either t *is* Γ*-anchored at* τ *or* t *is* Γ*-docked at* τ.

Proof. This follows by a progressive induction over the height $h(\nabla)$ of ∇. It is useful to do the case $h(\nabla) = 0$ first.

If $h(\nabla) = 0$ then ∇ is a Leaf and so is either an Axiom or a Projection. In the first case t is a constant, and in the second case t is an identifier declared in Γ. In either case t is anchored at depth 0 in Γ. Since $\Gamma \vdash t : \tau$ is derivable, t is Γ-anchored at τ.

For the induction step we first locate the principal rule of ∇, i.e. we find

$$(\nabla^-) \quad \Gamma^- \vdash t : \tau$$

where ∇ arises from ∇^- by a sequence of Weakenings. Here Γ^- is an initial part of Γ. (Of course, $\nabla^- = \nabla$ and $\Gamma^- = \Gamma$ can happen.) We have $h(\nabla^-) \leq h(\nabla)$.

If $h(\nabla^-) = 0$ then (as above) t is anchored at depth 0 in Γ and hence (since $\Gamma \vdash t : \tau$ is derivable) t is Γ-anchored at τ.

If $h(\nabla^-) \neq 0$ then ∇^- arises by a use of I or E. We look at these cases separately.

(I) We have

$$\nabla^- = \dfrac{\begin{matrix} R \\ \Gamma^-, y : \sigma \vdash r : \rho \end{matrix}}{\Gamma^- \vdash t : \tau}$$

for some subderivation R with $t = \lambda y : \sigma \,.\, r$ and $\tau = \sigma \to \rho$. The term r is normal (otherwise t is not normal), so t is Γ-docked at τ.

Observe that this case does not use the induction hypothesis.

(E) We have

$$\nabla^- = \dfrac{\begin{matrix} Q & P \\ \Gamma^- \vdash q : \pi \to \tau & \Gamma^- \vdash p : \pi \end{matrix}}{\Gamma^- \vdash t : \tau}$$

for subderivations Q, P with $t = qp$. Both q, p must be normal (otherwise t is not normal). Now $h(Q) < h(\nabla^-) \leq h(\nabla)$ so, by the induction hypothesis, one of

$$q \text{ is } \Gamma\text{-anchored at } \pi \to \tau \qquad q \text{ is } \Gamma\text{-docked at } \pi \to \tau$$

must hold. The second alternative can not arise (otherwise $t = qp$ is a redex). In the first alternative q is anchored at some depth in Γ^- and hence in Γ. Weakening gives

$$\Gamma \vdash q : \pi \to \tau \quad \Gamma \vdash p : \pi \qquad \text{where} \qquad q \text{ is anchored in } \Gamma \quad p \text{ is normal}$$

so t is anchored in Γ. Thus (since $\Gamma \vdash t : \tau$ is derivable) t is Γ-anchored at τ. □

Although this result gives us quite a lot of information, it by no means solves the normal inhabitation problem. The presence of constants will always cause extra work, but even without these all is not plain sailing. An example will illustrate what can happen.

6.20 EXAMPLE. Using the calculus $\mathbf{\lambda Add}$ let $\Gamma = x_1 : \mathcal{N}, \dots, x_k : \mathcal{N}$ and consider the normal inhabitation problem $(\Gamma, \mathcal{N})$. Any normal term t with $\Gamma \vdash t : \mathcal{N}$ is either Γ-anchored or Γ-docked in $\mathcal{N}$. Since $\mathcal{N}$ is an atom (and not an arrow type) no term cam be docked in $\mathcal{N}$. Thus t is Γ-anchored in $\mathcal{N}$ and hence $t = hp_1 \cdots p_m$ for some

$$\Gamma \vdash h : \pi_1 \to \cdots \to \pi_m \to \mathcal{N} \qquad \Gamma \vdash p_i : \pi_i \quad (1 \le i \le m)$$

with h a constant or declared in Γ and each p_i normal. The number m is also unknown.

When $h = x$ with x declared in Γ, we must have $m = 0$, for the displayed type of h must match that of x in Γ. Thus $t = x$.

When $h = \mathsf{0}$, we must have $m = 0$, for the displayed type of h must match that of the housing axiom of $\mathsf{0}$. Thus $t = \mathsf{0}$.

When $h = \mathsf{S}$, we must have $m = 1$ with $\pi_1 = \mathcal{N}$, for the displayed type of h must match that of the housing axiom of S. Thus $t = \mathsf{S}p$ where $\Gamma \vdash p : \mathcal{N}$ with p normal and we require a simpler solution p of $(\Gamma, \mathcal{N})$.

When $h = \mathsf{H}$, we must have $m = 2$ with $\pi_1 = \pi_2 = \mathcal{N}$, for the displayed type of h must match that of the housing axiom of H. Thus $t = \mathsf{H}qp$ where $\Gamma \vdash q : \mathcal{N}$ and $\Gamma \vdash p : \mathcal{N}$ with q, p normal and we require two simpler solutions q, p of $(\Gamma, \mathcal{N})$.

When $h = \mathsf{T}$, by the same reasoning we have $t = \mathsf{H}qp$ where $\Gamma \vdash q : \mathcal{N}$ and $\Gamma \vdash p : \mathcal{N}$ with q, p normal and we require two simpler solutions q, p of $(\Gamma, \mathcal{N})$.

This gives us a recursive way to generate all solutions. We must have

$$t = x \text{ or } t = \mathsf{0} \text{ or } t = \mathsf{S}p \text{ or } t = \mathsf{H}qp \text{ or } t = \mathsf{T}qp$$

where x is declared in Γ and q, p are simpler solutions. But, not all such terms are normal. When $t = \mathsf{H}qp$ or $t = \mathsf{T}qp$ we must have $\Gamma \vdash q : \mathcal{N}$ and $\Gamma \vdash p : \mathcal{N}$ with both q, p normal. By recursion we generate q and p except p can not be $\mathsf{0}$ and $\mathsf{S}r$ (for some r), since t must be normal and both the excluded p give reducible t. □

In general a normal inhabitant problem (never mind a general inhabitant problem), can be solved in full only under special circumstances.

EXERCISES

6.26 Consider the calculus $\mathbf{\lambda\emptyset}$, with no constants. Let X be a variable.

(a) Determine all normal inhabitants of

(i)	X	(ii)	X'	(iii)	$X \to X'$
(iv)	$X' \to X$	(v)	$X \to X \to X'$	(vi)	$X \to X' \to X$

in the empty context.

(b) Show that all normal inhabitants of

$$X'' \qquad X \to X' \to X'' \to X$$

(in the empty context) have the form

$$\lambda y : X', x : X \,.\, (y^m x) \qquad \lambda x : X, y : X', z : X'' \,.\, (z^n y)(y^m x)$$

for some $m, n \in \mathbb{N}$.

6.27 In the calculus $\boldsymbol{\lambda\emptyset}$ consider the types

$$\begin{array}{lll} \rho = X \to Z & \sigma = X \to Y & \tau = Y \to Z \\ \beta = \tau \to \sigma \to \rho & \gamma = \beta \to \delta & \delta = \sigma \to \tau \to \rho \end{array}$$

where X, Y, Z are distinct variables. Determine the normal inhabitants of β, γ, δ in the empty context. What happens if X, Y, Z are replaced by arbitrary types?

6.28 Show that if t is Γ-anchored at τ with witnessing derivation ∇ then there are normal derivations

$$(\nabla_0) \quad \Gamma \vdash h : \eta \qquad (\nabla_i) \quad \Gamma \vdash p_i : \pi_i \quad (1 \le i \le m)$$

where

$$\eta = \pi_1 \to \cdots \to \pi_m \to \tau \qquad h \text{ is a constant or an identifier} \qquad t = hp_1 \cdots p_m$$

(and $m = 0$ is possible). Describe ∇ in terms of $\nabla_0, \nabla_1, \ldots, \nabla_m$. Show how to generate all possible ∇.

6.7 Subject reduction

Suppose we wish to inhabit a type τ in some context Γ. We want to exhibit a derivation

$$(\nabla) \quad \Gamma \vdash t^- : \tau$$

with some inhabitant t^- of τ. Suppose we obtain such a derivation ∇ and then notice that t^- can be reduced to some term t^+. There ought to be a corresponding derivation of

$$\Gamma \vdash t^+ : \tau$$

and it should be routine to obtain this from ∇ (and the known reduction).

We know how to do this for $\boldsymbol{\lambda}$; now we lift those methods to $\boldsymbol{\lambda}\mathbf{Sig}$.

6.21 ALGORITHM. (Subject reduction) *An algorithm which, when supplied with*

$$(\nabla) \quad \Gamma \vdash t^- : \tau \qquad (\square) \quad t^- \rhd\!\!\!\rhd t^+$$

a derivation and a computation of indicated compatibility, will return a derivation

$$(\nabla \cdot \square) \quad \Gamma \vdash t^+ : \tau$$

the result of the action *of* $\square$ *on* ∇.

Construction. The algorithm proceeds by a double recursion over the two inputs ∇ and $\square$. At its base the algorithm calls on the recipe $\nabla \longmapsto \nabla \bullet 0$ given by the Specifics (for the reduction axiom) and the recipe $\nabla \longmapsto \nabla \bullet 1$ given by the Generalities (for redex removal).

The algorithm has two base clauses and five recursion clauses. Using the arboreal code these are as follows.

(Axiom)	$\nabla \cdot \mathsf{0} = \nabla \bullet \mathsf{0}$	(Abstr)	$R\uparrow \cdot \uparrow\mathsf{r} = (R \cdot \mathsf{r})\uparrow$
(Redex)	$\nabla \cdot \mathsf{1} = \nabla \bullet \mathsf{1}$	(Trans)	$\nabla \cdot (\mathsf{l} \circ \mathsf{r}) = (\nabla \cdot \mathsf{l}) \cdot \mathsf{r}$
(LAppl)	$(QP) \cdot \rfloor\mathsf{q} = (Q \cdot \mathsf{q})P$	(Weak)	$(S, x, \sigma)\downarrow \cdot \Box = (S \cdot \Box, x, \sigma)\downarrow$
(RAppl)	$(QP) \cdot \lfloor\mathsf{p} = Q(P \cdot \mathsf{p})$		

To use this algorithm we first check that the supplied pair $(\nabla, \Box)$ is compatible, i.e. the subjects of ∇ and $\Box$ are the same term. We then locate the appropriate rule to use. Usually this is determined by the shape of $\Box$.

Recall that the recipes $\nabla \longmapsto \nabla \bullet \mathsf{0}$ and $\nabla \longmapsto \nabla \bullet \mathsf{1}$ apply only to principal ∇. Thus to calculate $\nabla \cdot \mathsf{0}$ or $\nabla \cdot \mathsf{1}$ where ∇ is not principal, we first use W (perhaps several times) to locate the principal node of ∇. We can then use the appropriate recipe.

A case where there are alternative rules is when we are presented with a pair

$$\nabla = (S, x, \sigma)\downarrow \qquad \Box = \mathsf{l} \circ \mathsf{r}$$

for then both T and W are applicable. However, a use of T followed by two uses of W, and a use of W followed by T, give respectively

$$\nabla \cdot \Box = (\nabla \cdot \mathsf{l}) \circ \mathsf{r} = ((S \cdot \mathsf{l}) \cdot \mathsf{r}, x, \sigma)\downarrow \qquad \nabla \cdot \Box = (S \cdot \Box)\downarrow = ((S \cdot \mathsf{l}) \cdot \mathsf{r}, x, \sigma)\downarrow$$

so that both paths lead to the same result.

As a sensible strategy we should always use W whenever possible, and only use the other rules when ∇ is principal. In this way the algorithm is deterministic.

What is not so obvious is that the algorithm is well founded. Intuitively each recursion step which obtains $(\nabla, \Box)$ from $(\nabla', \Box')$ does simplify the problem, but we need to make this precise. □

How can we prove this algorithm is well founded? Given a pair $(\nabla, \Box)$ we use the height $h(\nabla)$ of ∇ and the width $w(\Box)$ of $\Box$ to produce the **cost**

$$(h, w) = (h(\nabla), w(\Box))$$

of evaluating $\nabla \cdot \Box$. We need to find a well founded, transitive, irreflexive comparison $\sqsubset$ on such pairs (of natural numbers) where each recursion step decreases the cost.

Consider first the laws LA, RA, Ab, W. In each of these the width of the two computations is the same, but the height of the derivation is reduced. Thus

$$k < h \Rightarrow (k, w) \sqsubset (h, w)$$

should hold. But what about T? Here we have a fixed derivation ∇ of height $h = h(\nabla)$. We have two component computations l, r with

$$l = w(\mathsf{l}) > 0 \quad r = w(\mathsf{r}) > 0 \quad \text{and then} \quad w = w(\Box) = l + r$$

is the width of the supplied computation. Note that $l < w$, $r < w$ (since both l, r are strictly positive). The rule first attacks (∇, l) at a cost of (h, l) to produce some derivation $\nabla \cdot \mathsf{l}$ with some unknown height $k = h(\nabla \cdot \mathsf{l})$. The rule then attacks $(\nabla \cdot \mathsf{l}, \mathsf{r})$ at a cost of (k, r) to produce the result. Thus both

$$l < w \Rightarrow (h, l) \sqsubset (h, w) \qquad r < w \Rightarrow (k, r) \sqsubset (h, w)$$

should hold. Here k is arbitrary.

In short we want

$$(h, w) \sqsubset (h + 1, w) \qquad (h, w) \sqsubset (0, w + 1)$$

to hold. Perhaps you can see how to achieve this?

There are some hidden subtleties in this algorithm, mostly concerning the renaming of identifiers. For instance, consider the redex removal law

$$\nabla \cdot \mathsf{1} = \nabla \bullet \mathsf{1}$$

which is an instruction to call on the redex removal algorithm 6.12. This tells us to decompose ∇ to produce

a derivation R a mutation $\mathcal{S}$

of a certain compatibility, and then take

$$\nabla \bullet \mathsf{1} = R \cdot \mathcal{S}$$

as the resulting derivation. This mutation has shaft $\mathfrak{s}$, and surely this replacement $\mathfrak{s}$ will be the replacement $\mathfrak{a}$ supplied by the original problem. No!

Remember that $\mathsf{1}$ may be embedded in a larger computation, and earlier parts of that may have changed the relevant bound variable. An example will illustrate this.

6.22 EXAMPLE. Using the context and terms

$$\Gamma = y : \theta', x : \theta \qquad r = yx \qquad s = \lambda x : \theta \,.\, r \qquad t = (\lambda y : \theta' \,.\, s)$$

consider the arboreal code of the following derivation.

$$\nabla'' = \Gamma[y]\Gamma[x] \qquad \nabla' = (\nabla''\uparrow\uparrow\downarrow\downarrow)\Gamma[y] \qquad \nabla = \nabla'\Gamma[x]$$

$$\dfrac{\dfrac{\dfrac{\dfrac{\dfrac{\dfrac{\dfrac{\Gamma \vdash y : \theta' \quad \Gamma \vdash x : \theta}{\Gamma \vdash r : \theta}}{\Gamma^{\uparrow} \vdash s : \theta'}(\uparrow)}{\vdash t : \theta''}(\uparrow)}{\Gamma^{\uparrow} \vdash t : \theta''}(\downarrow)}{\Gamma \vdash t : \theta''}(\downarrow) \quad \Gamma \vdash y : \theta'}{\Gamma \vdash ty : \theta'} \quad \Gamma \vdash x : \theta}{\Gamma \vdash tyx : \theta}
\qquad
\dfrac{\dfrac{\dfrac{\dfrac{\dfrac{\dfrac{\dfrac{\Gamma[y] \quad \Gamma[x]}{\nabla''}}{\bullet}(\uparrow)}{\bullet}(\uparrow)}{\bullet}(\downarrow)}{\bullet}(\downarrow) \quad \Gamma[y]}{\nabla'} \quad \Gamma[x]}{\nabla}$$

The root context of $\nabla''\uparrow\uparrow$ is $\emptyset$. The two Weakenings in ∇' build up another copy of Γ.

We have an informal reduction $tyx \mathrel{\rhd\!\!\!\rhd} sx \mathrel{\rhd\!\!\!\rhd} r$ which we organize into a computation. Using the replacements

$$\mathfrak{b} = [y \mapsto y]\mathfrak{i} \qquad '\mathfrak{b} = [x \mapsto x]\mathfrak{b} \qquad \mathfrak{a} = [x \mapsto x]\mathfrak{i}$$

(all of which have the same denotation as $\mathfrak{i}$) we have two reductions on the left

$$\begin{array}{l} ty \mathrel{\rhd\!\!\!\rhd} s \cdot \mathfrak{b} = \lambda x : \theta \,.\, (r \cdot '\mathfrak{b}) = s \\ sx \mathrel{\rhd\!\!\!\rhd} r \cdot \mathfrak{a} = r \end{array} \qquad (\Box = \downarrow\!1 \circ 1) \qquad \dfrac{\dfrac{\dfrac{ty \rhd s \cdot \mathfrak{b}}{ty \mathrel{\rhd\!\!\!\rhd} s}(1)}{tyx \mathrel{\rhd\!\!\!\rhd} sx}(\downarrow) \quad \dfrac{sx \rhd r \cdot \mathfrak{a}}{sx \mathrel{\rhd\!\!\!\rhd} r}(1)}{tyx \mathrel{\rhd\!\!\!\rhd} r}(\circ)$$

with the required computation on the right. (We could use $\mathfrak{i}$ in place of $\mathfrak{b}$, $'\mathfrak{b}$, $\mathfrak{a}$, but putting in the various updates makes it less confusing.)

Now let's hit ∇ with $\Box$ (to get ∇'', perhaps). We have

$$\nabla \cdot \Box = ((\nabla'\Gamma[x]) \cdot \downarrow\!\mathbf{1}) \cdot \mathbf{1} = ((\nabla' \cdot \mathbf{1})\Gamma[x]) \cdot \mathbf{1}$$

so we need to make two calls on the redex removal algorithm.

For the inner one we have

$$\nabla' \cdot \mathbf{1} = ((\nabla''\downarrow\downarrow\uparrow\uparrow)\Gamma[y]) \bullet \mathbf{1} = \nabla''\uparrow \cdot \mathfrak{y} = (\nabla'' \cdot '\mathfrak{y})\uparrow$$

where

$$(\mathfrak{y}) \quad \Gamma^{\uparrow} \xrightarrow{\;\mathfrak{y}\;} \Gamma \quad \text{with} \quad \mathfrak{y} = [y \mapsto y]\mathfrak{i}$$

$$('\mathfrak{y}) \quad \Gamma \xrightarrow{\;'\mathfrak{y}\;} \Lambda \quad \text{with} \quad '\mathfrak{y} = [x \mapsto u]\mathfrak{y} \quad \Lambda = \Gamma, u : \theta$$

where $u \notin \{y, x\}$ to ensure legality, and with

$$y \cdot \mathfrak{y} = \Gamma[y] \qquad y \cdot '\mathfrak{y} = \Gamma[y]\downarrow \qquad x \cdot '\mathfrak{y} = \Lambda(u)$$

as nominated derivations. In particular, we have

$$\nabla' \bullet \mathbf{1} = \nabla^{\times}\uparrow \quad \text{where} \quad \nabla^{\times} = (\Gamma[y]\downarrow)\Lambda[u]$$

to complete the first phase of the calculation. Notice that

$$(\nabla^{\times}) \quad \Lambda \vdash yu : \theta$$

(but perhaps not with the shape you expected), and although $\mathfrak{y} = \mathfrak{b}$ we have $'\mathfrak{y} \neq '\mathfrak{b}$.

Continuing we have

$$\nabla \cdot \Box = (\nabla^{\times}\uparrow)\Gamma[x] \bullet \mathbf{1} = \nabla^{\times} \cdot \mathfrak{X} = (y \cdot \uparrow\mathfrak{X})(x \cdot \mathfrak{X}) = (y \cdot \mathfrak{X})(x \cdot \mathfrak{X})$$

where

$$(\mathfrak{X}) \quad \Lambda \xrightarrow{\;\mathfrak{x}\;} \Gamma \quad \text{with} \quad \mathfrak{x} = [u \mapsto xy]\mathfrak{i}$$

with

$$y \cdot \mathfrak{X} = \Gamma(y) = \Gamma^{\uparrow}[y]\downarrow \qquad x \cdot \mathfrak{X} = \Gamma[x]$$

as nominated derivations.

This gives

$$\nabla \cdot \square = \Gamma(y)\Gamma(x)$$

which is *not* ∇'' (as perhaps expected), but a standardized version of this. Notice that $\mathfrak{x} \neq \mathfrak{a}$.

What is going on here? Both the substitution algorithm and the mutation algorithm require the renaming of (some) bound identifiers. The choice of a new identifier is more restrictive in the mutation case. When we set up the computation we didn't anticipate what might happen when we used a mutation.

With foresight we could have used the replacements

$$\mathfrak{y} = [y \mapsto y]\mathfrak{i} \qquad '\mathfrak{y} = [x \mapsto u]\mathfrak{y} \qquad \mathfrak{x} = [u \mapsto x]\mathfrak{i}$$

to get

$$ty \rhd s \cdot \mathfrak{y} = \lambda u : \theta \,.\, (r \cdot '\mathfrak{y}) = s' \qquad s'x \rhd r' \cdot \mathfrak{x} = r$$

where $r' = yu$, $s' = \lambda u : \theta \,.\, r'$ and then

$$(\square = \downarrow 1 \circ 1) \qquad \dfrac{\dfrac{\dfrac{ty \rhd s \cdot \mathfrak{y}}{ty \mathrel{\rhd\!\!\rhd} s'}(1)}{tyx \mathrel{\rhd\!\!\rhd} s'x}(\downarrow) \qquad \dfrac{s'x \rhd r \cdot \mathfrak{x}}{s'x \mathrel{\rhd\!\!\rhd} r}(1)}{tyx \mathrel{\rhd\!\!\rhd} r}(\circ)$$

organizes the reduction.

When we work only with terms there seems no good reason to change x to u only to change it back again. When we work with derivations we see why this is done. This also explains why the carrying replacement is not indicated in a redex removal 1. By the time that 1 is performed the required replacement may have changed. □

To conclude this section let's consider how we might normalize a derivation. Any given derivation

$$(\nabla) \quad \Gamma \vdash t^{-} : \tau$$

may contain abnormalities. These can be located and, with a little practice, we can set up a computation □ which will remove them. Isn't $\nabla \cdot \square$ the required normal derivation?

6.23 EXAMPLE. Within $\boldsymbol{\lambda\emptyset}$ consider the terms $r = \lambda x : \rho \,.\, yx$, $s = \lambda x : \rho \,.\, x$, and $t = \lambda y : \sigma \,.\, r$ where $\sigma = \rho'$ (and ρ is arbitrary). Note that $t^{-} = ts$ is a redex and there are no other reducible subterms (i.e. both t and s are normal). With the contexts

$$\Sigma = y : \sigma \qquad \Lambda = x : \rho \qquad \Gamma = y : \sigma,\, x : \rho \qquad \Pi = u : \rho \qquad \Delta = u : \rho,\, x : \rho$$

(where Π, Δ will be useful later) we see that

$$\dfrac{\dfrac{\dfrac{\dfrac{\Gamma \vdash y : \sigma \quad \Gamma \vdash x : \rho}{\Gamma \vdash yx : \rho}}{\Sigma \vdash r : \sigma}(\uparrow)}{\vdash t : \sigma'}(\uparrow) \qquad \dfrac{\Lambda \vdash x : \rho}{\vdash s : \sigma}(\uparrow)}{\vdash ts : \sigma}(*) \qquad\qquad \dfrac{\dfrac{\dfrac{\dfrac{\Gamma[y] \quad \Gamma[x]}{\bullet}}{R}}{L} \qquad \dfrac{\Lambda[x]}{S}}{\vdash ts : \sigma}(*)$$

gives a derivation

$$(\nabla) \quad \Gamma \vdash t^- : \sigma$$

(and is probably the derivation we first think of when asked to solve (Γ, t^-).) Thus we see that with

$$R = (\Gamma[y]\Gamma[x])\uparrow \quad L = R\uparrow \quad S = \Lambda[x]\uparrow$$

we have a derivation $\nabla = LS$.

How can we normalize ∇? This derivation has just one abnormality at $(*)$. This should be removed by using the right hand leg (above $(*)$) to witness the identifier y in the top left leaf.

With

$$\mathfrak{b} = [y \mapsto s]\mathfrak{i} \quad {}'\mathfrak{b} = [x \mapsto u]\mathfrak{b} \qquad \text{we have} \qquad t^- \rhd r \cdot \mathfrak{b} = \lambda u : \rho\,.\,((yx) \cdot {}'\mathfrak{b}) = t^0$$

with $t^0 = (\lambda u : \rho\,.\,su)$ and u suitably chosen. The computation

$$(1) \quad \dfrac{ts \rhd r \cdot \mathfrak{b}}{ts \mathrel{\rhd\!\!\rhd} t^0}(1)$$

organizes this reduction.

To remove the abnormality $(*)$ in ∇ we hit the derivation with $\mathsf{1}$. What does this give? We could argue a case for either of

$$\dfrac{\dfrac{\dfrac{\dfrac{\Lambda[x]}{\vdash s : \sigma}(\uparrow)}{\Pi \vdash s : \sigma}(\downarrow) \qquad \Pi[u]}{\Pi \vdash su : \rho}}{\vdash (\lambda u : \rho\,.\,su) : \sigma}(*) \qquad\qquad \dfrac{\dfrac{\dfrac{\Lambda[x]}{\Pi \vdash s : \sigma}(\uparrow) \qquad \Pi[u]}{\Pi \vdash su : \rho}}{\vdash (\lambda u : \rho\,.\,su) : \sigma}(*)$$

but which one does the algorithm actually produce? In either case the result is still not normal for an abnormality has been created (at $(*)$) which must be removed by hitting with some other computation. We expect this will produce $\Pi[u]\uparrow$ as the final result.

When we hit ∇ with $\mathsf{1}$ we get

$$\nabla \cdot \mathsf{1} = LS \cdot \mathsf{1} = ((R\uparrow)S) \bullet \mathsf{1} = R \cdot \mathfrak{B}$$

where

$$(\mathfrak{B}) \quad \Sigma \xrightarrow{\;\mathfrak{b}\;} \emptyset \qquad ({}'\mathfrak{B}) \quad \Gamma \xrightarrow{\;{}'\mathfrak{b}\;} \Pi$$

with

$$y \cdot \mathfrak{B} = S \qquad y \cdot {'\mathfrak{B}} = (S, u, \rho)\downarrow \quad x \cdot {'\mathfrak{B}} = \Pi[u]$$

as nominated mutations. Using these we have

$$\nabla \cdot \mathbf{1} = R \cdot \mathcal{B} = ((\Gamma[y]\Gamma[x]) \cdot {'\mathfrak{B}})\uparrow = ((S\downarrow)\Pi[u])\uparrow$$

which is the left hand alternative of the two suggested intermediate derivations. Note also that $u = x$ is possible.

Continuing with the global calculation we need to remove the abnormality in $\nabla \cdot \mathbf{1}$. With $\mathfrak{a} = [x \mapsto u]\mathrm{i}$ we have $su \rhd x \cdot \mathfrak{a} = u$ so that $t^0 \Rrightarrow \lambda u : \rho . u = t^+$ (say) and

$$(\uparrow\mathbf{1}) \quad \dfrac{\dfrac{su \rhd x \cdot \mathfrak{a}}{su \Rrightarrow u}\,(1)}{t^0 \Rrightarrow \lambda u : \rho . u}\,(\uparrow)$$

organizes this reduction. Here t^+ is normal and $\square = \mathbf{1} \circ \uparrow\mathbf{1}$ organizes the full reduction $t^- \Rrightarrow t^+$ (via t^0).

We hit $\nabla \cdot \mathbf{1}$ with $\uparrow\mathbf{1}$ to get

$$\nabla \cdot \square = ((S\downarrow)\Pi[u])\uparrow \cdot \uparrow\mathbf{1} = ((S\downarrow)\Pi[u] \bullet \mathbf{1})\uparrow = (\Lambda[x] \cdot \mathfrak{A})\uparrow = (x \cdot \mathfrak{A})\uparrow$$

for a mutation

$$(\mathfrak{A}) \ \Lambda \xrightarrow{\ \mathfrak{a}\ } \Pi \quad \text{with} \quad x \cdot \mathfrak{A} = \Pi[u]$$

as the sole nominated derivation.

This gives

$$\nabla \cdot \square = \Pi[u]\uparrow$$

as the expected final result.

You should observe how Weakening appears in this calculation. It occurs in neither the initial nor the final derivation, but is used in the intermediate step. The use of this rule makes the whole algorithm easier to operate. □

The removal of an abnormality

$$(\nabla, \square) \longmapsto \nabla \cdot \square$$

can create one or more new abnormalities. At first sight it is not obvious why $\nabla \cdot \square$ is simpler than ∇. It is true that every derivation can be normalized, and it doesn't matter in which order we remove abnormalities, but the proof of this requires a bit of subtlety.

EXERCISES

6.29 Using the terms r, s, t and derivation T of Exercise 6.3, and the computations $\square$ of Exercise 6.23, calculate $T \cdot \square$.

6.30 Consider the terms $\mathsf{B}, \mathsf{C}, \mathsf{D}$ as used in Exercises 6.2, 6.6, 6.17, 6.24.

(a) Using arboreal codes write down derivations as follows.

$$(B) \quad \vdash \mathsf{B} : \beta \qquad (C) \quad \vdash \mathsf{C} : \gamma \qquad (D) \quad \vdash \mathsf{D} : \delta$$

(b) Write down a computation as follows.

$$(\square) \quad \mathsf{CB} \rhd\!\!\!\rhd \mathsf{D}$$

(c) Show that

$$(CB) \cdot \square = D$$

holds (for the right choice of B, C, D).

You should describe all the mutations used in this calculation.

7
Multi-recursive Arithmetic

7.1 Introduction

When viewed as a whole an applied λ-calculus $\boldsymbol{\lambda}\mathbf{Sig}$ is quite a complicated object. To get a better understanding of $\boldsymbol{\lambda}\mathbf{Sig}$ it is useful to dissect it into smaller parts, understand these parts, and then understand how these parts fit together to form the whole. How should we do this? Each calculus $\boldsymbol{\lambda}\mathbf{Sig}$ has three interacting facilities: derivation, computation, and the mediating substitution. We could try to understand these separately and then put them together. This doesn't help much, for it is precisely the interactions which can cause the complications, so we need to look at bits of all three facilities together. What we will do is take 'slices' across all three facilities and build up the whole calculus in layers.

Each calculus $\boldsymbol{\lambda}\mathbf{Sig}$ has a crude layering given by the type structure. Intuitively the more complicated the types involved in a construction the more complicated that construction becomes. Suppose we restrict the family of types that can be used. Can we find a less coarse measure of complexity? One idea is to pick out certain crucial terms (constants) and look at the nesting of these in any term. The deeper the nesting the more complicated the term. Of course, when we do this we should take into account the computation mechanism of the calculus. In this and Chapter 9 we look at a particular calculus $\boldsymbol{\lambda G}$ which is almost tailor made for illustrating these ideas. We will see that $\boldsymbol{\lambda G}$ has many interesting subcalculi and these fit together to form a stratification of the whole calculus where each stratum has a restricted complexity.

The calculus $\boldsymbol{\lambda G}$ is a part of a calculus with the catchy name of Gödel's T. This is a more sophisticated kind of calculus which as well as all the features of an applied λ-calculus also has some equational reasoning facilities. Thus T can name certain arithmetical gadgets *and* demonstrate that certain pairs of names have the same intended interpretation. The calculus $\boldsymbol{\lambda G}$ is a simplified version of the naming facilities of T.

To define $\boldsymbol{\lambda G}$ it suffices to give its Specifics. We give some of these here, but leave the bulk of them until Section 7.2.

The calculus $\boldsymbol{\lambda G}$ has just one atom $\mathcal{N}$ which is intended to name the set $\mathbb{N}$ of natural numbers. The types are built from $\mathcal{N}$ without the use of variables (so that in $\boldsymbol{\lambda G}$ each type is molecular). Each type τ has an orthodox interpretation

$[\![\tau]\!]$ in the full type hierarchy over $\mathbb{N}$. Thus

$$[\![\mathcal{N}]\!] = \mathbb{N} \qquad [\![\sigma \to \rho]\!] = [\![\sigma]\!] \longrightarrow [\![\rho]\!]$$

generates these interpretations.

There are two constants with housing axioms $\mathsf{0} : \mathcal{N}$ and $\mathsf{S} : \mathcal{N}'$ where these are intended to name zero and the successor function, respectively. There are other constants (and housing axioms) but we need not worry about these just yet.

With these two constants we easily generate a canonical name for each natural number.

7.1 DEFINITION. For each $m \in \mathbb{N}$ we set $\ulcorner m \urcorner = \mathsf{S}^m\mathsf{0}$ to produce the **numeral** for m. Thus

$$\ulcorner 0 \urcorner = \mathsf{0} \qquad \ulcorner m' \urcorner = \mathsf{S}\ulcorner m \urcorner$$

is an iterative generation of these numerals. □

Using these constants we can produce some derivations which do not exist in $\boldsymbol{\lambda}\boldsymbol{\emptyset}$. In particular, we have $\vdash \ulcorner m \urcorner : \mathcal{N}$ and there is a unique such derivation. You should write down a description of this and observe its shape. There will be other inhabitants of $\mathcal{N}$ (in the empty context), but it will turn out that the numerals are the only normal inhabitants, and every non-normal inhabitant reduces to a numeral.

Of course, we can't prove these assertions just yet because they are concerned with the reduction mechanism of $\boldsymbol{\lambda G}$. We introduce a notion which relates to this mechanism, and which can be used to show how powerful (or weak) the mechanism is.

We are interested in k-placed numerical functions. To handle these we generate the 'first order' types together with their orthodox interpretations. Thus we set

$$\mathcal{N}[0] = \mathcal{N} \quad \mathcal{N}[k'] = \mathcal{N} \to \mathcal{N}[k] \qquad \mathbb{N}[0] = \mathbb{N} \quad \mathbb{N}[k'] = \mathbb{N} \longrightarrow \mathbb{N}[k]$$

for each $k < \omega$. The cases $k = 1$ and $k = 2$ will occur most frequently. For each type σ we set

$$\sigma' = \sigma \to \sigma \qquad \sigma^+ = \sigma \to \mathcal{N} \to \sigma \qquad \text{so that} \qquad \mathcal{N}[1] = \mathcal{N}' \quad \mathcal{N}[2] = \mathcal{N}^+$$

and $\mathcal{N}^+$ is the type of all binary numerical operations. (See Table 1.2 on page 7 for a discussion of the notation σ'. The more general notation σ^+ will be useful later.)

7.2 DEFINITION. A term $\ulcorner f \urcorner$ **represents** a k-placed function $f : \mathbb{N}[k']$ if

$$\vdash \ulcorner f \urcorner : \mathcal{N}[k'] \quad \text{and} \quad \ulcorner f \urcorner \ulcorner m_k \urcorner \cdots \ulcorner m_1 \urcorner \ggg \ulcorner f m_k \cdots m_1 \urcorner$$

for all $m_k, \ldots, m_1 \in \mathbb{N}$. □

For instance, for each $k \in \mathbb{N}$ the term

$$\lambda x : \mathcal{N} . \mathsf{S}^k x$$

represents the translation $x \longmapsto x + k$ on $\mathbb{N}$. We use the reflexive version of the reduction relation to avoid a couple of silly exceptions; in most cases ⊳⊳⊳ can be replaced by ⊳⊳.

As is shown in Exercise 7.4, with only the constants $\mathsf{0}$ and S no more than a limited range of rather uninteresting functions can be represented in $\boldsymbol{\lambda G}$. We need to add more constants to name other functions.

So far the only arithmetical operations we have used have been minor variants of the

Successor, Addition, Multiplication, Exponentiation, Stacking

operations, i.e. the operations S, A, M, E, B of 1, 2, 2, 2, and 3 arguments where

$$Sn = 1 + n \quad Anm = n + m \quad Mnm = n \times m \quad Enm = n^m \quad Bknm = \beth(k, n, m)$$

for all $m, n, k \in \mathbb{N}$. Intuitively these five operations are listed in order of increasing complexity, e.g. multiplication is more complicated than addition but less complicated than exponentiation. How can we make this idea of complexity precise, and is there anything beyond $\beth$? This is the central question of this and the remaining chapters. The calculus $\boldsymbol{\lambda G}$ is custom built to answer this kind of question.

The crucial idea is not to concentrate on the functions but on the operators which take us from one function to the next. We find that, apart from a slight hiccup at the beginning, there is a single operator which does this job.

For any set $\mathbb{S}$, any two functions of the type $\mathbb{S}'$ can be composed. In particular, for each $f : \mathbb{S}'$ and $m \in \mathbb{N}$, we can form the m-fold iterate

$$f^m = f \circ f \circ \cdots \circ f$$

of f (where there are m occurrences of f). We may evaluate this at any $s \in \mathbb{S}$ to produce $f^m s \in \mathbb{S}$, and any λ-calculus can describe this kind of compound.

Now observe that

$$Anm = S^m n \qquad Mnm = (An)^m 0 \qquad Enm = (Mn)^m 1 \qquad Bknm = (En)^m k$$

for all $m, n, k \in \mathbb{N}$. Here the 2-placed functions are viewed in curried form, that is they receive the arguments one after the other, not as a pair. In particular, An, Mn, En are all members of $\mathbb{N}'$ for each $n \in \mathbb{N}$. The functions A, M, E, B can be obtained from S by nested uses of iteration. Furthermore, the nesting depth is a measure of the intuitive complexity of the functions.

This suggests how to continue the sequence of functions. We use the construction that appeared in Exercise 2.12.

7.3 DEFINITION. For a 2-placed function $F : \mathbb{N}^+$ the function $F' : \mathbb{N}^+$ given by

$$F'nm = (Fn)^m 1$$

for all $m, n \in \mathbb{N}$ is called the Grzegorczyk jump of F. □

For instance $M' = E$, $E' = B1$ and it is easy to repeat this jump to produce new functions. More generally, for any function $F : \mathbb{N}^+$ we can form a chain of functions

$$\boldsymbol{F} = (F^{(i)} \mid i < \omega) \quad \text{by} \quad F^{(0)} = F \quad F^{(i')} = F^{(i)\prime}$$

for each $i < \omega$. In general F' is more complicated than F, so that $\boldsymbol{F}$ is a chain of 2-placed functions of ever increasing complexity. This $\boldsymbol{F}$ is called the 2-placed **Grzegorczyk hierarchy** on F. (Later we will construct a more amenable 1-placed version of this.)

The construction of $\boldsymbol{F}$ is just another use of iteration. We view the Grzegorczyk jump as an operator $\Gamma : \mathbb{N}^{+\prime}$ (where $F' = \Gamma F$) and then we have $F^{(i)} = \Gamma^i F$ for each $i < \omega$. We will use this kind of trick again and again to produce some very complicated functions.

The intention is that many functions can be represented in $\boldsymbol{\lambda G}$. To do that we certainly need more facilities. What other functions should be named by constants? It is tempting to throw in names for addition, multiplication, ... , but this won't get us very far. The insight gained from the Grzegorczyk hierarchy is that many functions can be generated from comparatively few functions using higher order gadgets which convert functions into other functions. We will concentrate on having names for these operators rather than names for more first order functions.

Exercises

7.1 For each $m \in \mathbb{N}$ write down the derivation $\nabla(m)$ of $\vdash \ulcorner m \urcorner : \mathcal{N}$ in abbreviated notation. Indicate the shape of $\nabla(m)$.

7.2 Find two different normal representations of the successor function on $\mathbb{N}$.

7.3 Given functions $\psi : \mathbb{N}[k']$ and $\theta_1, \dots, \theta_k : \mathbb{N}[l']$ we define the composite

$$\phi = \psi \circ (\theta_1, \dots, \theta_k) : \mathbb{N}[l'] \quad \text{by} \quad \phi \mathsf{x} = \psi(\theta_1 \mathsf{x}, \dots, \theta_k \mathsf{x})$$

for each sequence $\mathsf{x} = (x_1, \dots, x_k)$ from $\mathbb{N}$. Show that if $\psi, \theta_1, \dots, \theta_k$ are representable then so is ϕ.

7.4 Consider the minimal subcalculus of $\boldsymbol{\lambda G}$ which has $\mathsf{0}$ and S as the only constants with the corresponding housing axioms and no reduction axioms. Suppose a function $f : \mathbb{N}'$ is represented in this calculus by a normal term $\ulcorner f \urcorner$. What can the function f and the term $\ulcorner f \urcorner$ be?

7.2 The specifics of $\boldsymbol{\lambda G}$

The full calculus $\boldsymbol{\lambda G}$ and each of its subcalculi are determined by the corresponding Language, Derivation, and Computation specifics. We have seen some of these already, but now we can give a full description.

(**Language**) The only atom is $\mathcal{N}$ and variables are not used (so that each type is molecular). The constants are $\mathsf{0}, \mathsf{S}$, and I_σ for each of a selected family of types σ. We call I_σ the **iterator** over σ.

(Derivation) The housing axioms are $0 : \mathcal{N}$, $\mathsf{S} : \mathcal{N}'$ and $\mathsf{I}_\sigma : \iota_\sigma$ where $\iota_\sigma = \sigma' \to \sigma^+$ for each selected type σ.

(Computation) For each iterator I_σ there are two reduction axioms

$$\mathsf{I}_\sigma ts0 \rhd s \qquad \mathsf{I}_\sigma ts(\mathsf{S}u) \rhd t(\mathsf{I}_\sigma tsu)$$

for arbitrary terms t, s, u. Of course, these will be used only when these terms are suitably related (by type).

The full calculus $\boldsymbol{\lambda G}$ has an iterator I_σ for each type σ. Each subcalculus has a restricted class of iterators. In practice we almost always use I_σ where σ is one of $\mathcal{N}, \mathcal{N}', \mathcal{N}'', \mathcal{N}''', \ldots$, and in this chapter only the first three of these are needed.

We have yet to describe the computation recipe $\nabla \longmapsto \nabla \bullet 0$ associated with each reduction axiom. We do that at the end of this section. Before that let's look at some simple derivations and computations.

As explained in Section 7.1, the idea is that $\mathcal{N}$ names $\mathbb{N}$, and then 0 names 0 and S names the successor function. We want to use the calculus to name many other arithmetical gadgets. We already have names, the numerals, for natural numbers.

7.4 EXAMPLE. For each $m \in \mathbb{N}$ we have $\vdash \ulcorner m \urcorner : \mathcal{N}$ and there is a unique such derivation. For all terms t, s we have

$$(\square(I, m)) \quad \mathsf{I}_\sigma ts \ulcorner m \urcorner \mathrel{\rhd\!\!\!\rhd} t^m s$$

for any relevant iterator I_σ. □

These numerals play a distinguished role. Using the Type Inhabitation Lemma 6.19 as in Exercise 7.9, we obtain the following.

7.5 LEMMA. *Each normal term t with $\vdash t : \mathcal{N}$ is a numeral.*

What can we say about derivations $\vdash t : \mathcal{N}'$ or, more generally, $\vdash t : \mathcal{N}[k']$ with t normal? Not a lot, or almost anything, depending on your point of view. Let's look at some simple examples first.

7.6 EXAMPLE. For convenience let $\mathsf{I} = \mathsf{I}_\mathcal{N}$ and set

$$\mathsf{Add} = \mathsf{IS} \qquad \mathsf{Mlt} = \lambda v : \mathcal{N} \,.\, \mathsf{I}(\mathsf{Add}v)0$$

to produce two normal terms. The derivation

$$\dfrac{\dfrac{\dfrac{v : \mathcal{N} \vdash \mathsf{I} : \mathcal{N}' \to \mathcal{N}^+ \qquad \dfrac{\dfrac{v : \mathcal{N} \vdash \mathsf{I} : \mathcal{N}' \to \mathcal{N}^+ \quad v : \mathcal{N} \vdash \mathsf{S} : \mathcal{N}'}{v : \mathcal{N} \vdash \mathsf{Add} : \mathcal{N}^+} \qquad v : \mathcal{N} \vdash v : \mathcal{N}}{\bullet}}{\bullet} \qquad v : \mathcal{N} \vdash 0 : \mathcal{N}}{v : \mathcal{N} \vdash \mathsf{I}(\mathsf{Add}v)0 : \mathcal{N}'}}{\vdash \mathsf{Mlt} : \mathcal{N}^+}$$

indicates that $\vdash \mathsf{Add} : \mathcal{N}^+$ and $\vdash \mathsf{Mlt} : \mathcal{N}^+$. It is easy to generate computations

$$(\square(Add, n, m)) \ \mathsf{Add}\ulcorner n\urcorner\ulcorner m\urcorner \rhd\!\!\!\!\rhd \ulcorner n+m\urcorner \qquad (\square(Mlt, n, m)) \ \mathsf{Mlt}\ulcorner n\urcorner\ulcorner m\urcorner \rhd\!\!\!\!\rhd \ulcorner nm\urcorner$$

where the shape depends on both inputs. $\square$

What we have here are two terms Add, Mlt which represent the two operations *Add* and *Mlt* in the sense of Definition 7.2. It is quite easy to see that all the functions of the Grzegorczyk hierarchy are representable in $\boldsymbol{\lambda G}$ (where only $\mathsf{I} = \mathsf{I}_{\mathcal{N}}$ is needed for these representations). More complicated gadgets can also be named.

7.7 EXAMPLE. For convenience let $\mathsf{I} = \mathsf{I}_{\mathcal{N}}$, $\mathsf{I}^+ = \mathsf{I}_{\mathcal{N}^+}$, $1 = \ulcorner 1\urcorner$ and set

$$\mathsf{Grz} = \lambda w : \mathcal{N}^+, v : \mathcal{N} \,.\, \mathsf{I}(wv)1 \qquad \mathsf{GRZ} = \lambda w : \mathcal{N}^+, v : \mathcal{N} \,.\, \mathsf{I}^+\mathsf{Grz}wv$$

to produce two normal terms. Almost trivially we have both $\vdash \mathsf{Grz} : \mathcal{N}^{+\prime}$ and $\vdash \mathsf{GRZ} : \mathcal{N}^{++}$ (where you should write down appropriate derivations). If the term $\vdash \ulcorner F\urcorner : \mathcal{N}^+$ represents the 2-placed function $F : \mathbb{N}^+$, then we can generate computations

$$\begin{array}{ll} (\square(Grz, F, n, m)) & \mathsf{Grz}\ulcorner F\urcorner\ulcorner n\urcorner\ulcorner m\urcorner \rhd\!\!\!\!\rhd \ulcorner F'nm\urcorner \\ (\square(Grz, F, i, n, m)) & \mathsf{GRZ}\ulcorner F\urcorner\ulcorner i\urcorner\ulcorner n\urcorner\ulcorner m\urcorner \rhd\!\!\!\!\rhd \ulcorner F^{(i)}nm\urcorner \end{array}$$

for $m, n, i \in \mathbb{M}$. You should organize these computations. $\square$

To conclude this section we complete the missing details of the computation specifics, i.e. for each reduction axiom we describe a recipe $\nabla \longmapsto \nabla \bullet 0$ which moves across that reduction.

Consider first a derivation

$$(\nabla) \ \ \Gamma \vdash t^- : \xi$$

where $t^- = \mathsf{I}_\sigma tsr$ for some terms t, s, r, context Γ, and type ξ. What can ∇ look like? It doesn't take too long to see there is a sequence of shortenings $\Gamma, \Gamma^r, \Gamma^s, \Gamma^t$ of Γ and derivations

$$(I) \ \ \Gamma^t \vdash \mathsf{I}_\sigma : \iota_\sigma \qquad (T) \ \ \Gamma^t \vdash t : \phi \qquad (S) \ \ \Gamma^s \vdash s : \psi \qquad (R) \ \ \Gamma^r \vdash r : \theta$$

with

$$\nabla = (((((IT)\Downarrow)S)\Downarrow)R)\Downarrow$$

for appropriate Weakenings $\Downarrow$. From this we may check that the types involved are $\phi = \sigma'$, $\psi = \sigma$, $\theta = \mathcal{N}$, and $\xi = \sigma$ for some type σ. We now look at the two possible atomic reductions where $r = 0$ or $r = \mathsf{S}u$ for some term u.

When $r = 0$ we have $t^- = \mathsf{I}_\sigma ts0$ and $t^+ = s$ is the required result with

$$(\nabla) \ \ \Gamma \vdash t^- : \sigma \qquad (\nabla \bullet 0) \ \ \Gamma \vdash s : \sigma$$

as the two derivations. Thus $\nabla \bullet 0 = S\Downarrow$ will do (where the Weakenings $\Downarrow$ build up Γ from Γ^s).

When $r = \mathsf{S}u$ we have $t^- = \mathsf{I}_\sigma ts(\mathsf{S0})$ and $t^+ = t(\mathsf{I}_\sigma tsu)$ is the required result with

$$(\nabla)\ \Gamma \vdash t^- : \sigma \qquad (\nabla \bullet \mathsf{0})\ \Gamma \vdash t^+ : \sigma$$

as the two derivations. We find that

$$R = ((Suc)U)\Downarrow \quad \text{where} \quad (Suc)\ \Gamma^u \vdash \mathsf{S} : \mathcal{N}' \qquad (U)\ \Gamma^u \vdash u : \mathcal{N}$$

for some shortening Γ^u of Γ^r with a matching Weakening $\Downarrow$. We can Weaken each of I, T, S, U to produce

$$(I^\Downarrow)\ \Gamma \vdash \mathsf{I}_\sigma : \iota_\sigma \qquad (T^\Downarrow)\ \Gamma \vdash t : \sigma' \qquad (S^\Downarrow)\ \Gamma \vdash s : \sigma \qquad (U^\Downarrow)\ \Gamma \vdash u : \mathcal{N}$$

and then

$$\nabla \bullet \mathsf{0} = T^\Downarrow(I^\Downarrow T^\Downarrow S^\Downarrow U^\Downarrow)$$

will do. There are other possible choices for $\nabla \bullet \mathsf{0}$, for we may delay some of the Weakenings until nearer the root. The various effects of the different choices of $\nabla \bullet \mathsf{0}$ need not worry us here.

Exercises

7.5 Show how to generate a computation

$$(\Box(\mathsf{I}_\sigma, m))\ \ \mathsf{I}_\sigma ts\ulcorner m \urcorner \rhd\!\!\!\rhd t^m s$$

for arbitrary terms t, s. How does this depend on σ?

7.6 Let σ be an arbitrary type and consider the terms

$$r = \lambda x : \sigma \,.\, x \qquad s = \lambda y : \sigma^+ \,.\, \mathsf{I}(\lambda x : \sigma \,.\, yx\ulcorner 1 \urcorner) \qquad t = \mathsf{I}^+ s(\mathsf{I}r)\ulcorner 1 \urcorner$$

where $\mathsf{I} = \mathsf{I}_\sigma$ and $\mathsf{I}^+ = \mathsf{I}_{\sigma^+}$.

(a) Determine the type τ such that $\vdash t : \tau$ is derivable, and exhibit the shape of such a derivation ∇.

(b) Reduce t to a normal form t^* and write down, in abbreviated notation, a computation $\Box$ which organizes a reduction $t \rhd\!\!\!\rhd t^*$.

(c) Write down a derivation

$$(\nabla^*) \quad \vdash t^* : \tau$$

and compare its shape with that of ∇.

(d) Indicate how the calculation of $\nabla \cdot \Box$ (to value ∇^*) proceeds.

7.7 Referring to Example 7.6, show how to generate

$$(\Box(Add, n, m))\ \ \mathsf{Add}\ulcorner n \urcorner \ulcorner m \urcorner \rhd\!\!\!\rhd \ulcorner m + n \urcorner \qquad (\Box(Mlt, n, m))\ \ \mathsf{Mlt}\ulcorner n \urcorner \ulcorner m \urcorner \rhd\!\!\!\rhd \ulcorner mn \urcorner$$

(for each $m, n \in \mathbb{N}$).

7.8 Describe the computations $\square(Grz, F, n, m)$ and $\square(GRZ, F, i, n, m)$ given in Example 7.7. You may, of course, refer to computations

$$(\square(F, n, m)) \quad \ulcorner F \urcorner \ulcorner n \urcorner \ulcorner m \urcorner \rhd\!\!\rhd \ulcorner Fnm \urcorner$$

(which depend on F).

7.9 Prove Lemma 7.5, i.e. show that for each derivation $\vdash n : \mathcal{N}$ with a normal subject n, there is some $m \in \mathbb{N}$ with $n = \ulcorner m \urcorner$. Can you modify your argument to characterize those normal terms f such that $\vdash f : \mathcal{N}'$ is derivable?

7.10 The usual formulation of $\boldsymbol{\lambda G}$ uses a recursor with housing axiom $\mathsf{R}_\sigma : \sigma^{+\prime}$ (for arbitrary σ) in place of the iterator I_σ. The reduction axioms for R_σ are

$$\mathsf{R}_\sigma ws\mathsf{0} \rhd s \qquad \mathsf{R}_\sigma ws(\mathsf{S}r) \rhd w(\mathsf{R}_\sigma wsr)r$$

for all terms w, s, r. Find terms $\vdash \mathsf{B}_\sigma : \sigma^{+\prime} \to \iota_\sigma'$ and $\vdash \mathsf{A}_\sigma : \iota_\sigma$ and such that $\mathsf{B}_\sigma \mathsf{R}_\sigma \mathsf{A}_\sigma$ has the same computation properties as I_σ.

7.11 Suppose we have access to product, pairing, and projection gadgets. Arguing informally in the type hierarchy over $\mathbb{N}$, show that each use of the recursor R_σ can be replaced by a use of the iterator I_τ where $\tau = \sigma \times \mathbb{N}$.

7.3 Forms of recursion and induction

The calculus $\boldsymbol{\lambda G}$ is designed to name many different functions, both first order and higher order. Many of these functions $\phi : \mathbb{R} \longrightarrow \mathbb{S}$ will have a **recursive specification**, i.e. each value ϕr (for $r \in \mathbb{R}$) will be determined in a uniform way from certain 'earlier' values ϕr^- for certain $r^- \in \mathbb{R}$ 'smaller' than r. We will show how simpler forms of recursion can be used to simulate more complicated versions. To begin let's look at some of the more commonplace simple recursions.

Thus with the set $\mathbb{N}$ of natural numbers let $\mathbb{S}$ be a set of target values, let $\mathbb{P}$ be a set of parameters, and put $\mathbb{F} = \mathbb{P} \longrightarrow \mathbb{S}$ to form a function space. We are interested in constructions which produce functions $\phi : \mathbb{N} \longrightarrow \mathbb{F}$ out of various supplied 'simpler' data functions. We are particularly interested in such functions ϕ which are specified recursively over the distinguished first argument. Thus the value $\phi 0$ is given outright and then, for each $r \in \mathbb{N}$, the value $\phi r'$ is obtained from the previous values $\phi 0, \ldots, \phi r$ using some predetermined recipe. It is the nature of this recipe that we must look at.

7.8 **DEFINITION**. The function $\phi : \mathbb{N} \longrightarrow \mathbb{F}$ is obtained from the data functions

$$\theta : \mathbb{F} \qquad \psi : \mathbb{N} \longrightarrow \mathbb{S} \longrightarrow \mathbb{F} \qquad \kappa : \mathbb{N} \longrightarrow \mathbb{P}'$$

by an immediate use of a **body recursion** if

$$\phi 0 p = \theta p \qquad \phi r' p = \psi r s p \quad \text{where} \quad s = \phi r p^+ \quad \text{where} \quad p^+ = \kappa r p$$

holds for all $r \in \mathbb{N}, p \in \mathbb{P}$.

The function $\phi : \mathbb{N} \longrightarrow \mathbb{F}$ is obtained by an immediate use of a

head recursion   tail recursion

from the functions

$$\theta : \mathbb{F} \quad \psi : \mathbb{N} \longrightarrow \mathbb{S} \longrightarrow \mathbb{F} \qquad \theta : \mathbb{F} \quad \kappa : \mathbb{N} \longrightarrow \mathbb{P}'$$

respectively, if it is obtained by a body recursion using the projection

$$\kappa r p = p \qquad \psi r s p = s$$

as the third required data function. □

In the case where $\mathbb{P} = \mathbb{N}^k$ and $\mathbb{S} = \mathbb{N}$, a head recursion is just a primitive recursion and a body recursion is a primitive recursion with variation of parameters. When the types $\mathbb{P}$ or $\mathbb{S}$ are higher order these recursions have hidden power. Once we start to use higher types we can simplify the form of the recursion considerably.

7.9 DEFINITION. The function $\phi : \mathbb{N} \longrightarrow \mathbb{S}$ is obtained by an immediate use of a parameter-free head recursion, or a G(ödel) recursion for short, from the initial value and data function

$$s : \mathbb{S} \quad \psi : \mathbb{N} \longrightarrow \mathbb{S}' \quad \text{if} \quad \phi 0 = s \quad \phi r' = \psi r(\phi r)$$

holds for all $r \in \mathbb{N}$. □

This kind of recursion (and the previous kinds for that matter) should not be viewed as an action but as a higher order function $R_{\mathbb{S}}$ which when supplied with the input data s and ψ will return the function ϕ. What is the type of this function $R_{\mathbb{S}}$? The two arguments can be supplied in one of two ways. Here it is convenient to have

$$R_{\mathbb{S}} : (\mathbb{N} \longrightarrow \mathbb{S}') \longrightarrow \mathbb{S}^+ \quad \text{so that} \quad \phi = R_{\mathbb{S}}\psi s$$

as the constructed function. You will find it instructive to write down the value ϕr (for $r \in \mathbb{N}$) directly in terms of $R_{\mathbb{S}}$. (The other order is used in Exercise 7.10.)

This function $R_{\mathbb{S}}$ is called the recursor over $\mathbb{S}$. The usual formulation of Gödel's T uses such recursors (hence the name). It can be checked, as in Exercise 7.12, that such recursors are powerful enough to capture all body recursions (but there is a hidden cost when doing this). Here we do not need the full power of these recursors, so we use a simplified version.

The input data function ψ for $R_{\mathbb{S}}$ may be insensitive to its first argument. In this case we omit that argument and use a data function $\psi : \mathbb{S}'$ to produce ϕ where

$$\phi r = \psi^r s$$

for all $r \in \mathbb{N}$ (and $s \in \mathbb{S}$). Thus ϕr is just the r-fold iterate of ψ evaluated at s.

7.10 DEFINITION. The iterator over $\mathbb{S}$ is the function

$$I_{\mathbb{S}} : \mathbb{S}' \longrightarrow \mathbb{S}^{+} \qquad \text{given by} \qquad I_{\mathbb{S}}\psi s0 = s \quad I_{\mathbb{S}}\psi sr' = \psi(I_{\mathbb{S}}\psi sr)$$

for all $\psi : \mathbb{S}', s \in \mathbb{S}, r \in \mathbb{N}$. □

The full calculus $\boldsymbol{\lambda G}$ has a name for each possible iterator. More generally, $\boldsymbol{\lambda G}$ has many subcalculi each determined by those iterators that are named. If you look at the Specifics of $\boldsymbol{\lambda G}$ you will will see that for each type σ there is a constant $\mathsf{I}_\sigma : \iota_\sigma$ with reductions axioms that ensure

$$\mathsf{I}_\sigma t s \ulcorner m \urcorner \rhd\!\!\!\rhd t^m s$$

for terms s, t and $m \in \mathbb{N}$. The intention is that if σ names $\mathbb{S}$ ($[\![\sigma]\!] = \mathbb{S}$) then I_σ names $I_{\mathbb{S}}$.

We are interested in a subcalculus of $\boldsymbol{\lambda G}$ determined by a set Σ of types: the calculus where I_σ can be used only if $\sigma \in \Sigma$. In this chapter we will look at the case $\Sigma = \{\mathcal{N}, \mathcal{N}'\}$, and then at the more general case in Chapter 9.

The recursions considered so far have all been over a single recursion argument (which is a natural number). There are other kinds of recursion arguments.

7.11 DEFINITION. A multi-index of length s is a list

$$\mathsf{i} = (i_s, \ldots, i_0)$$

of $s+1$ natural numbers. Let $\mathbb{M}_s$ be the set of all such i for this s. □

Thus

$$\mathbb{M}_0 = \mathbb{N},\ \mathbb{M}_1 = \mathbb{N} \times \mathbb{N},\ \mathbb{M}_2 = \mathbb{N} \times \mathbb{N} \times \mathbb{N},\ \ldots$$

where we always add the next component at the left hand end. Each $\mathbb{M}_s$ carries a natural comparison $\leq_s$. To compare

$$\mathsf{i} = (i_s, \ldots, i_0) \qquad \mathsf{j} = (j_s, \ldots, j_0)$$

we read both from left to right to find the leftmost difference, and then we order according to that difference. Thus $\leq_0$ is just the natural ordering on $\mathbb{N}$. Furthermore, for indexes (i, i) and (j, j) in $\mathsf{M}_{s'}$ (with $\mathsf{i}, \mathsf{j} \in \mathbb{M}_s$) we have

$$(i, \mathsf{i}) \leq_{s'} (j, \mathsf{j}) \Longleftrightarrow i < j \text{ or } (i = j \text{ and } \mathsf{i} \leq_s \mathsf{j})$$

so we may generate $\leq_s$ by recursion on s. This construction can be used to show that $\leq_s$ is a linear comparison with other nice properties.

We can even use the set $\mathbb{M}$ of all multi-indexes of varying length. To compare two such indexes i, j we simply fill out the left hand end of one or other until we have indexes of the same length. We then compare these two indexes.

We often meet a function

$$F : \mathbb{M}_s \longrightarrow \mathbb{A}$$

$\mathbb{M}_3$-recursion

$$
\begin{array}{llcl}
(\bot) & F(0,0,0,0) & = & \text{an explicit value} \\
(0) & F(k,j,i,r') & = & \text{a compound of the values } F(k,j,i,0),\dots,F(k,j,i,r) \\
(1) & F(k,j,i',0) & = & \text{a value obtained from a compound of} \\
 & & & \text{the 1-placed functions} \quad F(k,j,0,\cdot),\dots,F(k,j,i,\cdot) \\
(2) & F(k,j',0,0) & = & \text{a value obtained from a compound of} \\
 & & & \text{the 2-placed functions} \quad F(k,0,\cdot,\cdot),\dots,F(k,j,\cdot,\cdot) \\
(3) & F(k',0,0,0) & = & \text{a value obtained from a compound of} \\
 & & & \text{the 3-placed functions} \quad F(0,\cdot,\cdot,\cdot),\dots,F(k,\cdot,\cdot,\cdot)
\end{array}
$$

Table 7.1: A typical multi-recursion scheme

for some $s \in \mathbb{N}$ and set $\mathbb{A}$. Many such functions can be specified by recursion over the multi-index arguments using the ordering $\leq_s$. The general idea can be seen from the case $s = 3$.

We are interested in a function $F : \mathbb{N}^4 \longrightarrow \mathbb{A}$ with typical value $F(k,j,i,r)$ for $r,i,j,k \in \mathbb{N}$. To specify such a function we use five clauses as given in Table 7.1. In general the $\mathbb{M}_s$-recursion scheme will use $s+2$ clauses to specify such a function $F : \mathbb{M}_s \longrightarrow \mathbb{A}$ of $s+1$ argument places each of which is a natural number. The reason for using the subscript 's' (rather than '$s+1$' or '$s+2$') will become clear in Chapter 9.

Such highly nested forms of recursion can produce some extremely complicated functions. Here we employ comparatively simple forms of multi-recursion, where the right hand side of each clause uses only a small part of the available data in a straightforward fashion. We will see that even this can produce functions of unexpected complexity.

You should be able to write down the $\mathbb{M}_s$-recursion scheme for an arbitrary s. Note the the $\mathbb{M}_0$-recursion scheme produces a function

$$
\phi : \mathbb{N} \longrightarrow \mathbb{A} \quad \text{where} \quad
\begin{array}{ll}
(\bot) & \phi 0 = \text{an explicit value} \\
(0) & \phi r' = \text{a compound of the values } \phi 0,\dots,\phi r
\end{array}
$$

which is the general form of recursion over $\mathbb{N}$ described at the beginning of this section. Just as this general $\mathbb{M}_0$-recursion has some important particular instances, so does the $\mathbb{M}_s$-recursion scheme.

When $\mathbb{A}$ is a numeric space $\mathbb{N}[k]$ for some k, the $\mathbb{M}_s$-recursion scheme can be restricted to what is known as s'-recursion. Thus 1-recursion is just primitive recursion. The 2-recursion scheme is a rather simple nested recursion more powerful than 1-recursion. We won't need the precise definition of s'-recursion here. We will meet some particular examples, and these should give a small glimpse of the larger picture.

Any recursion scheme has an associated form of induction. Consider a 4-placed

Abbreviations		Requirements	
$\langle k,j,i,r\rangle$	for $\cdots F(k,j,i,r)\cdots$	$[\bot]$	$\langle 0,0,0,0\rangle$
$\langle k,j,i\rangle$	for $(\forall r:\mathbb{N})\langle k,j,i,r\rangle$	$[0]$	$\langle k,j,i,0\rangle,\dots,\langle k,j,i,r\rangle \Rightarrow \langle k,j,i,r'\rangle$
$\langle k,j\rangle$	for $(\forall i:\mathbb{N})\langle k,j,i\rangle$	$[1]$	$\langle k,j,0\rangle,\dots\dots,\langle k,j,i\rangle \Rightarrow \langle k,j,i'\rangle$
$\langle k\rangle$	for $(\forall j:\mathbb{N})\langle k,j\rangle$	$[2]$	$\langle k,0\rangle,\dots\dots\dots,\langle k,j\rangle \Rightarrow \langle k,j'\rangle$
$\langle\rangle$	for $(\forall k:\mathbb{N})\langle k\rangle$	$[3]$	$\langle 0\rangle,\dots\dots\dots,\langle k\rangle \Rightarrow \langle k'\rangle$

Table 7.2: A typical multi-induction scheme

function F as specified in Table 7.1 and suppose we wish to verify a property

$$(\forall k,j,i,r:\mathbb{N})[\cdots F(k,j,i,r)\cdots]$$

where the body $\cdots F(k,j,i,r)\cdots$ says something about the indicated value of F. How can we prove such an assertion? As with 1-placed recursion and the associated induction, 4-placed induction tracks the parent recursion. At first sight this looks rather complicated, but with a little practice it can become quite routine.

It is convenient to introduce some abbreviations as on the left of Table 7.2. With these $\langle\rangle$ is the target assertion. To prove this it suffices to verify one instance and four implications as in the right of the table. Of course, these must be proved for arbitrary k,j,i,r. It is not immediately obvious that these suffice to prove $\langle\rangle$, so you need to think about this. If you are not familiar with this kind of induction then you should spend quite some time on the following exercises.

EXERCISES

7.12 Show that each body recursion can be achieved using a G-recursion. This involves re-formatting the data functions. Can you see any objections to this?

7.13 (i) Show that for each s the comparison $\leq_s$ on $\mathbb{M}_s$ is linear.

(ii) Show that each $\leq_s$ is a well ordering.

(iii) Describe the immediate successor of each $\mathsf{i}\in\mathbb{M}_s$.

(iv) Show that for $s>0$ there are infinitely many members of $\mathbb{M}_s$ which do not have an immediate predecessor.

7.14 Show that the $\mathbb{M}_3$-induction scheme does verify the target assertion $\langle\rangle$.

7.15 An $\mathbb{M}_3$-recursion often produces the clauses shown right rather than the official clauses $[\bot,0,1,2,3]$. Show that these new clauses $(0.\bot,\dots,0.3)$ collectively imply the old clauses.

$(0.\bot)$	$\langle 0,0,0,0\rangle$		
(0.0)	$\langle k,j,i,r\rangle$	$\Rightarrow$	$\langle k,j,i,r'\rangle$
(0.1)	$\langle k,j,i\rangle$	$\Rightarrow$	$\langle k,j,i',0\rangle$
(0.2)	$\langle k,j\rangle$	$\Rightarrow$	$\langle k,j',0,0\rangle$
(0.3)	$\langle k\rangle$	$\Rightarrow$	$\langle k',0,0,0\rangle$

The next six exercises use $\mathbb{M}_3$-recursion to construct functions with targets $\mathbb{N}, \mathbb{N}', \mathbb{N}'', \mathbb{N}'''$, ending with a general format for an arbitrary target set $\mathbb{A}$.

7.16 The function $F : \mathbb{N}^4 \longrightarrow \mathbb{N}$ is specified to the right. Show there are constants a, b, c, d, e such that

$$F(k,j,i,r) = ak + bj + ci + dr + e$$

holds for all $k, j, i, r \in \mathbb{N}$.

$$\begin{array}{lll}
(\bot) & F(0,0,0,0) & = 0 \\
(0) & F(k,j,i,r') & = F(k,j,i,r) + 1 \\
(1) & F(k,j,i',0) & = F(k,j,i,1) + 1 \\
(2) & F(k,j',0,0) & = F(k,j,1,0) + 2 \\
(3) & F(k',0,0,0) & = F(k,1,2,0)
\end{array}$$

7.17 Let $Zero : \mathbb{N}'$ be the function with $Zero\, x = 0$ for all $x \in \mathbb{N}$. For the function $F : \mathbb{N}^4 \longrightarrow \mathbb{N}'$ given to the right, show that

$$F(k,j,i,r)x = x^3k + x^2j + xi + r$$

(for $k, j, i, r, x \in \mathbb{N}$) taking care to set up and explain the induction steps.

$$\begin{array}{lll}
(\bot) & F(0,0,0,0) & = Zero \\
(0) & F(k,j,i,r') & = Suc \circ F(k,j,i,r) \\
(1) & F(k,j,i',0)x & = F(k,j,i,x)x \\
(2) & F(k,j',0,0)x & = F(k,j,x,0)x \\
(3) & F(k',0,0,0)x & = F(k,x,0,0)x
\end{array}$$

7.18 Given functions $f, g, h : \mathbb{N}'$, we specify $G : \mathbb{N}^4 \longrightarrow \mathbb{N}'$ as shown. Produce a function $H : \mathbb{N}^4 \longrightarrow \mathbb{N}'$ depending on h but not on f, g with

$$G(k,j,i,r)x = (f^e \circ g \circ h^e)x$$

for all k, j, i, r, x where $e = H(k,j,i,r)x$.

$$\begin{array}{lll}
(\bot) & G(0,0,0,0) & = g \\
(0) & G(k,j,i,r') & = f \circ G(k,j,i,r) \circ h \\
(1) & G(k,j,i',0)x & = G(k,j,i,1)x \\
(2) & G(k,j',0,0)x & = G(k,j,x,1)x \\
(3) & G(k',0,0,0)x & = G(k,x^2,x,2)x
\end{array}$$

7.19 For the function $L : \mathbb{N}^4 \longrightarrow \mathbb{N}''$ (where $f : \mathbb{N}$), show there is an auxiliary function $E : \mathbb{N}^4 \longrightarrow \mathbb{N}$ such that

$$L(k,j,i,r)f = f^e \quad \text{where} \quad e = E(k,j,i,r)$$

for all $k, j, i, r \in \mathbb{N}$. Describe E.

$$\begin{array}{lll}
(\bot) & L(0,0,0,0)f & = f \\
(0) & L(k,j,i,r')f & = L(k,j,i,r)f^2 \\
(1) & L(k,j,i',0)f & = L(k,j,i,2i)f^2 \\
(2) & L(k,j',0,0) & = L(k,j,j,0) \\
(3) & L(k',0,0,0) & = L(k,0,k,k)
\end{array}$$

7.20 Consider $\Phi : \mathbb{N}^4 \longrightarrow \mathbb{N}'''$ specified right for $k, j, i, r \in \mathbb{N}$ and for $F : \mathbb{N}''$, $f : \mathbb{N}'$, $x : \mathbb{N}$. Show that for each $F : \mathbb{N}''$ there are $F_0, F_1, F_2, F_3 : \mathbb{N}''$ such that

$$\Phi(k,j,i,r)F = F_0^r \circ F_1^i \circ F_2^j \circ F_3^k$$

holds for all $k, j, i, r \in \mathbb{N}$.

$$\begin{array}{lll}
(\bot) & \Phi(0,0,0,0) & = id_{\mathbb{N}''} \\
(0) & \Phi(k,j,i,r')F & = F \circ \Phi(k,j,i,r)F \\
(1) & \Phi(k,j,i',0)Ffx & = \Phi(k,j,i,x)Ffx \\
(2) & \Phi(k,j',0,0)Ffx & = \Phi(k,j,x,x)Ffx \\
(3) & \Phi(k',0,0,0)Ffx & = \Phi(k,x,x,x)Ffx
\end{array}$$

7.21 This exercise describes a rather general format which covers many (but not all) $\mathbb{M}_3$-recursions. A **3-structure**

$$\mathfrak{A} = (\mathbb{A},\, a,\, A,\, \mathcal{A}_0,\, \mathcal{A}_1,\, \mathcal{A}_2)$$

is determined by the following data.

- A carrying set $\mathbb{A}$
- A distinguished element $a : \mathbb{A}$
- A distinguished operation $A : \mathbb{A}'$
- Distinguished higher order operations $\mathcal{A}_l : (\mathbb{M}_l \to \mathbb{A}) \to \mathbb{A}$ for $0 \le l < 3$

A morphism between a pair of 3-structures

$$\mathfrak{B} = (\mathbb{B},\, b,\, B,\, \mathcal{B}_0,\, \mathcal{B}_1,\, \mathcal{B}_2) \qquad \mathfrak{A} = (\mathbb{A},\, a,\, A,\, \mathcal{A}_0,\, \mathcal{A}_1,\, \mathcal{A}_2)$$

is a function $\phi : \mathbb{B} \longrightarrow \mathbb{A}$ such that

$$\phi b = a \qquad \phi \circ B = A \circ \phi \qquad \phi \circ \mathcal{B}_l = \mathcal{A}_l \bullet \phi$$

(for $0 \le l < 3$). In the last clause

$$(\mathcal{A}_l \bullet \phi)p = \mathcal{A}_l(\phi \circ p)$$

for $p : \mathbb{M}_l \longrightarrow \mathbb{B}$. Each 3-structure $\mathfrak{A}$ induces a function $\mathfrak{A}(-) : \mathbb{M}_3 \longrightarrow \mathbb{A}$ specified as shown above for $k, j, i, r \in \mathbb{N}$.

$$\begin{array}{lll}
(\perp) & \mathfrak{A}(0,0,0,0) & = a \\
(0) & \mathfrak{A}(k,j,i,r') & = A\mathfrak{A}(k,j,i,r) \\
(1) & \mathfrak{A}(k,j,i',0) & = \mathcal{A}_0\mathfrak{A}(k,j,i,\cdot) \\
(2) & \mathfrak{A}(k,j',0,0) & = \mathcal{A}_1\mathfrak{A}(k,j,\cdot,\cdot) \\
(3) & \mathfrak{A}(k',0,0,0) & = \mathcal{A}_2\mathfrak{A}(k,\cdot,\cdot,\cdot)
\end{array}$$

(a) Fit as many as possible of the specifications of Exercises 7.16–7.20 into this format.

(b) Show that $\phi \circ \mathfrak{B}(-) = \mathfrak{A}(-)$ holds for each pair of 3-structures $\mathfrak{A}, \mathfrak{B}$ and morphism ϕ between these structures.

7.22 It is not immediately obvious that each instance of the $\mathbb{M}_3$-recursion scheme produces a unique function F. Prove this required uniqueness.

7.4 SMALL JUMP OPERATORS

How can we measure the complexity of a part of $\boldsymbol{\lambda G}$? Think of the set of numeric functions that can be represented in that part. To some extent this set reflects at least some of the complexity. We will take this as a good starting point, so our problem is to measure the complexity of the represented functions.

How can we measure the complexity of a numeric function? There are several ways, but here we concentrate on the one that is easiest to analyse. (That does not mean that it is the best method, but it is the one we should start with.)

We measure the complexity of a function by its **rate of growth**. For some functions this is not a very good measure (there are some very complicated slow growing functions), but is certainly quite good for the functions we are interested in.

We turn the standard comparison $\le$ on $\mathbb{N}$ into two comparisons between functions (one of which is also written $\le$).

7.12 DEFINITION. Let $\phi : \mathbb{N}[m]$ be an m-placed function and let $f : \mathbb{N}[1] = \mathbb{N}'$ be a 1-placed function. For $a \in \mathbb{N}$ we write

$$\phi \leq_a f \quad \text{and say} \quad \phi \text{ is } a\text{-dominated by } f \quad \text{if} \quad \phi(x_1, \ldots, x_m) \leq fx$$

holds for all $x_1, \ldots, x_m, x \in \mathbb{N}$ with $a \leq x$, $x_1 \leq x$, $\ldots$, $x_m \leq x$.

We write '$\leq$' for '$\leq_0$' and say ϕ is dominated by f if $\phi \leq f$.

We write $\phi \sqsubseteq f$ and say ϕ is eventually dominated by f if $\phi \leq_a f$ for some $a \in \mathbb{N}$. □

For the most part we use $\leq$ and $\sqsubseteq$. It would be nice if we could restrict our attention to domination, but at times we are forced to use eventual domination. It would also be nice if we could restrict our attention to 1-placed functions, but again sometimes we are forced to use multi-placed functions.

To make good use of these comparisons we need to extract a subset of $\mathbb{N}'$.

7.13 DEFINITION. A snake is a function $f : \mathbb{N}'$ which is strictly inflationary and monotone in the sense that both

$$\text{(inf)} \quad x + 1 \leq fx \qquad\qquad \text{(mon)} \quad x \leq y \Rightarrow fx \leq fy$$

hold for all $x, y \in \mathbb{N}$. □

We use the two comparisons $\phi \leq f$ and $\phi \sqsubseteq f$ only where f is a snake. In fact, most of the time we can replace ϕ by a snake. For any m-placed function ϕ let

$$\overline{\phi}x = \max\{x + 1, \phi(x_1, \ldots, x_m) \mid x_1 \leq x, \ldots, x_m \leq x\}$$

to obtain a 1-placed function $\overline{\phi}$. It is easy to check that $\overline{\phi}$ is a snake, and

$$\phi \leq f \Longleftrightarrow \overline{\phi} \leq f$$

holds for all snakes f. We call $\overline{\phi}$ the bounding snake of ϕ.

The composition of two snakes is a snake, and

$$\left.\begin{array}{l} f \leq g \\ h \leq k \end{array}\right\} \Rightarrow f \circ h \leq g \circ k$$

holds (for snakes f, g, h, k). The pointwise comparison $\leq$ partially orders snakes. Here are two other properties of snakes we will need later.

7.14 LEMMA. *For each snake f both*

$$(i) \quad r + s \leq f^r s \qquad\qquad (ii) \quad \left.\begin{array}{l} r \leq s \\ x \leq y \end{array}\right\} \Rightarrow f^r x \leq f^s y$$

hold for all $r, s, x, y \in \mathbb{N}$.

Given any snake f we may iterate to produce a chain

$$(\mathrm{Cmp}) \quad f \leq f^2 \leq f^3 \leq \cdots \leq f^{r+1} \leq \cdots \qquad (r < \omega)$$

of faster and faster snakes which dominate more and more functions. In this way we produce a rather fine stratification of those functions which do not grow too fast (relative to f). Of course, there will be some snakes which are even faster than those in (Cmp), and in time we would like to describe some of these. We say a snake f **captures** a function if that function is dominated by some member of (Cmp). In particular, if a function is eventually dominated by a snake then it is captured by that snake.

7.15 EXAMPLES. (0) If f is a non-trivial translation then f captures all translations, but does not capture $x \longmapsto 2x$.

(1) If f is linear and not a translation then f captures all linear functions, but does not capture $x \longmapsto x^2$.

(2) If f is a polynomial and not linear then f captures all polynomials, but does not capture $x \longmapsto 2^x$.

(3) If f is exponential then f captures all stacked exponentials, but does not capture $x \longmapsto \beth(x, 2, 1)$. □

How can we speed up, in a uniform fashion, any given snake so that it outstrips its composition hierarchy (Cmp)? Before we answer that let's hide some of the irrelevant features of (Cmp).

7.16 DEFINITION. For each snake f let $\mathcal{B}(f)$ be the set of all snakes captured by f, i.e. the set of all snakes g such that $g \leq f^{r+1}$ for some $r \in \mathbb{N}$. □

Almost trivially

$$g \leq r \Rightarrow \mathcal{B}(g) \subseteq \mathcal{B}(f) \qquad \mathcal{B}(f) = \mathcal{B}(f^2) = \mathcal{B}(f^3) = \cdots$$

(for snakes f, g). We have $\mathcal{B}(f) = \mathcal{B}(g)$ precisely when the composition hierarchies of f and g interlace. By concentrating on $\mathcal{B}(f)$ rather than on f we smooth out any compositional irregularities that f may have. A function ϕ is captured by a snake f exactly when $\overline{\phi} \in \mathcal{B}(f)$.

7.17 DEFINITION. A **basket** is a non-empty set $\mathcal{B}$ of snakes which is closed under composition and downward closed, i.e. satisfies

$$g, f \in \mathcal{B} \Rightarrow g \circ f \in \mathcal{B} \qquad g \leq f \in \mathcal{B} \Rightarrow g \in \mathcal{B}$$

for all snakes f, g. □

The intersection of any family of baskets is itself a basket, and so any set of snakes is included in a smallest basket. The smallest basket which contains f is $\mathcal{B}(f)$.

How can we escape from the basket $\mathcal{B}(f)$? We need a snake f' which is (eventually) faster than each member of (Cmp) to ensure that $\mathcal{B}(f) \subsetneq \mathcal{B}(f')$ holds. This leads us to a discussion of suitable constructions $f \longmapsto f'$. Once we have one of these we can repeat the construction to produce a chain

$$\mathcal{B}(f) \subsetneq \mathcal{B}(f') \subsetneq \mathcal{B}(f'') \subsetneq \mathcal{B}(f''') \subsetneq \cdots$$

of baskets which, we hope, will capture a substantial collection of functions.

It is convenient to isolate some of the properties of suitable constructions $f \longmapsto f'$. Here are the ones which seem most useful.

7.18 DEFINITION. An operator $F : \mathbb{N}''$ is **charming** if for all snakes f, g

(0) Ff is a snake (1) $f \leq Ff$ (2) $g \leq f \Rightarrow Fg \leq Ff$

(3) $Ff^2 \leq (Ff)^2$ (4) $f^2 \sqsubseteq Ff$

hold. □

You can probably see why (0,1,2) are useful, but what about (3) and (4)? For each snake f and charmer F the required inclusion $\mathcal{B}(f) \subseteq \mathcal{B}(Ff)$ is a consequence of (0,1). Using (3) a simple induction gives

$$Ff^{2^r} \leq (Ff)^{2^r}$$

for all $r < \omega$. Then using (2), for snakes f, g we have

$$g \in \mathcal{B}(f) \Rightarrow (\exists r)[g \leq f^{2^r}] \Rightarrow (\exists r)[Fg \leq Ff^{2^r} \leq (Ff)^{2^r}] \Rightarrow Fg \in \mathcal{B}(Ff)$$

and hence

$$\mathcal{B}(g) \subseteq \mathcal{B}(f) \Rightarrow \mathcal{B}(Fg) \subseteq \mathcal{B}(Ff) \quad \text{so that} \quad \mathcal{B}(g) = \mathcal{B}(f) \Rightarrow \mathcal{B}(Fg) = \mathcal{B}(Ff)$$

holds. This shows that the basket $\mathcal{B}(Ff)$ depends only on the basket $\mathcal{B}(f)$ and not on the particular generating snake f. Thus conditions (0–3) ensure we have an inflating operation on baskets, not just on snakes. So far we have not used (4), nor have we shown that $\mathcal{B}(f) \neq \mathcal{B}(Ff)$; that will come later.

Given a snake f and a charmer F we can iterate F to produce a chain

$$f \leq Ff \leq F^2f \leq \cdots \leq F^if \leq \cdots \quad (i < \omega)$$

of snakes and a corresponding chain

$$\text{(Eff)} \quad \mathcal{B}(f) \subseteq \mathcal{B}(Ff) \subseteq \cdots \subseteq \mathcal{B}(F^if) \subseteq \cdots \quad (i < \omega)$$

of baskets. Let

$$\mathcal{B}(F^*f) = \bigcup\{\mathcal{B}(F^if) \mid i < \omega\}$$

to herd into one basket all the snakes occurring at some step in the chain (Eff). You should be careful with this collection. We have *not* constructed an operation F^*; we have merely used a notation to suggest such an operation.

This outline gives us several questions we should investigate.

(?1) How can we construct charming operators F, and how can we ensure that the chain (Eff) is strictly increasing?

(?2) Can we characterize the functions captured by snakes in $\mathcal{B}(F^*f)$, and does the level $\mathcal{B}(F^i f)$ of first capture have any significance?

(?3) How can we escape from $\mathcal{B}(F^*f)$?

Furthermore, we should check that every gadget we use can be simulated within $\boldsymbol{\lambda G}$ or that part of $\boldsymbol{\lambda G}$ of immediate concern.

The chain (Cmp) gives a 2-placed function $(r, x) \longmapsto f^{r+1}x$ which we want to outpace. The standard trick is to take a section of this function by somehow connecting r and x. Here are some possibilities.

7.19 DEFINITION. The operators *pol*, *brw*, *rob*, *ack*, *exp* are given by

$$polfx = f^2x \quad brwfx = f^x 2 \quad robfx = f^{x+1}1 \quad ackfx = f^{x+1}x \quad expfx = f^{2^x}x$$

for all $f : \mathbb{N}'$ and $x \in \mathbb{N}$. □

The operator *ack* is due to Ackermann who used it to produce a recursive function that is not primitive recursive. This construction was simplified, independently, by R. M. Robinson and Rozsá Péter using *rob*. The operator *brw*, which for some jobs is easier to use, is part of folklore, but may have been first documented by Brewer. We refer to *ack*, *rob*, *brw* as the **standard jump operators**. The operators *pol* and *exp* will be used later to illustrate a point.

Any of these operators can be used to escape from $\mathcal{B}(f)$, but not all of them are charming in the strict sense of Definition 7.18. You will find that some of the comparisons don't quite work, but in each case it is not too hard to get round the problem. In any case, the properties (0–4) are merely the ones that seem most convenient in general, but there is nothing sacrosanct about them.

For the time being let us concentrate on the operator *ack*.

7.20 LEMMA. *The operator ack is charming.*

Proof. Consider any snake f.

Setting $r = s = x$ in 7.14(i) gives $x \leq 2x \leq f^x x$ and hence $x \leq fx \leq f^{x+1}x$ using the snake properties of f. Setting $r = x + 1$, $s = y + 1$ in 7.14(ii) shows that $ackf$ is monotone. This verifies (0,1).

For any snake $g \leq f$ we have $g^{r+1} \leq f^{r+1}$ for all $r < \omega$, and hence for $x \in \mathbb{N}$

$$ackgx = g^{x+1}x \leq f^{x+1}x = ackfx$$

to verify (2).

For (3) let $y = ackfy$ (for $x \in \mathbb{N}$). Setting $r = x + 1$, $s + x$ in 7.14(i) we have

$$x \leq 2x + 1 \leq ackfx = y$$

so that $2x + 2 \leq y + 1$. Setting $r = 2(x + 1)$, $s = y + 1$ in 7.14(ii) gives

$$ackf^2x = f^{2(x+1)}x \leq f^{y+1}y = ackfy = (ackf)^2x$$

as required.

Finally, if $x \geq 1$ then

$$f^2x \leq f^{x+1}x = ackfx$$

to verify (4). □

This gives one example of a charmer, but does it do the job it is supposed to?

7.21 LEMMA. *For all snakes f, g*

$$g \in \mathcal{B}(f) \Rightarrow g \sqsubseteq ackf \qquad ackf \notin \mathcal{B}(f) \qquad \mathcal{B}(f) \subsetneq \mathcal{B}(ackf)$$

hold.

Proof. If $g \in \mathcal{B}(f)$ then $g \leq f^{r+1}$ for some $r < \omega$, and hence

$$gx \leq f^{r+1}x \leq f^{x+1}x \leq ackfx$$

for all $x \geq r$.

If $ackf$ is in $\mathcal{B}(f)$ then so is the snake $g = Suc \circ (ackf)$, and hence

$$gx \leq ackfx = gx - 1$$

for all large x. This couldn't be more contradictory, hence $ackf \notin \mathcal{B}(ackf)$.

The strict inclusion is now immediate. □

This proof uses some of the special properties of *ack*. It does not generalize to an arbitrary charmer. To see this observe that *pol* is charming. Conditions (0,1,2) are immediate, and the comparisons (3,4) are equalities. In particular, $\mathcal{B}(polf) = \mathcal{B}(f)$. However, we can still use *pol*, or any charmer, to escape from $\mathcal{B}(f)$. We return to this shortly.

This partly answers (?1), but what about (?2)? Of course, any characterization will depend on the charmer F used, but there is a nice answer for $F = ack$ which is important historically.

Given a snake f, set

$$Ack_f(i, r) = (ack^i f)^{r+1}$$

to produce a doubly indexed family of snakes. If we fix i and let r increase then we move up the composition chain for $ack^i f$. If we set $r = 0$ and let i increase, then we move up the chain (Eff). The 3-placed function

$$(i, r, x) \longmapsto Ack_f(i, r)x$$

will encapsulate everything that is happening in these chains.

It is not hard to see that for each i, f the 2-placed function $Ack_f(i, \cdot)\cdot$ is primitive recursive in f. The following result, which is due to Grzegorczyk and is proved in Exercise 7.31, is a converse of this observation.

7.22 THEOREM. *If the numeric function ϕ is primitive recursive in the snake f, then ϕ is dominated by $Ack_f(i, r)$ for some $i, r < \omega$.*

This result gives a stratification of the functions which are primitive recursive in f. From the proof we see that the bounding indexes i, r measure the nesting depth of the use of primitive recursion followed by composition. Thus we have an answer to (?2) at least for the *ack* case.

The result can be viewed as a refinement of the earlier result due to Ackermann.

7.23 COROLLARY. *The 3-placed function Ack_f is not primitive recursive in f.*

Proof. By way of contradiction, suppose this function is primitive recursive in f. Then so is the function $\phi : \mathbb{N}'$ given by

$$\phi x = Ack_f(x, x)x + 1$$

(for $x \in \mathbb{N}$). By Theorem 7.22 this function ϕ is dominated by $Ack_f(i, r)$ for some $i, r \in \mathbb{N}$. But then, for all large x we have

$$\phi x \leq Ack_f(i, r)x < Ack_f(x, x)x = \phi x - 1$$

which is contradictory. □

This shows us how to escape from the grips of primitive recursion, but where have we escaped to? It can be shown (as in the exercises) that the 3-placed function Ack_f has a rather simple 2-recursive specification. We haven't escaped very far, and there are even nastier beasts waiting to swallow us.

Although the operators *brw* and *rob* are not quite charming, they can be used as replacements of *ack* to produce analogues of Theorem 7.22 and Corollary 7.23.

Our attention is now moving away from snakes to charmers.

7.24 LEMMA. *The composite of two charming operators is charming.*

Proof. Only the inflationary properties (3,4) are not immediate.

Let F, G be two charmers. Then, for each snake f, properties (3) for F, (2) for G, and (3) for G give

$$(G \circ f)f^2 = G(Fr^2) \leq G(Ff)^2 \leq (G(Ff))^2$$

to verify that $G \circ F$ has (3). Also property (1) for F and (2,4) for G give

$$f^2 \sqsubseteq Gf \leq G(Ff)$$

to verify that $G \circ F$ has (4). □

Each charmer F and each snake f gives us a chain (Eff) of baskets which we herd into one basket $\mathcal{B}(F^*f)$. How can we escape from this? This basket $\mathcal{B}(F^*f)$ is closed under F,

$$g \in \mathcal{B}(F^*f) \Rightarrow (\exists i)[g \leq F^i f] \Rightarrow (\exists i)[Fg \leq F^{i+1} f] \Rightarrow Fg \leq \mathcal{B}(F^*f)$$

so we can't use F to escape. We need a more powerful charmer.

7.25 DEFINITION. For each operator $F : \mathbb{N}''$ the **step-up** of F is the operator $F' : \mathbb{N}''$ given by

$$F'fx = F^x fx$$

for all $f : \mathbb{N}'$ and $x \in \mathbb{N}$. □

This is essentially the same trick as before; we have diagonalized along a chain.

7.26 LEMMA. *The step-up of a charming operator is charming.*

Proof. Let F be a charmer.

For each $x \geq 1$ the iterate F^x is charming and hence $f \leq F^x f$ to give $fx \leq F'fx$ for each snake f. This comparison is trivial for $x = 0$, and hence $f \leq F'f$. For $x \leq y$ we have $F^x f \leq F^y f$ so that $F'fx \leq F'fy$. This verifies the required properties (0,1).

For snakes $g \leq f$ and $x \geq 1$ we have $F^x g \leq F^x f$ (since F^x is charming). This gives $F'gx \leq F'fx$ for $x \geq 1$, and this comparison is trivial for $x = 0$, hence $F'g \leq F'f$, to verify (2).

To verify (3), for a snake f and $x \in \mathbb{N}$ let $y = F'fx = F^x fx$. Then $F^x f^2 \leq (F^x f)^2$ (since F^x is charming or $x = 0$). Thus

$$F'f^2x = F^x f^2 x \leq (F^x f)^2 x = F^x fy \leq F^y fy = F'fy = (F'f)^2 x$$

as required.

Finally, for large x we have

$$f^2 x \leq Ffx \leq \cdots \leq F^x fx = F'fx$$

to verify (4), as required. □

As an example consider the rather feeble charmer *pol*. A simple exercise shows that for each snake f and $r < \omega$

$$pol^r f = f^{2^r} \quad \text{so that} \quad pol' fx = f^{2^x} x = exp fx$$

i.e. $pol' = exp$. In particular, the step-up of *pol* is more powerful than *ack*, and in this form can be used to escape from a basket $\mathcal{B}(f)$.

So far we have not used property (4) of charmers. A simple induction gives

$$f^{2^r} \sqsubseteq F^r f$$

(for all charmers F, snakes f, and $r \in \mathbb{N}$). This gives a stepped-up analogue of Lemma 7.21.

7.27 LEMMA. *For each charming operator F and all snakes f, g*

$$g \in \mathcal{B}(F^* f) \Rightarrow g \sqsubseteq F'f \qquad F'f \notin \mathcal{B}(F^* f) \qquad \mathcal{B}(F^* f) \subsetneq \mathcal{B}(F'f)$$

hold.

Proof. If $g \in \mathcal{B}(F^* f)$ then $g \in \mathcal{B}(F^i)$ for some $i < \omega$ and hence $g \leq (F_i f)^{2^r}$ for some $r < \omega$. Thus, by the above observation

$$g \leq f^r(F^i f) = F^{r+i} f \quad \text{and hence} \quad gx \leq F^x fx = F' fx$$

for all $x \geq i + r$. Thus $g \sqsubseteq F' f$.

If $F' f$ is in $\mathcal{B}(F^* f)$, then so is the snake $g = Suc \circ F' f$, and then $g \sqsubseteq F' f$, which leads to a contradiction.

The strict inclusion is now immediate. □

You can probably see what we should now do, but before that let's look at some historically important hierarchies.

Exercises

7.23 Prove Lemma 7.14.

7.24 Show that if a function $\phi : \mathbb{N}[m]$ is eventually dominated by a snake $f : \mathbb{N}'$, then it is dominated by an iterate of f. Does this mean that eventual domination is not needed?

7.25 (a) Suppose

$$f_0 \leq f_1 \leq f_2 \leq \cdots \leq f_r \leq \cdots \quad (r < \omega)$$

is an ascending chain of snakes. Define the diagonal limit $g : \mathbb{N}'$ of this chain by $gx = f_x x$ (for $x \in \mathbb{N}$). Show that g is a snake which eventually dominates each f_r.

(b) What happens if in the given chain we have $\sqsubseteq$ in place of $\leq$?

7.26 The notion of a basket of snakes is introduced in Definition 7.17.

(a) Show that each basket $\mathcal{B}$ is directed, i.e. for each $f, g \in \mathcal{B}$ there is some $h \in \mathcal{B}$ with $f \leq h, g \leq h$.

(b) Show that $\mathcal{B}(f)$ is the least basket containing f.

(c) Show that for each set $\mathcal{S}$ of snakes there is a least basket which includes $\mathcal{S}$.

7.27 (a) Show that

$$f \leq polf \sqsubseteq brwf \leq robf \sqsubseteq ackf \leq expf$$

hold for all snakes f.

(b) Show that brw and rob are not quite charming. Does this matter?

7.28 Let $jmp : \mathbb{N}''$ be any of the jump operators pol, brw, rob, ack, exp. Show there is a term $\vdash \mathsf{jmp} : \mathcal{N}''$ such that if a function $f : \mathbb{N}'$ is represented by a term $\ulcorner f \urcorner$ then $\mathsf{jmp}\ulcorner f \urcorner$ represents $jmpf$. In each case locate the required iterator I_σ and its nesting depth.

7.29 Show that for each snake f the three functions

$$(i, r, x) \longmapsto (ack^i f)^{r+1} x \qquad (i, x) \longmapsto rob^i fx \qquad (i, x) \longmapsto brw^i fx$$

are 2-recursive in f.

7.30 For an arbitrary function $f : \mathbb{N}'$ consider the function $F : \mathbb{N}^2 \longrightarrow \mathbb{N}'$ specified by

$$\begin{array}{lrcl} (\perp) & F(0,0) & = & f \\ (0) & F(i,r') & = & F(i,b) \circ F(i,a) \quad \text{where } b = \lceil r/2 \rceil, a = \lfloor r/2 \rfloor \\ (1) & F(i',0)x & = & F(i,x)x \end{array}$$

where $\lceil \cdot \rceil$ and $\lfloor \cdot \rfloor$ are the ceiling and floor functions. Determine this known function F and verify your claim. Explain carefully the induction scheme you use.

7.31 This exercise contains a proof of Theorem 7.22. It gives more detailed information than Ackermann's original proof, but less than Grzegorczyk's version. We analyse how composition and primitive recursion interact with the a-indexed comparison $\leq_a$ of Definition 7.12.

(a) Suppose that ϕ is a composite of ψ and $\theta_1, \ldots, \theta_n$, i.e.

$$\phi \mathsf{x} = \psi(\theta_1 \mathsf{x}, \ldots, \theta_n \mathsf{x})$$

for each argument sequence x. Show that for snakes g, h

$$\psi \leq_a h \quad \theta_1, \ldots, \theta_n \leq_a g \qquad \Rightarrow \qquad \phi \leq_a h \circ g$$

holds.

(b) Suppose ϕ is obtained from ψ and θ by primitive recursion, i.e.

$$\phi 0 \mathsf{x} = \theta \mathsf{x} \qquad \phi r' \mathsf{x} = \psi r s \mathsf{x} \quad \text{where} \quad s = \phi r \mathsf{x}$$

for $r \in \mathbb{N}$ and sequence x of parameters. Show that if $\theta, \psi \leq_a g$ for a snake g, then $\phi r \leq_a g^{r+1}$ for all $r \in \mathbb{N}$. Hence show $\phi \leq_a ack\, g$.

(c) Show that if the function ϕ is primitive recursive in the snake f, then $\phi \leq_a ack^i f$ for some $i, a \in \mathbb{N}$. Hence prove Theorem 7.22.

7.5 The multi-recursive hierarchies

What we have done so far? Starting from any snake $f : \mathbb{N}'$ we can iterate to produce

$$(\text{Cmp}) \quad f \leq f^2 \leq f^3 \leq \cdots \leq f^{r+1} \leq \cdots \qquad (r < \omega)$$

a chain of faster and faster snakes. This is a rather fine hierarchy which is sensitive to function composition, and rather too fine for our purposes. We herd into one basket $\mathcal{B}(f)$ all those snakes which are dominated (or eventually dominated) by some member of (Cmp). This hides any insignificant properties of the initial snake f. We now concentrate on hierarchies of baskets rather than snakes. We have a nested collection of such hierarchies.

Let $jmp : \mathbb{N}''$ be a standard jump operator such as ack, rob, or brw, or any minor variant of these. For $i < \omega$ we set

$$\mathcal{G}_i(f) = \mathcal{B}(jmp^i f)$$

to produce the Grzegorczyk hierarchy of baskets

$$\text{(Grz)} \quad \mathcal{B}(f) = \mathcal{G}_0(f) \subseteq \cdots \subseteq \mathcal{G}_i(f) \subseteq \cdots \quad (i < \omega)$$

on f. We herd this into one basket

$$\mathcal{G}_\omega(f) = \mathcal{B}(jmp^* f) = \bigcup \{\mathcal{G}_i(f) \mid i < \omega\}$$

which is closed under jmp. From Theorem 7.22 (modified to this particular jmp) we see that $\mathcal{G}_\omega(f)$ captures all functions which are primitive recursive in f. This hierarchy (Grz) is the 1-placed analogue of the 2-placed hierarchy constructed in Section 7.1. This 1-placed version is easier to handle.

Each step

$$\mathcal{G}_i(f) \subseteq \mathcal{G}_{i+1}(f)$$

of (Grz) could be refined, at the snake level, by a composition hierarchy. In this way we produce a doubly indexed family

$$((jmp^i f)^{r+1} \mid i, r < \omega)$$

of snakes which gives a rather fine stratification of those functions primitive recursive in f. With hindsight, we can see this is precisely what Ackermann did, but it was first made explicit by Grzegorczyk.

The more delicate properties of (Grz) depend on the particular jump operator used. For instance, for some slow snakes f we have

$$\mathcal{B}(rob f) \neq \mathcal{B}(ack f)$$

and this discrepancy persists all the way up the two corresponding hierarchies. However, since

$$\mathcal{B}(rob^* f) = \mathcal{B}(ack^* f)$$

these differences are smoothed out by $\mathcal{G}_\omega(f)$. Just as the move $f \longmapsto \mathcal{B}(f)$ hides the insignificant properties of f, so the move to $\mathcal{B}(jmp^* f)$ hides the insignificant properties of jmp.

To escape from the herd $\mathcal{G}_\omega(f)$ we use the step-up $JMP = jmp'$ of jmp. We know $\mathcal{G}_\omega(f) \subsetneq \mathcal{B}(JMP f)$ and the function $JMP f$ is not 1-recursive in f but is 2-recursive. The new operator JMP is charming (or almost charming) so we may set

$$\mathcal{A}_i(f) = \mathcal{B}(JMP^i f)$$

to produce the Ackermann hierarchy of baskets on f.

$$\text{(Ack)} \quad \mathcal{B}(f) = \mathcal{A}_0(f) \subseteq \cdots \subseteq \mathcal{A}_i(f) \subseteq \cdots \quad (i < \omega)$$

We herd this into one basket

$$\mathcal{A}_\omega(f) = \bigcup \{\mathcal{A}_i(f) \mid i < \omega\}$$

which is closed under JMP. We know that $JMP f$ is 2-recursive in f, and hence each $JMP^i f$ is 2-recursive in f. It can be shown (but won't be here) that each function which is 2-recursive in f is caught by $\mathcal{A}_\omega(f)$.

Each step

$$\mathcal{A}_j(f) \subseteq \mathcal{A}_{j+1}(f)$$

of (Ack) may be refined by a Grzegorczyk hierarchy

$$\mathcal{A}_j(f) = \mathcal{G}_0(JMP^j f) \subseteq \ldots \subseteq \mathcal{G}_i(JMP^j f) \subseteq \ldots \subseteq \mathcal{G}_\omega(JMP^j f) \subsetneq \mathcal{A}_{j+1}(f)$$

(each step of which could be refined at the snake level by a composition hierarchy). Staying on the basket level this gives us a doubly indexed hierarchy

$$\Big(\mathcal{B}(jmp^i(JMP^j f)) \mid i, j < \omega\Big)$$

of baskets which stratifies the functions 2-recursive in f.

It is now clear what we should do next. We should use the second step-up jmp'' and then the third step-up jmp''', and so on. We need a better notation to handle this.

7.28 DEFINITION. Let $Pet : \mathbb{N}'''$ be the higher order operator given by

$$PetFfx = F^x fx$$

for all $F : \mathbb{N}''$, $f : \mathbb{N}$ and $x : \mathbb{N}$. □

Thus Pet converts each charmer (or near charmer) into its step-up. In particular, $JMP = Pet\,jmp$. For each $s < \omega$ let

$$jmp_s = Pet^s jmp$$

to produce a chain of charmers of increasing strength. For each s and snake f we have a chain of baskets

$$\mathcal{B}(f) \subseteq \mathcal{B}(jmp_s f) \subseteq \cdots \subseteq \mathcal{B}(jmp_s^i f) \subseteq \cdots \subseteq \mathcal{B}(jmp_s^* f) \subseteq \mathcal{B}(jmp_{s+1} f)$$

the analogue of (Eff) using $F = jmp_s$, together with the herded and stepped-up basket. Let

$$\mathcal{P}_s(f) = \mathcal{B}(jmp_s f)$$

to produce the Péter hierarchy of baskets on f

$$\text{(Pet)} \quad \mathcal{B}(f) \subseteq \mathcal{P}_0(f) \subseteq \cdots \subseteq \mathcal{P}_s(f) \subseteq \cdots \quad (s < \omega)$$

which we may herd into one basket

$$\mathcal{P}_\omega(f) = \bigcup\{\mathcal{P}_s(f) \mid s < \omega\}$$

like so. It can be shown that $jmp_s f$ is s'-recursive in f. It is known (but will not be shown here) that, in general, $jmp_s f$ is not s-recursive in f. Furthermore, provided f is representable in $\boldsymbol{\lambda G}$, then so is $jmp_s f$.

For the time being the basket $\mathcal{P}_\omega(f)$ is as far as we want to go.

The hierarchies we have set up give us a mechanism for measuring the rate of growth of a function to a quite precise degree. To measure the rate of growth of a function ϕ we compare its bounding snake $\overline{\phi}$ with a known snake f. We choose f as slow as possible with $\overline{\phi} \in \mathcal{P}_\omega(f)$. This gives some $s < \omega$ with $\overline{\phi} \in \mathcal{P}_s(f)$, and the least such s is the first, crude, measure of ϕ.

If $s = 0$, i.e. $\overline{\phi} \in \mathcal{B}(jmpf)$, then we should think of using a less powerful charmer, or a slower snake f, or going into the composition hierarchy. Let's assume $s \neq 0$. Thus, with a change of indexing, we have some $s < \omega$ with

$$\overline{\phi} \in \mathcal{P}_{s+1}(f) - \mathcal{P}_s(f)$$

and hence $\overline{\phi}$ occurs somewhere in

$$\mathcal{P}_s(f) = \mathcal{B}(jmp_s f) \subseteq \cdots \subseteq \mathcal{B}(jmp_s^{i+1} f) \subseteq \cdots \subseteq \mathcal{B}(jmp_s^* f) \subseteq \mathcal{B}(jmp_{s+1} f)$$

and not in $\mathcal{P}_s(f)$. If $\overline{\phi} \notin \mathcal{B}(jmp_s^* f)$, i.e.

$$\overline{\phi} \in \mathcal{P}_{s+1}(jmpf) - \mathcal{B}(jmp_s^* f)$$

then we take $jmp_{s+1} f$ as giving the upper bound for the rate of growth of ϕ. If $\overline{\phi} \in \mathcal{B}(jmp_s^* f)$ then we locate the least $i_s < \omega$ with

$$\overline{\phi} \in \mathcal{B}(jmp_s^{i_s+1} f) = \mathcal{B}(jmp_s(jmp_s^{i_s} f)) = \mathcal{B}(Pet\, jmp_t g)$$

where $g = jmp_s^{i_s} f$ and $t = s - 1$.

This means that $\overline{\phi}$ occurs somewhere in

$$\mathcal{P}_t(g) = \mathcal{B}(jmp_t g) \subseteq \cdots \subseteq \mathcal{B}(jmp_t^{i+1} g) \subseteq \cdots \subseteq \mathcal{B}(jmp_t^* g) \subseteq \mathcal{B}(jmp_s g)$$

and not in $\mathcal{P}_t(g)$. If $\overline{\phi} \notin \mathcal{B}(jmp_t^* g)$ then we take

$$jmp_s g = jmp_s^{i_s+1} f$$

as giving the upper bound for the rate of growth of ϕ. If $\overline{\phi} \in \mathcal{B}(jmp_t^* g)$ then we locate the least $i_t < \omega$ with

$$\overline{\phi} \in \mathcal{B}(jmp_t^{i_t+1} g) = \mathcal{B}(jmp_t(jmp_t^{i_t} g)) = \mathcal{B}(Pet\, jmp_u((jmp_t^{i_t} \circ jmp_s^{i_s}) f))$$

where $u = t - 1 = s - 2$. We now repeat this process using

$$h = (jmp_t^{i_t} \circ jmp_s^{i_s}) f$$

and then continue in this manner decreasing the number of iterates of Pet at each step.

In this way an upper bound for the rate of growth of ϕ can be measured using a multi-index. For each jump operator jmp and index $\mathsf{i} = (i_s, \ldots, i_0)$ of length s let

$$M_s \mathsf{i}\, jmp = jmp_0^{i_0} \circ \cdots \circ jmp_s^{i_s}$$

to obtain a compound charmer (or near charmer) built from jmp by an i-iteration of Pet. For each snake f let

$$\mathcal{M}_{\mathsf{i}}(f) = \mathcal{B}(M_s \mathsf{i}\, jmp\, f)$$

to obtain a multi-indexed array of baskets. It is not too hard to see that

$$\mathsf{i} \leq \mathsf{j} \Rightarrow \mathcal{M}_{\mathsf{i}}(f) \subseteq \mathcal{M}_{\mathsf{j}}(f)$$

and the process described above locates the least indexed i with $\overline{\phi} \in \mathcal{M}_{\mathsf{i}}(f)$. We take this index as the measure of ϕ (relative to f).

You may have recognized that the multi-indexes i are nothing more than concrete representations of the ordinals below ω^ω. Thus the 'value' we have attached to a rate of growth is just one of these ordinals. This suggests that we should reorganize this process to make the use of ordinals explicit, and then investigate how we can obtain larger ordinal rates of growth. This is the topic of Chapter 9.

EXERCISES

7.32 Let $f : \mathbb{N}'$ be arbitrary and recall that Solution 7.29 gives a spec of the function $F_0 : \mathbb{N} \longrightarrow \mathbb{N}^+$ where $F_0 irx = (ack^i f)^{r+1}x$ for all $i, r, x \in \mathbb{N}$. Recall that $ack_s = Pet^s ack$ for each $s < \omega$. Let $ACK = Pet\, ack$.

(a) Write down a spec of the function $F_1 : \mathbb{M}_1 \longrightarrow \mathbb{N}^+$ where

$$F_1(j,i)rx = (ack^i(ACK^j f))^{r+1}x$$

for all $(j,i) \in \mathbb{M}_1$ and $r, x \in \mathbb{N}$.

(b) Write down a spec of the function $F_3 : \mathbb{M}_3 \longrightarrow \mathbb{N}^+$ where

$$(*) \qquad F_3(l,k,j,i)rx = (ack_0^i(ack_1^j(ack_2^k(ack_3^l f))))^{r+1}x$$

for all $(l,k,j,i) \in \mathbb{M}_3$ and $r, x \in \mathbb{N}$. (You may find it useful to write down a spec of $F_2 : \mathbb{M}_2 \longrightarrow \mathbb{N}^+$ where

$$F_3(j,i) = F_2(0,j,i) \qquad F_2(k,j,i) = F_3(0,k,j,i)$$

for all $k, j, i \in \mathbb{N}$.)

(c) Is your spec for (b) an instance of $\mathbb{M}_3$-recursion?

(d) Set up an induction which proves $(*)$. (You need not give all the component verifications.)

7.33 Let $f : \mathbb{N}'$ be arbitrary, let $jmp = rob$ or brw, and recall that Solution 7.29 gives a spec of the function $G_0 : \mathbb{N} \longrightarrow \mathbb{N}'$ where

$$G_0 ix = (jmp^i f)x$$

for all $i, x \in \mathbb{N}$. (The specs for the two jmps are slightly different.) Recall that $jmp_s = Pet^s jmp$ for each $s < \omega$. Let $JMP = Pet\, jmp = jmp_1$.

(a) Write down a spec of the function $G_1 : \mathbb{M}_1 \longrightarrow \mathbb{N}'$ where

$$G_1(j,i) = jmp^i(JMP^j f)$$

for all $(j,i) \in \mathbb{M}_1$.

(b) Write down a spec of the function $G_3 : \mathbb{M}_3 \longrightarrow \mathbb{N}'$ where

$$(*) \qquad G_3(l,k,j,i) = (jmp_0^i \circ jmp_1^j \circ jmp_2^k \circ jmp_3^l)f$$

for all $(l,k,j,i) \in \mathbb{M}_3$.

(c) Is your spec for (b) an instance of $\mathbb{M}_3$-recursion?

(d) Set up an induction which proves $(*)$. (You need not give all the component verifications.)

7.34 (a) Using $\mathbb{M}_3$-recursion write down a spec of the function $\mathbb{G}_3 : \mathbb{M}_3 \longrightarrow \mathbb{N}'''$ which, when supplied with an arbitrary $\mathsf{i} \in \mathbb{M}_3$ and $jmp : \mathbb{N}''$ will return

$$jmp_0^i \circ jmp_1^j \circ jmp_2^k \circ jmp_3^l$$

where $\mathsf{i} = (l,k,j,i)$ and $jmp_s = Pet^s jmp$ (for $s < \omega$).

(b) The functions F_3 of Exercise 7.32 and G_3 of Exercise 7.33 use instances of $\mathbb{G}_3$. Why is it that $\mathbb{G}_3$ has an $\mathbb{M}_3$-recursive spec, but F_3 and G_3 do not?

7.6 The extent of $\boldsymbol{\lambda G}$

Which numeric functions are representable in $\boldsymbol{\lambda G}$? In more detail, which facilities are needed to represent a particular function, what is the syntactic complexity of the representing term, and how complicated are the associated computations?

We started this chapter with a look at the first few standard arithmetical operations *Suc*, *Add*, *Mlt*, *Exp*, ... and an operator $Grz : \mathbb{N}^{+\prime}$ which we can use to jump up any binary operation on $\mathbb{N}$. We have representations $\mathsf{S}, \mathsf{Add}, \mathsf{Mlt}, \dots$ of these operations. More generally, we have a term Grz with the property that if a term $\ulcorner F \urcorner$ represents a function $F : \mathbb{N}^{+}$, then the term $\mathsf{Grz}\ulcorner F \urcorner$ represents the jumped-up version $F' = Grz\,F$ of F. These terms use only the iterator $\mathsf{I} = \mathsf{I}_{\mathcal{N}}$, and application of *Grz* increases the nesting depth.

All this was by way of a preamble so that by Section 7.4 we decided to concentrate on hierarchies of 1-placed functions. In that section we introduced the three main small jump operators *ack*, *rob*, *brw* and it is not too hard, as in Exercise 7.28, to produce terms ack, rob, brw which name these operators. These terms have just one occurrence of $\mathsf{I} = \mathsf{I}_{\mathcal{N}}$ (and no other iterators).

By nesting these terms (and thereby nesting I) we easily climb up through the 1-placed version of the Grzegorczyk hierarchy based on *Suc*. By relativizing this process and diagonalizing out we produce an operator $f \longmapsto Ack_f$ (of type $\mathbb{N}''$) which, for a snake f, jumps outside the class of functions primitive recursive in f. Again it is not hard to produce a term $\vdash \mathsf{Ack} : \mathcal{N}''$ which names this operator. However, this term makes use of the iterator $\mathsf{J} = \mathsf{I}_{\mathcal{N}'}$.

More generally, we can produce a very simple term $\vdash \mathsf{Pet} : \mathcal{N}'''$ which names the stepping-up operator for jump operators. This term uses just one iterator J.

By nesting this term (and thereby nesting J) we can pass to any particular level of the Péter hierarchy. Given any such level s, there is an operator

$$M_s : \mathbb{M}_s \longrightarrow \mathbb{N}''' \quad \text{where} \quad M_s \mathsf{i}\, jmp = jmp_0^{i_0} \circ \cdots \circ jmp_s^{i_s}$$

for each $\mathsf{i} = (i_s, \ldots, i_0) \in \mathbb{M}_s$ and $jmp \in \mathbb{M}''$. This operator can be named using an appropriate nesting of $\mathsf{J} = \mathsf{I}_{\mathcal{N}'}$. The details are given in the exercises.

Using only $\mathsf{I} = \mathsf{I}_{\mathcal{N}}$ and $\mathsf{J} = \mathsf{I}_{\mathcal{N}'}$ we can represent some functions on any level of the multi-indexed hierarchy. Furthermore, the nesting structure of the representing term corresponds quite closely to the indexing of that level. But $\boldsymbol{\lambda G}$ has many more iterators such as $\mathsf{I}_{\mathcal{N}''}$, $\mathsf{I}_{\mathcal{N}'''}$, $\mathsf{I}_{\mathcal{N}''''}$, ... and others on more complicated types. What kind of functions can be represented using these? That is the central topic of Chapter 9.

Exercises

7.35 (a) Using the term ack of Solution 7.28, produce a derivation $\nabla(ack)$ of $\vdash \mathsf{ack} : \mathcal{N}''$ and indicate its shape.

(b) Suppose the derivation

$$\nabla(f) \quad \vdash \ulcorner f \urcorner : \mathcal{N}'$$

produces a term $\ulcorner f \urcorner$ which represents a function $f : \mathbb{N}$. Indicate the shape of

$$\nabla(ack, f, m) = \nabla(ack)\nabla(f)\nabla(m)$$

for arbitrary $m \in \mathbb{N}$.

(c) Organize a computation

$$(\Box(ack, f, m)) \quad \mathsf{ack}\ulcorner f \urcorner \ulcorner m \urcorner \mathrel{\rhd\!\!\!\rhd} \ulcorner ackfm \urcorner$$

(for $m \in \mathbb{N}$) to show that $\mathsf{ack}\ulcorner f \urcorner$ represents $ack\, f$.

(d) Show how the shape of $\nabla(ack, f, m)$ changes as it is hit by $\Box(ack, f, m)$.

7.36 Produce a derivation $\nabla(Ack)$ of a term $\vdash \mathsf{Ack} : \mathcal{N}''$ such that if a derivation $\nabla(f)$ gives a representation of a function $f : \mathbb{N}'$, then the derivation $\nabla(Ack)\nabla(f)$ gives a representation of the function Ack_f. Indicate how this representation is verified.

7.37 Let $\mathsf{J} = \mathsf{I}_{\mathcal{N}'}$. In what sense does the term

$$\mathsf{Pet} = \lambda w : \mathcal{N}'', v : \mathcal{N}', u : \mathcal{N} \,.\, \mathsf{J}wvuu$$

represent the stepping-up operator Pet?

7.38 For arbitrary $s < \omega$, produce a term with a derivation $\vdash \mathsf{M}_s : \mu_s$ which, in some sense, represents the operator M_s mentioned at the end of the section. What is the type μ_s? How does this term depend on s?

7.7 Naming in $\boldsymbol{\lambda G}$

The calculus $\boldsymbol{\lambda G}$ is an attempt at a formal description of a part of arithmetic or, more generally, an investigation of higher order numeric functions. In what sense can we claim the attempt is successful?

The numerals provide canonical names for the natural numbers. There is one such numeral $\ulcorner m \urcorner$ for each $m \in \mathbb{N}$ and this term $\ulcorner m \urcorner$ is normal, and correctly formed since $\vdash \ulcorner m \urcorner : \mathcal{N}$ holds. There are other terms which name natural numbers. We know (or have been told) that each term $\vdash t : \mathcal{N}$ reduces to a unique numeral (i.e. there is a unique $m \in \mathbb{N}$ such that $t = \ulcorner m \urcorner$ or $t \rhd\!\!\rhd \ulcorner m \urcorner$). We can think of such a non-normal term t as a non-canonical name for m.

Definition 7.2 gives the notion of representing a first order function over $\mathbb{N}$. Thus we say a term $\ulcorner f \urcorner$ represents a 1-placed function $f : \mathbb{N}'$ if

$$\vdash \ulcorner f \urcorner : \mathcal{N}' \qquad \ulcorner f \urcorner \ulcorner m \urcorner \rhd\!\!\rhd\!\!\rhd \ulcorner fm \urcorner$$

hold (for all $m \in \mathbb{N}$). In other words, the term $\ulcorner f \urcorner$ must be correctly formatted within the derivation system, and the computation mechanism can be used to calculate any particular value of f. Look more closely at this second point. Given the term $\ulcorner f \urcorner$, to calculate fm we form the term $\ulcorner f \urcorner \ulcorner m \urcorner$ using the canonical (normal) name $\ulcorner m \urcorner$ of m, and then we start to reduce until we meet a normal term s. This normal term s will be $\ulcorner n \urcorner$ for some $n \in \mathbb{N}$, and then $fm = n$.

There are some questions to be asked. How do we know that a reduction of $\ulcorner f \urcorner \ulcorner n \urcorner$ will eventually meet a normal term? How do we know that all the possible different reduction paths will meet the same term? How do we know that this term is a numeral? These are general questions about $\boldsymbol{\lambda G}$, and there are some general answers.

To describe these properties it is convenient to use the reflexive version $\rhd\!\!\rhd\!\!\rhd$ of the reduction relation. This is not essential, but it does make some of the statements rather more compact.

(**Normalization**) For each derivation

$$(\nabla) \quad \Gamma \vdash t : \tau \qquad \text{there is a computation} \qquad (\square) \quad t \rhd\!\!\rhd\!\!\rhd t^*$$

to a normal term t^* (provided we allow a 'do nothing' computation to deal with the case where t is already normal and so $t^* = t$).

(**Confluence**) For each divergent wedge

$$t \rhd\!\!\rhd\!\!\rhd t_1 \qquad t \rhd\!\!\rhd\!\!\rhd t_2$$

from a common source t, there is a convergent wedge

$$t_1 \rhd\!\!\rhd\!\!\rhd t' \qquad t_2 \rhd\!\!\rhd\!\!\rhd t'$$

to a common sink t'.

(Influence) For each derivation and computation

$$(\nabla)\quad \Gamma \vdash t : \tau \qquad\qquad (\square)\quad t \rhd\!\!\!> t^+$$

there is a derivation

$$(\nabla \cdot \square)\quad \Gamma \vdash t^+ : \tau$$

(the action of $\square$ on ∇).

The first two properties, Normalization and Confluence, are not proved in this book. The third, Influence, is the subject reduction property.

To calculate fm we form the term

$$(\nabla)\quad \Gamma \vdash \ulcorner f \urcorner \ulcorner m \urcorner : \mathcal{N}$$

and then start to reduce $\ulcorner f \urcorner \ulcorner m \urcorner$ in some fashion. The Normalization property ensures that at least one reduction path will eventually terminate in a normal term. The Confluence property ensures that all reduction paths will eventually terminate, and all will produce essentially the same normal term. Thus we may reduce at random until we meet the normal term s. By surveying what we have done we can organize a computation

$$(\square)\quad \ulcorner f \urcorner \ulcorner m \urcorner \rhd\!\!\!> s$$

and the Influence property ensures that

$$(\nabla \cdot \square)\quad \Gamma \vdash s : \mathcal{N}$$

holds. Finally, a use of the Type Inhabitation Lemma 6.19, as in Lemma 7.5 and Exercise 7.9, shows that s is a numeral, and so we produce a unique $n \in \mathbb{N}$ with $s = \ulcorner n \urcorner$. By assumption, the original term $\ulcorner f \urcorner$ represents f, and hence $fm = n$.

Thus, in principle, we can use the calculus $\boldsymbol{\lambda G}$ to evaluate certain functions $f : \mathbb{N}'$. In practice we will want efficient evaluations. This will involve developing strategies which produce the right kind of representing term $\ulcorner f \urcorner$ and the right kind of computation $\square$. We do not deal with these questions of strategy and efficiency in this book.

Similar remarks apply to the representation and evaluation of first order functions of type $\mathbb{N}[k]$ for any $k > 0$. But what about higher order functions?

For two examples consider the Ackermann jump operator $ack : \mathbb{N}''$ and the Péter step-up operator $Pet : \mathbb{N}'''$. In Exercises 7.36 and 7.37 we analysed how two particular terms

$$\vdash \mathsf{ack} : \mathcal{N}'' \qquad\qquad \vdash \mathsf{Pet} : \mathcal{N}'''$$

could 'represent' these operators. In the light of that analysis we could attempt to use a hierarchy of notions of representation.

(0) A term $\vdash \ulcorner f \urcorner : \mathcal{N}'$ is said to 0-represent a function $f : \mathbb{N}'$ if

$$\ulcorner f \urcorner \ulcorner m \urcorner \rhd\!\!\!>\!\!\!> \ulcorner fm \urcorner$$

holds for all $m \in \mathbb{N}$.

(1) A term $\vdash \ulcorner F \urcorner : \mathcal{N}''$ is said to 1-represent a function $F : \mathbb{N}''$ if

$$\ulcorner f \urcorner \text{ 0-represents } f \Rightarrow \ulcorner F \urcorner \ulcorner f \urcorner \text{ 0-represents } Ff$$

holds for all terms $\vdash \ulcorner f \urcorner : \mathcal{N}'$ and functions $f : \mathbb{N}'$.

(2) A term $\vdash \ulcorner \phi \urcorner : \mathcal{N}'''$ is said to 2-represent a function $\phi : \mathbb{N}'''$ if

$$\ulcorner F \urcorner \text{ 1-represents } F \Rightarrow \ulcorner \phi \urcorner \ulcorner F \urcorner \text{ 1-represents } \phi F$$

holds for all terms $\vdash \ulcorner F \urcorner : \mathcal{N}''$ and functions $F : \mathbb{N}''$.

There is an obvious extension to a notion of k-representability for arbitrary $k < \omega$.

There is a significant difference between the notion of a 0-representation (which is the notion of a representation used up to now) and the more general notion of a k'-representation. To determine which function a term $\vdash \ulcorner f \urcorner : \mathcal{N}'$ represents we test $\ulcorner f \urcorner$ on all *normal* inhabitants $\ulcorner m \urcorner$ of $\mathcal{N}$. To determine which operator a term $\vdash \ulcorner F \urcorner : \mathcal{N}''$ may 1-represent we test $\ulcorner F \urcorner$ on *all, not just normal*, terms $\vdash \ulcorner f \urcorner : \mathcal{N}'$. For each such term $\ulcorner f \urcorner$ we need to know which function f it represents, and then we need to determine which function $\ulcorner F \urcorner \ulcorner f \urcorner$ represents.

We have seen that even the normal inhabitants of $\mathcal{N}'$ are hard to classify. If we attempt to use all inhabitants of $\mathcal{N}'$, then we are heading for trouble.

The notion of 'representation' works well enough at the first order level (with types $\mathbb{N}[k]$) but is much too complicated at higher type levels.

We adopt a more radical solution, one that is nearer the original idea of the calculus. It is possible to set up a recipe which converts

$$(\nabla) \quad \Gamma \vdash t : \tau \quad \text{into} \quad [\![\Gamma]\!] \xrightarrow{[\![\nabla]\!]} [\![\tau]\!]$$

i.e. which converts a syntactic object ∇ into a set theoretic function $[\![\nabla]\!]$. We call $[\![\nabla]\!]$ the **meaning** of ∇, and say that t (or ∇) **names** this function $[\![\nabla]\!]$ in $\boldsymbol{\lambda G}$. (The recipe has a small, but crucial, ingredient which deals with the empty context. This need not worry us here.) This notion is concerned entirely with the derivation system, whereas the notion of representation is concerned with the computation mechanism. However, we find that this recipe is coherent in the following sense.

(Coherence) For each derivation and computation

$$(\nabla) \quad \Gamma \vdash t : \tau \qquad (\square) \quad t \rhd\!\!\!\rhd t^{+}$$

the action of $\square$ on ∇

$$[\![\nabla \cdot \square]\!] = [\![\nabla]\!]$$

doesn't change the meaning.

We won't look at the details of this recipe here, but we can give the idea behind it. The recipe simply makes precise (in a set theoretic way) the intuitive notions of 'application' and 'abstraction' for functions and arguments as used to generate derivations.

The two notions of 'naming' and 'representing' are related.

7.29 THEOREM. *Each derivation*

$$(\nabla) \quad \vdash \ulcorner f \urcorner : \mathcal{N}'$$

names a function $f : \mathbb{N}'$, *and then* $\ulcorner f \urcorner$ *represents* f.

This result is not as trivial as it looks. It makes use of the four general properties

Normalization Confluence Influence Coherence

which are the pillars on which applied λ-calculi like $\boldsymbol{\lambda G}$ stand.

To sum up, the notion of a representation in $\boldsymbol{\lambda G}$ is nice enough for first order functions, but for higher order functions we must transfer our affections to the notion of naming.

Exercises

7.39 Prove Theorem 7.29. You may use the four general properties of Normalization, Confluence, Influence, and Coherence.

8

ORDINALS AND ORDINAL NOTATIONS

8.1 INTRODUCTION

Although the usefulness of multi-indexes is not entirely exhausted, it is hard to take them much further without some rather intricate notation. We have reached a point where we must make the conceptual leap to the use of ordinals. You either know something about these or you don't.

If you do know something about the ordinals then you will have recognized that multi-indexes are nothing more than the ordinals smaller than ω^ω in disguise. You may have noticed also that by making the ordinals explicit much of Chapter 7 can be reworked in a neater form. The next chapter will extend such a reworking to a much longer stretch of ordinals.

If you don't know anything about ordinals then you can treat this chapter as an introduction to the topic via one particular use. Some of the chapter you won't understand, for ordinals can be quite tricky, but you will get the general idea.

In either case you will find that not everything in this chapter is cut and dried. Some things will not be proved; in particular, some of the ordinal manipulations will depend on unjustified (and sometimes unstated) facts.

EXERCISES

8.1 Refresh your knowledge of ordinal arithmetic. For this I prefer older textbooks such as [8], [18], and [25], but you may have your own favourites.

8.2 ORDINAL ARITHMETIC

What do we need to know about the ordinals?

Let $\mathbb{O}\mathrm{rd}$ be a reasonably long stretch of the ordinals. You can think of $\mathbb{O}\mathrm{rd}$ as the countable ordinals, although we won't need all of these. We assume that $\mathbb{O}\mathrm{rd}$ has no last element and is closed under all the countable suprema that we meet. In fact, we only need the ordinals up to ϵ_0, but it is useful to remember there are some more beyond that. (In case you don't know what ϵ_0 is, this section will explain it to you.)

We write

$$\alpha, \beta, \gamma, \dots, \mu, \nu, \dots, \zeta, \eta, \dots$$

for typical ordinals, although $\mu, \nu, \ldots$ will tend to have special properties.

The ordinals $\mathbb{O}\mathrm{rd}$ are linearly ordered by a comparison $\leq$. Thus

$$\beta \leq \alpha \quad \text{or} \quad \alpha \leq \beta$$

for all $\alpha, \beta \in \mathbb{O}\mathrm{rd}$. This is the unstrict, inclusive form of linearity, whereas

$$\beta < \alpha \quad \text{or} \quad \beta = \alpha \quad \text{or} \quad \alpha < \beta$$

is the strict, exclusive form. The comparison is a well ordering, and this leads to several important consequences.

There is a smallest ordinal 0. Thus $0 \leq \beta$ for all $\beta \in \mathbb{O}\mathrm{rd}$. Each ordinal α has a next largest ordinal α'. Thus $\alpha < \alpha'$ and

$$\beta \leq \alpha \quad \text{or} \quad \alpha' \leq \beta$$

for all $\beta \in \mathbb{O}\mathrm{rd}$. Notice that

$$\alpha < \beta \Longleftrightarrow \alpha' \leq \beta \qquad \beta \leq \alpha \Longleftrightarrow \beta < \alpha'$$

for all $\alpha, \beta \in \mathbb{O}\mathrm{rd}$. We call α' the successor of α, and this gives us a function $Suc : \mathbb{O}\mathrm{rd}'$ with $Suc\,\alpha = \alpha'$. (Note the two uses of $(\cdot)'$; one use for the successor of an ordinal, and one use for the successor of a concrete type. See Table 1.2 on page 7 for a discussion of this notation.)

The first few ordinals are

$$0,\ 0',\ 0'',\ 0''',\ \ldots,\ Suc^r 0,\ \ldots$$

which we write as

$$0,\ 1,\ 2,\ 3,\ \ldots,\ r,\ \ldots$$

and deliberately confuse this initial stretch of $\mathbb{O}\mathrm{rd}$ with $\mathbb{N}$. These are the finite ordinals, and their ordinal arithmetic (which we look at later) is precisely the same as natural number arithmetic.

For each countable subset $A \subseteq \mathbb{O}\mathrm{rd}$ there is an ordinal $\bigvee A \in A$ characterized by

$$\bigvee A \leq \beta \Longleftrightarrow (\forall \alpha \in A)[\alpha \leq \beta]$$

(for $\beta \in \mathbb{O}\mathrm{rd}$). We call $\bigvee A$ the **supremum** of A. Notice that

$$\beta < \bigvee A \Longleftrightarrow (\exists \alpha \in A)[\beta < \alpha]$$

(by a simple contrapositive). It can happen that $\bigvee A \in A$, but we are mostly concerned with those sets where $\bigvee A \notin A$, so $\bigvee A$ is strictly larger than every member of A.

The supremum of the set of finite ordinals is ω, the first infinite ordinal. So far we have used 'ω' merely as a notational convenience, now it is a particular ordinal.

A non-zero ordinal which is not a successor is a **limit ordinal**. Thus ω is the least limit ordinal. Later we will meet many more. Let μ be any limit ordinal and consider the set A of smaller ordinals.

$$\alpha \in A \Longleftrightarrow \alpha < \mu$$

$\mathbb{O}$rd-recursion		$\mathbb{O}$rd-induction	
(base)	$F0 =$ an explicit value	(base)	$\langle 0 \rangle$
(step)	$F\alpha' =$ a value obtained from $F\alpha$	(step)	$\langle \alpha \rangle \Rightarrow \langle \alpha' \rangle$
(leap)	$F\mu =$ a value obtained from the set $\{F\alpha \mid \alpha < \mu\}$ of earlier values	(leap)	$(\forall \alpha < \mu)\langle \alpha \rangle \Rightarrow \langle \mu \rangle$

Table 8.1: $\mathbb{O}$rd recursion and induction schemes

Let $\nu = \bigvee A$ so that

$$\nu \leq \mu \quad \text{i.e.} \quad \nu < \mu \quad \text{or} \quad \nu = \mu$$

and we wish to show $\nu = \mu$. Notice that

$$\alpha \in A \Rightarrow \alpha < \mu \Rightarrow \alpha' \leq \mu \Rightarrow \{\alpha' < \mu \text{ or } \alpha' = \mu\} \Rightarrow \alpha' \in A$$

(since μ is a limit ordinal and hence $\alpha' \neq \mu$). With $\alpha = \nu'$ we have

$$\nu < \mu \Rightarrow \nu \in A \Rightarrow \nu' \in A \Rightarrow \nu < \alpha \in A \Rightarrow \nu < \bigvee A = \nu$$

using the characterization of $\bigvee A$. Since $\nu \not< \nu$, we have $\nu \not< \mu$, so

$$\mu = \bigvee \{\alpha \in \mathbb{O}\mathrm{rd} \mid \alpha < \mu\}$$

which is a characteristic property of limit ordinals.

Each ordinal is either

(base)	the zero ordinal, 0
(step)	a successor ordinal, α'
(leap)	a limit ordinal, $\mu = \bigvee\{\alpha \mid \alpha < \mu\}$

and we have seen examples of all three. Shortly we will see many more examples. This crude classification gives us a scheme of $\mathbb{O}$rd-recursion by which we can specify functions

$$F : \mathbb{O}\mathrm{rd} \longrightarrow \mathbb{A}$$

(for arbitrary sets $\mathbb{A}$), and an associated principle of $\mathbb{O}$rd-induction by which we can verify properties of the spec. As with all recursions and inductions, these come in several forms. We concentrate on just one version which is the most useful here. You should compare these with $\mathbb{M}_s$-recursion and $\mathbb{M}_s$-induction.

To specify such a function F (as above) we use three clauses as given in the left hand part of Table 8.1. Here α is an arbitrary ordinal and μ is a limit ordinal. Of course, in any particular application there may be other parameters involved, and some of these may be ordinals.

Given such an ordinal specification we may wish to verify a property

$$(\forall \alpha : \mathbb{N})[\cdots F\alpha \cdots]$$

where the body $\cdots F\alpha \cdots$ says something about the indicated value of F. How can we prove such an assertion? Let

$$\langle\alpha\rangle \text{ abbreviate } \cdots F\alpha \cdots$$

so that $(\forall\alpha : \mathbb{O}\mathrm{rd})\langle\alpha\rangle$ is the target assertion. To prove this it suffices to verify one instance and two implications, as in the right hand part of Table 8.1. As before, α is an arbitrary ordinal and μ is a limit ordinal.

Let's illustrate this by setting up the basics of ordinal arithmetic.

The set $\mathbb{O}\mathrm{rd}$ carries three binary operations

$$(\beta,\alpha) \longmapsto \beta+\alpha \qquad (\beta,\alpha) \longmapsto \beta\times\alpha \qquad (\beta,\alpha) \longmapsto \beta^{\alpha}$$

of addition, multiplication, and exponentiation. In time we will write

$$\beta\alpha \quad \text{for} \quad \beta\times\alpha$$

but for now we stick to the official notation. Each such operation is specified by a recursion over the second argument, α, with the first argument, β, held fixed. Addition is specified in terms of Suc, multiplication is specified in terms of addition, and exponentiation in terms of multiplication.

$$\begin{array}{ccc}
\text{(base)} & \text{(step)} & \text{(leap)} \\
\beta+0=\beta & \beta+\alpha'=(\beta+\alpha)' & \beta+\mu=\bigvee\{\beta+\alpha \mid \alpha<\mu\} \\
\beta\times 0=0 & \beta\times\alpha'=(\beta\times\alpha)+\beta & \beta\times\mu=\bigvee\{\beta\times\alpha \mid \alpha<\mu\} \\
\beta^{0} \;=1 & \beta^{\alpha'} \;=(\beta^{\alpha})\times\beta & \beta^{\mu} \;=\bigvee\{\beta^{\alpha} \mid \alpha<\mu\}
\end{array}$$

Here, as usual, α is an arbitrary ordinal and μ is a limit ordinal. Also β is an arbitrary ordinal.

For finite ordinals α, β, the ordinal values of $\beta+\alpha$, $\beta\times\alpha$, β^{α} are just the natural number values. This is why we can use this notation, and let certain symbols mean two different things. However, once we start to use the infinite ordinals there are some important differences between ordinal and natural number arithmetic.

Exercises 8.2–8.4 will get you used to these operations, $\mathbb{O}\mathrm{rd}$-recursion, and $\mathbb{O}\mathrm{rd}$-induction.

These operations of addition, multiplication, and exponentiation can be constructed from Suc in a uniform fashion. Given any ordinal function $f : \mathbb{O}\mathrm{rd}'$, we construct the ordinal iterates $(f^{\alpha} \mid \alpha \in \mathbb{O}\mathrm{rd})$ of f by

$$f^{0} = id_{\mathbb{O}\mathrm{rd}} \qquad f^{\alpha'} = f \circ f^{\alpha} \qquad f^{\mu}\beta = \bigvee\{f^{\alpha}\beta \mid \alpha<\mu\}$$

for arbitrary ordinals α, β and limit ordinals μ. In particular

$$\beta+\alpha = Suc^{\alpha}\beta \qquad \beta\times\alpha = (\cdot+\beta)^{\alpha}0 \qquad \beta^{\alpha} = (\cdot\times\beta)^{\alpha}1$$

as with natural number arithmetic. We can even continue by setting

$$\beth(\gamma,\beta,\alpha) = (\beta^{\cdot})^{\alpha}\gamma$$

for arbitrary α, β, γ to produce the ordinal **stacking function**. The (base, step, leap) specification of this is

$$\beth(\gamma,\beta,0)=\gamma \quad \beth(\gamma,\beta,\alpha')=\beta^{\beth(\gamma,\beta,\alpha)} \quad \beth(\gamma,\beta,\mu)=\bigvee\{\beth(\gamma,\beta,\alpha)\,|\,\alpha<\mu\}$$

for arbitrary ordinals α, β, γ and limit ordinals μ. For finite α, β, γ the ordinal value $\beth(\gamma,\beta,\alpha)$ is just the natural number value. You might like to worry about the value $\beth(\omega,\omega,\omega)$ for a while.

Exercises

These exercises use the recursive construction of the ordinal arithmetical operations and appropriate properties of the comparison of ordinals. As you do the exercises, you should try to set down exactly what properties are needed.

8.2 (a) Using its recursive construction show that addition has a left neutral element, is associative, but is not commutative.

(b) The left cancellation law

$$\beta+\alpha=\beta+\gamma \Rightarrow \alpha=\gamma$$

is valid for ordinals. Investigate a proof of this (using induction over α).

8.3 (a) Using the recursive construction of multiplication show

$$\beta\times 1=\beta \quad 0\times\alpha=0 \quad 1\times\alpha=\alpha$$

and that multiplication is associative but not commutative.

(b) The left cancellation law

$$\beta\times\alpha=\beta\times\gamma \Rightarrow \alpha=\gamma$$

is valid provided $\beta \neq 0$. Investigate a proof of this (using induction over α).

(c) Investigate the distributive laws

$$\gamma\times(\beta+\alpha)=\gamma\times\beta+\gamma\times\alpha \qquad (\gamma+\beta)\times\alpha=\gamma\times\alpha+\beta\times\alpha$$

(for ordinals α, β, γ).

8.4 Using the recursive construction of exponentiation show that the three identities $\gamma^1=\gamma$, $\gamma^{\beta+\alpha}=\gamma^\beta\times\gamma^\alpha$, $(\gamma^\beta)^\alpha=\gamma^{\beta\times\alpha}$ hold for all ordinals α, β, γ.

8.3 Fundamental sequences

Because of the leap clause

$$f^\mu\beta=\bigvee\{f^\alpha\beta\,|\,\alpha<\mu\}$$

the use of ordinal iterates of a function $f : \mathbb{O}\mathrm{rd}'$ really only makes sense if the values $f^\alpha\beta$ are increasing with α (for a fixed β). There is a nice result which ensures this.

Given an ordinal function $f : \mathbb{O}\mathrm{rd}'$ we say

f is inflationary	if	for all β,	$\beta \le f\beta$
f is monotone	if	for all α, β,	$\beta\le\alpha \Rightarrow f\beta\le f\alpha$

(as with numeric functions).

8.1 LEMMA. *Suppose the function $f : \mathbb{O}\mathrm{rd}'$ is inflationary.*
(a) For each $\alpha \in \mathbb{O}\mathrm{rd}$ the iterate f^α is inflationary.
(b) For each $\beta \in \mathbb{O}\mathrm{rd}$ the function $\alpha \longmapsto f^\alpha\beta$ is monotone.

Proof. (a) Let

$$\langle\alpha\rangle \quad \text{abbreviate} \quad f^\alpha \text{ is inflationary}$$

so that $(\forall\alpha : \mathbb{O}\mathrm{rd})\langle\alpha\rangle$ is the target assertion. We proceed by induction on α.

(base) Since $f^0 = id_{\mathbb{O}\mathrm{rd}}$, the base case, $\langle 0\rangle$, is immediate.

(step) Assuming $\langle\alpha\rangle$, for each $\beta \in \mathbb{O}\mathrm{rd}$ we have

$$\beta \leq f^\alpha\beta \leq f(f^\alpha\beta) = f^{\alpha'}\beta$$

to deduce $\langle\alpha'\rangle$. Here the first comparison holds by the induction hypothesis $\langle\alpha\rangle$, the second holds since f is inflationary, and the final equality holds by the (step) clause of the spec of f.

(leap) For each limit ordinal μ we have $1 < \mu$ and then

$$\beta \leq f\beta = f^1\beta \leq f^\mu\beta$$

for each $\beta \in \mathbb{O}\mathrm{rd}$ to verify $\langle\mu\rangle$. Here the last comparison uses the (leap) clause of the spec of f. Notice that this argument does not use the induction hypothesis $(\forall\alpha < \mu)\langle\alpha\rangle$ which is available.

(b) For any fixed β let

$$\langle\alpha\rangle \quad \text{abbreviate} \quad (\forall\gamma : \mathbb{O}\mathrm{rd})[\gamma \leq \alpha \Rightarrow f^\gamma\beta \leq f^\alpha\beta]$$

so that $(\forall\alpha : \mathbb{O}\mathrm{rd})\langle\alpha\rangle$ is the target assertion. We proceed by induction on α.

(base) We have

$$\gamma \leq 0 \Rightarrow \gamma = 0 \Rightarrow f^\gamma\beta = \beta = f^0\beta$$

to verify $\langle 0\rangle$, as required.

(step) Assuming $\langle\alpha\rangle$, for each ordinal γ we have

$$\begin{aligned}
\gamma \leq \alpha' \quad &\Rightarrow \quad \begin{cases} \gamma \leq \alpha \\ \text{or} \\ \gamma = \alpha' \end{cases} \\
&\Rightarrow \quad \begin{cases} f^\gamma\beta \leq f^\alpha\beta \\ \text{or} \\ f^\gamma\beta = f^{\alpha'}\beta \end{cases} \\
&\Rightarrow \quad \begin{cases} f^\gamma\beta \leq f^\alpha\beta \leq f(f^\alpha\beta) \leq f^{\alpha'}\beta \\ \text{or} \quad f^\gamma\beta = f^{\alpha'}\beta \end{cases} \quad \Rightarrow \quad f^\gamma\beta \leq f^{\alpha'}\beta
\end{aligned}$$

to deduce $\langle\alpha'\rangle$. Here the second comparison holds by the induction hypothesis, and the third uses the given inflationary property of f.

(leap) For each limit ordinal μ we have

$$f^\mu\beta = \bigvee\{f^\gamma\beta \mid \gamma < \mu\}$$

so that

$$\begin{aligned}\gamma \le \mu \quad &\Rightarrow \quad \gamma < \mu \text{ or } \gamma = \mu \\ &\Rightarrow \quad f^{\gamma}\beta \le f^{\alpha}\beta \text{ or } f^{\gamma}\beta = f^{\mu}\beta \quad \Rightarrow \quad f^{\gamma}\beta \le f^{\alpha}\beta\end{aligned}$$

to verify $\langle\mu\rangle$. Notice that this argument does not use the induction hypothesis. □

The (leap) clause of an iteration

$$f^{\mu}\beta = \bigvee\{f^{\alpha}\beta \mid \alpha < \mu\}$$

uses all the previous values $f^{\alpha}\beta$ for $\alpha < \mu$. It is often convenient to use just a selection of these values. We say a set A of ordinals is **cofinal** in a limit ordinal μ if both

$$\alpha \in A \Rightarrow \alpha < \mu \qquad \gamma < \mu \Rightarrow (\exists \alpha \in A)[\gamma \le \alpha]$$

hold for all ordinal α, γ.

8.2 LEMMA. *Suppose $f : \mathbb{O}\mathrm{rd}'$ is inflationary. Then*

$$f^{\mu}\beta = \bigvee\{f^{\alpha}\beta \mid \alpha \in A\}$$

holds for each ordinal β, limit ordinal μ, and set A cofinal in μ.

Proof. Let

$$\nu = \bigvee\{f^{\alpha}\beta \mid \alpha \in A\}$$

so that $\nu \le f^{\mu}\beta$ since both values are suprema and the right hand one is calculated over a larger set of values.

Consider any $\gamma < \mu$. There is some $\alpha \in A$ with $\gamma \le \alpha$ and hence

$$f^{\gamma}\beta \le f^{\alpha}\beta \le \nu$$

since $f^{\bullet}\beta$ is monotone (by Lemma 8.1). Thus

$$f^{\mu}\beta = \bigvee\{f^{\gamma}\beta \mid \gamma < \mu\} \le \nu$$

as required. □

In this book we always calculate with a limit ordinal via a selected cofinal set of smaller ordinal. To do this efficiently we introduce an important idea.

8.3 DEFINITION. A **fundamental sequence** for a limit ordinal μ is a function

$$\mu[\cdot] : \mathbb{N} \longrightarrow \mathbb{O}\mathrm{rd}$$

which is monotone and with a range that is cofinal in μ, i.e. such that

- $(\forall r, s : \mathbb{N})[r \le s \Rightarrow \mu[r] \le \mu[s]]$
- $(\forall r : \mathbb{N})[\mu[r] < \mu]$
- $(\forall \gamma < \mu)(\exists r : \mathbb{N})[\gamma \le \mu[r]]$

hold. □

We always calculate with a limit ordinal using a fundamental sequence. For example, given an inflationary $f : \mathbb{O}\mathrm{rd}'$ we use (base,step,leap) clauses

$$f^0 = id_{\mathbb{N}} \qquad f^{\alpha'} = f \circ f^\alpha \qquad f^\mu \beta = \bigvee\{f^{\mu[r]}\beta \mid r < \omega\}$$

to specify the iterates of f. In the (leap) clause $\mu[\cdot]$ is a fundamental sequence for the limit ordinal μ. This spec produces the same value as before since the range of $\mu[\cdot]$ is cofinal in μ, so Lemma 8.2 can be used. Later on we produce the selected fundamental sequence of each limit ordinal. Before that let's try to meet as many ordinals as possible.

Exercises

8.5 Show that

$$\beta < \alpha \Rightarrow \omega^\beta + \omega^\alpha = \omega^\alpha$$

holds for all ordinals α, β. For this you may state without proof some monotone properties of the comparison.

8.6 Using

$$\beta = \omega^{\omega+2}3 + \omega^\omega + \omega + 7 \qquad \alpha = \omega^{\omega+1}2 + \omega^\omega + 3$$

calculate $\beta \times \alpha$ in a canonical form. Use this example to discover some useful rules of manipulation for ordinal arithmetic.

8.7 (a)What is the value of 0^α for arbitrary α?
(b) Can you produce a discriminator function $\ell : \mathbb{O}\mathrm{rd}'$ where

$$\ell\alpha = \begin{cases} \alpha & \text{if } \alpha \text{ is a limit ordinal} \\ 0 & \text{if } \alpha \text{ is not limit ordinal} \end{cases}$$

(for $\alpha \in \mathbb{O}\mathrm{rd}$)?

8.4 Some particular ordinals

How can we write down particular ordinals? Initially $\mathbb{O}\mathrm{rd}$ looks like $\mathbb{N}$.

$$0,\ 1,\ 2,\ \ldots,\ r,\ \ldots \quad (r < \omega)$$

The supremum of this set is ω (as indicated), the first non-finite ordinal or least limit ordinal. After that we can continue by repeatedly taking successors. This gives

$$\omega,\ \omega+1,\ \omega+2,\ \ldots,\ \omega+r,\ \ldots \quad (r < \omega)$$

and now we need the supremum of this sequence. The spec of multiplication gives

$$\begin{aligned} \omega 0 &= 0 \\ \omega 1 &= \omega 0' = \omega 0 + \omega = \omega \\ \omega 2 &= \omega 1' = \omega 1 + \omega = \omega + \omega \end{aligned}$$

and then

$$\omega + \omega = \bigvee\{\omega + r \mid r < \omega\}$$

by the spec of addition. Thus the supremum of the second sequence is

$$\omega + \omega = \omega 2$$

and we can write either as is convenient. We are now using the standard notation and so we write $\beta\alpha$ for the product of two ordinals α, β in the order indicated (which can be read as 'α copies of β').

Next consider a sequence

$$\omega i,\ \omega i + 1,\ \omega i + 2,\ \ldots,\ \omega i + r,\ \ldots \quad (r < \omega)$$

for some fixed, but arbitrary, i. Since $\omega 0 = 0$ the case $i = 0$ gives the first sequence, and since $\omega 1 = \omega$ the case $i = 1$ gives the second sequence. The specs of multiplication and addition give

$$\omega i' = \omega i + \omega = \bigvee\{\omega i + r \mid r < \omega\}$$

so the general sequence has supremum $\omega i'$.

This gives us an array

$$\begin{array}{l}
0\ ,\quad 1,\quad 2, \ldots,\quad r, \ldots,\quad \omega \\
\omega\ ,\ \omega + 1,\ \omega + 2, \ldots,\ \omega + r, \ldots,\ \omega + \omega \\
\omega 2\ , \omega 2 + 1\ , \omega 2 + 2, \ldots, \omega 2 + r\ , \ldots, \omega 2 + \omega \\
\vdots \\
\omega i\ ,\ \omega i + 1, \omega i + 2, \ldots, \omega i + r\ , \ldots, \omega i + \omega \\
\omega i'\ , \omega i' + 1,\ \ldots \\
\vdots
\end{array}$$

where each line is continued on the next (at the same value). How do we compare two such ordinals?

$$\omega i + r \qquad \omega j + s$$

If $i \neq j$ then these ordinals lie on different lines and they are ordered by i, j. If $i = j$ then they lie on the same line and are ordered by r, s. Thus

$$\omega i + r \leq \omega j + s \Longleftrightarrow (i < j \quad \text{or} \quad i = j \text{ and } r \leq s)$$

which should remind you of something.

What is the supremum of the whole array? We have

$$\bigvee\{\omega i + r \mid i, r < \omega\} = \bigvee\{\omega i \mid i < \omega\}$$

since every member of the right hand set is in the left hand set, and every member of the left hand set is bounded above by a member of the right hand set. Thus

$$\bigvee\{\omega i + r \mid i, r < \omega\} = \bigvee\{\omega i \mid i < \omega\} = \omega\omega = \omega^2$$

using the spec of multiplication and exponentiation. The array is precisely the set of ordinals $\alpha < \omega^2$. We continue in the obvious manner to generate all ordinals

$$\omega^2 j + \omega i + r$$

for $j, i, r < \omega$. This stretch has supremum ω^3. Thus in turn we generate

$$\begin{array}{llllll} \text{all} & r & \text{for} & r < \omega & \text{with supremum} & \omega \\ \text{all} & \omega i + r & \text{for} & i, r < \omega & \text{with supremum} & \omega^2 \\ \text{all} & \omega^2 j + \omega i + r & \text{for} & j, i, r < \omega & \text{with supremum} & \omega^3 \\ \text{all} & \omega^3 k + \omega^2 j + \omega i + r & \text{for} & k, j, i, r < \omega & \text{with supremum} & \omega^4 \\ \vdots & & & & & \end{array}$$

and eventually all

$$\omega^s i_s + \omega^{s-1} i_{s-1} + \cdots + \omega^2 i_2 + \omega i_1 + i_0$$

for $i_s, \ldots, i_0 \in \mathbb{N}$ with varying s. If this doesn't remind you of something then you haven't been paying attention.

Remember the set $\mathbb{M}$ of all multi-indexes introduced in Section 7.3. These have just reappeared in clerical dress.

8.4 DEFINITION. The **indexing function**

$$\text{ind} : \mathbb{M} \longrightarrow \mathbb{O}\text{rd}$$

is given by

$$\text{ind}\,\mathsf{i} = \omega^s i_s + \cdots + \omega^2 i_2 + \omega i_1 + i_0$$

for all $\mathsf{i} = (i_s, \ldots, i_0) \in \mathbb{M}$. □

In other words, by using ind the multi-indexes give us all the ordinals in an initial stretch of $\mathbb{O}$rd. Notice also that

$$\text{ind}\,\mathsf{i} \le \text{ind}\,\mathsf{j} \iff \mathsf{i} \le \mathsf{j}$$

i.e. the comparison between multi-indexes is precisely that between the corresponding ordinals.

What is the least ordinal that is not generated in this way? We have

$$\text{ind}\,\mathsf{i} \le \omega^{s+1}$$

where s is the length of i. Thus

$$\bigvee\{\text{ind}\,\mathsf{i} \mid \mathsf{i} \in \mathbb{M}\} = \bigvee\{\omega^{s+1} \mid s < \omega\} = \bigvee\{\omega^s \mid s < \omega\} = \omega^\omega$$

using the spec of exponentiation.

8.5 THEOREM. *The indexing function* ind *isomorphically matches* $\mathbb{M}$ *with the initial stretch* $\mathbb{O}\text{rd}(\omega^\omega)$ *of ordinals* $\alpha < \omega^\omega$.

Multi-indexes give us a nice way of naming (and calculating with) smallish ordinals, but what about larger ordinals? To answer that we state a result without proof. At some stage you should look at the proof of this (in a text devoted to a more detailed analysis of ordinals), but for now we simply accept it as a fact.

For each ordinal α there is a descending sequence

$$\alpha \geq \alpha(a) \geq \alpha(a-1) \geq \cdots \geq \alpha(2) \geq \alpha(1) > 0$$

of ordinals with

$$\alpha = \omega^{\alpha(a)} + \omega^{\alpha(a-1)} + \cdots + \omega^{\alpha(2)} + \omega^{\alpha(1)} + r$$

for some $r \in \mathbb{N}$.

The case $a = 0$ gives the finite ordinals. With $\alpha(a)$ finite (for arbitrary a) we get all ordinals $< \omega^\omega$. Using some infinite exponents we get larger and larger ordinals.

To name α it suffices to name the exponents $\alpha(a), \ldots, \alpha(1)$ and use the expansion above. Some of these exponents may be infinite, but each one has an expansion which produces a second level of exponents. Some of these may be infinite, but each can be expanded to produce some third level exponents, and so on. At each level the exponents are getting smaller, so eventually this procedure will terminate. Here is an example of the kind of expansion generated.

$$\omega^{(\omega^{(\omega^\alpha + \beta)} + \gamma)} + \omega^{(\omega^\theta + \psi)} + \omega^\phi + \xi$$

Here $\alpha, \beta, \gamma, \theta, \psi, \phi, \xi$ are ordinals $< \omega^\omega$. The exponents have been punctuated to show the structure more clearly.

In this way it appears that we can use a whole battery of multi-indexes to name all ordinals. Except there is a catch.

In the expansion

$$\alpha = \omega^{\alpha(a)} + \cdots$$

we can ensure only that $\alpha \geq \alpha(a)$, *not* that $\alpha > \alpha(a)$. If we always have a strict comparison then the procedure outlined above does terminate. But if $\alpha(a) = \alpha$ then the search for an expansion starts to cycle, and we are stymied.

An ordinal θ such that $\omega^\theta = \theta$ is said to be critical. The existence of such ordinals is one of the surprises of ordinal arithmetic.

Consider the sequence $(\epsilon[r] \mid r < \omega)$ of ordinals generated by

$$\epsilon[0] = \omega \qquad \epsilon[r'] = \omega^{\epsilon[r]}$$

(for $r < \omega$). We have

$$\epsilon[0] < \epsilon[1] < \epsilon[2] < \cdots < \epsilon[r] < \cdots$$

so that

$$\epsilon_0 = \bigvee \{\epsilon[r] \mid r < \omega\}$$

is a way of topping this chain.

8.6 LEMMA. *The ordinal ϵ_0 is the smallest critical ordinal.*

Proof. Since

$$\omega^{\epsilon_0} = \bigvee\{\omega^{\epsilon[r]} \mid r < \omega\} = \bigvee\{\epsilon[r'] \mid r < \omega\} = \bigvee\{\epsilon[r] \mid r < \omega\} = \epsilon_0$$

we see that ϵ_0 is critical.

Suppose θ is critical, i.e. $\omega^\theta = \theta$. Since $0 \leq \omega$ we have

$$\begin{array}{ccccccc} 1 & = & \omega^0 & \leq & \omega^\theta & = & \theta \\ \omega & = & \omega^1 & \leq & \omega^\theta & = & \theta \end{array}$$

and then $\epsilon[r] \leq \theta$ follows by induction on r. Thus $\epsilon_0 \leq \theta$, as required. □

The ordinal ϵ_0 (mentioned at the beginning of this section) is the supremum of

$$\omega,\ \omega^\omega,\ \omega^{\omega^\omega},\ \omega^{\omega^{\omega^\omega}},\ \ldots$$

and so is quite large (at least in comparison with the ordinals we have met so far). It turns out that ϵ_0 has some natural closure properties which correspond to those of $\boldsymbol{\lambda G}$. We thus restrict out attention to ordinals $\alpha < \epsilon_0$.

8.7 DEFINITION. For each ordinal $\alpha < \epsilon_0$ there is a unique descending sequence

$$\alpha > \alpha(a) \geq \alpha(k-1) \geq \cdots \geq \alpha(2) \geq \alpha(1) > 0$$

of ordinals and a unique $r \in \mathbb{N}$ where

$$\alpha = \omega^{\alpha(a)} + \omega^{\alpha(a-1)} + \cdots + \omega^{\alpha(2)} + \omega^{\alpha(1)} + r$$

holds. Iteration of this procedure on the exponents $\alpha(a), \ldots, \alpha(1)$ leads to the **canonical expansion** for α. □

This gives us a way of writing down any particular ordinal $\alpha < \epsilon_0$. This canonical expansion is a kind of exponential polynomial in ω. Furthermore, that expression reflects many of the structural properties of α.

What does ordinal arithmetic look like using ordinal expansion? Consider

$$\mu = \omega^{\mu(m)} + \cdots + \omega^{\mu(1)} \qquad \alpha = \omega^{\alpha(a)} + \cdots + \omega^{\alpha(1)} + r$$

in canonical expansion where μ is a limit ordinal. We have

$$\mu + \alpha = \omega^{\mu(m)} + \cdots + \omega^{\mu(1)} + \omega^{\alpha(a)} + \cdots + \omega^{\alpha(1)} + r$$

but this need *not* be the canonical expansion for $\mu + \alpha$.

For instance

$$(\omega^\omega + 1) + (\omega + 1) = \omega^\omega + (1 + \omega) + 1 = \omega^\omega + \omega + 1$$

since $1 + \omega = \omega$.

Returning to the general example, if at the join $\mu(1) < \alpha(a)$ then the final expression is not a canonical form. Some of the right hand end of μ can be absorbed into α. Doing this will produce the canonical expansion for $\mu + \alpha$.

8.8 DEFINITION. Given a limit ordinal μ and an ordinal α with a canonical expansion as above, we say μ **meshes with** α and write

$$\mu \gg \alpha$$

if $\mu(1) \geq \alpha(a)$. □

Notice that

$$\omega^\gamma \gg \omega^\beta \Longleftrightarrow \gamma \geq \beta$$

so, in the canonical expansion for α we have

$$\omega^{\alpha(a)} \gg \omega^{\alpha(a-1)} \gg \cdots \gg \omega^{\alpha(2)} \gg \omega^{\alpha(1)}$$

i.e. the notation is a **meshing sum** of powers of ω.

When we add together two ordinals, $\beta + \alpha$, both given in canonical expansion, sometimes the tail of β can be absorbed into α. In extreme cases the whole of β can be absorbed into α. Thus

$$1 + \omega = \omega \qquad (\omega 7 + 3) + \omega^2 = \omega^2 \qquad (\text{indi}) + \omega^\omega = \omega^\omega$$

are example of this.

A similar thing happens with the product, $\beta \times \alpha$, of ordinals. The resulting canonical notation can be quite a lot simpler than expected at first sight. Exercise 8.6 gave an example of this.

I indicated earlier that we always calculate with a limit ordinal μ using a fundamental sequence $\mu[\cdot]$. However, I haven't yet told you how to produce such a sequence. This is dealt with in detail in the next section, but a few words are in order here.

Consider first a limit ordinal $\eta = \omega^\alpha$ which is a power of ω (so $\alpha \neq 0$). We first choose a fundamental sequence for each such η. Once we have these we produce the fundamental sequence for an arbitrary limit ordinal μ in a uniform way.

Each such limit ordinal μ which is not a power of ω can be decomposed as a meshing sum

$$\mu = \zeta + \eta$$

where $\zeta \gg \eta$ and η is a power of ω. We then set

$$\mu[x] = \zeta + \eta[x]$$

for all $x \in \mathbb{N}$. Further information on this is contained in Exercises 8.8 and 8.9.

We now have an idea of the ordinals we are interested in, the ordinals $\alpha < \epsilon_0$ (although as yet, we have no indication of why these are relevant for the analysis of $\boldsymbol{\lambda G}$). We have a canonical expansion which names each such ordinal, and we know (in principle) how to perform ordinal arithmetic using this expansion. In Chapter 7 we saw how, via the indexing function, the ordinals $\alpha < \omega^\omega$ index longish iterations. We now want to extend this to produce much longer iterations.

At this point I have to come clean; the ordinals are not good enough for what we want to do. The original idea is to index iterations using ordinals, and then show that the arithmetic of composites of iterations is reflected by the arithmetic of the indexing ordinals. Unfortunately, ordinals are not fine enough to do this job.

Intuitively, given an operator F and an ordinal α, we write

$$F^{\alpha} \quad \text{to indicate} \quad \text{a sequence of } \alpha \text{ applications of } F$$

(where the right hand side is made precise in some way). But then we want

$$F^{\alpha} \circ F^{\beta} = F^{\beta+\alpha}$$

i.e. both sides indicate β applications of F followed by α applications of F. If we can make this work then, since $1+\omega = \omega$, we have

$$F^{\omega} \circ F = F^{1+\omega} = F^{\omega}$$

which is simply false.

EXERCISES

8.8 (a) Assuming $\omega^{t+1}[x] = \omega^t x$ for all $t, x \in \mathbb{N}$, show how to calculate the fundamental sequence

$$\mathrm{ind}(\mathsf{I}, i', 0, 0)[\cdot]$$

for a multi-index of the indicated form.

(b) Write down a spec of the function which converts a multi-index into the canonical expansion for its ordinal value.

8.9 (a) Show that

$$\begin{array}{llll} \mu \gg \nu \geq \alpha & \Rightarrow \ \mu \gg \alpha & \quad \mu \gg \nu & \Rightarrow \ \mu \gg \nu[x] \\ \mu \gg \nu \gg \alpha & \Rightarrow \ \mu \gg \alpha & \quad \mu \gg \nu & \Rightarrow \ (\mu+\nu)[x] = \mu + \nu[x] \end{array}$$

hold for all limit ordinals μ, ν, ordinals α, and $x \in \mathbb{N}$.

(b) Show, by example, that in general none of

$$\mu \geq \nu \gg \alpha \Rightarrow \mu \gg \alpha \qquad \mu \gg \mu[x] \qquad (\mu+\nu)[x] = \mu + \nu[x]$$

hold for arbitrary limit ordinals μ, ν, ordinals α, and $x \in \mathbb{N}$.

(c) Describe the limit ordinals μ for which $\mu \gg \mu$.

(d) Let μ, ν be limit ordinals with $\mu \gg \nu \gg \mu$. Describe the relationship between μ and ν.

8.5 ORDINAL NOTATIONS

The problem is that the arithmetic of ordinals is just too coarse to reflect all the properties of iteration. With experience we find that it is not the ordinals we need but the *notations for ordinals*. Each notation has an ordinal value but the passage from notation to value loses something. Each of

$$F^{\omega}, F^{1+\omega}, F^{2+\omega}, F^{3+\omega}, \ldots$$

uses a notation for the first infinite ordinal, but in general the iterates are all different.

As I said earlier, not everything in this section (and chapter) is cut and dried. A watertight account of this topic, without the loopholes and pitfalls, would need a fully developed formal system of ordinal notations. That can't be done here (because it would make the book too long) so we will have to make do with a fudge.

However, we can give some of the flavour of such a formal development.

8.9 DEFINITION. The set $\mathbb{O}$ of **canonical notations** is generated as follows.

(Zero) 0 is a canonical notation.
(Omega) ω is a canonical notation.
(Step) if α is a canonical notation, then so is α'.
(Exp) if α is a canonical notation, then so is ω^α.
(Mesh) If ζ, η are canonical notations, then so is $\zeta + \eta$.

Each canonical notation α has an associated ordinal value $|\alpha|$. We routinely confuse α with $|\alpha|$. Use of (Mesh) is restricted to limit ordinals ζ, η with $\zeta \gg \eta$. □

We ought to make a clear distinction between

$\mathbb{O}$ – the set of ordinal notations $\qquad$ $\mathbb{O}\mathrm{rd}$ – the set of ordinals

even to the point of setting up a semantic function $[\![\cdot]\!] : \mathbb{O} \longrightarrow \mathbb{O}\mathrm{rd}$. Here most of the time we will behave as though $\mathbb{O}$ and $\mathbb{O}\mathrm{rd}$ are the same thing, but every now and then I will point out where this leads to problems.

In a proper account of ordinal notations the restriction in the (Mesh) clause of Definition 8.9 is dropped. This, for instance, allows us to write $1+\omega$ and distinguish it from ω.

In calculations we often meet a sum

$$\omega^\alpha + \cdots + \omega^\alpha$$

of the same powers of ω. We write this as

$$\omega^\alpha m$$

where m is the number of terms. In fact, ordinal arithmetic gives

$$\omega^\alpha r' = \omega^\alpha r + \omega^\alpha$$

so we aren't doing anything new here. Notice that

$$\omega^\alpha 0 \le \omega^\alpha 1 \le \omega^\alpha 2 \le \cdots \le \omega^\alpha r \le \cdots$$

and the supremum of this is

$$\bigvee\{\omega^\alpha r \mid r < \omega\} = \omega^\alpha \omega = \omega^{\alpha+1}$$

which leads us nicely to the next point.

In terms of notations a fundamental sequence of a limit ordinal μ is a function

$$\mu[\cdot] : \mathbb{N} \longrightarrow \mathbb{O}$$

such that

$$\mu = \bigvee\{\mu[r] \mid r < \omega\}$$

holds. We want to calculate with μ via a selected fundamental sequence $\mu[\cdot]$. How can we produce such a $\mu[\cdot]$? The first job is to decide on $\omega[\cdot]$.

8.10 DEFINITION. The selected fundamental sequence for ω is

$$\omega[\cdot] = id_{\mathbb{N}}$$

the identity function on $\mathbb{N}$. □

Any inflationary and monotone function on $\mathbb{N}$ such as

$$id_{\mathbb{N}} \qquad Suc \qquad r \longmapsto 2^r$$

could be used as a fundamental sequence for ω. We have selected $id_{\mathbb{N}}$ for convenience. Later we will see that the fundamental sequence we select can have a considerable impact on the arithmetic of notations. In fact, if we did the job properly, then we would have to consider all possible fundamental sequences. We don't go that far here.

Once we have the selected fundamental sequence for ω we can generate a fundamental sequence for all limit ordinals $\mu < \epsilon_0$ using the canonical expansion. The canonical expansion of a limit ordinal μ has one of the four forms

$$\omega \qquad \omega^{\alpha'} \qquad \omega^{\nu} \qquad \zeta + \eta$$

where α is an arbitrary ordinal, and ν, ζ, η are limit ordinals with $\zeta \gg \eta$. We give a rule for attaching a fundamental sequence to each of these shapes.

8.11 DEFINITION. Each limit ordinal μ has a selected fundamental sequence

$$\mu[\cdot] : \mathbb{N} \longrightarrow \mathbb{O}$$

generated by recursion over the canonical notation for μ using the following clauses.

(Omega)	$\omega[x]$	$= x$
(Step)	$\omega^{\alpha'}[x]$	$= \omega^{\alpha}x$
(Leap)	$\omega^{\nu}[x]$	$= \omega^{\nu[x]}$
(Mesh)	$(\zeta + \eta)[x]$	$= \zeta + \eta[x]$

Here α is an arbitrary ordinal, and ν, ζ, η are limit ordinals with $\zeta \gg \eta$, and $x \in \mathbb{N}$. For convenience we have repeated the construction of $\omega[\cdot]$. □

Given a limit ordinal μ and $m \in \mathbb{N}$, how do we calculate $\mu[m]$? We take the canonical notation for μ, use the appropriate clause of Definition 8.9, and then transform that into a canonical notation. This may involve some arithmetic and further references to Definition 8.9. The result may turn out to be a limit ordinal, in which case we can calculate $\mu[m][n]$ for $n \in \mathbb{N}$. This process can be quite long.

8.12 EXAMPLE. Consider the limit $\mu = \omega^{\omega^\omega}$. What is $\mu[3]$? Using the (Leap) clause of Definition 8.11 twice followed by (Omega) we have

$$\mu[3] = \omega^{(\omega^\omega)[3]} = \omega^{\omega^{\omega[3]}} = \omega^{\omega^3}$$

and this is still a limit ordinal so it has its own fundamental sequence. We find that a use of (Leap) then (Step) produces

$$\mu[3][3] = \omega^{\omega^3[3]} = \omega^{\omega^2 3} = \omega^{\omega^2 2 + \omega^2}$$

which is yet another limit ordinal with its own fundamental sequence. We find that

$$\mu[3][3][3] = \omega^{(\omega^2 2+\omega^2)[3]} = \omega^{\omega^2 2+\omega^2[3]} = \omega^{\omega^2 2+\omega 3} = \omega^{\omega^2 2+\omega 2+\omega} = \omega^{\nu+\omega}$$

where $\nu = \omega^2 2 + \omega 2$.

Continuing, and using a self explanatory notation, we have

$$\begin{aligned}
\mu[3]_4 &= \omega^{\nu+3} \\
\mu[3]_5 &= \omega^{\nu+2}3 &&= \omega^{\nu+2}2 + \omega^{\nu+2} \\
\mu[3]_6 &= \omega^{\nu+2}2 + \omega^{\nu+1}3 &&= \omega^{\nu+2}2 + \omega^{\nu+1}2 + \omega^{\nu+1} \\
\mu[3]_7 &= \omega^{\nu+2}2 + \omega^{\nu+1}2 + \omega^{\nu}3 &&= \omega^{\nu+2}2 + \omega^{\nu+1}2 + \omega^{\nu}2 + \omega^{\nu} \\
\mu[3]_8 &= \omega^{\nu+2}2 + \omega^{\nu+1}2 + \omega^{\nu}2 + \omega^{\nu[3]} &&= \omega^{\nu+2}2 + \omega^{\nu+1}2 + \omega^{\nu}2 + \omega^{\omega^2 2+\omega+3} \\
&\;\;\vdots
\end{aligned}$$

and, as you may be able to imagine this could go on for quite a long time. □

This calculation is based on $\omega[\cdot] = id_{\mathbb{N}}$. We can change this selected fundamental sequence of ω and retain the (Step, Leap, Mesh) clauses of Definition 8.11 to produce a different selected fundamental sequence for each limit ordinal. Even a small change in the fundamental sequence for ω can have a considerable impact.

8.13 EXAMPLE. Suppose we use $\omega^+[\cdot] = Suc$ as the fundamental sequence of ω. We can re-do the calculation of Example 8.12 using $\omega^+[\cdot]$ in place of $\omega[\cdot]$. Let $\mu^+[\cdot]$ be the resulting fundamental sequence for μ, i.e.

$$\mu^+[x] = (\omega^{\omega^\omega})^+[x] = \omega^{(\omega^\omega)^+[x]} = \omega^{\omega^{(\omega^+[x])}} = \omega^{\omega^{1+x}}$$

for each $x \in \mathbb{N}$. Thus

$$\mu^+[3] = \omega^{\omega^4} \quad \text{whereas} \quad \mu[3] = \omega^{\omega^3}$$

and we immediately see a difference. Continuing we have

$$\begin{array}{rclcl} \mu^+[3]_2 &=& \omega^{\omega^3 3} &=& \omega^{\omega^3 2+\omega^3} \\ \mu^+[3]_3 &=& \omega^{\omega^3 2+\omega^2 3} &=& \omega^{\omega^3 2+\omega^2 2+\omega^2} \\ \mu^+[3]_4 &=& \omega^{\omega^3 2+\omega^2 2+\omega 3} &=& \omega^{\omega^3 2+\nu+\omega 3} \end{array}$$

with $\nu = \omega^2 2 + \omega 2$ as before, then

$$\begin{array}{rclcl} \mu^+[3]_5 &=& \omega^{\omega^3 2+\nu+\omega^+[3]} &=& \omega^{\omega^3 2+\nu+4} \\ \mu^+[3]_6 &=& \omega^{\omega^3 2+\nu+3} 3 &=& (\omega^{\omega^3 2+\nu+3})2 + (\omega^{\omega^3 2+\nu+3}) \\ &\vdots&&& \\ \mu^+[3]_9 &&&=& \omega^{\eta+3}2 + \omega^{\eta+2}2 + \omega^{\eta+1}2 + \omega^{\eta}2 + (\omega^{\omega^3 2+\nu}) \\ \mu^+[3]_{10} &&&=& \omega^{\eta+3}2 + \omega^{\eta+2}2 + \omega^{\eta+1}2 + \omega^{\eta}2 + (\omega^{\omega^3 2+\nu^+[3]}) \\ &\vdots&&& \end{array}$$

where

$$\eta = \omega^3 2 + \nu = \omega^3 2 + \omega^2 2 + \omega 2$$

so that

$$\nu^+[3] = \omega^2 2 + \omega + \omega^+[3] = \omega^2 2 + \omega + 4$$

and we can see how that extra 1 begins to build up to cause quite a difference. In particular, by comparing

$$\begin{array}{rcl} \mu[3]_8 &=& \omega^{\nu+2}2 + \omega^{\nu+1}2 + \omega^{\nu}2 + \omega^{\omega^3 2+\omega^2 2+\omega+4} \\ \mu^+[3]_{10} &=& \omega^{\eta+3}2 + \omega^{\eta+2}2 + \omega^{\eta+1}2 + \omega^{\eta}2 + \omega^{\omega^2 2+\omega+3} \end{array}$$

and remembering that $\eta = \omega^3 2 + \nu$, we see the difference is quite dramatic. □

The reasoning behind the (Step) clause of Definition 8.9 is that

$$\omega^{\alpha'} = \omega^{\alpha}\omega \quad \text{hence} \quad \omega^{\alpha'}[x] = (\omega^{\alpha}\omega)[x] = \omega^{\alpha}(\omega[x]) = \omega^{\alpha}x$$

(since $\omega[\cdot] = id_{\mathbb{N}}$). If we use $\omega^+[\cdot] = Suc$ then logically we should change the (Step) clause to

$$(\omega^{\alpha'})^+[x] = \omega^{\alpha}(x+1)$$

(for $x \in \mathbb{N}$). If we carry this through in Example 8.13 then we get

$$\mu^+[3][3] = \omega^{\omega^2 4}$$

which leads to an even more dramatic difference.

Throughout the next chapter we use the selected fundamental sequence given by Definition 8.9. This will be just about good enough for what we do there. However, if we try to go much further then these fixed choices become a liability and a more sophisticated approach is necessary. Every now and then we will see a hint of these problems (and solutions).

Exercises

8.10 Here is a recipe for finding the canonical expansion of an ordinal $\alpha < \epsilon_0$.

- We may suppose $\omega \leq \alpha$.
- Find the largest ordinal β with $\omega^\beta \leq \alpha$.
- With this β, find the largest ordinal γ with $\omega^\beta + \gamma \leq \alpha$.
- Observe that $\beta, \gamma < \alpha$ and $\omega^\beta \gg \gamma$, so we may proceed by recursion.

Fill in the details of this recipe, giving the necessary justifications.

8.11 The canonical expansion of an ordinal $\alpha < \epsilon_0$ can be described as an exponential polynomial in ω. Let $\{\alpha\}x$ be the exponential polynomial in x obtained from the expansion of α by replacing each occurrence of 'ω' by 'x'. We can view $\{\alpha\}$ as a function of type $\mathbb{N}'$.

(a) Write down a recursive specification of the indexed family

$$(\{\alpha\} \mid \alpha \in \mathbb{O})$$

of exponential polynomials.

(b) Show that

$$\{\mu[x]\}x = \{\mu\}x$$

for all limit ordinals $\mu < \epsilon_0$ and $x \in \mathbb{N}$.

(c) Write down a recursive specification of the indexed family

$$(\{\alpha\} \mid \alpha \in \mathbb{O})$$

of exponential polynomial functions. This really does mean you should re-do part (a). However, now you should concentrate on generating functions rather than syntactic expressions. You will find that by using (b) you can produce a much neater specification than for (a).

(d) Describe the sequence

$$(\{\epsilon[r]\} \mid r < \omega)$$

of functions, and suggest a possible limit of this sequence.

8.12 (a) An ordinal θ is said to be **additively critical** if $\theta \neq 0$ and

$$\beta < \theta \Rightarrow \beta + \theta = \theta$$

holds for all β. Show that θ is additively critical if and only if $\theta = \omega^\alpha$ for some α.

(b) An ordinal θ is said to be **multiplicatively critical** if θ is infinite and

$$\beta < \theta \Rightarrow \beta \times \theta = \theta$$

holds for all non-zero β. Show that θ is multiplicatively critical if and only if $\theta = \omega^{\omega^\alpha}$ for some α.

8.13 Show that if θ is critical, i.e. $\omega^\theta = \theta$, then $\alpha^\theta = \theta$ for all $2 \leq \alpha < \theta$.

8.14 Show that $\beth(\omega, \omega, \omega) = \epsilon_0$, and calculate $\beth(\alpha, \beta, \omega)$ for $\alpha, \beta < \epsilon_0$.

9
HIGHER ORDER RECURSION

9.1 INTRODUCTION

We review what we did in Chapter 7 and then describe a way to continue.

Starting from any function $f : \mathbb{N}'$ we iterate

$$(\boldsymbol{f}) \quad f^0, f, f^2, f^3, \dots, f^r, \dots \quad (r < \omega)$$

to produce a chain of functions which, for snake-like f, have increasing rates of growth. This chain can be used as a standard against which other functions can be compared (by domination or eventual domination). For instance, when $fx = 2^x$ we have $f^r x = \beth(x, 2, r)$ which is useful for measuring combinatorial problems.

To produce a faster function we take the diagonal limit $x \longmapsto f^x x$ of the sequence $(\boldsymbol{f})$. Certainly, for suitable f, this new function eventually dominates each member of $\boldsymbol{f}$. In Chapter 7 we used slightly modified versions of this limit. This was partly for historical reasons and partly to help with certain calculations. In all cases we obtain an operator $F : \mathbb{N}''$ which, for each suitable $f : \mathbb{N}'$, produces $Ff : \mathbb{N}'$ which leaps above each member of $\boldsymbol{f}$.

Starting from any suitable pair $F : \mathbb{N}'', f : \mathbb{N}'$, the iterates of F give a chain

$$(\boldsymbol{F}, \boldsymbol{f}) \quad f, Ff, F^2 f, F^3 f, \dots, F^i f, \dots \quad (i < \omega)$$

of functions where now at each step the increase in rate of growth is much larger, at least as much as the increase along the whole of $\boldsymbol{f}$. This process can be simulated in the bottom layer of $\boldsymbol{\lambda G}$. We may take $f = Suc$ and use $\mathsf{I}_{\mathcal{N}}$ to name a suitable jump operator jmp. In the bottom layer we can represent some quite fast functions. The results of Ackermann and Grzegorczyk show that these rates of growth are precisely those obtained from the primitive recursive functions. Furthermore, the chain $(\boldsymbol{F}, \boldsymbol{f})$ is related to the structural complexity of primitive recursive specifications.

To produce a function which outstrips all of $(\boldsymbol{F}, \boldsymbol{f})$ we take the next diagonal limit $x \longmapsto F^x f x$ of the sequence. For suitable f and F this eventually dominates each $F^i f$. This is the definition of the operator $Pet : \mathbb{N}'''$, but we can easily conceive of other operators $P : \mathbb{N}'''$ which do a similar job.

Using a suitable triple $P : \mathbb{N}''', F : \mathbb{N}'', f : \mathbb{N}'$, the iterates of P give a chain

$$(\boldsymbol{P}, \boldsymbol{F}, \boldsymbol{f}) \quad f, Ff, PFf, P^2 Ff, \dots, P^j Ff, \dots \quad (j < \omega)$$

of functions with larger and larger leaps in rates of growth. Using just $\mathsf{I}_{\mathcal{N}}$ and $\mathsf{I}_{\mathcal{N}'}$ this process can be simulated in a small part of $\boldsymbol{\lambda G}$. We know (i.e. it was stated in

Chapter 7 but not proved) that these rates of growth arise from functions specified by multi-recursion. This is as far as Chapter 7 gets.

How can we outstrip all of $(\boldsymbol{P}, \boldsymbol{F}, \boldsymbol{f})$? The obvious thing to try is the diagonal limit $x \longmapsto P^x F f x$ of the sequence. This is precisely what we do, and then we repeat the trick again and again at higher and higher levels. In this way we produce a layering of the spine of $\boldsymbol{\lambda G}$, that part in which

$$\mathsf{I}_{\mathcal{N}}, \mathsf{I}_{\mathcal{N}'}, \mathsf{I}_{\mathcal{N}''}, \ldots, \mathsf{I}_{\mathcal{N}^{(r)}}, \ldots \quad (r < \omega)$$

are the only iterators used.

To be more precise, consider the sequence $\boldsymbol{O} = (O_r \mid r < \omega)$ of operators where

$$\begin{array}{lll} O_0 : \mathbb{N}''' & \text{is given by} & O_0 F f n = F^n f n \\ O_1 : \mathbb{N}^{(4)} & \text{is given by} & O_1 \phi F f n = \phi^n F f n \\ O_2 : \mathbb{N}^{(5)} & \text{is given by} & O_2 \Phi \phi F f n = \Phi^n \phi F f n \\ \vdots & & \end{array}$$

for $n \in \mathbb{N}$, $f : \mathbb{N}'$, $F : \mathbb{N}''$, $\phi : \mathbb{N}'''$, $\Phi : \mathbb{N}^{(4)}$, etc. Thus O_0 is *Pet* and O_1 is the new operator just suggested, and O_2 is the one after that, etc.

We must be careful with this notation. Here's how to keep it under control.

9.1 DEFINITION. For $r < \omega$ let $\mathbb{N}(r) = \mathbb{N}^{(r+1)}$ and let $O_r : \mathbb{N}(r)''$ be given by

$$O_r t s \phi_r \cdots \phi_2 \phi_1 x = t^x s \phi_r \cdots \phi_2 \phi_1 x$$

where

$$(*) \quad t : \mathbb{N}(r)',\ s : \mathbb{N}(r),\ \phi_r : \mathbb{N}^{(r)}, \ldots \phi_2 : \mathbb{N}'',\ \phi_1 : \mathbb{N}',\ x : \mathbb{N}$$

are the consumed arguments. □

The sequence $\boldsymbol{O}$ uses higher and higher types. This is not a problem in $\boldsymbol{\lambda G}$. Mimicking Definition 9.1 let $\mathcal{N}(r) = \mathcal{N}^{(r+1)}$ for each $r < \omega$, so that $[\![\mathcal{N}(r)]\!] = \mathbb{N}(r)$. Let

$$\mathsf{O}_r = \lambda z : \mathcal{N}(r)', y : \mathcal{N}(r) \,.\, \lambda v_r : \mathcal{N}^{(r)}, \ldots, v_1 : \mathcal{N}', x : \mathcal{N} \,.\, \mathsf{I}_{\mathcal{N}(r)} z y x v_r \cdots v_1 x$$

to produce a term with $\vdash \mathsf{O}_r : \mathcal{N}(r)''$. Almost trivially O_r names O_r. Later in this chapter we will produce a different name for each O_r. This will allow a direct measure of the increasing strength of the sequence $\boldsymbol{O}$.

Using $\boldsymbol{O}$ we produce a sequence $\boldsymbol{E} = (E_r \mid r < \omega)$ of operators each of type $\mathbb{N}'''$.

$$E_0 = O_0,\ E_1 = O_1 O_0,\ E_2 = O_2 O_1 O_0,\ \ldots,\ E_r = O_r O_{r-1} \cdots O_1 O_0,\ \ldots$$

Each such E_r is named by

$$\mathsf{O}_r \mathsf{O}_{r-1} \cdots \mathsf{O}_1 \mathsf{O}_0$$

and this term is built from the iterators I_σ for $\sigma \in \{\mathcal{N}(0), \mathcal{N}(1), \ldots, \mathcal{N}(r)\}$. There are many other operators obtainable from $\boldsymbol{O}$. For instance

$$\begin{array}{ll} \text{of type } \mathbb{N}(2) \ - & O_0^2 \quad O_1 O_0^2 \quad O_1^2 O_0 \quad O_1^2 O_0^2 \quad O_2 O_1^2 O_0 \quad O_2 O_1^2 O_0^2 \quad \cdots \\ \text{of type } \mathbb{N}(3) \ - & O_1^2 \quad O_2 O_1^2 \quad O_2^2 O_1 \quad O_2^2 O_1^2 \quad \cdots \\ \text{of type } \mathbb{N}(4) \ - & O_2^2 \quad \cdots \end{array}$$

and other such compounds. We will describe these operators in a uniform manner to allow a direct comparison of the relative strengths. In short we attach to each such compound a measure which determines that compound and facilitates the required comparisons.

How can we measure the power of an operator $F : \mathbb{N}''$? In Chapter 7 we used multi-indexes. Much of the presentation was in terms of $\mathbb{M}_3$, but it should be clear how to use an arbitrary $\mathbb{M}_s$. Now we want to use the whole of $\mathbb{M}$.

Each multi-index $\mathsf{i} \in \mathbb{M}$ is a list $\mathsf{i} = (i_s, \dots, i_0)$ of natural numbers thought of as extending from right to left as the subscripts indicate. In fact, we can think of such a multi-index as an infinite list which is eventually zero. Two such indexes $\mathsf{i} = (i_s, \dots, i_0)$ and $\mathsf{j} = (i_t, \dots, j_0)$ are compared by first filling out with zeros the left hand end of one or other (or both) to make the lengths equal, and then looking for the leftmost difference. In this way $\mathbb{M}$ becomes a linearly ordered (in fact, well ordered) set. The left hand zeros make no difference to the position of a multi-index in this ordering.

We often want to concentrate on one or two adjacent components a multi-index. To do this we display it as

$$(\mathsf{l}, k, j, \mathsf{r})$$

where k and j are the relevant components and l and r are the left hand and right hand ends. In some circumstances one or other of l, r may be empty.

There is an $\mathbb{M}$-recursion scheme and an associated $\mathbb{M}$-induction scheme. These are best illustrated by a particular example. The details will be explained after the definition.

9.2 DEFINITION. The Péter function $\mathfrak{P} : \mathbb{M} \longrightarrow \mathbb{N}'''$ is specified by

$$\begin{array}{lll} \text{(base)} & \mathfrak{P}0 & = Id_{\mathbb{N}'} \\ \text{(step)} & \mathfrak{P}(\mathsf{l}, i')F & = F \circ \mathfrak{P}(\mathsf{l}, i)F \\ \text{(leap)} & \mathfrak{P}(\mathsf{l}, i', 0, 0)Ffx & = \mathfrak{P}(\mathsf{l}, i, x, 0)Ffx \end{array}$$

for all $F : \mathbb{N}'', f : \mathbb{N}', x : \mathbb{N}$ and multi-indexes of the indicated form. Here $Id_{\mathbb{N}'} : \mathbb{N}'''$ satisfies $Id_{\mathbb{N}'}Ff = f$ for all $F : \mathbb{N}'', f : \mathbb{N}'$, i.e. $Id_{\mathbb{N}'}F = id_{\mathbb{N}'}$. □

Thus $\mathfrak{P}$ is a higher order function which consumes a multi-index i to return a value $\mathfrak{P}\mathsf{i} : \mathbb{N}'''$. This, in turn, consumes some $F : \mathbb{N}''$, then some $f : \mathbb{N}'$, and finally some $x : \mathbb{N}$, to return an eventual natural number value $\mathfrak{P}\mathsf{i}Ffx$.

In the (base) clause 0 is the zero multi-index $(\dots, 0, 0, 0)$, the least member of $\mathbb{M}$. The successor i' of a multi-index i is obtained by increasing i_0 to $i_0 + 1$. In the (step) clause (l, i') is the successor of (l, i) and this clause can be rephrased as

$$\mathfrak{P}\mathsf{i}'F = F \circ \mathfrak{P}\mathsf{i}F$$

(for $\mathsf{i} \in \mathbb{M}$). Any non-zero, non-successor multi-index must satisfy $i_{t+1} \neq 0$ and $i_t = i_{t-1} = \cdots = i_0 = 0$ for some position t. Such a multi-index is displayed, as in the (leap) clause, as

$$(\mathsf{l}, i', 0, 0)$$

where $i_{t+1} = i'$ and $i_t = 0$ with an arbitrary left hand end l and a zero right hand end $\mathsf{0}$ (which is empty if $t = 0$). This clause pulls the eventual last argument x into this index.

The specification of $\mathfrak{P}$ is an example of $\mathbb{M}$-recursion. It is by no means the most general kind, but it is the one we use most often. There is a corresponding principle of $\mathbb{M}$-induction. Again we illustrate this with an example.

Given any $F : \mathbb{N}''$ we may use $Pet : \mathbb{N}'''$ to obtain $F_k = Pet^k F$ for each $k < \omega$. Thus

$$F_0 = F \qquad F_{k'}fx = F_k^x fx$$

for all $f : \mathbb{N}', x : \mathbb{N}$. Each multi-index $\mathsf{i} = (i_s, \ldots, i_0)$ gives us a compound member

$$F_0^{i_0} \circ F_1^{i_1} \circ \cdots \circ F_s^{i_s}$$

of $\mathbb{N}''$. The function $\mathfrak{P}$ encapsulates this multi-iterated use of Pet.

9.3 LEMMA. *For each $F : \mathbb{N}''$ and $\mathsf{i} = (i_s, \ldots, i_0) \in \mathbb{M}$*

$$\mathfrak{P}\mathsf{i}F = F_0^{i_0} \circ F_1^{i_1} \circ \cdots \circ F_s^{i_s}$$

holds.

Proof. We proceed by an $\mathbb{M}$-induction in three phases as follows.

(base) Verify the identity for the particular case $\mathsf{i} = \mathsf{0}$.

(step) Show that, for arbitrary i, the assertion for i implies the assertion for i'.

(leap) Show that for each non-zero, non-successor multi-index i, the assertion for all previous $\mathsf{j} < \mathsf{i}$ implies the assertion for i.

The details are left for Exercise 9.1. □

To go beyond the family $(\mathfrak{P}\mathsf{i} \mid \mathsf{i} \in \mathbb{M})$ of operators (of type $\mathbb{N}'''$) we need to extend the linearly ordered set $\mathbb{M}$ of indexes. This can be done in an ad hoc fashion, but eventually this becomes too cumbersome. We need a uniform approach to indexing long iterations. We make explicit use of ordinals. So far we have used a disguised form of the ordinals below ω^ω, but now we go beyond this.

The intuitive idea is that for suitable sets $\mathbb{S}$, operations $F : \mathbb{S}''$, and ordinals α, we compose F with itself α times to produce a new operator $F^\alpha : \mathbb{S}''$, the αth iterate of F. We know how to do this for $\alpha < \omega$. Using the Péter function we can do this for $\alpha < \omega^\omega$ provided $\mathbb{S} = \mathbb{N}'$. We now continue the process to much longer ordinals, first for $\mathbb{S} = \mathbb{N}'$, and then for arbitrary $\mathbb{S}$.

This is why we reviewed the relevant properties of ordinals in Chapter 8.

Exercises

9.1 Prove Lemma 9.3 using the three suggested phases.

9.2 As in Definition 9.1 let $\mathbb{N}(r) = \mathbb{N}^{(r+1)}$, so that $O_r : \mathbb{N}(r)$. For each $t : \mathbb{N}(r)'$ let $t_k = O_r^k t$ for each $k \in \mathbb{N}$ (as used in Lemma 9.3 for the case $r = 0$).

(a) Write down the spec of a function $\mathfrak{P}_r : \mathbb{M} \longrightarrow \mathbb{N}(r)''$ such that

$$\mathfrak{P}_r(i_s, \dots, i_0)t = t_0^{i_0} \circ \cdots \circ t_s^{i_s}$$

for all $t : \mathbb{N}(r)'$ and $\mathsf{i} = (i_s, \dots, i_0) \in \mathbb{M}$. You need not give the proof of correctness.

(b) Express O_r in the form $\mathfrak{P}_r\mathsf{i}$ for a suitable r.

(c) Using the array of functions $\mathfrak{P}_\bullet$, describe the function $(k, t) \longmapsto t_k$ (of type $\mathbb{N} \longrightarrow \mathbb{N}(r)''$) for arbitrary r.

9.3 Given any function $f : \mathbb{N}'$ we can produce a chain $(F_s \mid s < \omega)$ of functions

$$F_s : \mathbb{M}_s \longrightarrow \mathbb{N} \longrightarrow \mathbb{N}' \quad \text{by} \quad F_s\mathsf{i}rx = (ack_0^{i_0}(ack_1^{i_1}(\cdots(ack_s^{i_s} f)\cdots)))^{r+1}x$$

for each $\mathsf{i} = (i_s, \dots, i_0) \in \mathbb{M}_s$ and $r, x \in \mathbb{N}$. Here $ack_k = Pet^k ack$ for each $k < \omega$. In particular, $F_0 = Ack_f$ as used in Section 7.4 and specified in Exercise 7.29 (and in Exercise 7.30). Write down a spec of F_s for arbitrary s. You need not justify this spec, but you should relate it to other known constructions.

9.4 Given any function $f : \mathbb{N}'$ and any jump operator $jmp : \mathbb{N}''$ we can produce a chain $(G_s \mid s < \omega)$ of functions $G_s : \mathbb{M}_s \longrightarrow \mathbb{N}'$ by

$$G_s\mathsf{i}x = (jmp_0^{i_0}(jmp_1^{i_1}(\cdots(jmp_s^{i_s} f)\cdots)))x$$

for each $\mathsf{i} = (i_s, \dots, i_0) \in \mathbb{M}_s$ and $x \in \mathbb{N}$. Here $jmp_k = Pet^k jmp$ for each $k < \omega$.

(a) Write down a spec of G_s for arbitrary s for the cases $jmp = rob$ and $jmp = brw$.

(b) Can you write down a spec of G_s for the case $jmp = ack$?

9.2 The long iterator

The Péter function $\mathfrak{P} : \mathbb{M} \longrightarrow \mathbb{N}'''$ encapsulates much of what is done in Chapter 7. From that chapter we know that the (base, step, leap) specification of $\mathfrak{P}$ describes a highly nested battery of iterations along ω. The indexing function $\text{ind} : \mathbb{M} \longrightarrow \mathbb{O}(\omega^\omega)$ allows us to replace this battery by a single iteration along ω^ω. Furthermore, the nesting structure of the battery is reflected by the structure of ω^ω.

We can iterate along much larger ordinals.

9.4 **DEFINITION**. The **long iterator** $\mathbb{G} : \mathbb{O} \longrightarrow \mathbb{N}'''$ is specified by

$$\text{(base)}\ \mathbb{G}0 = Id_{\mathbb{N}'} \qquad \text{(step)}\ \mathbb{G}\alpha' F = F \circ \mathbb{G}\alpha F \qquad \text{(leap)}\ \mathbb{G}\mu F f x = \mathbb{G}\mu[x] F f x$$

for all ordinals α, limit ordinals μ, and $F : \mathbb{N}'', f : \mathbb{N}', x : \mathbb{N}$. Here $Id_{\mathbb{N}'} : \mathbb{N}'''$ is as in Definition 9.2. □

The (leap) clause makes use of the selected fundamental sequence $\mu[\cdot]$ of μ. The value $\mathbb{G}\alpha F f$ is quite sensitive to this selection.

9.5 EXAMPLE. It is routine to show $\mathbb{G}mFf = F^m f$ for all $m < \omega$, and hence

$$\mathbb{G}\omega Ffx = \mathbb{G}\omega[x]Ffx = F^{\omega[x]}fx$$

for all $F : \mathbb{N}'', f : \mathbb{N}', x : \mathbb{N}$. We selected $\omega[\cdot] = id_{\mathbb{N}}$ as the fundamental sequence for ω, but for many purposes $\omega^+[\cdot] = Suc$ is neater. These two selections give

$$\mathbb{G}\omega Ffx = F^x fx \qquad \mathbb{G}\omega^+ Ffx = F^{x+1}fx$$

so that $\mathbb{G}\omega = O_0 = Pet$ for our preferred selection with $\mathbb{G}\omega^+ F = PetF \circ F$ for the other. A large ordinal has many occurrences of ω nested inside it, and then the difference is quite dramatic. □

We often write F^α for $\mathbb{G}\alpha F$ (for an ordinal α). This example gives

$$F^\omega = \mathbb{G}\omega F \neq \mathbb{G}\omega^+ F = F^{\omega +}$$

in fact $F^{\omega +} = F^\omega \circ F$. The value F^α is sensitive to the fundamental sequences used in α. The operator $\mathbb{G}\alpha$ is not determined solely by the ordinal value of α, but by the *notation* for that ordinal. We should keep in mind this distinction between ordinal and notation.

The indexing function makes precise the similarity between $\mathbb{G}$ and $\mathfrak{P}$.

9.6 THEOREM. $\mathfrak{P} = \mathbb{G} \circ \mathrm{ind}$

Proof. We show $\mathfrak{P}\mathsf{i} = \mathbb{G}(\mathrm{ind}\,\mathsf{i})$ for all $\mathsf{i} \in \mathbb{M}$ by $\mathbb{M}$-induction. We use

$$\mathrm{ind}\,\mathsf{0} = 0 \qquad \mathrm{ind}\,\mathsf{i}' = (\mathrm{ind}\,\mathsf{i})' \qquad \mathrm{ind}\,(\mathsf{l}, i', 0, 0)[\cdot] = \mathrm{ind}\,(\mathsf{l}, i, \cdot, 0)$$

for the (base, step, leap) clauses, respectively, as given in Solution 8.8(b). Only the (leap) case is not immediate.

Consider any $F : \mathbb{N}'', f : \mathbb{N}', x : \mathbb{N}$. Using, in turn, the (leap) clause of the spec of $\mathfrak{P}$, the induction hypothesis, the description of a fundamental sequence $\mathrm{ind}\,(\cdot)[\cdot]$, and the (leap) clause of the spec of $\mathbb{G}$, we have

$$\begin{aligned} \mathfrak{P}(\mathsf{l}, i', 0, 0)Ffx &= \mathfrak{P}(\mathsf{l}, i, x, 0)Ffx \\ &= \mathbb{G}\mathrm{ind}\,(\mathsf{l}, i, x, 0)Ffx \\ &= \mathbb{G}(\mathrm{ind}\,(\mathsf{l}, i, \cdot, 0)[x])Ffx \ = \ \mathbb{G}\mathrm{ind}\,(\mathsf{l}, i', 0, 0)Ffx \end{aligned}$$

to give the required result. □

With this we can extend the hierarchies of Chapter 7 much, much further. How much further? For now let's see what the first new step is.

9.7 EXAMPLE. For each small ordinal

$$\alpha = \omega^s i_s + \cdots + \omega i_1 + i_0 \quad \text{we have} \quad \mathbb{G}\alpha F = \mathfrak{P}(i_s, \ldots, i_0)F = F_0^{i_0} \circ \cdots \circ F_s^{i_s}$$

in the notation of Lemma 9.3. Since $\omega^\omega[x] = \omega^x$ this gives

$$\mathbb{G}\omega^\omega Ffx = \mathbb{G}\omega^x Ffx = F_x fx$$

so that $\mathbb{G}\omega^\omega$ is the operator E_1. □

We will describe $\mathbb{G}\alpha$ for all $\alpha < \epsilon_0$. In particular, we show $E_r = \mathbb{G}\epsilon[r]$ for all $r < \omega$. After that we attach to each ordinal $\alpha < \epsilon_0$ a term $\langle\alpha\rangle$ of $\boldsymbol{\lambda G}$ which names $\mathbb{G}\alpha$. Thus some extremely fast functions can be represented in $\boldsymbol{\lambda G}$.

There are several standard long function hierarchies $\boldsymbol{f} = (f_\alpha \mid \alpha \in \mathbb{O})$ all generated in the form

$$f_0 = \text{explicit function} \qquad f_{\alpha'} = \text{modified version of } f_\alpha \qquad f_\mu x = f_{\mu[x]}x$$

for the usual α, μ, x. These can be obtained using $\mathbb{G}$. Let's look at the hierarchies given in Example 2.16 of [9].

9.8 EXAMPLES. (Fast) For each $f : \mathbb{N}'$ and a standard jump operator jmp set

$$f_0 = f \qquad f_{\alpha'} = jmp\, f_\alpha \qquad f_\mu x = f_{\mu[x]}x$$

for the usual α, μ, x to obtain $\boldsymbol{f}$. This produces a constructor

$$\text{Fast} : \mathbb{N}'' \longrightarrow \mathbb{N}' \longrightarrow \mathbb{O} \longrightarrow \mathbb{N}'$$

with $\boldsymbol{f} = \text{Fast}\, jmp f$. The two Fast-Growing hierarchies $B_\bullet$ and $F_\bullet$ of [9] are $\text{Fast}\, polSuc$ and $\text{Fast}\, ackSuc$ respectively. Notice that $\text{Fast}\, Ff\alpha = \mathbb{G}\alpha Ff$, i.e. Fast is just a rearranged version of $\mathbb{G}$.

(Slow) For each $g : \mathbb{N}'$ set

$$g_0 = \mathsf{zero} \qquad g_{\alpha'} = g \circ g_\alpha \qquad g_\mu x = g_{\mu[x]}x$$

for the usual α, μ, x to obtain $\boldsymbol{g}$. This produces a constructor

$$\text{Slow} : \mathbb{N}' \longrightarrow \mathbb{O} \longrightarrow \mathbb{N}'$$

with $\boldsymbol{g} = \text{Slow}\, g$. The Slow-Growing hierarchy $G_\bullet$ of [9] is $\text{Slow}\, Suc$. Each $g : \mathbb{N}'$ produces a $G : \mathbb{N}''$ such that $\text{Slow}\, g\alpha = \mathbb{G}\alpha G\mathsf{zero}$ (for $\alpha \in \mathbb{O}$).

(Hardy) For each $h : \mathbb{N}'$ set

$$h_0 = id_{\mathbb{N}} \qquad h_{\alpha'} = h_\alpha \circ h \qquad h_\mu x = h_{\mu[x]}x$$

for the usual α, μ, x to obtain $\boldsymbol{h}$. This produces a constructor

$$\text{Hard} : \mathbb{N}' \longrightarrow \mathbb{O} \longrightarrow \mathbb{N}'$$

with $\boldsymbol{h} = \text{Hard}\, h$. The Hardy hierarchy $H_\bullet$ of [9] is $\text{Hard}\, Suc$. Each $h : \mathbb{N}'$ produces a $H : \mathbb{N}''$ such that $\text{Hard}\, h\alpha = \mathbb{G}\alpha H id_{\mathbb{N}}$ (for $\alpha \in \mathbb{O}$).

(Laurel) For each $g, h, l : \mathbb{N}'$ set

$$l_0 = l \qquad l_{\alpha'} = g \circ l_\alpha \circ h \qquad l_\mu x = l_{\mu[x]}x$$

for the usual α, μ, x to obtain $\boldsymbol{l}$. This produces a constructor

$$\text{Laur} : \mathbb{N}' \longrightarrow \mathbb{N}' \longrightarrow \mathbb{N}' \longrightarrow \mathbb{O} \longrightarrow \mathbb{N}'$$

with $\boldsymbol{l} = \text{Laur}\, ghl$. Notice that

$$\text{Slow}\, g = \text{Laur}\, g id_{\mathbb{N}}\mathsf{zero} \qquad \text{Hard}\, h = \text{Laur}\, id_{\mathbb{N}} h id_{\mathbb{N}}$$

and hence this construction encompasses both the Slow and the Hardy constructions. Each $g, h : \mathbb{N}'$ produces some $L : \mathbb{N}''$ such that $\text{Laur}\, ghl\alpha = \mathbb{G}\alpha Ll$ (for $\alpha \in \mathbb{O}$). □

Many of the rather complicated specifications we saw in Chapter 7 can be rephrased as ordinal iterations using $\mathbb{G}$. Some of these are given in the exercises.

EXERCISES

9.5 (a) Describe constructors $g \longmapsto G$, $h \longmapsto H$, and $(g,h) \longmapsto L$ where

$$\mathrm{Slow}\, g = \mathbb{G}(\cdot)G\mathsf{zero} \quad \mathrm{Hard}\, g = \mathbb{G}(\cdot)Hid_{\mathbb{N}} \quad \mathrm{Laur}\, gh = \mathbb{G}(\cdot)L$$

for $g, h, l : \mathbb{N}'$.

(b) Show that for each $g, h, l : \mathbb{N}'$ and $\alpha \in \mathbb{O}, x \in \mathbb{N}$ we have

$$\mathrm{Laur}\, ghl\alpha x = (g^y \circ l \circ h^y)x \quad \text{where } y = e_\alpha x$$

where the sequence $(e_\alpha \mid \alpha \in \mathbb{O})$ of functions is determined by h.

9.6 Exercise 8.11 gives a family $(\{\alpha\} \mid \alpha \in \mathbb{O})$ of exponential polynomial functions (where $\{\alpha\}$ codes α). Relate this to the standard hierarchies of Example 9.8.

9.7 Show there is a family $(p_\alpha \mid \alpha \in \mathbb{O})$ of exponential polynomial functions such that $g_\alpha x = g^y 0$ where $g_\bullet = \mathrm{Slow}\, g$ and $y = p_\alpha x$ for each $g : \mathbb{N}'$ and $\alpha \in \mathbb{O}, x \in \mathbb{N}$.

9.8 Exercise 9.3 attaches to each $f : \mathbb{N}'$ a chain of functions F_s of varying type.

(a) Write down the spec of a single generating function $F : \mathbb{O} \longrightarrow \mathbb{N} \longrightarrow \mathbb{N}'$ such that $F(\mathsf{ind}\,\mathsf{i}) = F_s\mathsf{i}$ for each $\mathsf{i} \in \mathbb{M}_s$ and $s < \omega$.

(b) Indicate how

$$F\alpha r x = (\mathbb{G}\alpha ack f)^{r+1} x$$

follows by an appropriate induction (for arbitrary α, r, x).

(c) What are the functions $F\omega^\omega 0$ and $F\omega^{\omega^\omega} 0$?

9.9 Exercise 9.4 attaches to each $f : \mathbb{N}'$ a chain of functions G_s of varying type.

(a) Write down the spec of a single generating function $G : \mathbb{O} \longrightarrow \mathbb{N}'$ such that $G(\mathsf{ind}\,\mathsf{i}) = G_s\mathsf{i}$ for each $\mathsf{i} \in \mathbb{M}_s$ (for each $s < \omega$).

(b) How is G related to $\mathbb{G}$?

(c) What are the function $G\omega^\omega$ and $G\omega^{\omega^\omega}$?

9.3 LIMIT CREATION AND LIFTING

We use the long iterator $\mathbb{G}$ to describe in a uniform fashion the members of the sequence $\boldsymbol{O}$ and various compounds of these. To do that we bring out two kinds of parametric properties of the construction.

Definition 9.1 produces an operator $O_{\mathbb{S}} : \mathbb{S}''$ for each one of a certain family of sets $\mathbb{S}$. The type of $\mathbb{S}$ is a parameter of the construction. In Definition 9.1 this is coded by the subscript r. Later we make this parametric behaviour more explicit.

Each operator O_r is used to form the diagonal limit of a certain kind of ω-sequence (where the kind is determined by r). In some vague sense this is the same trick used at different levels. In this section we make this notion less vague.

For each set $\mathbb{S}$ or type σ we use

$$\mathcal{L}(\mathbb{S}) = (\mathbb{N} \longrightarrow \mathbb{S}) \longrightarrow \mathbb{S} \qquad \mathcal{L}(\sigma) = (\mathcal{N} \to \sigma) \to \sigma$$

as convenient abbreviations. Note that $[\![\mathcal{L}(\sigma)]\!] = \mathcal{L}(\mathbb{S})$ whenever $[\![\sigma]\!] = \mathbb{S}$.

9.9 DEFINITION. For each set $\mathbb{S}$ a **limit creator** on $\mathbb{S}$ is a function $L : \mathcal{L}(\mathbb{S})$ which converts each sequence $p : \mathbb{N} \longrightarrow \mathbb{S}$ in $\mathbb{S}$ into an element of $\mathbb{S}$. □

Some sets $\mathbb{S}$ carry 'natural' and useful limit creators and others don't. For instance, $\mathbb{N}$ does not. (Can you think of an interesting way of converting an *arbitrary* function $p : \mathbb{N}'$ into a natural number?) Here are some examples of immediate interest.

9.10 EXAMPLE. (a) Let $\mathbb{S} = (\mathbb{N} \longrightarrow \mathbb{T})$ where $\mathbb{T}$ is arbitrary. The function $D : \mathcal{L}(\mathbb{S})$ where

$$Dpu = puu$$

is the **diagonal limit creator** for $\mathbb{T}$. The diagonal limit creator Δ for $\mathbb{N}$ inhabits $\mathcal{L}(\mathbb{N}')$.

(b) Let L be a limit creator on $\mathbb{S}$. Let $\mathbb{S}^{\mathbb{R}} = (\mathbb{R} \longrightarrow \mathbb{S})$ for an arbitrary set $\mathbb{R}$. The function $L^{\mathbb{R}}$ where

$$L^{\mathbb{R}}qr = Lp \quad \text{where } pu = qur \text{ for } u \in \mathbb{N}$$

(for $q : \mathbb{N} \longrightarrow \mathbb{S}^{\mathbb{R}}$) is the **lift** of L from $\mathbb{S}$ to $\mathbb{S}^{\mathbb{R}}$. □

Later we will concentrate on a uniform family of limit creators on $(\mathbb{N}(r) \mid r < \omega)$, but we need not be so particular just yet.

9.11 DEFINITION. For each set $\mathbb{S}$ let $\Uparrow_{\mathbb{S}} : \mathcal{L}(\mathbb{S}) \longrightarrow \mathcal{L}(\mathbb{S}')$ be given by

$$\Uparrow_{\mathbb{S}} Lqs = Lp \quad \text{where} \quad px = qxs$$

for $L : \mathcal{L}(\mathbb{S})$, $q : \mathbb{N} \longrightarrow \mathbb{S}'$, $s : \mathbb{S}$ and $x \in \mathbb{N}$. We call $\Uparrow_\sigma$ the **limit lifter** for $\mathbb{S}$. This is the special case $\mathbb{R} = \mathbb{S}$ of Example 9.10(b).

For each type σ let

$$\uparrow_\sigma = \lambda l : \mathcal{L}(\sigma), q : \mathcal{N} \to \sigma', x : \sigma \,.\, l(\lambda u : \mathcal{N} \,.\, qux)$$

to produce a term $\vdash \uparrow_\sigma : \mathcal{L}(\sigma) \to \mathcal{L}(\sigma')$. Clearly $[\![\uparrow_\sigma]\!] = \Uparrow_{\mathbb{S}}$ whenever $[\![\sigma]\!] = \mathbb{S}$. □

It is convenient to write

$$L' \quad \text{for} \quad \Uparrow_{\mathbb{S}} L \qquad \text{to match} \qquad \mathbb{S}' \quad \text{for} \quad (\mathbb{S} \longrightarrow \mathbb{S})$$

and refer to L' as *the* lift of L. Lifting can be iterated to produce

$$\begin{array}{lcll}
L' & = & \Uparrow_{\mathbb{S}} L & \text{with } L' : \mathcal{L}(\mathbb{S}') \\
L'' & = & \Uparrow_{\mathbb{S}'} (\Uparrow_{\mathbb{S}} L) & \text{with } L'' : \mathcal{L}(\mathbb{S}'') \\
L''' & = & \Uparrow_{\mathbb{S}''} (\Uparrow_{\mathbb{S}'} (\Uparrow_{\mathbb{S}} L) & \text{with } L''' : \mathcal{L}(\mathbb{S}''') \\
 & \vdots & & \\
L^{(r')} & = & \Uparrow_{\mathbb{S}^{(r)}} L^{(r)} \qquad = L^{(r)\prime} & \\
 & \vdots & &
\end{array}$$

where the uniformity of this construction is at the level of *types* not inhabitants. We have $L^{(r)} : \mathcal{L}(\mathbb{S}^{(r)})$, but this notation omits some information.

We need to know how to evaluate an iterated lift $L^{(r)}$ of a limit creator L.

9.12 LEMMA. *For each set* $\mathbb{A}$ *and* $L : \mathcal{L}(\mathbb{A}')$

$$L^{(r)}pa_r \cdots a_1 = Lp_r \quad \text{where } p_r u = pua_r \cdots a_1 \text{ for } u \in \mathbb{N}$$

for $a_r : \mathbb{A}^{(r)}, \ldots, a_1 : \mathbb{A}'$ *and* $p : \mathbb{N} \longrightarrow \mathbb{A}^{(r+1)}$.

Proof. We proceed by induction on r. The base case, $r = 0$, is trivial, and the first case, $r = 1$, is just the definition of L'. For the induction step, $r \mapsto r'$, consider

$$q : \mathbb{N} \longrightarrow \mathbb{A}^{(r+2)} \qquad b : \mathbb{A}^{(r+1)}, a_r : \mathbb{A}^{(r)}, \ldots, a_1 : \mathbb{A}'$$

and recall that $L^{(r')} = L^{(r)\prime}$. Then with $pu = qub$ for $u \in \mathbb{N}$ we have

$$L^{(r')}qba_r \cdots a_1 = L^{(r)}pa_r \cdots a_1 = Lq_r \quad \text{where} \quad q_r u = pua_r \cdots a_1 = quba_r \cdots a_1$$

for $u \in \mathbb{N}$, to give the required result. □

The (leap) clause of the spec of the long iterator $\mathbb{G}$ can be written

$$\mathbb{G}\mu Ff = \Delta p \quad \text{where } p = \mathbb{G}\mu[\cdot]Ff$$

where $\Delta : \mathcal{L}(\mathbb{N}')$ is the diagonal limit creator. We generalize this construction. Using

$$\Delta^{(0)} = \Delta \qquad \Delta^{(r')} = \Delta^{(r)\prime}$$

we generate a sequence of limit creators $\Delta^{(r)} : \mathcal{L}(\mathbb{N}(r))$ (for $r < \omega$). The limit creator Δ can be named in $\boldsymbol{\lambda G}$. This term with the terms $\uparrow_\sigma$ (for appropriate σ) shows that each $\Delta^{(r)}$ can be named in $\boldsymbol{\lambda G}$ (without the use of iterators).

Exercises

9.10 Show that for each standard jump operator jmp there is an operator $J : \mathbb{N}'^{+}$ such that $jmp = \Delta \circ J$. Write down a term which names J.

9.11 (a) Write down a term δ which names the diagonal limit creator $\Delta : \mathcal{L}(\mathbb{N}')$.
(b) The sequence $(\delta_r \,|\, r < \omega)$ is generated by $\delta_0 = \delta$ and $\delta_{r'} = \delta_r{}'$ using the appropriate limit lifter at each stage. Reduce each δ_r to normal form.

9.4 Parameterized ordinal iterators

The long iterator $\mathbb{G} : \mathbb{O} \longrightarrow \mathbb{N}'''$ has a selected target, $\mathbb{N}'''$, and uses a particular limit creator, Δ. We turn these two components into parameters of the construction, and show how the ability to vary these gives a quite flexible method of calculating with $\mathbb{G}$.

9.13 DEFINITION. For each set $\mathbb{S}$ the iterator $\mathbb{G}_\mathbb{S} : \mathbb{O} \longrightarrow \mathcal{L}(\mathbb{S}) \longrightarrow \mathbb{S}''$ is specified by

$$\mathbb{G}_\mathbb{S} 0L = Id_\mathbb{S} \qquad \mathbb{G}_\mathbb{S}\alpha' Lt = t \circ \mathbb{G}_\mathbb{S}\alpha Lt \qquad \mathbb{G}_\mathbb{S}\mu Lts = Lp \ \text{ where } pu = \mathbb{G}_\mathbb{S}\mu[u]Lts$$

for all ordinals α, limit ordinals μ, and $L : \mathcal{L}(\mathbb{S}), t : \mathbb{S}', s : \mathbb{S}$ (and $u : \mathbb{N}$). Here $Id_\mathbb{S} : \mathbb{S}''$ satisfies $Id_\mathbb{S} ts = s$ for all $t : \mathbb{S}', s : \mathbb{S}$, i.e. $Id_\mathbb{S} t = id_\mathbb{S}$. □

The long iterator uses a particular instance of this construction. A comparison with Definition 9.4 immediately gives the following.

9.14 THEOREM. *Using the parameters $\mathbb{S} = \mathbb{N}'$ and $L = \Delta$ we have $\mathbb{G}\alpha = \mathbb{G}_{\mathbb{S}}\alpha L$ for each ordinal $\alpha < \epsilon_0$.*

Here the parameters $\mathbb{S}$ and L are specialized. Later we will describe the operators $O_r : \mathbb{N}(r)$ and $E_r : \mathbb{N}''$ using different parameters. We begin with an example.

9.15 EXAMPLE. We show that for each set $\mathbb{S}$

$$\mathbb{G}_{\mathbb{S}}\omega Lts = Lp \quad \text{where } pu = t^u s$$

for all $L : \mathcal{L}(\mathbb{S}), t : \mathbb{S}', s : \mathbb{S}, u : \mathbb{N}$. Almost trivially for each $u \in \mathbb{N}$ we have

$$\mathbb{G}_{\mathbb{S}}uLts = t^u s = pu \quad \text{and hence} \quad \mathbb{G}_{\mathbb{S}}\omega[u]Lts = \mathbb{G}_{\mathbb{S}}uLts = pu$$

since $\omega[\cdot] = id_{\mathbb{N}}$. The limit clause of Definition 9.13 gives the required result. □

This with Lemma 9.12 gives the description of O_r.

9.16 THEOREM. *For each $r < \omega$, using the parameters $\mathbb{S} = \mathbb{N}(r)$ and $L = \Delta^{(r)}$ we have $O_r = \mathbb{G}_{\mathbb{S}}\omega L$.*

Proof. Consider any sequence $(*)$ of arguments

$$t : \mathbb{N}(r)',\ s : \mathbb{N}(r),\ \phi_r : \mathbb{N}^{(r)},\ \ldots ,\phi_1 : \mathbb{N}',\ x : \mathbb{N}$$

as in the definition of O_r. Using the suggested $\mathbb{S}$ and L, by Example 9.15 we have

$$\mathbb{G}_{\mathbb{S}}\omega Lts = \Delta^{(r)}p \quad \text{where } pu = t^u s$$

and hence

$$\mathbb{G}_{\mathbb{S}}\omega Lts\phi_r \cdots \phi_1 = \Delta p_r \quad \text{where } p_r u = t^u s\phi_r \cdots \phi_1$$

using Lemma 9.12. Thus,

$$\mathbb{G}_{\mathbb{S}}\omega Lts\phi_r \cdots \phi_1 x = \Delta p_r x = O_r ts\phi_r \cdots \phi_1 x$$

as required. □

This description brings out the uniformity of the construction of the O_r. What about the operator E_r? From its definition we have

$$E_r = (\mathbb{G}_{\mathbb{N}(r)}\omega\Delta^{(r)}) \cdots (\mathbb{G}_{\mathbb{N}(0)}\omega\Delta^{(0)})$$

and our problem is to convert the right hand side into $\mathbb{G}_{\mathbb{N}'}\alpha_r\Delta = \mathbb{G}\alpha_r$ for some ordinal α_r. To calculate this ordinal we need some compositional properties of $\mathbb{G}_\bullet$ which enable us to collapse the compound to a single term $\mathbb{G}_{\mathbb{S}}\alpha L$ for some $\mathbb{S}, L$, and α.

9.17 LEMMA. *For each set* $\mathbb{S}$

$$\begin{array}{llll}
(i) & \mathbb{G}_{\mathbb{S}}(\mu+\alpha)Lt & = \mathbb{G}_{\mathbb{S}}\alpha Lt \circ \mathbb{G}_{\mathbb{S}}\mu Lt & \text{provided } \mu \gg \alpha \\
(ii) & \mathbb{G}_{\mathbb{S}}(\mu m)Lts & = (\mathbb{G}_{\mathbb{S}}\mu Lt)^m s & \text{where } \mu = \omega^\alpha \\
(iii) & \mathbb{G}_{\mathbb{S}}\omega^{\alpha+1}L & = \mathbb{G}_{\mathbb{S}}\omega L \circ \mathbb{G}_{\mathbb{S}}\omega^\alpha L & \\
(iv) & \mathbb{G}_{\mathbb{S}}\omega^\alpha L & = (\mathbb{G}_{\mathbb{S}'}\alpha L')(\mathbb{G}_{\mathbb{S}}\omega L) &
\end{array}$$

hold for all ordinals α, *limit ordinals* μ, *and* $L : \mathcal{L}(\mathbb{S}), t : \mathbb{S}', s : \mathbb{S}, m : \mathbb{N}$.

To keep the flow going we relegate the proofs to Exercise 9.14. We use property (iv) to calculate the ordinal α_r for E_r. For this we need a more extensive class of operators.

For each $r, s \in \mathbb{N}$ let

$$E_{(s,r)} = O_{s+r} \cdots O_{s+1} O_s$$

to produce $E_{(s,r)} : \mathbb{N}^{(s+3)}$. In particular $E_r = E_{(0,r)}$. Note that

$$E_{(s,0)} = O_s \qquad E_{(s,r')} = E_{(s',r)} O_s$$

is a neat construction of $E_{(\cdot,\cdot)}$ (by recursion on r with variation of s).

9.18 THEOREM. *For each* $s < \omega$, *using the parameters* $\mathbb{S} = \mathbb{N}(s)$ *and* $L = \Delta^{(s)}$ *we have* $E_{(s,r)} = \mathbb{G}_{\mathbb{S}}\epsilon[r]L$ *for each* $r < \omega$.

Proof. We proceed by induction on r with variation of s. Since $\epsilon[0] = \omega$, the base case, $r = 0$, is just Theorem 9.16. For the induction step, $r \mapsto r'$, fix s and use the suggested $\mathbb{S}, L$. We have

$$E_{(s',r)} = \mathbb{G}_{\mathbb{S}'}\epsilon[r]L' \qquad O_s = \mathbb{G}_{\mathbb{S}}\omega L$$

by the induction hypothesis and Theorem 9.16. Thus, since $\epsilon[r'] = \omega^{\epsilon[r]}$ we have

$$E_{(s,r')} = E_{(s',r)} O_s = (\mathbb{G}_{\mathbb{S}'}\epsilon[r]L')(\mathbb{G}_{\mathbb{S}}\omega L) = \mathbb{G}_{\mathbb{S}}\omega^{\epsilon[r]}L = \mathbb{G}_{\mathbb{S}}\epsilon[r']L$$

as required. Here the third step uses property (iv) of Lemma 9.17. □

With this we have the uniform description of the operators E_r.

9.19 COROLLARY. *Using the parameters* $\mathbb{S} = \mathbb{N}'$ *and* $L = \Delta$ *for each* $r < \omega$ *we have* $E_r = \mathbb{G}_{\mathbb{S}}\epsilon[r]L = \mathbb{G}\epsilon[r]$.

Proof. With $\mathbb{S} = \mathbb{N}(0) = \mathbb{N}'$ and $L = \Delta^{(0)} = \Delta$, Theorems 9.18 and 9.14 give

$$E_r = E_{(0,r)} = \mathbb{G}_{\mathbb{S}}\epsilon[r]L = \mathbb{G}\epsilon[r]$$

as required. □

The operators in the sequence $\boldsymbol{O}$ can be combined to form a whole battery of operators of type $\mathbb{N}'''$ (and higher types, for that matter). Using the composition properties of Lemma 9.17 many, but not all, of these can be described as $\mathbb{G}\alpha$

for certain ordinals α. This allows a direct comparison of the strengths of these operators; the larger the ordinal, the stronger the operator. We know that

$$\mathbb{G}\alpha = \mathbb{G}_{\mathbb{S}}\alpha L \quad \text{where } \mathbb{S} = \mathbb{N}', L = \Delta$$

and, almost trivially, Δ can be named in $\boldsymbol{\lambda G}$. Thus, to name the compound operator it suffices to name $\mathbb{G}_{\mathbb{S}}\alpha : \mathcal{L}(\mathbb{S}) \longrightarrow \mathbb{S}''$ in $\boldsymbol{\lambda G}$ or a part of $\boldsymbol{\lambda G}$. We do this in the next section.

It is possible to name all, not just some, of the compound operators in the form $\mathbb{G}\alpha$ provided we use *notations* for ordinals, not just ordinals. A detailed study of the notations is beyond the scope of this book, but a hint of what can happen is given in Exercise 9.18.

Exercises

9.12 For each set $\mathbb{S}$ and known $L : \mathcal{L}(\mathbb{S})$ it is convenient to write F^α for $(\mathbb{G}_{\mathbb{S}}\alpha L)F$ for each $F : \mathbb{S}'$ and $\alpha : \mathbb{O}$.

(a) Write down the four composition properties (of Lemma 9.17) in this notation.

(b) Calculate $(F^\omega)^\omega$, $(F^{\omega^\omega})^\omega$ and $(F^\omega)^{\omega^\omega}$ in the form F^α for some α.

9.13 For each set $f : \mathbb{N}'$ the ordinal iterates $(f^\alpha \mid \alpha \in \mathbb{O})$ of f are obtained by

$$f^0 = id_{\mathbb{N}} \quad f^{\alpha'} = f \circ f^\alpha \quad f^\mu x = f^{\mu[x]}x$$

for the usual α, μ and $x \in \mathbb{N}$.

(a) Show how to generate these iterates using $\mathbb{G}$.

(b) Show $(f^\omega)^\omega = f^{\omega^2}$.

9.14 Prove the four composition properties of Lemma 9.17.

9.15 Recall the notion of a 3-structure, as given in Exercise 7.21. A **limit structure** $\mathfrak{A} = (\mathbb{A}, a, A, \mathcal{A})$ is carried by a set $\mathbb{A}$ which is furnished with three distinguished attributes $a : \mathbb{A}, A : \mathbb{A}', \mathcal{A} : \mathcal{L}(\mathbb{A})$. A morphism between two such structures $\mathfrak{B}$ and $\mathfrak{A}$ is a function $\phi : \mathfrak{B} \longrightarrow \mathfrak{A}$ (between the carriers) satisfying

$$\phi b = a \quad \phi \circ B = A \circ \phi \quad \phi \circ \mathcal{B} = \mathcal{A} \bullet \phi$$

using the distinguished attributes of $\mathfrak{B}$ and $\mathfrak{A}$. Each limit structure produces a function

$$\mathfrak{A}(-) : \mathbb{O} \longrightarrow \mathbb{A} \quad \text{generated by} \quad \mathfrak{A}0 = a \quad \mathfrak{A}\alpha' = A(\mathfrak{A}\alpha) \quad \mathfrak{A}\mu = \mathcal{A}(\mathfrak{A}\mu[\cdot])$$

for the usual α, μ.

(a) What is the connection between limit structures and ordinal iterators?

(b) Show that each limit structure gives a 3-structure, but not conversely.

(c) Show that the composite of two morphisms is a morphism.

9.16 Using the operators $O_0, O_1, O_2, \ldots$ describe $\mathbb{G}\alpha Ff$ for the ordinal

$$\alpha = \omega^{\omega^{\omega^\omega}} + \omega^{\omega^2} + \omega^{\omega 2} + \omega^3 + 7$$

and arbitrary $F : \mathbb{N}'', f : \mathbb{N}'$.

9.17 Let $\mathbb{S}$ and $L : \mathcal{L}(\mathbb{S})$ be arbitrary.

(a) Assuming $\mu \gg \alpha$, calculate $\mathbb{G}_\mathbb{S}\omega^\alpha L \circ \mathbb{G}_\mathbb{S}\omega^\mu L$ in the form $\mathbb{G}_\mathbb{S}\beta L$ for some β.

(b) For each $m \in \mathbb{N}$ let $(m+\omega)$ be a notation for the ordinal ω with fundamental sequence given by $(m+\omega)[u] = m + u$. Show that

$$\mathbb{G}_\mathbb{S}\omega Lt \circ \mathbb{G}_\mathbb{S}mLt = \mathbb{G}_\mathbb{S}(m+\omega)Lt$$

holds for all $t : \mathbb{S}'$.

(c) Calculate $\mathbb{G}_\mathbb{S}\omega^\omega L \circ \mathbb{G}_\mathbb{S}\omega L$ as best you can.

9.18 For an arbitrary $\mathbb{S}$ and $L : \mathcal{L}(\mathbb{S})$ let $P = \mathbb{G}_\mathbb{S}\omega L$, $Q = \mathbb{G}_{\mathbb{S}'}\omega L'$, $R = \mathbb{G}_{\mathbb{S}''}\omega L''$ and consider the following compounds, each of type $\mathbb{S}''$.

(i) RQP	(ii) R^2QP	(iii) $(RQ)^2P$	(iv) $(RQP)^2$
(v) $(RQ)(QP)$	(vi) RQ^2P	(vii) RQP^2	(viii) $R(QP)^2$

Express each of these in the form $\mathbb{G}_\mathbb{S}\alpha L$ where α is a notation for an ordinal. For some of these it is important to distinguish between a notation and its ordinal value. Use these representations to rank the operators (i–viii) in order of strength.

9.5 How to name ordinal iterates

We wish to name in $\boldsymbol{\lambda G}$ the operators $\mathbb{G}_\mathbb{S}\alpha : \mathcal{L}(\mathbb{S}) \longrightarrow \mathbb{S}''$ for $\mathbb{S} = \mathbb{N}'$ and ordinals $\alpha < \epsilon_0$. We allow the parameter $\mathbb{S}$ to vary and use the composition properties of Lemma 9.17, especially property (iv). We attach to each ordinal $\alpha < \epsilon_0$ and type σ a term

$$\alpha_\sigma = \lambda l : \mathcal{L}(\sigma), y : \sigma', x : \sigma \,.\, A_\sigma(l, y, x)$$

where only the identifiers l, y, x are free in $A_\sigma(l, y, x)$. This body term satisfies

$$l : \mathcal{L}(\sigma), y : \sigma', x : \sigma \vdash A_\sigma(l, y, x) : \sigma \quad \text{and hence} \quad \vdash \alpha_\sigma : \mathcal{L}(\sigma) \to \sigma''$$

so that α_σ has the required type. We will show that α_σ names $\mathbb{G}_\mathbb{S}\alpha$.

To construct α_σ it is convenient to abbreviate the prefix

$$\lambda l : \mathcal{L}(\sigma), y : \sigma', x : \sigma \quad \text{as} \quad \boldsymbol{\lambda}_\sigma l, y, x \quad \text{so that} \quad \alpha_\sigma = \boldsymbol{\lambda}_\sigma l, y, x \,.\, A_\sigma(l, y, x)$$

for the required body $A_\sigma(l, y, x)$.

You can probably make a good guess at what m_σ is for each $m < \omega$ (but you will be wrong). The crucial step is to produce ω_σ, and this is how the iterators are brought in.

9.20 DEFINITION. For each type σ we set

$$\omega_\sigma = \boldsymbol{\lambda}_\sigma l, y, x \,.\, l(\lambda u : \mathcal{N} \,.\, \mathsf{I}_\sigma yxu)$$

using the iterator I_σ attached to σ. □

This term is normal. In general the term α_σ is compact but not normal.

9.21 LEMMA. *For each type σ, if $[\![\sigma]\!] = \mathbb{S}$ then ω_σ names $\mathbb{G}_\mathbb{S}\omega$.*

Proof. We know that I_σ names $I_\mathbb{S}$ and $\vdash \omega_\sigma : \mathcal{L}(\sigma) \to \sigma''$. Let ω_σ name the function $W : \mathcal{L}(\mathbb{S}) \longrightarrow \mathbb{S}''$. Unravelling the construction of ω_σ we have

$$WLts = Lp \quad \text{where } pm = I_\mathbb{S}tsm = t^m s$$

for all $L : \mathcal{L}(\mathbb{S}), t : \mathbb{S}', s : \mathbb{S}, m \in \mathbb{N}$. Also, since $\omega[u] = u$ we have

$$\mathbb{G}_\mathbb{S}Lts = Lp \quad \text{where } pm = \mathbb{G}_\mathbb{S}mts = t^m s$$

so that $W = \mathbb{G}_\mathbb{S}$ as required. □

The fact that ω_σ names $\mathbb{G}_\mathbb{S}\omega$ depends on the selected fundamental sequence $\omega[\cdot] = id_\mathbb{N}$. If we decide to change this selection, say to $\omega^+[\cdot] = Suc$, then we must change the naming term. (For ω^+ the crucial part of the body is $\mathsf{I}_\sigma yx(\mathsf{S}u)$.) In a more extensive analysis the complexity of the selected fundamental sequence comes into play.

With this we can give the full construction.

9.22 DEFINITION. To each ordinal $\alpha < \epsilon_0$ and type σ the term α_σ is obtained by recursion over the canonical notation for α using the following clauses.

$$\begin{array}{lll}
\text{(Zero)} & 0_\sigma & = \boldsymbol{\lambda}_\sigma l, y, x \,.\, x \\
\text{(Omega)} & \omega_\sigma & = \boldsymbol{\lambda}_\sigma l, y, x \,.\, l(\lambda u : \mathcal{N} \,.\, \mathsf{I}_\sigma yxu) \\
\text{(Step)} & (\alpha')_\sigma & = \boldsymbol{\lambda}_\sigma l, y, x \,.\, y(\alpha_\sigma lyx) \\
\text{(Exp)} & (\omega^\alpha)_\sigma & = \boldsymbol{\lambda}_\sigma l, y, x \,.\, (\alpha_{\sigma'} l')(\omega_\sigma l)yx \\
\text{(Mesh)} & (\zeta + \eta)_\sigma & = \boldsymbol{\lambda}_\sigma l, y, x \,.\, (\eta_\sigma ly)(\zeta_\sigma lyx)
\end{array}$$

In the (Exp) clause we have $l' = \uparrow_\sigma l$. The (Mesh) clause is used only when $\zeta \gg \eta$ where both are limit ordinals. For convenience we have repeated the construction of ω_σ. □

To construct α_σ we first write down the canonical notation for α, and then build up α_σ using the relevant clauses. Let's look at some examples.

9.23 EXAMPLE. (a) For $m < \omega$ the term m_σ is obtained from 0_σ by m uses of (Step).

$$\begin{array}{rcl}
0_\sigma & = & \boldsymbol{\lambda}_\sigma l, y, x \,.\, x \\
1_\sigma & = & \boldsymbol{\lambda}_\sigma l, y, x \,.\, y(0_\sigma lyx) \\
 & = & \boldsymbol{\lambda}_\sigma l, y, x \,.\, y((\boldsymbol{\lambda}_\sigma l, y, x \,.\, x)lyx) \\
2_\sigma & = & \boldsymbol{\lambda}_\sigma l, y, x \,.\, y(1_\sigma lyx) \\
 & = & \boldsymbol{\lambda}_\sigma l, y, x \,.\, y((\boldsymbol{\lambda}_\sigma l, y, x \,.\, y(0_\sigma lyx))lyx) \\
 & = & \boldsymbol{\lambda}_\sigma l, y, x \,.\, y((\boldsymbol{\lambda}_\sigma l, y, x \,.\, y((\boldsymbol{\lambda}_\sigma l, y, x \,.\, x)lyx))lyx) \\
 & \vdots &
\end{array}$$

For $m > 0$ the term m_σ is not normal. A simple induction shows that

$$m_\sigma \mathrel{\rhd\!\!\!\rhd} \boldsymbol{\lambda}_\sigma l, y, x \,.\, y^m x$$

which is probably the term you first thought of.

(b) As ordinals $\omega^0 = 1$ and $\omega^1 = \omega$ whereas

$$(\omega^0)_\sigma = \boldsymbol{\lambda} l, y, x \,.\, (0_{\sigma'} l')(\omega_\sigma l) yx \qquad (\omega^1)_\sigma = \boldsymbol{\lambda} l, y, x \,.\, (1_{\sigma'} l')(\omega_\sigma l) yx$$

which are not 1_σ and ω_σ. However

$$(\omega^0)_\sigma \mathrel{\rhd\!\!\!\rhd} \boldsymbol{\lambda}_\sigma l, y, x \,.\, yx \qquad 1_\sigma \mathrel{\rhd\!\!\!\rhd} \boldsymbol{\lambda}_\sigma l, y, x \,.\, yx \qquad (\omega^1)_\sigma \mathrel{\rhd\!\!\!\rhd} \omega_\sigma$$

so that $[\![\omega^0_\sigma]\!] = [\![1_\sigma]\!]$ and $[\![\omega^1_\sigma]\!] = [\![\omega_\sigma]\!]$ which is what we want.

(c) The syntactic complexity of α_σ is determined by α, but doesn't always increase with the size of α; it sometimes decreases. For instance, $(\omega^\omega)_\sigma$ is simpler than $(\omega^{17}16 + \omega^{16}35 + \cdots + \omega^8 + 9)_\sigma$. For some special collections of ordinals it is possible to find simpler naming terms than the ones Definition 9.22 generates. Exercise 9.20 does this for the collection $(\epsilon[r] \mid r < \omega)$. □

It is time to prove the semantics of the terms α_σ are correct.

9.24 THEOREM. *For each $\alpha < \epsilon_0$ and each type σ, if $[\![\sigma]\!] = \mathbb{S}$ then α_σ names $\mathbb{G}_\mathbb{S}\alpha$.*

Proof. We proceed by induction over the canonical notation for α (with variation of σ), *not* over the size of α. The two base cases, $\alpha = 0$ and $\alpha = \omega$, have been dealt with already. We look at the three induction steps in turn.

($\alpha \mapsto \alpha'$) Let α_σ name A and let $(\alpha')_\sigma$ name B. For all $L : \mathcal{L}(\mathbb{S}), t : \mathbb{S}', s : \mathbb{S}$ the construction of $(\alpha')_\sigma$ gives $BLts = t(ALts)$ and the induction hypothesis gives $A = \mathbb{G}_\mathbb{S}\alpha$. Thus

$$BLt = t \circ ALt = t \circ \mathbb{G}_\mathbb{S}\alpha Lt = \mathbb{G}_\mathbb{S}\alpha' Lt$$

using the step clause of the spec of $\mathbb{G}_\mathbb{S}$, to give the required result.

($\alpha \mapsto \omega^\alpha$) Let $\alpha_{\sigma'}$ name A, let ω_σ name W, and let $(\omega^\alpha)_\sigma$ name M. For all $L : \mathcal{L}(\mathbb{S}), t : \mathbb{S}', s : \mathbb{S}$ the construction of $(\omega^\alpha)_\sigma$ gives $MLts = (AL')(WL)ts$, the induction hypothesis gives $A = \mathbb{G}_{\mathbb{S}'}\alpha$, and Lemma 9.21 gives $W = \mathbb{G}_\mathbb{S}\omega$. Thus

$$ML = (AL')(WL) = (\mathbb{G}_{\mathbb{S}'}\alpha L')(\mathbb{G}_\mathbb{S}\omega L) = \mathbb{G}_\mathbb{S}\omega^\alpha L$$

using Lemma 9.17(iv), to give the required result.

($(\zeta, \eta) \mapsto \zeta + \eta$) Here both ζ and η are limit ordinals with $\zeta \gg \eta$. Let ζ_σ name Z, let η_σ name H, and let $(\zeta + \eta)_\sigma$ name S. For all $L : \mathcal{L}(\mathbb{S}), t : \mathbb{S}', s : \mathbb{S}$ the construction of $(\zeta + \eta)_\sigma$ gives $SLts = (HLt)(ZLts)$, and the induction hypothesis gives $H = \mathbb{G}_\mathbb{S}\eta$ and $Z = \mathbb{G}_\mathbb{S}\zeta$. Thus, since $\zeta \gg \eta$, we have

$$SLt = HLt \circ ZLt = \mathbb{G}_\mathbb{S}\eta Lt \circ \mathbb{G}_\mathbb{S}\zeta Lt = \mathbb{G}_\mathbb{S}(\zeta + \eta)Lt$$

using Lemma 9.17(i), to give the required result. □

For any particular pair α, σ only a part of $\boldsymbol{\lambda G}$ is needed to construct α_σ. Given σ let

$$\sigma(0) = \sigma,\ \sigma(1) = \sigma',\ \sigma(2) = \sigma'',\ \ldots \qquad \text{i.e.} \quad \sigma(0) = \sigma \qquad \sigma(r') = \sigma(r)'$$

for each $r < \omega$. Notice that $\sigma(r') = \sigma'(r)$.

9.25 THEOREM. *For each $r < \omega$ and ordinal $\alpha < \epsilon[r]$, the term α_σ is constructed from iterators $\mathsf{I}_{\sigma(i)}$ for $i < r$. In particular, for $\alpha < \omega$, no iterators at all are needed.*

We leave the proof of this as Exercise 9.21. A more delicate analysis shows that the nesting properties of the I_ρ in α_σ precisely match the 'nesting' structure of α. A more extensive study of ordinal notations would investigate these 'nesting' properties.

The methods of this chapter have been described mainly for ordinals $\alpha < \epsilon_0$. It is possible to go a little further without too much extra work. We have a fundamental sequence $\epsilon[\cdot]$ for ϵ_0, so

$$\mathbb{G}_{\mathbb{S}}\epsilon_0 Lts = Lp \quad \text{where } pu = \mathbb{G}_{\mathbb{S}}\epsilon[u]Lts$$

is an instance of the limit clause of the spec of $\mathbb{G}_{\mathbb{S}}$. With $\mathbb{S} = \mathbb{N}', L = \Delta, F : \mathbb{N}''$, and $f : \mathbb{N}$ we have $pu = E_u F f$ so that, for each $x \in \mathbb{N}$

$$\mathbb{G}\epsilon_0 F f x = \Delta(u \mapsto E_u F f)x = E_x F f x$$

and hence $\mathbb{G}\epsilon_0$ is the obvious limit of $(E_r \mid r < \omega)$.

For $\alpha < \epsilon_0$ consider the functions $S_\alpha = \mathbb{G}\alpha ackSuc$. By Theorems 9.14 and 9.24 we have $\mathbb{G}\alpha = \mathbb{G}_{\mathbb{N}'}\alpha\Delta$ and hence $\mathbb{G}\alpha$ can be named in $\boldsymbol{\lambda G}$. Thus S_α is representable in $\boldsymbol{\lambda G}$. Furthermore, by looking at the structure of α we can determine that part of $\boldsymbol{\lambda G}$ needed for this representation. This long sequence $(S_\alpha \mid \alpha < \epsilon_0)$ exhausts $\boldsymbol{\lambda G}$.

9.26 THEOREM. *The function $\mathbb{G}\epsilon_0 ackSuc$ is not representable in $\boldsymbol{\lambda G}$.*

The proof of this requires a more detailed analysis of $\boldsymbol{\lambda G}$ than is possible here. Such an analysis uncovers a tight connection between $\boldsymbol{\lambda G}$ and the structural properties of ϵ_0.

EXERCISES

9.19 Verify $\vdash \alpha_\sigma : \mathcal{L}(\sigma) \to \sigma''$ for each ordinal $\alpha < \epsilon_0$ and type σ. You will not find this difficult, but you should explain how your induction works. Exhibit the shape of the derivation for each α.

9.20 For each type σ and term l let

$$\omega[\sigma, l] = \lambda y : \sigma', x : \sigma \,.\, l(\lambda u : \mathcal{N} \,.\, \mathsf{I}_\sigma yxu)$$

(so that $\omega_\sigma = \lambda l : \mathcal{L}(\sigma) \,.\, \omega[\sigma.l]$). Also set

$$E[\sigma, 0, l] = \omega[\sigma, l] \qquad E[\sigma, r', l] = E[\sigma', r.l']\omega[\sigma, l]$$

for each $r < \omega$. Show that

$$\epsilon[r]_\sigma \rhd\!\!\!\rhd \boldsymbol{\lambda} l, y, x \,.\, E[\sigma, r, l]yx$$

(for each type σ and $r < \omega$).

9.21 Prove Theorem 9.25.

9.6 THE GODS

The Sunday school humour of the section header might seem a bit diabolical, but it does underline an important idea. The GODS are the four components used to measure and compare type systems.

(G) The long iterator $\mathbb{G} : \mathbb{O}\mathrm{rd} \longrightarrow \mathbb{N}'''$, which organizes the whole method.

(O) The ordinals or ordinal notations $\mathbb{O}$ used to index length.

(D) A diagonalization operator $D : \mathbb{N}''$ which we iterate along the ordinals using $\mathbb{G}$.

(S) A base function $S : \mathbb{N}$ on top of which we build a long tower.

Of these the first component $\mathbb{G}$ is fixed. We know that for each $\alpha \in \mathbb{O}$

$$\mathbb{G}\alpha = \mathbb{G}_{\mathbb{S}}\alpha L \quad \text{where} \quad \mathbb{S} = \mathbb{N}', L = \Delta$$

and the 3-argument gadget $\mathbb{G}_{\bullet}(\cdot, \cdot)$ is a useful computational device for calculations involving $\mathbb{G}$.

The second component $\mathbb{O}$ is an initial stretch of ordinals which is long enough for the job in hand. Of course, hidden in the choice of $\mathbb{O}$ is the problem of selecting a fundamental sequence for each limit ordinal in $\mathbb{O}$. We have seen that this is not entirely straight forward, and have decided not to face that problem here. In this book we are concerned in the main with canonical notations for ordinals $\alpha < \epsilon_0$. A fuller development would take seriously the distinction between ordinal and notation, would use non-canonical notations, and would go at least a little way beyond ϵ_0.

The third component D is the one for which we have greatest freedom of choice. The most common choice is one of *ack*, *rob*, *brw* (or something similar). We find that the relative differences between the values of

$$\mathbb{G}\alpha ack \quad \mathbb{G}\alpha rob \quad \mathbb{G}\alpha brw$$

become less and less as α becomes larger. When we are interested in small α we can choose D to be something like the operators G, H, L, of Example 9.8 which generate the Slow, Hardy, and Laurel hierarchies. This gives us a finer structure to work with. In this section we concentrate on $D = ack$.

The final component S is almost always the successor function. Occasionally, when we want a relativized version, we can replace S by some faster growing snake.

With these controlling GODS we set

$$S_\alpha = \mathbb{G}\alpha\, ack\, Suc = \mathbb{G}_{\mathbb{N}'}\alpha\Delta\, ack\, Suc = ack^\alpha Suc$$

for each $\alpha < \epsilon$ (as at the end of Section 9.5). This produces an $\mathbb{O}$-indexed sequence

$$\boldsymbol{S} = (S_\alpha \mid \alpha \in \mathbb{O})$$

of snakes of ever increasing rates of growth. This sequence is generated directly by

$$S_0 = Suc \quad S_{\alpha'} = ack\, S_\alpha \quad S_\mu x = S_{\mu[x]} x$$

for all ordinals α, limit ordinals μ, and $x \in \mathbb{N}$. The informal description

$$S_\alpha = ack^\alpha Suc$$

indicates how we should think of $\boldsymbol{S}$, and the description

$$S_\alpha = \mathbb{G}_{\mathbb{S}} \alpha L\, ack\, Suc \quad \text{where} \quad \mathbb{S} = \mathbb{N}', L = \Delta$$

provides a way of calculating with $\boldsymbol{S}$.

It has been found that for many type systems $\mathcal{TS}$ there is an ordinal $\alpha = |\mathcal{TS}|$ such that, in some sense, S_α encapsulate the complexity of $\mathcal{TS}$. This is true for many natural parts of $\boldsymbol{\lambda G}$, and the hierarchy $\boldsymbol{S}$ provides a neat stratification of $\boldsymbol{\lambda G}$. (The same idea can work with type systems more complicated that $\boldsymbol{\lambda G}$, provided we use larger ordinals.)

Look at Table 9.1. The left hand column lists the ordinals and indicates some of the important steps and leaps along the way to ϵ_0. The right hand column gives for certain ordinals α a system of arithmetic with 'complexity S_α'. These named systems occur mostly for smaller ordinals. This is not because systems with larger ordinals are not important, merely because they don't have standard names. The strength of $\boldsymbol{\lambda G}$ is measured by ϵ_0, and the internal structure of this ordinal corresponds to some extent to the internal structure of $\boldsymbol{\lambda G}$.

Various other functions and hierarchies that we have met, either in the examples or the exercises, can be compared with $\boldsymbol{S}$. Some of these are listed in the central columns.

We began the chapter by producing a chain $(E_r \mid r < \omega)$ of operators. Using these we may set

$$A_r = E_r\, ack\, Suc$$

for each $r < \omega$. In Corollary 9.19 we saw that $E_r = \mathbb{G}\epsilon[r]$ and hence $A_r = S_{\epsilon[r]}$ for each $r < \omega$. The diagonal limit A_ω of this chain $(A_r \mid r < \omega)$ is given by

$$A_\omega x = A_x x = \mathbb{G}\epsilon[x] ack\, Suc\, x = \mathbb{G}\epsilon_0 ack\, Suc\, x$$

and hence by Theorem 9.18 the function A_ω is not representable in $\boldsymbol{\lambda G}$.

In Exercises 9.3 and 9.4 we produced two chains

$$\boldsymbol{F} = (F_s \mid s < \omega) \qquad \boldsymbol{G} = (G_s \mid s < \omega)$$

of multi-argument functions starting from an arbitrary 1-placed function f. Consider the case $f = Suc$ and set $B_s = F_s$ or $B_s = G_s$ as you wish to produce a chain $\boldsymbol{B}$. These functions are positioned in the next column. You should make sure you know why B_s has the same complexity as $S_{\omega^{s+1}}$ (and not, for instance, as S_{ω^s}). The number of arguments of B_s increases with s, so it is not immediately clear what the diagonal limit B_ω of $\boldsymbol{B}$ could be. However, any sensible definition

Ord	$\boldsymbol{A}$	$\boldsymbol{B}$	$\boldsymbol{C}$	Nature of S_α	Bounding snake of
0			C_0	$x \mapsto x+1$	
1			C_ω	$x \mapsto 2x$	
2			$C_{\omega 2}$	$x \mapsto 2^x$	$\boldsymbol{\lambda}$, Feasible arithmetic
3			$C_{\omega 3}$	$x \mapsto \beth(x,2,1)$	Rudimentary arithmetic
$\vdots$					
r			$C_{\omega r}$		r^{th} Grzegorczyk level
$\vdots$					
ω	A_0	B_0	C_{ω^2}	Ackermann	Primitive recursive arithmetic
$\omega 2$					
$\omega 3$					
$\vdots$					
ω^2		B_1	C_{ω^3}		2-recursive arithmetic
$\vdots$					
ω^3		B_2	C_{ω^4}		3-recursive arithmetic
$\vdots$					
ω^s		B_{s-1}	$C_{\omega^{s+1}}$		s-recursive arithmetic
$\vdots$					
ω^{s+1}		B_s	$C_{\omega^{s+2}}$		s'-recursive arithmetic
$\vdots$					
ω^ω	A_1	B_ω			Multi-recursive arithmetic
$\vdots$					
ω^{ω^ω}	A_2				
$\vdots$					
$\epsilon[r]$	A_r				
$\vdots$					
ϵ_0	A_ω				$\boldsymbol{\lambda G}$
$\vdots$					

Table 9.1: Mount Olympus

will produce a 1-placed function on level ω^ω, as indicated. The details of this are discussed in Exercise 9.22.

There are other hierarchies which are comparable with $\boldsymbol{S}$. Each ordinal α has the form $\alpha = \mu + r$ for some unique limit ordinal μ and $r < \omega$. Let us set $|\alpha| = \mu$ so that the sandwich $|\alpha| \leq \alpha < |\mu| + \omega$ determines this limit ordinal. Now set

$$C_0 = Suc \quad C_{\alpha'} = C_{|\alpha|} \circ C_\alpha \quad C_\mu x = C_{\mu[x]} x$$

for each ordinal α and limit ordinal μ. It can be shown that $C_{\omega\alpha} = S_\alpha$ for all $\alpha < \omega^\omega$ as indicated in the table. This produces a hierarchy that is finer than the initial part of $\boldsymbol{S}$; each step of $\boldsymbol{S}$ corresponds to ω steps of $\boldsymbol{C}$. We can continue

$\boldsymbol{C}$ beyond ω^ω, but the correspondence with $\boldsymbol{S}$ is more complicated. The details of this are discussed in Exercise 9.23.

Perusing the table and reading these remarks may suggest to you several questions.

($\uparrow$) Is it possible to go beyond ϵ_0, and are there any sensible systems which these levels classify? It is quite easy to go a little way beyond ϵ_0. We already have a fundamental sequence $\epsilon[\cdot]$ for ϵ_0, and using this we can produce notations for

$$\epsilon_0 + \epsilon_0 \quad \epsilon_0 \times \omega \quad \epsilon_0^\omega \quad \epsilon_0^{\epsilon_0} \quad \beth(3, \epsilon_0, \epsilon_0)$$

etc. In fact using ϵ_0 we can produce critical ordinals $\epsilon_0, \epsilon_1, \epsilon_2, \ldots, \epsilon_\alpha, \ldots$ for all 'previously constructed' ordinals α. These ordinals can not be named in $\boldsymbol{\lambda G}$, but we can still use them outside the system. However, if we are going to take this step then we should develop a formal system of ordinal notations (and do this chapter properly). That would enable us to go much, much further along the ordinals. These higher levels do classify natural system using more powerful principles.

($\leftrightarrow$) What do the various levels of $\boldsymbol{S}$ tell us about the complexity of subsystems of $\boldsymbol{\lambda G}$? In particular, how do the multi-recursion hierarchies fit into this scheme? I will say something about this shortly.

($\downarrow$) We have seen that for the lower levels $\boldsymbol{C}$ gives a finer stratification than $\boldsymbol{S}$. How much finer can we go? The Slow-Growing hierarchy $\boldsymbol{g}$ of Example 9.8(Slow) has the form $g_\alpha = \mathbb{G}\alpha G\, zero$ for a certain $G : \mathbb{N}''$. Using Exercises 8.11 and 9.6 for $r < \omega$ we have

$$g_{\epsilon[r]}(x) = \beth(r, x, x) \quad \text{so that} \quad g_\epsilon(x) = \beth(x, x, x)$$

which is around the level $3\frac{1}{2}$ of $\boldsymbol{S}$ (i.e. g_ϵ is faster than S_3 but slower than S_4). This kind of extremely fine hierarchy can be used to classify combinatorial properties.

In Section 7.5 we produced several function hierarchies, such as the Grzegorczyk hierarchy, the Ackermann hierarchy, the Péter hierarchy, and the all embracing multi-index hierarchy. Because of the use of multi-indexes $\mathbb{M}$ rather than ordinals $\mathbb{O}$, these hierarchies were severely limited in length. We can now take them much further. Also, we find that the steps in these hierarchies correspond to simple ordinal operations.

(Suc) For each snake f the Grzegorczyk hierarchy $(ack^i f \mid i < \omega)$ is obtained by repeatedly applying the small jump operator $ack : \mathbb{N}''$. This hierarchy stratifies those functions which are 1-recursive (primitive recursive) in f. For the case $f = Suc$ this hierarchy corresponds precisely to the finite levels $(S_i \mid i < \omega)$ of $\boldsymbol{S}$. Thus we may think of $\boldsymbol{S}$ as the long Grzegorczyk hierarchy. In particular, each successor increase, $\alpha \mapsto \alpha'$, in ordinal index corresponds to a Grzegorczyk jump in complexity.

(Add) For each snake f the Ackermann hierarchy $(ACK^i f \mid i < \omega)$ is obtained by repeatedly applying the middling jump operator $ACK : \mathbb{N}''$. This hierarchy stratifies those functions which are 2-recursive in f. We have

$$ACK = ack_1 = Pet\, ack = ack^\omega \quad \text{so that} \quad ACK^i f = ack^{\omega i} f$$

for each $i < \omega$. For the case $f = Suc$ this hierarchy corresponds precisely to the limit levels $(S_{\omega i} \mid i < \omega)$ of $\boldsymbol{S}$. This selection of limit ordinals can be continued throughout $\mathbb{O}$rd to produce the long Ackermann hierarchy. In particular, each additive leap to the next limit ordinal, $\alpha \mapsto \alpha + \omega$, corresponds to an Ackermann jump in complexity.

(Mlt) For each snake f the Péter hierarchy $(Pet^s ack\, f \mid s < \omega)$ is obtained by repeatedly applying the higher order operator $Pet : \mathbb{N}'''$ to ack and then applying the resulting jump operator to f. This hierarchy stratifies those functions which are multi-recursive in f. In fact, for each s the function $Pet^s ack\, f$ eventually dominates each function which is s-recursive in f, and itself is 'minimally' s'-recursive in f. For the case $f = Suc$ this hierarchy corresponds to the limit-limit levels $(S_{\omega^s} \mid s < \omega)$ of $\boldsymbol{S}$. (This is equivalent to the chain $\boldsymbol{B}$ of Table 9.1 with a slightly different indexing.) This selection of limit-limit ordinals can be continued throughout $\mathbb{O}$rd to produce the long Péter hierarchy. In particular, each multiplicative leap to the next limit-limit ordinal, $\alpha \mapsto \alpha \times \omega$, corresponds to a Péter jump in complexity. Observe that the essential feature of this construction is a use of a higher type object $Pet : \mathbb{N}'''$. In a way the base function $f : \mathbb{N}$ is not very important.

(Exp) At the beginning of this chapter we produced a hierarchy $(E_r \mid r < \omega)$ of operators of type $\mathbb{N}'''$. The essential feature of this construction is the use of higher and higher types. In fact, the doubly indexed family $(E_{(s,r)} \mid s, r < \omega)$ of operators $E_{(s,r)} : \mathbb{N}^{(s+3)}$ constructed in Section 9.4 gives a better picture of the process. We have seen that

$$E_r = \mathbb{G}\epsilon[r] \quad \text{so that} \quad E_r\, ack\, suc = S_{\epsilon[r]}$$

and we obtain a rather coarse stratification of the functions representable in $\boldsymbol{\lambda G}$. (This is just the chain $\boldsymbol{A}$ of the table.) The jumps between the levels of the hierarchy correspond to an exponential leap, $\alpha \mapsto \omega^\alpha$, in ordinal index.

This is as far as we can go in this book but, of course, it is not the end of the story. Along the way we have omitted many important topics. Nowhere have we seen formal versions of propositional connectives other than $\rightarrow$, and nowhere have we seen any formal quantifiers. We have seen no equational reasoning, no predicate logic, and no enrichment of the constructive systems in the direction of boolean logic. All of these facilities can be added to $\boldsymbol{\lambda G}$, in various combinations, to produce much more detailed analyses of number theoretic functions of finite type.

Beyond that we can strengthen the type system in various ways. A rather modest extension produces a formal system of ordinal notations (which is still an applied λ-calculus) which turns out to have enormous power. Such extensions very quickly lead to unexplored (and sometimes impenetrable) regions, as well as bringing in methods of other parts of mathematics.

This is as far as we go in this book, but it is not the end of the story, it is only the beginning.

Exercises

9.22 Show that in Table 9.1, in terms of complexity the functions B_s are correctly positioned on level ω^{s+1}.

9.23 As in the construction of chain $\boldsymbol{C}$ of Table 9.1, for each ordinal α let $|\alpha|$ be the unique non-successor ordinal such that $|\alpha| \leq \alpha < |\alpha| + \omega$ holds. For an arbitrary function $f : \mathbb{N}'$, set

$$f_0 = f \qquad f_{\alpha'} = f_{|\alpha|} \circ f_\alpha \qquad f_\mu x = f_{\mu[x]} x$$

for each ordinal α and limit ordinal μ.

(a) Compare the two hierarchies $(f_\alpha \mid \alpha < \omega^\omega)$ and $(ack^\alpha f \mid \alpha < \omega^\omega)$ by showing that one refines the other.

(b) What happens in the extensions of these hierarchies for ordinals $\alpha \geq \omega^\omega$?

Part II
Solutions

A
Derivation Systems

A.1 Introduction

1.1 The three trees below parse β, γ, and δ, as indicated. $\square$

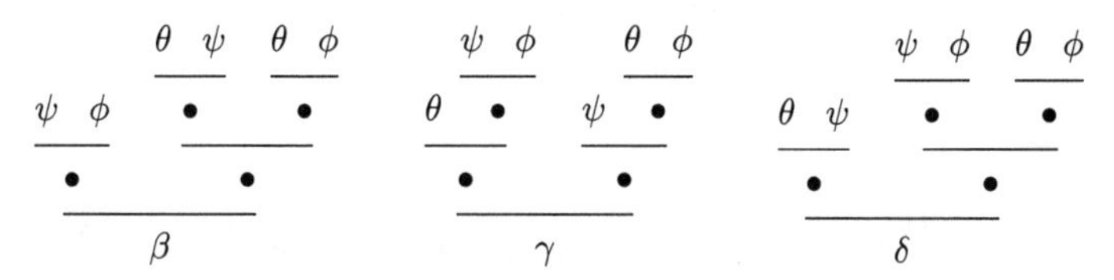

1.2 Each of (i,ii,iii) follows by induction over T. Part (ii) is the most complicated.
For the base case, $T = \bullet$, we have $s(T) = 1$, $h(T) = 0$, so that

$$s(T) + 1 = 2 \qquad h(T) + 1 = 1$$

and hence the required comparison holds for this T.

The induction step has two subcases.

Suppose $T = \frac{C}{\bullet}$ where $s(C) = s$ and $h(C) = h$ satisfy $s + 1 \leq 2^{h+1}$. Then we have $s(T) = s + 1$ and $h(T) = h + 1$ so that

$$s(T) + 1 = s + 2 \leq 2(s + 1) \leq 2 \times 2^{h+1} \leq 2^{h+2} = 2^{h(T)+1}$$

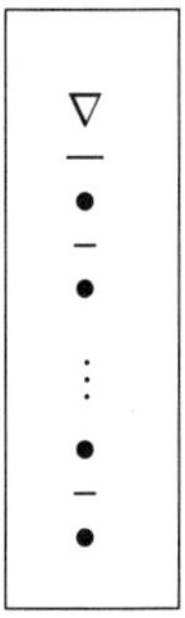

as required.

Suppose $T = \frac{L\,R}{\bullet}$ where the four measures $s(L) = s_l$, $s(R) = s_r$, and $h(L) = h_l$, $h(R) = h_r$ satisfy the assumed comparisons. For convenience let $m = \max(h_l, h_r)$, so that we have $h(T) = m + 1$ and

$$s_l + 1, s_r + 1 \leq 2^{m+1} = 2^{h(T)}$$

and hence

$$s(T) + 1 = s_l + s_r + 2 \leq (s_l + 1) + (s_r + 1) \leq 2^{m+1} + 2^{m+1} = 2^{h(T)+1}$$

as required.

The other two comparisons follow in a similar fashion.

Finally, given any tree ∇ of a specified width, we may use the template shown on the previous page to grow a tree of the same width but arbitrary height. □

1.3 We first obtain either of the two identities

$$T(u, H(v,s)) = T(u+v, s) \qquad T(u, H(v,s)) = H(u+v, s)$$

and then set $v = 0$ in the left hand one or $u = 0$ in the right hand one. These identities are proved by induction on u or v, respectively. □

A.2 GENERALITIES

1.4 (β) For convenience let

$$\rho = \theta \to \phi \qquad \sigma = \theta \to \psi \qquad \tau = \psi \to \phi$$

so that $\beta = \tau \to \sigma \to \rho$. Either of the trees

$$\dfrac{\dfrac{\dfrac{\dfrac{\dfrac{\dfrac{\theta \vdash \theta \quad \theta,\psi \vdash \psi}{\theta,\sigma \vdash \psi}\,(\mathrm{LI}\oplus) \qquad \theta,\sigma,\phi \vdash \phi}{\theta,\sigma,\tau \vdash \phi}\,(\mathrm{LI})}{\tau,\sigma,\theta \vdash \phi}\,(\mathrm{X}\cdots\mathrm{X})}{\tau,\sigma \vdash \rho}\,(\mathrm{RI})}{\tau \vdash \sigma \to \rho}\,(\mathrm{RI})}{\vdash \beta}\,(\mathrm{RI})$$

$$\dfrac{\dfrac{\dfrac{\dfrac{\dfrac{\dfrac{\theta,\tau \vdash \theta \qquad \dfrac{\dfrac{\theta,\psi \vdash \psi \quad \theta,\psi,\phi \vdash \phi}{\theta,\psi,\tau \vdash \phi}\,(\mathrm{LI}\oplus)}{\theta,\tau,\psi \vdash \phi}\,(\mathrm{X})}{\theta,\tau,\sigma \vdash \phi}\,(\mathrm{LI}\oplus)}{\sigma,\psi,\tau \vdash \phi}\,(\mathrm{LI}\oplus)}{\tau,\sigma,\theta \vdash \phi}\,(\mathrm{X}\cdots\mathrm{X})}{\tau,\sigma \vdash \rho}\,(\mathrm{RI})}{\tau \vdash \sigma \to \rho}\,(\mathrm{RI})}{\vdash \beta}\,(\mathrm{RI})$$

obtains $\vdash \beta$ in G+. Here X$\cdots$X is a number of uses of X. Either of the trees

$$\dfrac{\dfrac{\dfrac{\dfrac{\dfrac{\dfrac{\theta \vdash \theta \quad \psi \vdash \psi}{\theta,\sigma \vdash \psi}\,(\mathrm{LI}\otimes) \qquad \phi \vdash \phi}{\theta,\sigma,\tau \vdash \phi}\,(\mathrm{LI}\otimes)}{\tau,\sigma,\theta \vdash \phi}\,(\mathrm{X}\cdots\mathrm{X})}{\tau,\sigma \vdash \rho}\,(\mathrm{RI})}{\tau \vdash \sigma \to \rho}\,(\mathrm{RI})}{\vdash \beta}\,(\mathrm{RI})$$

$$\dfrac{\dfrac{\dfrac{\dfrac{\dfrac{\theta \vdash \theta \qquad \dfrac{\psi \vdash \psi \quad \phi \vdash \phi}{\psi,\tau \vdash \phi}\,(\mathrm{LI}\otimes)}{\theta,\sigma,\tau \vdash \phi}\,(\mathrm{LI}\otimes)}{\tau,\sigma,\theta \vdash \phi}\,(\mathrm{X}\cdots\mathrm{X})}{\tau,\sigma \vdash \rho}\,(\mathrm{RI})}{\tau \vdash \sigma \to \rho}\,(\mathrm{RI})}{\vdash \beta}\,(\mathrm{RI})$$

derives $\vdash \beta$ in G$\times$. Again X$\cdots$X indicates a number of uses of X.

You should compare these derivations with those displayed in Table 1.3.

(γ) For convenience set

$$\rho = \theta \to \phi \qquad \sigma = \theta \to \psi \qquad \tau = \psi \to \phi \qquad \lambda = \theta \to \tau \qquad \mu = \psi \to \rho$$

so that $\gamma = \lambda \to \mu$. The two derivations

$$\dfrac{\dfrac{\dfrac{\dfrac{\dfrac{\psi,\theta\vdash\theta \qquad \dfrac{\psi,\theta\vdash\psi \quad \psi,\theta,\phi\vdash\phi}{\psi,\theta,\rho\vdash\phi}\,(\text{LI}\oplus)}{\psi,\theta,\lambda\vdash\phi}\,(\text{LI}\oplus)}{\lambda,\psi,\theta\vdash\phi}\,(\text{XX})}{\lambda,\psi\vdash\sigma}\,(\text{RI})}{\lambda\vdash\mu}\,(\text{RI})}{\vdash\gamma}\,(\text{RI}) \qquad \dfrac{\dfrac{\dfrac{\dfrac{\theta\vdash\theta \qquad \dfrac{\dfrac{\psi\vdash\psi \quad \phi\vdash\phi}{\psi,\rho\vdash\phi}\,(\text{LI}\otimes)}{\rho,\psi\vdash\phi}\,(\text{X})}{\psi,\theta,\lambda\vdash\phi}\,(\text{LI}\otimes)}{\lambda,\psi\vdash\sigma}\,(\text{RI})}{\lambda\vdash\mu}\,(\text{RI})}{\vdash\gamma}\,(\text{RI})$$

obtain $\vdash \gamma$ in $\mathsf{G}+$ and $\mathsf{G}\times$, respectively. Here XX indicates two uses of X.

(δ) For convenience let

$$\rho = \theta \to \phi \qquad \sigma = \theta \to \psi \qquad \tau = \psi \to \phi$$

so that $\delta = \sigma \to \tau \to \rho$. The derivations

$$\dfrac{\dfrac{\dfrac{\dfrac{\dfrac{\dfrac{\theta\vdash\theta \quad \theta,\psi\vdash\psi}{\theta,\sigma\vdash\psi}\,(\text{LI}\oplus) \qquad \theta,\sigma,\phi\vdash\phi}{\theta,\sigma,\tau\vdash\phi}\,(\text{LI})}{\sigma,\tau,\theta\vdash\phi}\,(\text{XX})}{\sigma,\tau\vdash\rho}\,(\text{RI})}{\sigma\vdash\tau\to\rho}\,(\text{RI})}{\vdash\delta}\,(\text{RI}) \qquad \dfrac{\dfrac{\dfrac{\dfrac{\dfrac{\dfrac{\theta\vdash\theta \quad \psi\vdash\psi}{\theta,\sigma\vdash\psi}\,(\text{LI}\otimes) \qquad \theta,\sigma,\phi\vdash\phi}{\theta,\sigma,\tau\vdash\phi}\,(\text{LI})}{\sigma,\tau,\theta\vdash\phi}\,(\text{XX})}{\sigma,\tau\vdash\rho}\,(\text{RI})}{\sigma\vdash\tau\to\rho}\,(\text{RI})}{\vdash\delta}\,(\text{RI})$$

obtain $\vdash \delta$ in $\mathsf{G}+$ and $\mathsf{G}\times$, respectively. Here XX indicates two uses of X.

1.5 The templates

$$\dfrac{\dfrac{\Gamma^l\vdash\theta\to\phi}{\begin{array}{c}\vdots\\ \text{W+X}\\ \vdots\\ \hline \Gamma^l,\Gamma^r\vdash\theta\to\phi\end{array}} \qquad \dfrac{\Gamma^r\vdash\theta}{\begin{array}{c}\vdots\\ \text{W+X}\\ \vdots\\ \hline \Gamma^l,\Gamma^r\vdash\theta\end{array}}}{\Gamma^l,\Gamma^r\vdash\phi}\,(\text{MP}\oplus) \qquad\qquad \dfrac{\Gamma\vdash\theta\to\phi \qquad \Gamma\vdash\theta}{\begin{array}{c}\Gamma,\Gamma\vdash\theta\\ \hline \vdots\\ \text{C+X}\\ \vdots\\ \hline \Gamma\vdash\phi\end{array}}\,(\text{MP}\otimes)$$

give simulations of

$$\text{MP}\otimes \text{ from MP}\oplus \text{ with W and X} \qquad \text{MP}\oplus \text{ from MP}\otimes \text{ with C and X}$$

respectively. □

1.6 Consider first the passage from $\mathsf{G}\times$ to $\mathsf{G}+$. The problem is to simulate W and LI$\otimes$ in $\mathsf{G}+$. The way we handle W is important. For contexts Γ, Σ we say

$$\Gamma \text{ thins to } \Sigma$$

if Σ as a list can be obtained from Γ by inserting new elements at appropriate positions. Thus if we split Γ

$$\Gamma = \Gamma^l, \Gamma^r$$

and we use any formula θ, then

$$\Gamma^+ = \Gamma^l, \theta, \Gamma^r$$

is a 1-step thinning of Γ. All thinnings of Γ can be obtained by a succession of such 1-step thinnings where at each stage we may split the context at a different position.

We describe an algorithm which, when supplied with

$$(\nabla)\ \ \Gamma \vdash \phi \qquad (A)\ \ \Gamma \text{ thins to } \Sigma$$

a derivation ∇ in $\mathsf{G}\times$ and an explicit multiple thinning A, will return a derivation

$$(\nabla \cdot A)\ \ \Sigma \vdash \phi$$

in $\mathsf{G}+$. Once we have this the particular case with $\Sigma = \Gamma$ gives the required translation. However, building thinning into the algorithm in this way makes it easier to handle.

The algorithm proceeds by recursion on the construction of ∇.

Because of the way we have set up the algorithm, the step across W is trivial.

The crucial step is the one across a use of LI$\otimes$. Thus suppose the root of the given derivation is

$$\dfrac{\Gamma^l \vdash \theta \quad \psi, \Gamma^r \vdash \phi}{\Gamma \vdash \phi}$$

where $\Gamma = \Gamma^l, \theta \to \psi, \Gamma^r$ and both

$$\Gamma^l \vdash \theta \qquad \psi, \Gamma \vdash \phi$$

are derivable in $\mathsf{G}\times$. With the given thinning Σ of Γ, notice that

$$\Gamma^l \text{ thins to } \Sigma \qquad \psi, \Gamma^r \text{ thins to } \psi, \Sigma$$

and, furthermore, $\theta \to \psi$ occurs in Σ (since $\theta \to \psi$ occurs in Γ). After two recursion calls on the algorithm we obtain derivations

$$\Sigma \vdash \theta \qquad \psi, \Sigma \vdash \phi$$

in $\mathsf{G}+$, and then an application of LI$\oplus$ gives a derivation of

$$\Sigma, \theta \to \psi \vdash \phi$$

$$\Gamma, \psi \vdash \phi$$

$$\vdots \quad X \quad \vdots$$

$$\dfrac{\Gamma \vdash \theta \quad \psi, \Gamma \vdash \phi}{\Gamma, \theta \to \psi, \Gamma \vdash \phi}\ (\mathrm{LI}\otimes)$$

$$\vdots \quad X \quad \vdots$$

$$\Gamma, \Gamma, \theta \to \psi \vdash \phi$$

$$\vdots \quad C + X \quad \vdots$$

$$\Gamma, \theta \to \psi \vdash \phi$$

in G+. Several uses of X (to move the position of $\theta \to \psi$) followed by one use of C gives the required result.

The passage from G+ to G× is easy. For instance, the template on the previous page simulates LI⊗. □

A.3 THE SYSTEMS H AND N

1.7 (β) For the given θ, ψ, ϕ let

$$\rho = \theta \to \phi \qquad \sigma = \theta \to \psi \qquad \tau = \psi \to \phi \qquad \nu = \sigma \to \rho$$

so that $\beta = \tau \to \sigma \to \rho$ is the target formula. Let

$$\begin{array}{lll} \kappa_2 = \tau \to \theta \to \tau & \alpha = \theta \to \tau & \sigma_2 = \alpha \to \sigma \to \rho \\ \kappa_1 = \sigma_2 \to \tau \to \sigma_2 & \mu = \tau \to \sigma_2 & \sigma_1 = \mu \to \kappa_2 \to \beta \end{array}$$

so that κ_1 and κ_2 are axioms. By unravelling we have

$$\begin{array}{l} \sigma_1 = (\tau \to \alpha \to \nu) \to (\tau \to \alpha) \to (\tau \to \nu) \\ \sigma_2 = (\theta \to \psi \to \phi) \to (\theta \to \psi) \to (\theta \to \phi) \end{array}$$

so these too are axioms. (The original σ need not be an axiom.) Let $\Sigma = \sigma, \tau, \theta$. With these

$$\dfrac{\dfrac{\vdash \sigma_1 \quad \dfrac{\vdash \kappa_1 \quad \vdash \sigma_2}{\vdash \mu}}{\vdash \kappa \to \beta} \quad \vdash \kappa_2}{\vdash \beta} \qquad\qquad \dfrac{\dfrac{\dfrac{\dfrac{\Sigma \vdash \tau \quad \dfrac{\Sigma \vdash \sigma \quad \Sigma \vdash \theta}{\Sigma \vdash \psi}}{\Sigma \vdash \phi}}{\tau, \sigma \vdash \rho}\,(\uparrow)}{\tau \vdash \sigma \to \rho}\,(\uparrow)}{\vdash \beta}\,(\uparrow)$$

$$\emptyset[\sigma_1](\emptyset[\kappa_1]\emptyset[\sigma_2])\emptyset[\kappa_2] \qquad\qquad \Sigma[2](\Sigma[1]\Sigma[0])\uparrow\uparrow\uparrow$$

are an H-derivation and an N-derivation with arboreal codes as indicated.

(γ) For the given ρ, σ, τ let

$$\beta = \tau \to \sigma \to \rho \qquad \delta = \sigma \to \tau \to \rho$$

so that $\gamma = \beta \to \delta$ is the target formula. With

$$\xi = \tau \to \sigma \qquad \chi = \tau \to \rho \qquad \eta = \xi \to \chi$$

let

$$\begin{array}{lll} \kappa_4 = \sigma \to \tau \to \sigma & \omega = \beta \to \kappa_4 & \sigma_4 = \pi \to \omega \to \gamma \\ \kappa_3 = \kappa_4 \to \beta \to \kappa_4 & \pi = \beta \to \kappa_4 \to \delta & \sigma_3 = \beta \to \xi \to \chi \\ \epsilon = \kappa_4 \to \delta & \beta_2 = \eta \to \epsilon & \beta_1 = \beta_2 \to \sigma_3 \to \pi \end{array}$$

so that κ_3 and κ_4 are axioms. By unravelling we have

$$\begin{aligned}
\sigma_3 &= (\tau \to \sigma \to \rho) \to (\tau \to \sigma) \to (\tau \to \rho)\\
\sigma_4 &= (\beta \to \kappa_4 \to \delta) \to (\beta \to \kappa_4) \to (\beta \to \delta)\\
\beta_1 &= (\eta \to \epsilon) \to (\beta \to \eta) \to (\beta \to \epsilon)\\
\beta_2 &= (\xi \to \chi) \to (\sigma \to \xi) \to (\sigma \to \chi)
\end{aligned}$$

so that σ_3, σ_4 are axioms and β_1, β_2 are derivable by part (β). (Of course, the original σ need not be an axiom.) Let $\Gamma = \beta, \sigma, \tau$. With these

$$\dfrac{\dfrac{\vdash \sigma_4 \qquad \dfrac{\dfrac{\vdash \beta_1 \quad \vdash \beta_2}{\vdash \sigma_3 \to \pi} \qquad \vdash \sigma_3}{\vdash \pi}}{\vdash \omega \to \gamma} \qquad \dfrac{\vdash \kappa_3 \quad \vdash \kappa_4}{\vdash \omega}}{\vdash \gamma}
\qquad\qquad
\dfrac{\dfrac{\dfrac{\dfrac{\dfrac{\Gamma \vdash \beta \quad \Gamma \vdash \tau}{\Gamma \vdash \sigma \to \rho} \qquad \Gamma \vdash \sigma}{\Gamma \vdash \rho}\,(\uparrow)}{\beta, \sigma \vdash \tau \to \rho}\,(\uparrow)}{\beta \vdash \delta}\,(\uparrow)}{\vdash \gamma}$$

$$\emptyset[\sigma_4](B_1 B_2 \emptyset[\sigma_3])(\emptyset[\kappa_3]\emptyset[\kappa_4]) \qquad\qquad (\Gamma[2]\Gamma[0]\Gamma[1])\uparrow\uparrow\uparrow$$

are an H-derivation and an N-derivation with arboreal codes as indicated. Here B_1, B_2 are derivations of β_1, β_2 with shapes as given by part (β).

(δ) As in part (β), for the given θ, ψ, ϕ let

$$\rho = \theta \to \phi \qquad \sigma = \theta \to \psi \qquad \tau = \psi \to \phi$$

so that $\delta = \sigma \to \tau \to \rho$ is the target formula, and $\beta = \tau \to \sigma \to \rho$ is derivable. With

$$\xi = \tau \to \sigma \qquad \chi = \tau \to \rho \qquad \eta = \xi \to \chi \qquad \zeta = \sigma \to \eta$$

let

$$\begin{aligned}
\kappa_5 &= \eta \to \sigma \to \eta & \qquad \sigma_5 &= \zeta \to \kappa_6 \to \delta\\
\kappa_6 &= \sigma \to \tau \to \sigma & \qquad \sigma_6 &= \beta \to \xi \to \chi
\end{aligned}$$

so that κ_1 and κ_2 are axioms. By unravelling we have

$$\begin{aligned}
\sigma_5 &= (\sigma \to \xi \to \chi) \to (\sigma \to \zeta) \to (\sigma \to \chi)\\
\sigma_6 &= (\tau \to \sigma \to \rho) \to (\tau \to \sigma) \to (\tau \to \rho)
\end{aligned}$$

so these too are axioms. (The original σ is not an axiom.) Let $\Delta = \sigma, \tau, \theta$. With these

$$\dfrac{\dfrac{\vdash \sigma_5 \qquad \dfrac{\vdash \kappa_5 \qquad \dfrac{\vdash \sigma_6 \quad \begin{matrix}\vdots\\ \vdash \beta\end{matrix}}{\vdash \eta}}{\vdash \zeta}}{\vdash \kappa_6 \to \delta} \qquad \vdash \kappa_6}{\vdash \delta}
\qquad\qquad
\dfrac{\dfrac{\dfrac{\dfrac{\Delta \vdash \tau \qquad \dfrac{\Delta \vdash \sigma \quad \Delta \vdash \theta}{\Delta \vdash \psi}}{\Delta \vdash \phi}\,(\uparrow)}{\sigma, \tau \vdash \rho}\,(\uparrow)}{\sigma \vdash \tau \to \rho}\,(\uparrow)}{\vdash \delta}$$

$$(\emptyset[\sigma_5](\emptyset[\kappa_5](\emptyset[\sigma_6]B)))\emptyset[\kappa_6] \qquad\qquad (\Delta[1](\Delta[2]\Delta[0]))\uparrow\uparrow\uparrow$$

are an H-derivation and an N-derivation with arboreal codes as indicated. Here B is a derivation of β. □

1.8 We look at each formula in turn.

(ϵ) For convenience let

$$\alpha = \psi \to \phi \quad \xi = \zeta \to \alpha \quad \mu = \theta \to \zeta \quad \chi = \mu \to \theta \to \alpha \quad \eta = \psi \to \zeta \to \phi$$

so that $\epsilon = \eta \to \chi$ is the target formula. Let

$$\begin{aligned} \gamma &= \eta \to \xi &&= (\psi \to \zeta \to \phi) \to (\zeta \to \psi \to \phi) \\ \beta_2 &= \xi \to \chi &&= (\zeta \to \alpha) \to (\theta \to \zeta) \to (\theta \to \alpha) \\ \beta_1 &= \beta_2 \to \gamma \to \epsilon &&= (\xi \to \chi) \to (\eta \to \xi) \to (\eta \to \chi) \end{aligned}$$

to produce an axiom γ and two derivable formulas β_1, β_2. Let $\Gamma = (\eta, \mu, \theta, \psi)$ to produce a context. With these

$$\dfrac{\dfrac{\begin{matrix}\vdots \\ \vdash \beta_1\end{matrix} \quad \begin{matrix}\vdots \\ \vdash \beta_2\end{matrix}}{\vdash \gamma \to \epsilon} \quad \begin{matrix}\vdots \\ \vdash \gamma\end{matrix}}{\vdash \epsilon} \qquad\qquad \dfrac{\dfrac{\Gamma \vdash \eta \quad \Gamma \vdash \psi}{\Gamma \vdash \zeta \to \phi} \quad \dfrac{\Gamma \vdash \mu \quad \Gamma \vdash \theta}{\Gamma \vdash \zeta}}{\Gamma \vdash \phi}$$

$$B_1 B_2 C \qquad\qquad ((\Gamma[3]\Gamma[0])(\Gamma[2]\Gamma[1]))\uparrow\uparrow\uparrow\uparrow$$

are two H-derivations. When the right hand tree is concluded with four uses of Introduction we obtain $\vdash \epsilon$ in N. The arboreal codes are shown below the trees. (Here B_1, B_2, C are codes from Solution 1.7.)

(ω) For convenience let

$$\iota = \theta \to \theta \quad \lambda = \theta \to \phi \quad \mu = \theta \to \lambda \quad \nu = \mu \to \iota$$

so that $\omega = \mu \to \lambda$ is the target formula. Let

$$\begin{aligned} \kappa &= \iota \to \nu &&= \iota \to \mu \to \iota \\ \sigma_2 &= \mu \to \iota \to \lambda &&= (\theta \to \theta \to \phi) \to (\theta \to \theta) \to (\theta \to \phi) \\ \sigma_1 &= \sigma_2 \to \nu \to \omega &&= (\mu \to \iota \to \lambda) \to (\mu \to \iota) \to (\mu \to \lambda) \end{aligned}$$

to produce three axioms. Let $\Gamma = (\mu, \theta)$ to produce a context. With these

$$\dfrac{\dfrac{\vdash \sigma_1 \quad \vdash \sigma_2}{\vdash \nu \to \omega} \quad \dfrac{\vdash \kappa \quad \vdash \iota}{\vdash \nu}}{\vdash \omega} \qquad\qquad \dfrac{\dfrac{\Gamma \vdash \mu \quad \Gamma \vdash \theta}{\Gamma \vdash \lambda} \quad \Gamma \vdash \theta}{\Gamma \vdash \phi}$$

$$(\emptyset[\sigma_1]\emptyset[\sigma_2])(\emptyset[\kappa]\emptyset[\iota]) \qquad\qquad (\Gamma[1]\Gamma[0]\Gamma[0])\uparrow\uparrow$$

are two H-derivations. When the right hand tree is concluded with two uses of Introduction we obtain $\vdash \omega$ in N. The arboreal codes are shown below each tree.

(μ) For convenience let

$$\begin{array}{llll} \alpha = \psi \to \phi & \eta = \theta \to \alpha & \chi = \rho \to \eta & \omega = \zeta \to \chi \\ \lambda = \theta \to \psi & \xi = \lambda \to \theta \to \phi & \pi = \rho \to \xi & \epsilon = \chi \to \pi \end{array}$$

so that $\mu = \omega \to \zeta \to \pi$ is the target formula. Let

$$\begin{aligned} \sigma &= \eta \to \xi = (\theta \to \psi \to \phi) \to (\theta \to \psi) \to (\theta \to \phi) \\ \beta_1 &= \epsilon \to \mu = (\chi \to \pi) \to (\zeta \to \chi) \to (\zeta \to \pi) \\ \beta_2 &= \sigma \to \epsilon = (\eta \to \xi) \to (\rho \to \eta) \to (\rho \to \xi) \end{aligned}$$

to produce an axiom σ and two derivable formulas β_1, β_2. Let $\Gamma = (\omega, \zeta, \rho, \lambda, \theta)$ to produce a context. With these

$$\dfrac{\dfrac{\vdots}{\vdash \beta_1} \quad \dfrac{\dfrac{\vdots}{\vdash \beta_2} \quad \vdash \sigma}{\vdash \epsilon}}{\vdash \mu} \qquad \qquad \dfrac{\dfrac{\dfrac{\dfrac{\Gamma \vdash \omega \quad \Gamma \vdash \zeta}{\Gamma \vdash \chi} \quad \Gamma \vdash \rho}{\Gamma \vdash \eta} \quad \Gamma \vdash \theta}{\Gamma \vdash \alpha} \quad \dfrac{\Gamma \vdash \lambda \quad \Gamma \vdash \theta}{\Gamma \vdash \psi}}{\Gamma \vdash \phi}$$

$$B_1(B_2\emptyset[\sigma]) \qquad \qquad ((\Gamma[4]\Gamma[3]\Gamma[2]\Gamma[0])(\Gamma[1]\Gamma[0]))\uparrow\uparrow\uparrow\uparrow\uparrow$$

are two H-derivations. When the right hand tree is concluded with five uses of Introduction we obtain $\vdash \mu$ in N. The arboreal codes are shown below each tree.

(ν) For convenience let

$$\eta = \theta \to \psi \to \phi \quad \chi = \rho \to \eta \quad \lambda = \theta \to \psi \quad \xi = (\theta \to \psi) \to (\theta \to \phi)$$

so that $\nu = (\rho \to \eta) \to (\rho \to \xi)$ is the target formula. Let

$$\begin{aligned} \sigma_2 &= \eta \to \xi && = (\theta \to \psi \to \phi) \to (\theta \to \psi) \to (\theta \to \phi) \\ \sigma_1 &= (\rho \to \sigma_2) \to \nu && = (\rho \to \eta \to \xi) \to (\rho \to \eta) \to (\rho \to \xi) \\ \kappa &= \sigma_2 \to \rho \to \sigma_2 \end{aligned}$$

to produce three axioms. Let $\Gamma = (\chi, \rho, \lambda, \theta)$ to produce a context. With these

$$\dfrac{\vdash \sigma_1 \quad \dfrac{\vdash \kappa \quad \vdash \sigma_2}{\vdash \rho \to \sigma_2}}{\vdash \nu} \qquad \qquad \dfrac{\dfrac{\dfrac{\Gamma \vdash \chi \quad \Gamma \vdash \rho}{\Gamma \vdash \eta} \quad \Gamma \vdash \theta}{\Gamma \vdash \psi \to \phi} \quad \dfrac{\Gamma \vdash \lambda \quad \Gamma \vdash \theta}{\Gamma \vdash \psi}}{\Gamma \vdash \phi}$$

$$\emptyset[\sigma_1](\emptyset[\kappa]\emptyset[\sigma_2]) \qquad \qquad ((\Gamma[3]\Gamma[2]\Gamma[0])(\Gamma[1]\Gamma[0]))\uparrow\uparrow\uparrow\uparrow$$

are two H-derivations. When the right hand tree is concluded with four uses of Introduction we obtain $\vdash \mu$ in N. The arboreal codes are shown below each tree.

(τ) for convenience let

$$\alpha = \psi \to \phi \quad \eta = \theta \to \alpha \quad \omega = \zeta \to \theta \quad \lambda = \zeta \to \psi \quad \epsilon = \zeta \to \phi \quad \sigma = \zeta \to \alpha$$

so that $\tau = \eta \to \omega \to \lambda \to \epsilon$ is the target formula. Let

$$\begin{aligned} \beta &= \eta \to \omega \to \sigma = (\theta \to \alpha) \to (\zeta \to \theta) \to (\zeta \to \alpha) \\ \mu &= \beta \to \tau \qquad\; = (\eta \to \omega \to \zeta \to \psi \to \phi) \to \eta \to \omega \to (\zeta \to \psi) \to (\zeta \to \phi) \end{aligned}$$

to produce two derivable formulas. Let $\Gamma = (\eta, \omega, \lambda, \zeta)$ to produce a context. With these

$$\dfrac{\begin{matrix}\vdots & \vdots \\ \vdash \mu & \beta\end{matrix}}{\vdash \tau} \qquad \dfrac{\dfrac{\Gamma \vdash \eta \qquad \dfrac{\Gamma \vdash \omega \quad \Gamma \vdash \zeta}{\Gamma \vdash \theta}}{\Gamma \vdash \psi \to \phi} \qquad \dfrac{\Gamma \vdash \lambda \quad \Gamma \vdash \zeta}{\Gamma \vdash \psi}}{\Gamma \vdash \phi}$$

$$MB \qquad\qquad ((\Gamma[3](\Gamma[2]\Gamma[0]))(\Gamma[1]\Gamma[0]))\uparrow\uparrow\uparrow\uparrow$$

are two H-derivations. When the right hand tree is concluded with four uses of Introduction we obtain $\vdash \tau$ in N. The arboreal codes are shown below each tree. (Here M and B are codes from above.) □

1.9 We make use of the derivable formulas γ, δ of Exercise 1.7. We use the same abbreviations throughout all five parts. Given θ, ψ, ϕ let

$$\begin{array}{lllll} \eta = \theta \to \psi & \xi = \psi \to \phi & \chi = \theta \to \phi & & \nu = \eta \to \psi = \theta^{\psi} \\ \rho = \xi \to \chi & \lambda = \theta^{\phi} & \mu = \nu^{\phi} & \zeta = \lambda \to \tau & \tau = \xi \to \phi = \psi^{\phi} \end{array}$$

so that

$$\omega(1) = \theta \to \nu \quad \omega(2) = \theta \to \mu \quad \omega(3) = \eta \to \zeta \quad \omega(4) = \lambda \to \mu \quad \omega(5) = \theta \to \mu$$

are the target formulas.

(i) Using the derivable formula

$$\begin{aligned} \gamma &= (\eta \to \eta) \to \omega(1) \\ &= (\eta \to \theta \to \psi) \to (\theta \to \eta \to \psi) \end{aligned}$$

and a context $\Gamma = \theta, \eta$ the two derivations

$$C\emptyset[\iota] \qquad (\Gamma[1]\Gamma[0])\uparrow\uparrow$$

shown to the right will do.

$$\dfrac{\begin{matrix}\vdots & \\ \vdash \gamma & \vdash \eta \to \eta\end{matrix}}{\vdash \omega(1)} \qquad \dfrac{\dfrac{\dfrac{\Gamma \vdash \eta \quad \Gamma \vdash \theta}{\Gamma \vdash \psi}}{\theta \vdash \nu}\,(\uparrow)}{\vdash \omega(1)}\,(\uparrow)$$

(ii) Using a variant of $\omega(1)$, a derivable formula δ, and two contexts Σ, Γ

$$\begin{aligned} &\omega = \nu \to \mu = \nu \to \nu^{\phi} \\ &\delta\; = \omega(1) \to \omega \to \omega(2) \\ &\Sigma = \theta, \nu \to \phi \qquad \Gamma = \Sigma, \eta \end{aligned}$$

the derivations $DW(1)W$ and

$$(\Sigma[0]((\Gamma[0]\Gamma[2])\uparrow))\uparrow\uparrow$$

shown to the right will do.

$$\dfrac{\dfrac{\begin{matrix}\vdots & \vdots \\ \vdash \delta & \vdash \omega(1)\end{matrix}}{\bullet} \qquad \begin{matrix}\vdots \\ \vdash \omega\end{matrix}}{\vdash \omega(2)} \qquad \dfrac{\dfrac{\dfrac{\Sigma \vdash \nu \to \phi \qquad \dfrac{\dfrac{\Gamma \vdash \eta \quad \Gamma \vdash \theta}{\Gamma \vdash \psi}}{\Sigma \vdash \nu}\,(\uparrow)}{\Sigma \vdash \phi}}{\theta \vdash \mu}\,(\uparrow)}{\vdash \omega(2)}\,(\uparrow)$$

(iii) Using four derivable formulas and two contexts

$$\begin{array}{ll} \delta_{12} = \eta \to \rho & \delta_{23} = \rho \to \zeta \\ \epsilon = \delta_{23} \to \omega(3) & \delta_{13} = \delta_{12} \to \epsilon \\ \Sigma = \eta, \lambda, \xi & \Gamma = \Sigma, \theta \end{array}$$

the derivations

$$D_{13}D_{12}D_{23} \qquad (\Sigma[1]((\Gamma[1](\Gamma[3]\Gamma[0]))\uparrow))\uparrow\uparrow\uparrow$$

shown to the right will do.

$$\dfrac{\dfrac{\dfrac{\dfrac{\Sigma \vdash \lambda \qquad \dfrac{\dfrac{\Gamma \vdash \xi \qquad \dfrac{\Gamma \vdash \eta \quad \Gamma \vdash \theta}{\Gamma \vdash \psi}}{\Gamma \vdash \phi}}{\Sigma \vdash \chi}(\uparrow)}{\Sigma \vdash \phi}}{\eta, \lambda \vdash \tau}(\uparrow)}{\eta \vdash \zeta}(\uparrow)}{\vdash \omega(3)}(\uparrow)$$

(iv) With a variant of $\omega(3)$ and three contexts

$$\omega(1) \to \omega(4) = (\theta \to \nu) \to (\theta^\phi \to \nu^\phi)$$
$$\Sigma = \lambda, \nu \to \phi \qquad \Delta = \Sigma, \theta \qquad \Gamma = \Delta, \eta$$

the derivations $W(3)W(1)$ and

$$(\Sigma[1]((\Delta[1]((\Gamma[0]\Gamma[1])\uparrow))\uparrow))\uparrow\uparrow$$

will do. The N-derivation is shown to the right.

$$\dfrac{\dfrac{\dfrac{\Sigma \vdash \lambda \qquad \dfrac{\dfrac{\Delta \vdash \nu \to \phi \qquad \dfrac{\dfrac{\Gamma \vdash \eta \quad \Gamma \vdash \theta}{\Gamma \vdash \psi}}{\Delta \vdash \nu}(\uparrow)}{\Delta \vdash \phi}}{\Sigma \vdash \chi}(\uparrow)}{\Sigma \vdash \phi}}{\lambda \vdash \mu}(\uparrow)}{\vdash \omega(4)}(\uparrow)$$

(v) The variant $\omega = \theta \to \lambda = \theta \to \theta^\phi$ of $\omega(1)$ and the derivable formula

$$\begin{aligned} \delta &= \omega \to \omega(4) \to \omega(5) \\ &= (\theta \to \lambda) \to (\lambda \to \mu) \to (\theta \to \mu) \end{aligned}$$

produce the H-derivation shown. Any H-derivation can be turned into an N-derivation. We replace each leaf by an appropriate N-derivation. When we do this for $\omega(5)$ we get an N-derivation that is much bigger than the one in part (ii). □

$$\dfrac{\dfrac{\begin{matrix}\vdots \\ \vdash \delta\end{matrix} \qquad \begin{matrix}\vdots \\ \vdash \omega\end{matrix}}{\bullet} \qquad \begin{matrix}\vdots \\ \vdash \omega(4)\end{matrix}}{\vdash \omega(5)}$$

$$(DW(1)W(4))$$

1.10 The three derivations have shapes shown to the right. You should make sure you can fill in the missing components. □

$$\dfrac{\bullet}{\Gamma \vdash \iota}(\uparrow) \qquad \dfrac{\dfrac{\bullet}{\bullet}(\uparrow)}{\Gamma \vdash \kappa}(\uparrow) \qquad \dfrac{\dfrac{\dfrac{\dfrac{\dfrac{\bullet\ \bullet}{\bullet} \quad \dfrac{\bullet\ \bullet}{\bullet}}{\bullet}}{\bullet}(\uparrow)}{\bullet}(\uparrow)}{\Gamma \vdash \sigma}(\uparrow)$$

A.4 SOME ALGORITHMS ON DERIVATIONS

1.11 The effect of $(\cdot)_N$ is to grow extra branches above each Axiom, as in Solution 1.10. The largest increase in dimensions is caused by a use of the s-Axiom. A simple induction shows that

$$h(\nabla_H) \le h(H) + 5 \qquad w(\nabla_H) \le 4w(H) \qquad s(\nabla_H) \le s(H) + 9 \times 2^{h(H)}$$

hold. □

1.12 We show

$$h(\nabla_D) \le 2h(\nabla) + 1$$

by induction over ∇. By inspection, when ∇ is a leaf we have $h(\nabla_D) \le 1$, which gives the base cases. For $\nabla = QP$ let $m = \max(h(Q), h(P))$ so that $h(\nabla) = m+1$. Using the induction hypothesis at the comparison we have

$$h(\nabla_D) = h(SQ_DP_D) = \max(2, h(Q_D) + 2, h(P_D) + 1) \le 2m + 3 = 2h(\nabla) + 1$$

to obtain the required result. □

1.13 With

$$\rho = \theta \to \phi \qquad \sigma = \theta \to \psi \qquad \tau = \psi \to \phi$$

we have

$$\nabla = \Gamma[2](\Gamma[1]\Gamma[0])$$

where $\Gamma = \tau, \sigma, \theta$ (and σ need not be an axiom). Let $\Sigma = \tau, \sigma$. Then we have

$$\begin{array}{rcllcl}
\Gamma[2]_D & = & \Sigma[\kappa_2]\Sigma[1] & \text{where} & \kappa_2 & = \tau \to \theta \to \tau \\
\Gamma[1]_D & = & \Sigma[\kappa_3]\Sigma[0] & \text{where} & \kappa_3 & = \sigma \to \theta \to \sigma \\
\Gamma[0]_D & = & \Sigma[\iota] & \text{where} & \iota & = \theta \to \theta
\end{array}$$

using the leaf clauses of the algorithm. With the axioms

$$\sigma_2 = (\theta \to \tau) \to \sigma \to \rho \qquad \sigma_3 = (\theta \to \sigma) \to \iota \to \sigma$$

we have

$$\begin{aligned}
\nabla_D &= \Sigma[\sigma_2]\Gamma[2]_D(\Gamma[1]\Gamma[0])_D = \Sigma[\sigma_2](\Sigma[\kappa_2]\Sigma[1])(\Sigma[\sigma_3]\Gamma[1]_D\Gamma[0]_D) \\
&= \Sigma[\sigma_2](\Sigma[\kappa_2]\Sigma[1])(\Sigma[\sigma_3](\Sigma[\kappa_3]\Sigma[0])\Sigma[\iota])
\end{aligned}$$

to complete the first calculation. This derivation has shape

$$\dfrac{\dfrac{\Sigma \vdash \sigma_2 \quad \dfrac{\Sigma \vdash \kappa_2 \quad \Sigma \vdash \tau}{\bullet}}{\bullet} \quad \dfrac{\dfrac{\Sigma \vdash \sigma_3 \quad \dfrac{\Sigma \vdash \kappa_3 \quad \Sigma \vdash \sigma}{\bullet}}{\bullet} \quad \Sigma \vdash \iota}{\bullet}}{\Sigma \vdash \rho}$$

which is bigger than ∇.

To calculate ∇_{DD} notice that of the leaves of ∇_D only $\Sigma \vdash \sigma$ is a Gate Projection. Also, there are six uses of Elimination. Thus ∇_{DD} is quite large (with 37 nodes). After this we must compute ∇_{DDD} which will be very much larger and nothing like the standard derivation of $\vdash \beta$. □

1.14 With $Q' = Q_{HD}$ and $P' = P_{HD}$ we have

$$(QP)\uparrow_H = (QP)_{HD} = (Q_H P_H)_D = SQ'P'$$

where S is the use of an s-Axiom. After two more uses of s-Axioms we obtain

$$(RQP)_D = S(RQ)_D P_D = S(SR_D Q_D)P_D$$

and hence

$$(QP)\uparrow\uparrow_H = (QP)\uparrow_{HD} = (SQ'P')_D = S(SS_D Q'')P'' = S(S(KS)Q'')P''$$

where $Q'' = Q_{HDD}, P' = P_{HDD}$, K is the use of a k-Axiom, and each S is the use of an s-Axiom. These first two calculations produce the following shapes.

$$\dfrac{\dfrac{S \quad Q'}{\bullet} \quad P'}{\bullet} \qquad\qquad \dfrac{\dfrac{S \quad \dfrac{\dfrac{S \quad \dfrac{K \quad S}{\bullet}}{\bullet} \quad Q''}{\bullet}}{\bullet} \quad P''}{\bullet}$$

Continuing in this way we find that $(QP)\uparrow\uparrow\uparrow_H$ produces a shape

$$\dfrac{\dfrac{S \quad \dfrac{\dfrac{S \quad S_D}{\bullet} \quad \dfrac{\dfrac{S \quad \dfrac{\dfrac{S \quad S_D}{\bullet} \quad S_{DD}}{\bullet}}{\bullet} \quad Q'''}{\bullet}}{\bullet}}{\bullet} \quad P'''}{\bullet}$$

where S_D and S_{DD} must be expanded. After this the calculation gets rather out of hand. □

1.15 (i) With $\iota = \theta \to \theta$ and $I = \emptyset[\iota]$ we have $I_N = \Theta[0]\uparrow$ where $\Theta = \theta$ (a context of length 1). Thus

$$I_{NH} = \Theta[0]\uparrow_H = \Theta[0]_{HD} = \Theta[0]_D = I$$

so we do get back to where we started from.

(k) With $\kappa = \psi \to \theta \to \psi$, $K = \emptyset[\kappa]$ we have $K_N = \Gamma[1]\uparrow\uparrow$ where $\Gamma = \psi, \theta$. Thus

$$K_{NH} = \Gamma[1]\uparrow_{HD} = \Gamma[1]_{HDD} = \Gamma[1]_{DD} = (\Psi[\kappa]\Psi[0])_D$$

where Ψ is the obvious 1-element context and the algorithm requires $\kappa = \psi \to \theta \to \psi$, the original formula! With this

$$K_{NH} = S\Psi[\kappa]_D\Psi[0]_D = S(K'K)I$$

$$\dfrac{\dfrac{\vdash \sigma \qquad \dfrac{\vdash \kappa' \quad \vdash \kappa}{\vdash \psi \to \kappa}}{\bullet} \qquad \vdash \iota}{\vdash \kappa}$$

for suitable Axioms S, K', I. In full this is shown above right. Here we have

$$\iota = \psi \to \psi \qquad \kappa' = \kappa \to \psi \to \kappa \qquad \mu = \theta \to \psi \qquad \sigma = (\psi \to \psi \to \mu) \to \iota' \to \kappa$$

so that $\kappa = \psi \to \mu$. This is a pretty daft derivation. It uses an axiom $\vdash \kappa$ to derive itself. Thus something must be lost as we translate between the two systems.

(s) With $\sigma = (\theta \to \psi \to \phi) \to (\theta \to \psi) \to (\theta \to \phi)$, $S = \emptyset[\sigma]$ and

$$\lambda = \theta \to \phi \qquad \mu = \theta \to \psi \qquad \nu = \theta \to \psi \to \phi$$

we have

$$S_N = ((\Gamma[2]\Gamma[0])(\Gamma[1]\Gamma[0]))\uparrow\uparrow\uparrow$$

where $\Gamma = \nu, \mu, \theta$. Since

$$S_{NH} = ((\Gamma[2]\Gamma[0])(\Gamma[1]\Gamma[0]))_{DDD}$$

we find, as in Solution 1.14, that this is quite large. In fact, it has 79 leaves

This illustrates that there is much more to all this than meets the eye. □

1.16 We proceed by induction on ∇.

When ∇ is a Projection we have $\nabla_H = \nabla$ and $h(\nabla) = 0$, so the comparison is immediate. This gives the base case.

Suppose $\nabla = R\uparrow$. Using the result of Exercise 1.12 we have

$$h(\nabla_H) + 1 = h(R_{HD}) + 1 \leq 2(h(R_H) + 1) \leq 2 \times 2^{h(R)+1} = 2^{h(R)+2} = 2^{h(\nabla)+1}$$

where the second comparison uses the induction hypothesis. This gives the induction step across a use of Introduction.

Suppose $\nabla = QP$ with $m = \max(h(Q), h(P))$. Then

$$\begin{aligned} h(\nabla_H) + 1 &= h(Q_H P_H) + 1 \\ &= \max(h(Q_H) + 1, h(P_H) + 1) + 1 \leq 2^{m+1} \leq 2^{m+2} = 2^{h(\nabla)+1} \end{aligned}$$

where the first comparison uses the induction hypothesis. This gives the induction step across a use of Elimination.

With a little more effort we can show that $h(\nabla_H) + 3 \leq (\sqrt{2})^{h(\nabla)+3}$ holds. □

1.17 (a) With ρ, σ, τ as suggested let

$$\Pi = \sigma, \tau \qquad \Gamma = \beta, \Pi \qquad \Sigma = \tau, \sigma, \theta$$

and

$$\Gamma(C) = \Gamma[2]\Gamma[0]\Gamma[1] \quad C = \Gamma(C)\uparrow\uparrow\uparrow \qquad \Sigma(B) = \Sigma[2](\Sigma[1]\Sigma[0]) \quad B = \Sigma(B)\uparrow\uparrow\uparrow$$

so that $\nabla = CB$ is the new derivation of $\vdash \delta$. In full this is

$$\dfrac{\left(\dfrac{\dfrac{\dfrac{\dfrac{\dfrac{\beta, \Pi \vdash \beta \quad \Gamma \vdash \tau}{\Gamma \vdash \sigma \to \rho} \qquad \Gamma \vdash \sigma}{\Gamma \vdash \rho}}{\beta, \sigma \vdash \tau \to \rho}\,(\uparrow)}{\beta \vdash \delta}\,(\uparrow)}{\vdash \beta \to \delta}\,(\uparrow)\right) \left(\dfrac{\dfrac{\dfrac{\dfrac{\dfrac{\Sigma \vdash \sigma \quad \Sigma \vdash \theta}{\Sigma \vdash \tau} \qquad \Sigma \vdash \psi}{\Sigma \vdash \phi}}{\tau, \sigma \vdash \rho}\,(\uparrow)}{\tau \vdash \sigma \to \rho}\,(\uparrow)}{\vdash \beta}\,(\uparrow)\right)}{\vdash \delta}\,(*)$$

where the subderivations C and B have been set in parentheses.

Since

$$\nabla = CB = (\Gamma(C)\uparrow\uparrow\uparrow)B$$

this has an abnormality at $(*)$, as indicated. To remove this we graft B to the top left hand leaf of C to take care of the hypothesis β, as indicated.

$$\left(\dfrac{\begin{array}{c}\dfrac{\dfrac{\overset{\downarrow}{\beta}, \Pi \vdash \beta \quad \Gamma \vdash \tau}{\Gamma \vdash \sigma \to \rho} \qquad \Gamma \vdash \sigma}{\bullet} \\ \vdots \end{array}}{\beta \vdash \delta}\,(\uparrow)\right) * \left(\begin{array}{c} \vdots \\ \vdash \beta \end{array}\right)$$

Thus, using the arboreal code, to remove this abnormality we pass to

$$\nabla' = \Gamma(C)\uparrow\uparrow * B = (\Gamma(C)\uparrow * B\downarrow)\uparrow = (\Gamma(C) * B\downarrow\downarrow)\uparrow\uparrow$$

for a Weakened version $B\downarrow\downarrow$ of B. The nominated component of Γ is β in position 2. Let $\Xi = \Pi, \Sigma$ and

$$\Xi(B) = \Xi[2](\Xi[1]\Xi[0])$$

so that $B\downarrow\downarrow = \Xi(B)\uparrow\uparrow\uparrow$. Now with $B' = B\downarrow\downarrow$ we have

$$\Gamma[2] * B' = B' \quad \Gamma[1] * B' = \Pi[1] \quad \Gamma[0] * B' = \Pi[0]$$

so that

$$\Gamma(C) * B' = B'\Pi[0]\Pi[1] \quad \text{and hence} \quad \nabla' = ((B\downarrow\downarrow)\Pi[0]\Pi[1])\uparrow\uparrow$$

is the result of the first removal.

In full this is the derivation to the right. This has an abnormality at $(*)$ (which has been created by the first removal). The problem is with

$$B'\Pi[0] = \Xi(B)\uparrow\uparrow\uparrow\Pi[0]$$

which must replaced by

$$\begin{aligned}\Xi(B)\uparrow\uparrow * \Pi[0] &= (\Xi(B)\uparrow * \Pi[0]\downarrow)\uparrow \\ &= (\Xi(B) * \Pi[0]\downarrow\downarrow)\uparrow\uparrow\end{aligned}$$

for a Weakened version $\Pi[0]\downarrow\downarrow$ of $\Pi[0]$. This grafting must be done with some care.

$$\dfrac{\dfrac{\dfrac{\dfrac{\dfrac{\dfrac{\dfrac{\dfrac{\Xi \vdash \tau \qquad \dfrac{\Xi \vdash \sigma \quad \Xi \vdash \theta}{\Xi \vdash \psi}}{\Xi \vdash \phi}}{\Pi, \tau, \sigma \vdash \rho}(\uparrow)}{\Pi, \tau \vdash \sigma \to \rho}(\uparrow)}{\Pi \vdash \beta}(\uparrow) \qquad \Pi \vdash \tau}{\Pi \vdash \sigma \to \rho}(*) \qquad \Pi \vdash \sigma}{\Pi \vdash \rho}}{\sigma \vdash \tau \to \rho}(\uparrow)}{\vdash \delta}(\uparrow)$$

We have contexts

$$\Xi = \Pi, \Sigma = \sigma, \tau, \tau, \sigma, \theta$$

and it is the *rightmost* occurrence of τ, in position 2, which is nominated. We must not graft onto the other position.

Let

$$\Lambda = \Pi, \sigma, \theta$$

i.e. Λ is Ξ with the nominated component removed. With this we have

$$\Pi[0]\downarrow\downarrow = \Lambda[2] \qquad \Xi[2] * \Lambda[2] = \Lambda[2] \qquad \Xi[1] * \Lambda[2] = \Lambda[1] \qquad \Xi[0] * \Lambda[2] = \Lambda[0]$$

to give

$$\Xi(B) * \Pi[0]\downarrow\downarrow = \Lambda[2](\Lambda[1]\Lambda[0]) = \Lambda(B) \qquad \text{(say)}$$

and hence

$$\nabla'' = ((\Lambda(B)\uparrow\uparrow)\Pi[1])\uparrow\uparrow$$

is the result of the second removal.

In full this is the derivation to the right. This has an abnormality at $(*)$ (which has been created by the second removal). We have

$$\Lambda = \Pi, \sigma, \theta = \sigma, \tau, \sigma, \theta$$

and the problem is with

$$(\Lambda(B)\uparrow\uparrow)\Pi[1]$$

which must be replaced by

$$\Lambda(B)\uparrow * \Pi[1] = (\Lambda(B) * \Pi[1]\downarrow)\uparrow$$

using the σ in position 1 of Λ as the nominated component.

$$\dfrac{\dfrac{\dfrac{\dfrac{\dfrac{\dfrac{\Lambda \vdash \tau \qquad \dfrac{\Lambda \vdash \sigma \quad \Lambda \vdash \theta}{\Lambda \vdash \psi}}{\Lambda \vdash \phi}}{\Pi, \sigma \vdash \rho}(\uparrow)}{\Pi \vdash \sigma \to \rho}(\uparrow) \qquad \Pi \vdash \sigma}{\Pi \vdash \rho}(*)}{\sigma \vdash \tau \to \rho}(\uparrow)}{\vdash \delta}(\uparrow)$$

Let

$$\Delta = \sigma, \tau, \theta$$

i.e. Δ is Λ with the nominated component removed. With this we have

$$\Pi[1]\!\downarrow = \Delta[2] \qquad \Lambda[2] * \Delta[2] = \Delta[1] \quad \Lambda[1] * \Delta[2] = \Delta[2] \quad \Lambda[0] * \Delta[2] = \Delta[2]$$

to give

$$\Lambda(B) * \Pi[1] = \Delta[1](\Delta[2]\Delta[0]) = \Delta(D) \quad \text{(say)}$$

and hence

$$\nabla''' = \Delta(D)\!\uparrow\uparrow\uparrow$$

is the result of the third removal.

This is shown in full to the right. The derivation is normal and precisely the known derivation of $\vdash \delta$ in H.

$$\dfrac{\dfrac{\dfrac{\dfrac{\Delta \vdash \tau \qquad \dfrac{\Delta \vdash \sigma \quad \Delta \vdash \theta}{\Delta \vdash \psi}}{\Delta \vdash \phi}}{\Pi \vdash \rho}(\uparrow)}{\sigma \vdash \tau \to \rho}(\uparrow)}{\vdash \delta}(\uparrow)$$

This example illustrates how the present notation for derivations and algorithms is not very good. It is far too easy to graft at the wrong occurrence of a repeated hypothesis. We need to develop better bookkeeping techniques. This is done in the later chapters.

(b) There is a standard 'normalization' algorithm which will transform the new H-derivation of $\vdash \delta$ into the old one. However, this cannot be described without a decent notation for derivations and algorithms, and this has not yet been developed (but will be later). □

1.18 Informally any use of a structural rule can be pushed towards the leaves where it is easily absorbed. Formally we can deal with all the structural rules in unison.

Given contexts Γ, Σ let

$$(A) \quad \Gamma \sqsubseteq \Sigma$$

mean that Σ is a structurally modfied version of Γ, i.e. Σ can be obtained from Γ by a sequence of interchanges, deletions of repetitions, and insertions, in some order. We can think of A as a recipe for producing Σ from Γ. Each structural rule has the form

$$\dfrac{\Gamma \vdash \phi}{\Sigma \vdash \phi}$$

for some particular Σ with $\Gamma \sqsubseteq \Sigma$. Furthermore, for any $\Gamma \sqsubseteq \Sigma$ and any formula ϕ there is a derivation

$$\begin{array}{c} \Gamma \vdash \phi \\ \hline \\ \vdots \\ \hline \Sigma \vdash \phi \end{array}$$

which uses only structural rules.

We describe an algorithm which, when supplied with a pair

$$(\nabla) \quad \Gamma \vdash \phi \qquad (A) \quad \Gamma \sqsubseteq \Sigma$$

will return a derivation

$$(\nabla \cdot A) \quad \Sigma \vdash \phi$$

where ∇ and $\nabla \cdot A$ are in the same system. Thus if a judgement $\Gamma \vdash \phi$ is derivable in a structurally enriched version of H or N, then it is derivable in the unenriched version.

The algorithm proceeds by recursion on ∇.

If

$$(\nabla) \quad \Gamma \vdash \phi$$

is a leaf, an Axiom or a Projection, then so is

$$\Sigma \vdash \phi$$

so we take this as $\nabla \cdot A$.

Suppose $\nabla = R\uparrow$ where

$$(R) \quad \Gamma, \theta \vdash \psi$$

with $\phi = \theta \to \psi$. The given A easily converts into some

$$(\downarrow A) \quad \Gamma, \theta \sqsubseteq \Sigma, \theta$$

and then

$$(R \cdot \downarrow A) \quad \Sigma, \theta \vdash \psi$$

is obtained by a recursive call on the algorithm. Thus we set

$$R\uparrow \cdot A = (R \cdot \downarrow A)\uparrow$$

to pass across a use of I.

Using

$$QP \cdot A = (Q \cdot A)(P \cdot A)$$

the step across a use of E is immediate.

The shape of $\nabla \cdot A$ is exactly the same as that of ∇. Only the contexts in the judgements are different. □

1.19 (a) From Solution 1.8(τ) with a change of notation we have a derivation

$$(\Gamma[T]) \quad \Gamma \vdash \tau$$

in H for each context Γ. Then

$$\Gamma[\iota]_H = \Gamma[\iota] \quad \Gamma[\kappa]_H = \Gamma[\kappa] \quad \Gamma[\tau]_H = \Gamma[T] \quad (QP)_H = Q_H P_H$$

translates each T-derivation ∇ into an H-derivation ∇_H with the same root. (This, of course, is not the same as Algorithm 1.14.)

(b) Given θ, ψ, ϕ let $\zeta = \theta$, $\lambda = \psi \to \phi$, $\mu = \psi$, $\nu = \phi$, $\iota = \iota(\lambda)$ and consider $\tau(\zeta; \lambda, \mu, \nu)$. Note that $\lambda \to \mu \to \nu = \iota'$ and then $\tau = \iota' \to \sigma$, which leads to a derivation

$$(\Gamma[S]) \quad \Gamma \vdash \sigma$$

in T for each context Γ. Then

$$\Gamma[\iota]_T = \Gamma[\iota] \quad \Gamma[\kappa]_T = \Gamma[\kappa] \quad \Gamma[\sigma]_T = \Gamma[S] \quad (QP)_T = Q_T P_T$$

translates each H-derivation ∇ into an T-derivation ∇_T with the same root. □

1.20 (a) Let

$$(Z_0) \quad \Gamma \vdash \xi_0$$

be the given T-derivation. Observe first that $\tau_r = \xi_r \to \xi_{r+1}$. Then

$$Z_{r+1} = \Gamma[\tau_r] Z_r$$

generates derivations

$$(Z_r) \quad \Gamma \vdash \xi_r$$

for all relevant r. The derivation Z_r is an ascending staircase of height r.

(b) Given T-derivations

$$(Z_0) \quad \Gamma \vdash \xi_0 \qquad (L_r) \quad \Gamma \vdash \lambda_r \qquad (M_r) \quad \Gamma \vdash \mu_r$$

we use the derivations Z_r of part (a) to set $N_r = Z_r L_r M_r$ to generate derivations

$$(N_r) \quad \Gamma \vdash \nu_r$$

for all relevant r. The derivation Z_r is an ascending staircase of height r with a couple of basement steps. □

1.21 (a) Suppose we have a T-derivation

$$(\nabla) \quad \Gamma, \theta_r, \dots, \theta_1 \vdash \phi$$

whose concluding rule is a use of E. Thus $\nabla = QP$ where

$$(Q) \quad \Gamma, \theta_r, \dots, \theta_1 \vdash \psi \to \phi \qquad (P) \quad \Gamma, \theta_r, \dots, \theta_1 \vdash \psi$$

are the left and right legs for some formula ψ.

Suppose we have T-derivations

$$(L_r) \quad \Gamma \vdash \theta_r \to \cdots \to \theta_1 \to \psi \to \phi \qquad (M_r) \quad \Gamma \vdash \theta_r \to \cdots \to \theta_1 \to \psi$$

(which we may obtain by recursive calls on the algorithm under construction). Let

$$\lambda = (\psi \to \phi) \quad \mu = \psi \quad \nu = \phi$$

and consider the formulas λ_r, μ_r, ν_r of Exercise 1.20. Note that ξ_0 is $\iota(\lambda, \lambda)$ and

$$\begin{aligned} \lambda_r &\text{ is } \theta_r \to \cdots \to \theta_1 \to \psi \to \phi \\ \mu_r &\text{ is } \theta_r \to \cdots \to \theta_1 \to \psi \\ \nu_r &\text{ is } \theta_r \to \cdots \to \theta_1 \to \phi \end{aligned}$$

and so we have derivations

$$(Z_0)\ \ \Gamma \vdash \xi_0 \qquad (L_r)\ \ \Gamma \vdash \lambda_r \qquad (M_r)\ \ \Gamma \vdash \mu_r$$

where the first is an Axiom. Thus, using Exercise 1.20, we obtain a T-derivation of

$$\Gamma \vdash \theta_r \to \cdots \to \theta_1 \to \phi$$

as required. This is considerably more regular and smaller than the corresponding derivation obtained using H.

(b) For instance, the translation of

$$\nabla = (QP)\uparrow\uparrow\uparrow\uparrow$$

is shown to the right. Here Q'''', P'''' are the translations of Q, P and T, I indicate a use of the axioms. This shows how translating from N to T is easier than translating to H. □

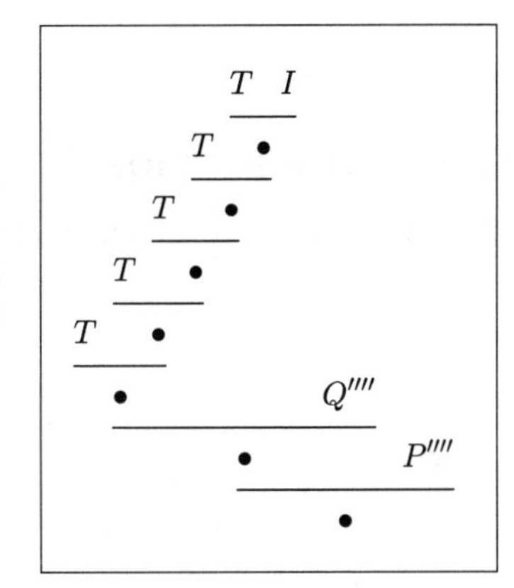

B
Computation Mechanisms

B.1 Introduction

2.1 (a) Instances of $\dot{\rhd}$ can be generated from instances of $\rhd$ using the rules

$$\text{(Axiom)}\quad \frac{t^- \rhd t^+}{t^- \dot{\rhd} t^+}(A) \qquad \text{(Left application)}\quad \frac{q^- \dot{\rhd} q^+}{\bullet q^- p \dot{\rhd} \bullet q^+ p}(L) \qquad \text{(Right application)}\quad \frac{p^- \dot{\rhd} p^+}{\bullet q p^- \dot{\rhd} \bullet q p^+}(R)$$

where each $\bullet$ is $+$ or $\times$. These rules take an axiom $l \rhd r$ and then build up matching terms around l and r.

(b) There are two ways to generate instances of $\rhd\!\!\rhd$. Both need a rule of transitive composition ($\circ$) as shown top right. To this we can either add a rule which moves from $\dot{\rhd}$ to $\rhd\!\!\rhd$ as shown bottom right, or we can repeat the rules above with $\dot{\rhd}$ replaced by $\rhd\!\!\rhd$. Both sets of rules generate the same relation $\rhd\!\!\rhd$, but they do *not* produce the same computations. The first alternative set of rules produces computations in a kind of standard form; all applications are done before compositions.

$$\frac{t^- \rhd\!\!\rhd t^0 \quad t^0 \rhd\!\!\rhd t^+}{t^- \rhd\!\!\rhd t^+}(\circ) \qquad \frac{t^- \dot{\rhd} t^+}{t^- \rhd\!\!\rhd t^+}$$

(c) For each $m \in \mathbb{N}$ we set

$$\ulcorner m \urcorner = \mathsf{S}^m 0$$

i.e. we generate these terms by recursion

$$\ulcorner 0 \urcorner = 0 \qquad \ulcorner m' \urcorner = \mathsf{S}\ulcorner m \urcorner$$

(where $m' = 1 + m$). These are the **numerals**. Each reduction axiom requires an instance of $+$ or $\times$, so the numerals are normal. Part (d) shows these are all the normal terms.

(d) We first use induction on m to show that

$$+\ulcorner n \urcorner \ulcorner m \urcorner \rhd\!\!\rhd \ulcorner n + m \urcorner \qquad \times \ulcorner n \urcorner \ulcorner m \urcorner \rhd\!\!\rhd \ulcorner n \times m \urcorner$$

hold. The second uses the first. For instance, to obtain the induction step we argue

$$\times \ulcorner n \urcorner \ulcorner m' \urcorner = \times \ulcorner n \urcorner (\mathsf{S}\ulcorner m \urcorner) \rhd\!\!\rhd \ulcorner n \urcorner (\times \ulcorner n \urcorner \ulcorner m \urcorner) \rhd\!\!\rhd \ulcorner n \urcorner \ulcorner n \times m \urcorner \rhd\!\!\rhd \ulcorner n + n \times m \urcorner = \ulcorner n \times m' \urcorner$$

using an axiom, the induction hypothesis, and the the first part.

The full result now follows by induction over the construction of terms. For instance, for a term $t = \times sr$, the induction hypothesis gives

$$s \gg \ulcorner n \urcorner \qquad r \gg \ulcorner m \urcorner$$

(for some $n, m \in \mathbb{N}$) and then

$$t \gg \ulcorner n \times m \urcorner$$

follows by the preliminary result.

Notice that this result is nothing more than a version of the well-foundedness of the usual recursive description of addition and multiplication.

(e) Given a divergent wedge

$$t_0 \dot{\rhd} t_1 \qquad t_0 \dot{\rhd} t_2$$

we produce the required convergent wedge by a double recursion on the heights of the two slim trees that witness these reductions. For each tree the root rule is either an Axiom, a Left Application, or a Right Application, so that, roughly speaking, there are $3^2 = 9$ cases to consider. However, some of the cases come in several different kinds, and there are some symmetries involved. Let's look at some typical cases.

(A,A) Each term can match the left hand side of at most one axiom. Thus if both trees are obtained by axioms then $t_1 = t_2$.

(A,L) There are four possible axioms. For the most complicated case the two trees will be

$$\frac{t_0 \rhd +s(\times sr)}{t_0 \dot{\rhd} +s(\times sr)}\,(\mathrm{A}) \qquad \frac{\begin{array}{c}\vdots\\ s \dot{\rhd} s_2\end{array}}{t_0 \dot{\rhd} \times s_2(\mathsf{S}r)}\,(\mathrm{L})$$

where $t_0 = \times s(\mathsf{S}r)$ is the common source. The two trees

$$\frac{\dfrac{\begin{array}{c}\vdots\\ s \dot{\rhd} s_2\end{array}}{t_1 \dot{\rhd} +s_2(\times sr)}\,(\mathrm{L}) \qquad \dfrac{\dfrac{\begin{array}{c}\vdots\\ s \dot{\rhd} s_2\end{array}}{\times sr \dot{\rhd} \times s_2 r}\,(\mathrm{L})}{+s_2(\times sr) \dot{\rhd} t_3}\,(\mathrm{R})}{t_1 \gg t_3}\,(\circ) \qquad \frac{t \rhd t_3}{t \dot{\rhd} t_3}\,(\mathrm{A})$$

show that $t_3 = +s_2(\times sr)$ is a common target. Notice the use of transitive composition in the left hand tree. This is because two subterms s have to be dealt with.

(L,L) Here we must have $t_0 = \bullet sr$ where $\bullet$ is $+$ or $\times$. The two trees will end

$$\frac{s \dot{\rhd} s_1}{t_0 \dot{\rhd} t_1}\,(\mathrm{L}) \qquad \frac{s \dot{\rhd} s_2}{t_0 \dot{\rhd} t_2}\,(\mathrm{L})$$

where $t_i = \bullet s_i r$ for $i = 1, 2$. By recursion we obtain $s_i \ggg s_3$ and then uses of (L) produce $t_3 = \bullet s_3 r$ as a common target.

(L,R) Here we must have $t = \bullet sr$ where $\bullet$ is $+$ or $\times$. The two trees will end

$$\frac{s \mathrel{\dot{\rhd}} s_1}{t_0 \mathrel{\dot{\rhd}} \bullet s_1 r}\,(\mathrm{L}) \qquad \frac{r \mathrel{\dot{\rhd}} r_2}{t_0 \mathrel{\dot{\rhd}} \bullet s r_2}\,(\mathrm{R})$$

where $t_1 = \bullet s_1 r$ and $t_2 = \bullet s r_2$. Two obvious trees show that $\bullet s_1 r_1$ is a common target.

Observe that many of the cases do not require a recursion call on the algorithm.

(f) Part (e) cannot be improved in the suggested way. This is because some cases, such as (A,L), need a use of transitive composition.

(g,h) Let us say a term t is **unusual** if there is a divergent wedge

$$t \mathrel{\rhd\!\!\!\rhd\!\!\!\rhd} \ulcorner m \urcorner \qquad t \mathrel{\rhd\!\!\!\rhd\!\!\!\rhd} \ulcorner n \urcorner$$

for some distinct $m, n \in \mathbb{N}$. By (d) it suffices to show that there are no unusual terms.

By way of contradiction suppose there is at least one unusual term t. Consider an example wedge, as above. Neither of these reductions can be a (syntactic) equality, so we have

$$t \mathrel{\dot{\rhd}} r \mathrel{\rhd\!\!\!\rhd\!\!\!\rhd} \ulcorner m \urcorner \qquad t \mathrel{\dot{\rhd}} s \mathrel{\rhd\!\!\!\rhd\!\!\!\rhd} \ulcorner n \urcorner$$

for some terms r, s. We show that at least one of these is unusual.

By (e) and (d) there is some $k \in \mathbb{N}$ such that

$$r \mathrel{\rhd\!\!\!\rhd\!\!\!\rhd} \ulcorner k \urcorner \qquad s \mathrel{\rhd\!\!\!\rhd\!\!\!\rhd} \ulcorner k \urcorner$$

hold. Since $m \neq n$, we must have $k \neq m$ or $k \neq n$. If $k \neq m$ then r is unusual, and if $k \neq n$ then s is unusual.

This shows that if t is unusual then $t \mathrel{\dot{\rhd}} t'$ for some unusual t'. Thus, by iteration, if t is unusual then there is an infinite chain

$$t = t_0 \mathrel{\dot{\rhd}} t_1 \mathrel{\dot{\rhd}} t_2 \mathrel{\dot{\rhd}} \cdots \mathrel{\dot{\rhd}} t_i \mathrel{\dot{\rhd}} \cdots$$

of (unusual) terms. To complete the proof it suffices to observe that $\dot{\rhd}$ is well founded, i.e. there are no such infinite chains of reductions.

This argument is known as **Newman's Lemma**. The fact that $\dot{\rhd}$ is well founded is not obvious (it is more than a mere observation), but we don't deal with such matters here.

(i) Each term t has a value $\llbracket t \rrbracket \in \mathbb{N}$ generated by recursion over the construction of t.

$$\llbracket 0 \rrbracket = 0 \quad \llbracket \mathsf{S}t \rrbracket = 1 + \llbracket t \rrbracket \quad \llbracket +sr \rrbracket = \llbracket s \rrbracket + \llbracket r \rrbracket \quad \llbracket \times sr \rrbracket = \llbracket s \rrbracket \times \llbracket r \rrbracket$$

In particular, $\llbracket \ulcorner m \urcorner \rrbracket = m$ for each $m \in \mathbb{N}$. The 1-step reductions are such that

$$t^- \mathrel{\dot{\rhd}} t^+ \Rightarrow \llbracket t^- \rrbracket = \llbracket t^+ \rrbracket \quad \text{and hence} \quad t^- \mathrel{\rhd\!\!\!\rhd\!\!\!\rhd} t^+ \Rightarrow \llbracket t^- \rrbracket = \llbracket t^+ \rrbracket$$

follows by induction over reduction trees.

Now consider a divergent wedge

$$t \mathrel{\rhd\!\!\!\rhd\!\!\!\rhd} \ulcorner m \urcorner \qquad t \mathrel{\rhd\!\!\!\rhd\!\!\!\rhd} \ulcorner n \urcorner$$

where $m, n \in \mathbb{N}$. Then $m = \llbracket \ulcorner m \urcorner \rrbracket = \llbracket t \rrbracket = \llbracket \ulcorner n \urcorner \rrbracket = n$ to give (e, f, g, h).

Notice that this argument does *not* prove that $\dot{\rhd}$ is well founded. $\square$

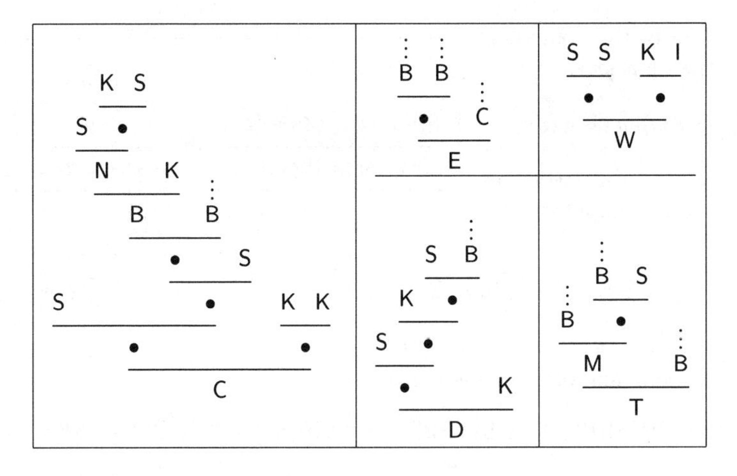

Table B.1: Some example parsing trees

B.2 Combinator terms

2.2 The trees of Table B.1 contain the parsing information for $\mathsf{B}, \mathsf{C}, \dots, \mathsf{T}$. □

B.3 Combinator reduction

2.3 It can be shown that the following will do.

(i)	(b′)	$\mathsf{B}z \rhd\!\!\!\rhd \mathsf{S}(\mathsf{K}z)$	where	b'	$= 0 \circ \rfloor 0$
	(b)	$\mathsf{B}zyx \rhd\!\!\!\rhd z(yx)$	where	b	$= \rfloor\rfloor \mathsf{b}' \circ \mathsf{b}'$
(ii)	(c′)	$\mathsf{C}z \rhd\!\!\!\rhd \mathsf{B}(\mathsf{S}z)\mathsf{K}$	where	c'	$= 0 \circ \rfloor \mathsf{b} \circ \lfloor 0$
	(c)	$\mathsf{C}zyx \rhd\!\!\!\rhd zxy$	where	c	$= \rfloor(\rfloor \mathsf{c}' \circ \mathsf{b}) \circ (0 \circ \lfloor 0)$
(iii)	(d′)	$\mathsf{D}zy \rhd\!\!\!\rhd \mathsf{B}yz$	where	d'	$= \lfloor 0 \circ \rfloor\rfloor 0 \circ 0 \circ \lfloor 0$
	(d)	$\mathsf{D}zyx \rhd\!\!\!\rhd y(zx)$	where	d	$= \rfloor \mathsf{d}' \circ \mathsf{b}$
(iv)	(e′)	$\mathsf{E}a \rhd\!\!\!\rhd \mathsf{B}(\mathsf{C}a)$	where	e'	$= \mathsf{b}$
	(e)	$\mathsf{E}azyx \rhd\!\!\!\rhd (ax)(zy)$	where	e	$= \rfloor\rfloor\rfloor \mathsf{b} \circ \rfloor \mathsf{b} \circ \mathsf{c}$
(v)	(w′)	$\mathsf{W}y \rhd\!\!\!\rhd (\mathsf{S}y)(\mathsf{KI}y)$	where	w'	$= 0 \circ \lfloor 0$
	(w)	$\mathsf{W}yx \rhd\!\!\!\rhd yxx$	where	w	$= \rfloor \mathsf{w}' \circ 0 \circ \lfloor 0$
(vi)	(m′)	$\mathsf{M}cba \rhd\!\!\!\rhd \mathsf{S}(cba)$	where	m'	$= \rfloor \mathsf{b} \circ \mathsf{b}$
	(m)	$\mathsf{M}abcxy \rhd\!\!\!\rhd (cbax)(yx)$	where	m	$= \rfloor\rfloor \mathsf{m}' \circ 0$
(vii)	(n′)	$\mathsf{N}az \rhd\!\!\!\rhd \mathsf{S}(az)$	where	n'	$= 0 \circ \rfloor 0 = \mathsf{b}'$
	(n)	$\mathsf{N}azyx \rhd\!\!\!\rhd (azx)(yx)$	where	n	$= \rfloor\rfloor \mathsf{n}' \circ 0$
(viii)	(t′)	$\mathsf{T}az \rhd\!\!\!\rhd \mathsf{S}(\mathsf{B}az)$	where	t'	$= \mathsf{m}'$
	(t)	$\mathsf{T}azyx \rhd\!\!\!\rhd a(zx)(yx)$	where	t	$= \rfloor\rfloor \mathsf{t}' \circ 0 \circ \rfloor \mathsf{b}$

Some of the later codes are not fully bracketed; any bracketed version will do.

(i) Firstly the computation to the right gives

$$(\mathsf{b}')\quad \mathsf{S}(\mathsf{K}w)vu \mathrel{\rhd\!\!\rhd} w(vu)$$

for any u, v, w. In particular we have three computations

$$\dfrac{\dfrac{\mathsf{S}(\mathsf{K}w)vu \rhd (\mathsf{K}wu)(vu)}{\mathsf{S}(\mathsf{K}w)vu \mathrel{\rhd\!\!\rhd} (\mathsf{K}wu)(vu)}(0) \qquad \dfrac{\dfrac{\mathsf{K}wv \rhd w}{\mathsf{K}wv \mathrel{\rhd\!\!\rhd} w}(0)}{(-)(vu) \mathrel{\rhd\!\!\rhd} w(vu)}(\downarrow)}{\mathsf{S}(\mathsf{K}w)vu \mathrel{\rhd\!\!\rhd} w(vu)}(\circ)$$

$$(\mathsf{b}')\ \mathsf{B}z \mathrel{\rhd\!\!\rhd} \mathsf{S}(\mathsf{K}z) \qquad (\downarrow\downarrow\mathsf{b}')\ \mathsf{B}zyx \mathrel{\rhd\!\!\rhd} \mathsf{S}(\mathsf{K}z)yx \qquad (\mathsf{b}')\ \mathsf{S}(\mathsf{K}z)yx \mathrel{\rhd\!\!\rhd} z(yx)$$

so that $\mathsf{b} = \downarrow\downarrow\mathsf{b}' \circ \mathsf{b}'$ gives the final result.

(ii) A compound of the three reductions

$$(0)\ \mathsf{C}z \mathrel{\rhd\!\!\rhd} (\mathsf{BBS}z)(\mathsf{KK}z) \quad (\downarrow\mathsf{b})\ (\mathsf{BBS}s)(-) \mathrel{\rhd\!\!\rhd} \mathsf{B}(\mathsf{S}z)(-) \quad (\lfloor 0)\ (-)(\mathsf{KK}z) \mathrel{\rhd\!\!\rhd} (-)\mathsf{K}$$

gives c', where the three components can be put together in two ways. Then

$$\begin{array}{lllll}
(\downarrow\mathsf{c}') & \mathsf{C}zy & \mathrel{\rhd\!\!\rhd} \mathsf{B}(\mathsf{S}z)\mathsf{K}y & (\mathsf{b}) & \mathsf{B}(\mathsf{S}z)\mathsf{K}y \mathrel{\rhd\!\!\rhd} (\mathsf{S}z)(\mathsf{K}y) \\
(\downarrow\downarrow\mathsf{c}') & \mathsf{C}zyx & \mathrel{\rhd\!\!\rhd} \mathsf{B}(\mathsf{S}z)\mathsf{K}yx & (\downarrow\mathsf{b}) & \mathsf{B}(\mathsf{S}z)\mathsf{K}yx \mathrel{\rhd\!\!\rhd} (\mathsf{S}z)(\mathsf{K}y)x \\
(0) & (\mathsf{S}z)(\mathsf{K}y)x & \mathrel{\rhd\!\!\rhd} (zx)(\mathsf{K}yx) & (\lfloor 0) & (zx)(\mathsf{K}yx) \mathrel{\rhd\!\!\rhd} zxy
\end{array}$$

show that suitably bracketed versions of either of

$$\mathsf{c} = \downarrow(\downarrow\mathsf{c}' \circ \mathsf{b}) \circ 0 \circ \lfloor 0 \qquad \mathsf{c} = \downarrow\downarrow\mathsf{c}' \circ \downarrow\mathsf{b} \circ 0 \circ \lfloor 0$$

will do. There is a further discussion of this reduction at the end of this solution.

(iii) A composite of the reductions

$$\begin{array}{llll}
(\downarrow 0) & \mathsf{D}zy \mathrel{\rhd\!\!\rhd} (\mathsf{K}(\mathsf{SB})z)(\mathsf{K}z)y & (\downarrow\downarrow 0) & (\mathsf{K}(\mathsf{SB})z)(-)y \mathrel{\rhd\!\!\rhd} \mathsf{SB}(-)y \\
(0) & \mathsf{SB}(\mathsf{K}z)y \mathrel{\rhd\!\!\rhd} (\mathsf{B}y)(\mathsf{K}zy) & (\lfloor 0) & (-)(\mathsf{K}zy) \mathrel{\rhd\!\!\rhd} \mathsf{B}yz
\end{array}$$

gives the first phase. Notice also that $\mathsf{d}' = \downarrow(0 \circ \downarrow 0) \circ 0 \circ \lfloor 0$ will do. With this

$$(\downarrow\mathsf{d}')\ \mathsf{D}zyx \mathrel{\rhd\!\!\rhd} \mathsf{B}yzx \qquad (\mathsf{b})\ \mathsf{B}yzx \mathrel{\rhd\!\!\rhd} y(zx)$$

gives d.

(iv) Since $\mathsf{E} = \mathsf{BBC}$, the first part follows from (i). A composite of

$$(\downarrow\downarrow\downarrow\mathsf{b})\ \mathsf{E}azyx \mathrel{\rhd\!\!\rhd} \mathsf{B}(\mathsf{C}a)zyx \quad (\downarrow\mathsf{b})\ \mathsf{B}(\mathsf{C}a)zyx \mathrel{\rhd\!\!\rhd} \mathsf{C}a(zy)x \quad (\mathsf{c})\ \mathsf{C}a(zy)x \mathrel{\rhd\!\!\rhd} (ax)(zy)$$

gives e.

(v) The full computation for w' is to the right and then a composite of the reductions

$$\begin{array}{ll}
(\downarrow\mathsf{w}') & \mathsf{W}yx \mathrel{\rhd\!\!\rhd} \mathsf{S}y\mathsf{I}x \\
(0) & \mathsf{S}y\mathsf{I}x \mathrel{\rhd\!\!\rhd} (yx)(\mathsf{I}x) \\
(\lfloor 0) & (yx)(\mathsf{I}x) \mathrel{\rhd\!\!\rhd} yxx
\end{array}$$

$$\dfrac{\dfrac{\mathsf{SS}(\mathsf{KI})y \rhd (\mathsf{S}y)(\mathsf{KI}y)}{\mathsf{W}y \mathrel{\rhd\!\!\rhd} (\mathsf{S}y)(\mathsf{KI}y)}(0) \qquad \dfrac{\dfrac{\mathsf{KI}y \rhd \mathsf{I}}{\mathsf{KI}y \mathrel{\rhd\!\!\rhd} \mathsf{I}}(0)}{(\mathsf{S}y)(\mathsf{KI}y) \mathrel{\rhd\!\!\rhd} \mathsf{S}y\mathsf{I}}(\downarrow)}{\mathsf{W}y \mathrel{\rhd\!\!\rhd} \mathsf{S}y\mathsf{I}}(\circ)$$

gives w.

(vi) Firstly

$$(\mathsf{b})\quad \mathsf{B}(\mathsf{BS})cb \rhd\!\!\!\rhd \mathsf{BS}(cb) \qquad (\downharpoonleft\mathsf{b})\quad \mathsf{M}cba \rhd\!\!\!\rhd \mathsf{BS}(cb)a \qquad (\mathsf{b})\quad \mathsf{Bs}(cb)a \rhd\!\!\!\rhd \mathsf{S}(cba)$$

gives m' and then

$$(\downharpoonleft\downharpoonleft\mathsf{m}')\quad \mathsf{M}cbayx \rhd\!\!\!\rhd \mathsf{S}(cba)yx \qquad (\mathsf{0})\quad \mathsf{S}(cba)yx \rhd\!\!\!\rhd (cbax)(yx)$$

gives m.

(vii) The two reductions

$$\mathsf{N}az \rhd\!\!\!\rhd (\mathsf{KS}z)(az) \rhd\!\!\!\rhd \mathsf{S}(az) \qquad \mathsf{N}azyx \rhd\!\!\!\rhd \mathsf{S}(az)yx \rhd\!\!\!\rhd (azx)(yx)$$

are given by $\mathsf{n}' = \mathsf{0} \circ \downharpoonleft\mathsf{0}$ and $\downharpoonleft\downharpoonleft\mathsf{n}' \circ \mathsf{0}$, respectively.

(viii) Particular cases from (vi) and (i) give

$$(\mathsf{m}')\quad \mathsf{MB}az \rhd\!\!\!\rhd \mathsf{S}(\mathsf{B}az) \qquad (\mathsf{0})\quad \mathsf{S}(\mathsf{b}az)yx \rhd\!\!\!\rhd (\mathsf{B}azx)(yx) \qquad (\mathsf{b})\mathsf{B}azx \rhd\!\!\!\rhd a(zx)$$

so that the indicated t' and t will do.

In some case there are some different computations which achieve the same result.

For instance, consider the combinator C and the reductions to the right. We obtain the first one, b', from (i) and this gives the second and third. Then setting

$$\mathsf{c}'' = \downharpoonleft\downharpoonleft\downharpoonright\downharpoonleft\mathsf{b}' \circ \mathsf{0} \circ \downharpoonright\mathsf{0} \qquad \mathsf{Z} = \mathsf{S}(\mathsf{KB})\mathsf{S}z$$

(b')	$\mathsf{BB} \rhd\!\!\!\rhd \mathsf{S}(\mathsf{KB})$
$(\downharpoonleft\downharpoonright\downharpoonleft\mathsf{b}')$	$\mathsf{C} \rhd\!\!\!\rhd \mathsf{S}(\mathsf{S}(\mathsf{KB})\mathsf{S})(\mathsf{KK})$
$(\downharpoonleft\downharpoonleft\downharpoonright\downharpoonleft\mathsf{b}' \circ \mathsf{0})$	$\mathsf{C}z \rhd\!\!\!\rhd (\mathsf{S}(\mathsf{KB})\mathsf{S}z)(\mathsf{KK}z)$
(c'')	$\mathsf{C}z \rhd\!\!\!\rhd \mathsf{ZK}$
(b')	$\mathsf{Z} \rhd\!\!\!\rhd \mathsf{B}(\mathsf{S}z)$
$(\mathsf{c}'' \circ \downharpoonleft\mathsf{b}')$	$\mathsf{C}z \rhd\!\!\!\rhd \mathsf{B}(\mathsf{S}z)\mathsf{K}$

gives the fourth. A second use of b' produces $\mathsf{c}'' \circ \downharpoonleft\mathsf{b}'$ as an alternative to c'.

These examples suggest there is an algebra of 'equivalent' computations lurking beneath the informal notation. We need more experience before we can make this precise.

The combinators $\mathsf{B}, \mathsf{C}, \mathsf{D}, \mathsf{E}, \mathsf{W}, \mathsf{M}, \mathsf{N}, \mathsf{T}$ code the shapes of the H-derivations of the formulas $\beta, \gamma, \delta, \epsilon, \omega, \mu, \nu, \tau$. Later, when we enrich the combinators with formula/type information, we get a precise description of the derivations. □

2.4 (a) We have $\mathsf{Y} = \mathsf{XX} = \mathsf{B}(\mathsf{SI})\mathsf{AX} \rhd\!\!\!\rhd \mathsf{SI}(\mathsf{AX}) \rhd\!\!\!\rhd \mathsf{SI}(\mathsf{XX})$ so that $\mathsf{Y} \rhd\!\!\!\rhd \mathsf{SIY}$. But then $\mathsf{Y}f \rhd\!\!\!\rhd \mathsf{SIY}f \rhd\!\!\!\rhd \mathsf{I}f(\mathsf{Y}f) \rhd\!\!\!\rhd f(\mathsf{Y}f)$.

(b) We have $\mathsf{Z}f = \mathsf{SXX}f \rhd\!\!\!\rhd (\mathsf{X}f)(\mathsf{X}f) = aa$ where $a = \mathsf{X}f = \mathsf{CBA}f \rhd\!\!\!\rhd \mathsf{B}f\mathsf{A}$ so that with $g = aa$ and $h = fg$ we see that

$$g \rhd\!\!\!\rhd \mathsf{B}f\mathsf{A}a \rhd\!\!\!\rhd f(\mathsf{A}a) \rhd\!\!\!\rhd f(aa) = fg = h$$

holds. Thus $\mathsf{Z}f \rhd\!\!\!\rhd g \rhd\!\!\!\rhd h$ and hence $f(\mathsf{Z}f) \rhd\!\!\!\rhd fg = h$ as required. □

2.5 A routine computation shows that $\mathsf{LA}yz \rhd\!\!\!\rhd \mathsf{A}y$ for all A, y, z. We then find that $\mathsf{K}^i\mathsf{A}x_1 \cdots x_i \rhd\!\!\!\rhd \mathsf{A}$ and $\mathsf{L}^j\mathsf{A}yz_1 \cdots z_j \rhd\!\!\!\rhd \mathsf{A}y$ and hence

$$\mathsf{K}^i(\mathsf{L}^j\mathsf{A})x_1 \cdots x_iyz_1 \cdots z_j \rhd\!\!\!\rhd \mathsf{A}y$$

for all $\mathsf{A}, x_1, \ldots, x_i, y, z_1, \ldots, z_j$. The cases $i = 0$ or $j = 0$ are allowed.

Thus, given $1 \leq k \leq n$, with $i = k-1$ and $j = n-k$, we see that the compound $\binom{n}{k} = \mathsf{K}^i(\mathsf{L}^j\mathsf{I})$ has the required projection property. □

2.6 Let us say a combinator A is special if it behaves like $\binom{n}{k}$ for some $1 \leq k \leq n$ where $k = n$ or $n = k+1$. Thus A selects the ultimate or penultimate component of a list of arguments (of a specified length). We call the pair (k, n) the index of A. A simple induction over m shows that $\mathsf{K}^m\mathsf{I}$ and $\mathsf{K}^m\mathsf{K}$ are special with indexes $(m+1, m+1)$ and $(m+1, m+2)$ respectively. We show that every combinator built up from K and I is special, and hence there are projection combinators, such as $\binom{3}{1}$, which cannot be built using only K and I.

Both K and I are special (with $k = 1, n = 2$ and $k = n = 1$, respectively). Any other combinator built from K and I has the form $\mathsf{A} = \mathsf{LR}$ where, by the induction hypothesis, both L and R are special. The behaviour of L is fixed by a pair (k, n) where $1 \leq k \leq n$ with $k = n$ or $n = k+1$. We consider the possible values of k and n.

When $k = n = 1$ we have $\mathsf{A} = \mathsf{LR} \rhd \mathsf{R}$ so A behaves like R.

When $k = 1, n = 2$ we have $\mathsf{A}x = \mathsf{LR}x \rhd \mathsf{R}$ so that, using the index (j, m) of R, we have $\mathsf{A}xx_1 \cdots x_m \mathsf{R}x_1 \cdots x_m \gg x_j$ and hence A is special with index $(j+1, m+1)$.

For all other cases we have $2 \leq k \leq n$ so that $\mathsf{A}x_2 \cdots x_n = \mathsf{LA}x_2 \cdots x_n \gg x_k$ and hence A is special with index $(k-1, n-1)$. □

2.7 The crucial property of T is $\mathsf{T}axyz \gg a(xz)(yz)$ for all a, x, y, z. The required result follows by an induction over n. For the induction step, since $\mathsf{T}^{n+1}\mathsf{I} = \mathsf{T}(\mathsf{T}^n\mathsf{I})$

$$\mathsf{T}^{n+1}\mathsf{I}vuxx_1 \cdots x_n \gg (\mathsf{T}^n\mathsf{I})(vx)(ux)x_1 \cdots x_n \gg (vxx_1 \cdots x_n)(uxx_1 \cdots x_n)$$

where the last reduction uses the induction hypothesis. □

2.8 (a) By Exercise 2.5, the combinator $\mathsf{A} = \binom{n}{k}$ satisfies $\mathsf{A}x_1 \cdots x_n \gg x_k$ (for all $1 \leq k \leq n$). Similarly, for each combinator Z, the compound $\mathsf{A} = \binom{n+1}{1}\mathsf{Z}$ satisfies $\mathsf{A}x_1 \cdots x_n \gg \mathsf{Z}$ (both for all $x_1, \ldots, x_n$). This gives the required result for 'atomic' expressions X. All other expressions have the form $X = qp$ for smaller expressions p, q. By recursion we have combinators Q, P satisfying $\mathsf{Qx} \gg q$ and $\mathsf{Px} \gg p$ where the sequence $x_1, \ldots, x_n$ has been abbreviated to x. Now let $\mathsf{A} = (\mathsf{T}^n\mathsf{I})\mathsf{QP}$ so that $\mathsf{Ax} \gg (\mathsf{Qx})(\mathsf{Px}) \gg qp = X$ as required.

(b) For convenience set $\mathsf{G} = \mathsf{T}^3\mathsf{I}$.

(i) We require $\mathsf{A} = \mathsf{Gab}$ where $\mathsf{a}xyz \gg zx$ and $\mathsf{b}xyz \gg y$ (for all x, y, z). Thus we can set $\mathsf{b} = \binom{3}{2}$ and $\mathsf{a} = \mathsf{Gcd}$ where $\mathsf{c}xyz \gg z$ and $\mathsf{d}xyz \gg x$ (for all x, y, z). Both these are projections, so we see that $\mathsf{A} = \mathsf{G}(\mathsf{G}\binom{3}{3}\binom{3}{1})\binom{3}{2}$ will do the job.

(ii) A similar construction shows that $\mathsf{A} = \mathsf{G}(\mathsf{G}\binom{3}{3}\binom{3}{2})\binom{3}{1}$ will do.

(iii) In this case we set $\mathsf{A} = \mathsf{Ga}\binom{3}{1}$ where a must satisfy $\mathsf{a}xyz \gg \mathsf{S}(z(\mathsf{K}y))$ (for all x, y, z). For this we set $\mathsf{a} = \mathsf{Gbc}$ where $\mathsf{b}xyz \gg \mathsf{S}$ and $\mathsf{c}xyz \gg z(\mathsf{K}y)$ are the required properties. Thus we set $\mathsf{b} = \binom{4}{1}\mathsf{S}$ and $\mathsf{c}xyz = \mathsf{Gde}$ where $\mathsf{d}xyz \rhd\!\!\rhd z$ and $\mathsf{e}xyz \rhd\!\!\rhd \mathsf{K}y$ are the required properties. The first of these is a projection and then we find that $\mathsf{e} = \mathsf{G}(\binom{4}{1}\mathsf{K})\binom{3}{2}$ by a further unravelling. Putting all these together we see that

$$\mathsf{A} = \mathsf{G}(\mathsf{G}(((\tbinom{4}{1}\mathsf{S})((\tbinom{3}{3})(\mathsf{G}((\tbinom{4}{1}\mathsf{K})(\tbinom{3}{2})))))(\tbinom{3}{1})$$

will do the job. □

2.9 We have

$$\overline{0}yx = \mathsf{KI}yx \rhd\!\!\rhd \mathsf{I}x \rhd\!\!\rhd x = y^0x$$

for all y, x. Suppose $\overline{m}yx \rhd\!\!\rhd y^m x$ for some $m \in \mathbb{N}$. Since $\overline{m'} = (\mathsf{SB})^{m'}\overline{0} = \mathsf{SB}\overline{m}$ we have

$$\overline{m'}yx = \mathsf{SB}\overline{m}yx \rhd\!\!\rhd (\mathsf{B}y)(\overline{m}y)x \rhd\!\!\rhd y(\overline{m}yx) \rhd\!\!\rhd y(y^m x) = y^{m'}x$$

to provide the step for a proof by induction over m.

It is *not* true that $\overline{\mathrm{Suc}}\overline{m} \rhd\!\!\rhd \overline{m+1}$ since, in fact, these two terms are the same (so that $\overline{\mathrm{Suc}}\overline{m} \rhd\!\!\rhd\!\!\rhd \overline{m+1}$ does hold). □

2.10 We have

$$\overline{\mathrm{Add}}\,\overline{m}\,\overline{n}\,yx = \mathsf{TB}\,\overline{m}\,\overline{n}\,yx \rhd\!\!\rhd \mathsf{B}(\overline{m}y)(\overline{n}y)x \rhd\!\!\rhd (\overline{m}y)(\overline{n}yx) \rhd\!\!\rhd y^m(y^n x) = y^{n+m}x$$

using the known properties of the combinators and the numerals.

Next we have

$$\overline{\mathrm{Mlt}}\,\overline{m}\,\overline{n}\,yx = \mathsf{B}\,\overline{m}\,\overline{n}\,yx \rhd\!\!\rhd \overline{m}(\overline{n}y)x \rhd\!\!\rhd (\overline{n}y)^m x \rhd\!\!\rhd y^{n\times m}x$$

where the last step follows by a simple induction over m.

Finally we have

$$\overline{\mathrm{Exp}}\,\overline{m}\,\overline{n}\,yx = \mathsf{I}\,\overline{m}\,\overline{n}\,yx \rhd\!\!\rhd \overline{m}\,\overline{n}\,yx \rhd\!\!\rhd \overline{n}^m yx \rhd\!\!\rhd y^{n^m}x$$

where the last step follows by a not so simple induction over m. You should make sure you can do this induction; it will crop up again. □

2.11 We show that

$$\mathsf{Add}\,\overline{n}\,\overline{m} \rhd\!\!\rhd \overline{n+m} \qquad \mathsf{Mlt}\,\overline{n}\,\overline{m} \rhd\!\!\rhd \overline{n\times m} \qquad \mathsf{Exp}\,\overline{n}\,\overline{m} \rhd\!\!\rhd \overline{n^m} \qquad \mathsf{Bth}\,\overline{k}\,\overline{n}\,\overline{m} \rhd\!\!\rhd \overline{\beth(k,n,m)}$$

for all $m, n, k \in \mathbb{N}$.

(Add) Remembering that $\mathsf{SB} = \overline{\mathrm{Suc}}$ is used to generate the numerals we have

$$\mathsf{Add}\,\overline{n}\,\overline{m} = \mathsf{Swp}(\mathsf{SB})\,\overline{n}\,\overline{m} \rhd\!\!\rhd \overline{m}(\mathsf{SB})\overline{n} \rhd\!\!\rhd (\mathsf{SB})^m\overline{n} = \overline{n+m}$$

as required.

(Mlt) We have

$$\mathsf{Mlt}\,\overline{n}\,\overline{m} = \mathsf{Jmp}\overline{0}\mathsf{Add}\,\overline{n}\,\overline{m} \rhd\!\!\rhd \overline{m}(\mathsf{Add}\overline{n})\overline{0} \rhd\!\!\rhd (\mathsf{Add}\overline{n})^m\overline{0} \rhd\!\!\rhd \overline{n\times m}$$

where the last step follows by a simple induction over m using part (Add).

(Exp) We have

$$\mathsf{Exp}\,\overline{n}\,\overline{m} = \mathsf{Jmp}\overline{1}\mathsf{Mlt}\,\overline{n}\,\overline{m} \rhd\!\!\rhd \overline{m}(\mathsf{Mlt}\overline{n})\overline{1} \rhd\!\!\rhd (\mathsf{Mlt}\overline{n})^m\overline{1} \rhd\!\!\rhd \overline{n^m}$$

where the last step follows by a simple induction over m using part (Mlt).

(Bth) Since $\mathsf{Bth} = \mathsf{CJmpExp}$ we have

$$\mathsf{Bth}\,\overline{k}\,\overline{n}\,\overline{m} \rhd\!\!\rhd \mathsf{Jmp}\overline{k}\mathsf{Exp}\,\overline{n}\,\overline{m} \rhd\!\!\rhd \overline{m}(\mathsf{Exp}\overline{n})\overline{k} \rhd\!\!\rhd (\mathsf{Exp}\overline{n})^m\overline{k} \rhd\!\!\rhd \overline{\beth(k,n,m)}$$

where the last step follows by a simple induction over m using part (Exp). □

2.12 We transform any 2-placed function $F : \mathbb{N} \longrightarrow \mathbb{N} \longrightarrow \mathbb{N}$ into another 2-placed function $F' : \mathbb{N} \longrightarrow \mathbb{N} \longrightarrow \mathbb{N}$ by $F'nm = (Fn)^m 1$ where $(\cdot)^m$ indicates the mth iterate of the 1-placed function Fn. This function F' is called the Grzegorczyk jump of F. For instance $\mathrm{Mlt}' = \mathrm{Exp}$, $\mathrm{Exp}' = \beth(\cdot,\cdot,1)$. By iterating this jump, $F^{(0)} = F$, $F^{(i')} = F^{(i)\prime}$, we produce the Grzegorczyk hierarchy on F. In general, the value $F^{(i)}nm$ is very large.

Suppose F represents F. Then

$$\mathsf{Jmp}\overline{1}\mathsf{F}\,\overline{n}\,\overline{m} \rhd\!\!\rhd \overline{m}(\mathsf{F}\overline{n})\overline{1} \rhd\!\!\rhd (\mathsf{F}\overline{n})^m\overline{1} \rhd\!\!\rhd \overline{F'nm}$$

by a simple induction on m. Thus $\mathsf{Jmp}\overline{1}\mathsf{F}$ represents F'. Next

$$\mathsf{Grz}\,\mathsf{F}\,\overline{i}\,\overline{n}\,\overline{m} = \mathsf{Swp}(\mathsf{Jmp}\overline{1})\mathsf{F}\,\overline{i}\,\overline{n}\,\overline{m} \rhd\!\!\rhd \overline{i}(\mathsf{Jmp}\overline{1}\mathsf{F}\,\overline{n}\,\overline{m} \rhd\!\!\rhd (\mathsf{Jmp}\overline{1})^i\mathsf{F}\,\overline{n}\,\overline{m} \rhd\!\!\rhd \overline{F^{(i)}nm}$$

by an induction on i. Thus GrzF represents the Grzegorczyk hierarchy on F as a 3-placed function. The size of the computation

$$(\square)\quad \mathsf{Grz}\,\mathsf{F}\,\overline{i}\,\overline{n}\,\overline{m} \rhd\!\!\rhd \overline{F^{(i)}nm}$$

will be comparable with $F^{(i)}nm$, i.e. very large in general. □

B.4 λ-TERMS

2.13 Consider the two trees to the right. The left hand one is the parsing tree of the term with the set of free identifiers indicated for each subterm. This is always $\{x\}$ or $\emptyset$. There are four uses of 'x' as indicated by the four leaves. Replacing the first three uses by u, v, w we obtain the right hand tree which parses the term

$$(\lambda w\,.\,((\lambda v\,.\,((\lambda u\,.\,u)v))w))x$$

$$\cfrac{\cfrac{\cfrac{\cfrac{\cfrac{\cfrac{\{x\}}{\emptyset}\,(x) \quad \{x\}}{\{x\}}}{\emptyset}\,(x) \quad \{x\}}{\{x\}}}{\emptyset}\,(x) \quad \{x\}}{\{x\}} \qquad\qquad \cfrac{\cfrac{\cfrac{\cfrac{\cfrac{\cfrac{u}{\bullet}\,(u) \quad v}{\bullet}}{\bullet}\,(v) \quad w}{\bullet}}{\bullet}\,(w) \quad x}{\bullet}$$

and this is a more readable variant of the original term. □

2.14 Comparing the result of $(\lambda x\,.\,x)[x := x]$ with the standard substitution term $(\lambda y\,.\,r)[x := s]$ we see that in this case

$$\mathcal{U} = (\partial(x) - \{x\}) \cup \partial(x) \cup \{x\} = \{x\}$$

is the unsafe set. With $u \notin \mathcal{U}$, i.e. with $u \neq x$, we have

$$r' = x[x := u] = u \quad \text{and then} \quad (\lambda x \,.\, x)[x := x] = \lambda u \,.\, (u[x := x]) = \lambda u \,.\, u$$

is the substituted version. Notice how this forces a change of bound identifier even though there is no danger of capture. Some versions of the substitution algorithm have a release clause to prevent such unnecessary renaming. □

2.15 There is plenty of scope for identifier capture. The set of unsafe identifiers is

$$\mathcal{U} = (\partial(yx) - \{y\}) \cup \partial(s) \cup \{x\} = \{x\} \cup \partial(s) = \begin{cases} \{x\} & \text{if } s = x \text{ or } s = t \\ \{x, y\} & \text{if } s = y \text{ or } s = xy \\ \{x, y, z\} & \text{if } s = xyz \end{cases}$$

and we require $v \notin \mathcal{U}$ with $v = y$ when possible. With such a v we have

$$(yx)' = (yx)[y := v] = vx \quad \text{and hence} \quad t[y := s] = \lambda v \,.\, ((vx)[x := s]) = \lambda v \,.\, vs$$

since $v \neq x$ in all cases. Thus we get

$$\lambda y \,.\, yx \qquad \lambda v \,.\, vy \qquad \lambda v \,.\, v(xy) \qquad \lambda v \,.\, v(xyz) \qquad \lambda y \,.\, yt$$

for the five cases of s. □

2.16 We find there are three cases, as indicated right. We show that

(1)	$x = z$	$b = w$
(2)	$x \neq z, x = y$	$b = u$
(3)	$x \neq z, x \neq y$	$b = s$

$$t[x := s] = \lambda u \,.\, ((\lambda w \,.\, wb)u)$$

where b is as shown and where $u = y$ and $w = z$ unless we are forced to rename either of these identifiers.

For convenience let $q = \lambda z \,.\, zx$ and $r = qy$ so that $t = \lambda y \,.\, r$ with

$$\partial r = \partial q \cup \{y\} = \begin{cases} \{y\} & (1) \\ \{x, y\} & (2,3) \end{cases}$$

depending on the case. With $\mathcal{S} = \partial s$ we have

$$\mathcal{U} = (\partial(r) - \{y\}) \cup \mathcal{S} \cup \{x\} = \begin{cases} \{x\} \cup \mathcal{S} & (1) \\ \{x, y\} \cup \mathcal{S} & (2,3) \end{cases}$$

as the first relevant unsafe set. We take $u \notin \mathcal{U}$ with $u = y$ if possible. Note that $u \neq x$ in all cases. With this we have $t[x := s] = \lambda u \,.\, r'[x := s]$ where $r' = r[y := u] = q'u$ where $q' = q[y := u]$ and then $r'[x := s] = q''u$ where $q'' = q'[x := s]$ since $u \neq x$. This gives $t[x := s] = \lambda u \,.\, q''$ so we must calculate $q'' = (\lambda z \,.\, zx)[y := u][x := s]$ which is an iterated substitution.

To calculate q' we have

$$\mathcal{V} = (\partial(zx) - \{z\}) \cup \{u\} \cup \{y\} = \begin{cases} \{u, y\} & (1) \\ \{u, x, y\} & (2,3) \end{cases}$$

as the second relevant unsafe set. We take $v \notin \mathcal{V}$ with $v = z$ if possible. Note that $v \neq u$ and $v \neq y$ in all cases. With this we have

$$\begin{aligned} q' &= \lambda v\,.\,((zx)[z := v][y := u]) \\ &= \begin{Bmatrix} \lambda v\,.\,(vv)[y := u] & (1) \\ \lambda v\,.\,(vx)[y := u] & (2) \\ \lambda v\,.\,(vx)[y := u] & (3) \end{Bmatrix} = \begin{cases} \lambda v\,.\,(vv)[y := u] & (1) \\ \lambda v\,.\,(vu)[y := u] & (2) \\ \lambda v\,.\,(vu)[y := u] & (3) \end{cases} \end{aligned}$$

for the three cases.

To calculate q'' we have

$$\mathcal{W} = (\partial(vp) - \{v\}) \cup \mathcal{S} \cup \{x\} = \begin{cases} \{x\} \cup \mathcal{S} & (1) \\ \{u, x\} \cup \mathcal{S} & (2) \\ \{x\} \cup \mathcal{S} & (3) \end{cases}$$

as the third relevant unsafe set. We take $w \notin \mathcal{W}$ with $w = v$ if possible. Thus $w = v = z$ whenever possible. Since $w \neq x$ and $u \neq x$ we have

$$q'' = \lambda w\,.\,((vp)[v := w][x := s]) = \begin{Bmatrix} \lambda w\,.\,((ww)[x := s]) & (1) \\ \lambda w\,.\,((wu)[x := s]) & (2) \\ \lambda w\,.\,((wx)[x := s]) & (3) \end{Bmatrix} = \lambda w\,.\,wb$$

where b is defined above, and hence

$$t[x := s] = \lambda u\,.\,q''u = \lambda u\,.\,((\lambda w\,.\,wb)u)$$

as required. □

B.5 λ-REDUCTION

2.17 We have

$$\begin{array}{lll} ts \mathrel{\rhd\!\!\rhd} sx \mathrel{\rhd\!\!\rhd} xx & st \mathrel{\rhd\!\!\rhd} tx \mathrel{\rhd\!\!\rhd} xx & rt \mathrel{\rhd\!\!\rhd} ty \mathrel{\rhd\!\!\rhd} yx \\ tr \mathrel{\rhd\!\!\rhd} rx \mathrel{\rhd\!\!\rhd} xy & sr \mathrel{\rhd\!\!\rhd} rx \mathrel{\rhd\!\!\rhd} xy & rs \mathrel{\rhd\!\!\rhd} sy \mathrel{\rhd\!\!\rhd} yx \end{array}$$

so that

$$\begin{array}{llll} tsr \mathrel{\rhd\!\!\rhd} xxr & t(sr) \mathrel{\rhd\!\!\rhd} srx \mathrel{\rhd\!\!\rhd} xyx & trs \mathrel{\rhd\!\!\rhd} xys & t(rs) \mathrel{\rhd\!\!\rhd} rsx \mathrel{\rhd\!\!\rhd} yxx \\ str \mathrel{\rhd\!\!\rhd} yxr & s(tr) \mathrel{\rhd\!\!\rhd} trx \mathrel{\rhd\!\!\rhd} xyx & srt \mathrel{\rhd\!\!\rhd} xyt & s(rt) \mathrel{\rhd\!\!\rhd} rtx \mathrel{\rhd\!\!\rhd} yxx \\ rts \mathrel{\rhd\!\!\rhd} yxs & r(ts) \mathrel{\rhd\!\!\rhd} tsy \mathrel{\rhd\!\!\rhd} xyy & rst \mathrel{\rhd\!\!\rhd} yxt & r(st) \mathrel{\rhd\!\!\rhd} sty \mathrel{\rhd\!\!\rhd} yxy \end{array}$$

are the required reductions. □

2.18 The parsing tree of the term t is given to the right. Here we use the abbreviations

$$\begin{array}{llll} (0) = x & (1) = [0]x & (2) = [1]x & (3) = [2]x \\ [0] = \lambda x\,.\,(0) & [1] = \lambda x\,.\,(1) & [2] = \lambda x\,.\,(2) & \end{array}$$

and hence $t = (3)$. Each of (1), (2), (3) is a redex and

$$(3) \rhd (2) \qquad (2) \rhd (1) \qquad (1) \rhd (0)$$

are 1-step reductions. The second computation $\downarrow\uparrow 0$ shown

$$\dfrac{\dfrac{\dfrac{\dfrac{\dfrac{\dfrac{(0)}{[0]}\,(x) \quad x}{(1)}}{[1]}\,(x) \quad x}{(2)}}{[2]}\,(x) \quad x}{(3)}$$

below right converts (1) $\rhd$ (0) into (2) $\rhd\!\!\!\rhd$ (1), and in a similar way we may convert any

$$(\mathsf{a})\quad (2) \rhd\!\!\!\rhd (1) \quad \text{into} \quad (\downarrow\uparrow\mathsf{a})\quad (3) \rhd\!\!\!\rhd (2)$$

from which we generate several component reductions. Using any of the choices

$$[3,2] = \begin{cases} \mathsf{0} \\ \downarrow\uparrow\mathsf{0} \\ \downarrow\uparrow\downarrow\uparrow\mathsf{0} \end{cases} \qquad [2,1] = \begin{cases} \mathsf{0} \\ \downarrow\uparrow\mathsf{0} \end{cases} \qquad [1,0] = \mathsf{0}$$

we have

$$([3,2])\ (3) \rhd\!\!\!\rhd (2) \qquad ([2,1])\ (2) \rhd\!\!\!\rhd (1) \qquad ([1,0])\ (1) \rhd\!\!\!\rhd (0)$$

$$(\downarrow\uparrow\mathsf{0})\quad \cfrac{\cfrac{\cfrac{(1) \rhd (0)}{(1) \rhd\!\!\!\rhd (0)}\,(0)}{[1] \rhd\!\!\!\rhd [0]}\,(\uparrow)}{(2) \rhd\!\!\!\rhd (1)}\,(\downarrow)$$

and then any bracketed version of

$$\square = [3,2] \circ [2,1] \circ [1,0]$$

organizes $t \rhd\!\!\!\rhd x$. This gives $3 \times 2 \times 1 = 6$ compounds which when bracketed give 12 different computations. □

2.19 (a) Let $\Phi = \lambda x\,.\,f(xx)$, $F = \Phi\Phi$, and $G = \mathsf{Y}f$ so that $\Omega[y := f] = \Phi$ to give

$$G \rhd\!\!\!\rhd \Phi\Phi = F \rhd\!\!\!\rhd f(\Phi\Phi) = fF \qquad fG \rhd\!\!\!\rhd fF$$

as required.

(b) Parts (i,ii,iv) are the cases $n = 1$, $n = 2$, $n = 24$ of (iii). Writing $\Omega(i)$ to indicate where each 'Ω' has come from with

$$\mathsf{Y} = \Omega(0)\Omega(1)\cdots\Omega(n) \quad \text{we have} \quad \mathsf{Y} \rhd\!\!\!\rhd \lambda y\,.\,y(\Omega(i(1))\cdots\Omega(i(n+1))y)$$

so that

$$\mathsf{Y}f \rhd\!\!\!\rhd f(\Omega(i(1))\cdots\Omega(i(n+1))f) = f(\mathsf{Y}f)$$

as required. □

2.20 Many of the calculations here are similar to those in the solutions to Exercises 2.10 and 2.11. The terms $\overline{m}$ are called the untyped Church numerals (for the λ-calculus).

(a) We have

$$\overline{\mathrm{Suc}}\,\overline{m} \rhd\!\!\!\rhd \lambda y, x\,.\,y(\overline{m}yx) \rhd\!\!\!\rhd \lambda y, x\,.\,y(y^m x) \rhd\!\!\!\rhd \lambda y, x\,.\,y^{1+m}x = \overline{1+m}$$

so that $\overline{\mathrm{Suc}}$ encodes the successor function.

Next we have

$$\overline{\mathrm{Add}}\,\overline{n}\,\overline{m} \rhd\!\!\!\rhd \lambda y, x\,.\,(\overline{m}y)(\overline{n}yx) \rhd\!\!\!\rhd \lambda y, x\,.\,(\overline{m}y)(y^n x) \rhd\!\!\!\rhd \lambda y, x\,.\,y^{m+n}x = \overline{n+m}$$

by a simple induction over m. Thus $\overline{\mathrm{Add}}$ encodes addition.

For $\overline{\mathrm{Mlt}}$ we have

$$\overline{\mathrm{Mlt}}\,\overline{n}\,\overline{m} \rhd\!\!\!\rhd \lambda y, x\,.\,\overline{m}(\overline{n}y)x \rhd\!\!\!\rhd \lambda y, x\,.\,(\overline{n}y)^m x \rhd\!\!\!\rhd \lambda y, x\,.\,y^{n\times m}x = \overline{n\times m}$$

by a routine induction over m. Thus $\overline{\mathrm{Mlt}}$ encodes multiplication.

Finally

$$\overline{\mathrm{Exp}}\,\overline{n}\,\overline{m} \rhd\!\!\!\rhd \lambda y, x\,.\,\overline{m}\,\overline{n}yx \rhd\!\!\!\rhd \lambda y, x\,.\,\overline{n}^m yx \rhd\!\!\!\rhd \lambda y, x\,.\,y^{n^m}x = \overline{n^m}$$

by a not so simple induction. Thus $\overline{\mathrm{Exp}}$ encodes exponentiation.

(b) Let

$$\overline{\mathrm{Bth}} = \lambda w, v, u\,.\,u(\overline{\mathrm{Exp}}v)w$$

using the term $\overline{\mathrm{Exp}}$. For all $m, n, k \in \mathbb{N}$ we have

$$\overline{\mathrm{Bth}}\,\overline{k}\,\overline{n}\,\overline{m} \rhd\!\!\!\rhd \overline{m}(\overline{\mathrm{Exp}}\overline{n})\overline{k} \rhd\!\!\!\rhd (\overline{\mathrm{Exp}}\overline{n})^m\overline{k} \rhd\!\!\!\rhd \overline{\beth(k, n, m)}$$

by a routine induction over m. □

B.6 INTERTRANSLATABILITY

2.21 The obvious translation $\mathsf{Z}_\lambda = \lambda x_1, \dots, x_n.X_\lambda$ works. □

2.22 For convenience let

$$\alpha = \mathsf{S}(\mathsf{KS})(\mathsf{KI}) \qquad \beta = \mathsf{S}(\mathsf{KK})\mathsf{I} \qquad \gamma = \mathsf{S}(\mathsf{KS})\beta \qquad \delta = \mathsf{S}(\mathsf{KS})(\mathsf{S}\alpha\beta)$$

so that

$$[x](xy) = (\mathsf{SI})(\mathsf{K}y) \quad [y][x](xy) = \mathsf{S}\alpha\beta \quad [x](yx) = \mathsf{S}(\mathsf{K}y)\mathsf{I} \quad [y][x](yx) = \mathsf{S}\gamma(\mathsf{KI})$$

hold. Next

$$[y](\mathsf{S}([x](xy))) = \mathsf{S}([y]\mathsf{S})([y][x](xy)) = \delta$$

so that with $W = (xy)(yx)$, we have

$$[x]W = (\mathsf{S}([x](xy)))([x]](yx)) \quad \text{hence} \quad [y][x]W = \mathsf{S}\delta([y][x](yx)) = \mathsf{S}\delta(\mathsf{S}\gamma(\mathsf{KI}))$$

for the last translation. This term contains 21 constants. □

2.23 For the combinator term, since $[x]t$ is x-free we have

$$([x]t)s = (([x]t)x)[x := s] \rhd\!\!\!\rhd t[x := s]$$

and then, for the λ-term

$$\langle(\lambda x.t)s\rangle = \langle(\lambda x.t)\rangle\langle s\rangle = ([x]\langle t\rangle)\langle s\rangle \rhd\!\!\!\rhd \langle t\rangle[x := \langle s\rangle]$$

as required. □

2.24 This is proved by induction on the structure of t. The only non-trivial case is for $t = \lambda x\,.\,r$ and even for this the subcases where r is an identifier are trivial.

For the remaining subcases we have $\langle r\rangle = qp$ for some $q, p \in \mathbb{C}omb$. But then with $P = ([x]p)_\lambda$ and $Q = ([x]q)_\lambda$ we see that

$$\langle t\rangle_\lambda = ([x]\langle r\rangle)_\lambda = ([x](qp))_\lambda = (\mathsf{S}([x]q)([x]p))_\lambda = \mathsf{S}_\lambda QP \mathrel{\rhd\!\!\!\rhd} \lambda x\,.\,(Qx)(Px)$$

holds. Lemma 2.14 gives

$$Qx = ([x]q)_\lambda x = ([x]qx)_\lambda \mathrel{\rhd\!\!\!\rhd} q_\lambda \qquad Px = ([x]p)_\lambda x = ([x]px)_\lambda \mathrel{\rhd\!\!\!\rhd} p_\lambda$$

so that

$$\langle t\rangle_\lambda \mathrel{\rhd\!\!\!\rhd} \lambda x\,.\,q_\lambda p_\lambda \mathrel{\rhd\!\!\!\rhd} \lambda x\,.\,(qp)_\lambda = \lambda x\,.\,\langle r\rangle_\lambda \mathrel{\rhd\!\!\!\rhd} \lambda x\,.\,r = t$$

where the final reduction uses the induction hypothesis. Notice that there has been a judicious choice of bound identifier in this argument. □

2.25 In the following computations '$\dot{=}$' indicates a use of the new clause. Firstly we have

$$\langle \mathsf{I}_\lambda\rangle = \langle \lambda x.x\rangle = [x]\langle x\rangle = [x]x = \mathsf{I}$$

and

$$\langle \mathsf{K}_\lambda\rangle = \langle \lambda y, x.y\rangle = [y]\langle \lambda x.y\rangle = [y]([x]\langle y\rangle) = [y](\mathsf{K}y) \doteq \mathsf{K}$$

for the first two constants. Next note that

$$\langle \lambda x.(zx)(yx)\rangle = \mathsf{S}([x](zx))([x](yx)) \doteq \mathsf{S}zy$$

and

$$\langle \lambda y, x.(zx)(yx)\rangle = [y]((\mathsf{S}z)y) \doteq \mathsf{S}z$$

so that

$$\langle \mathsf{S}_\lambda\rangle = \langle \lambda z, y, x.(zx)(yx)\rangle = [z](\mathsf{S}z) \doteq \mathsf{S}$$

for the third constant. The full result now follows by an easy induction.

With this translation the comparison $\langle t\rangle_\lambda \mathrel{\rhd\!\!\!\rhd} t$ no longer holds. □

B.7 Confluence and normalization

2.26 The general outline is the same as in Solution 2.1(e). The two reductions

$$t_0 \mathrel{\dot{\rhd}} t_1 \qquad t_0 \mathrel{\dot{\rhd}} t_2$$

are witnessed by slim computations (of width 1). Each is a use of a single 1-step reduction with various term forming constructs wrapped around. We consider the possible root rules of the two trees and proceed by recursion.

($\mathbb{C}omb$) This is quite easy (but tedious), partly because of the nature of the reduction axioms.

(Ax,Ax) Each term can match the left hand side of at most one axiom. (This property is built into the definition of $\rhd$. We could easily think of reduction relations where this isn't true.)

(Ap,Ap) If neither of the two root rules is an axiom, i.e. if both are a $\downharpoonleft$ or $\downharpoonright$ (possible mixed), then the construction is easy. Some of these cases need recursion calls on the algorithm.

(Ax,Ap) There are several cases here. The most complicated is where

$$t_0 = \mathsf{S}tsr \qquad t_1 = (tr)(sr) \qquad t = qp$$

where either

$$\text{(LA)} \quad \mathsf{S}ts \,\dot{\rhd}\, q,\ p = r \qquad\qquad \text{(RA)} \quad q = \mathsf{S}ts,\ r \,\dot{\rhd}\, p$$

hold.

The right hand case (RA) is straightforward, but requires a use of transitive composition to deal with the two occurrences of r. Of course, $t_3 = (tp)(sp)$ for this case.

For the left hand case (LA), what can the root rule of this computation be? It cannot be an axiom, but could be either LA or RA. Thus we have two subcases. Unravelling in this way we find we need

$$\left.\begin{array}{r} t \rhd\!\!\rhd\!\!\rhd t' \\ s \rhd\!\!\rhd\!\!\rhd s' \\ r \rhd\!\!\rhd\!\!\rhd r' \end{array}\right\} \Longrightarrow \mathsf{S}tsr \rhd\!\!\rhd\!\!\rhd (t'r')(s'r')$$

to hold. This is easy to verify. Notice this also covers the first case.

($\mathbb{L}amb$) For these terms each 1-step reduction is achieved by a redex removal, and there is an extra rule of Abstraction to consider. How can these rules be matched, and what is needed to handle these matchings? We find we need

$$\left.\begin{array}{r} s \rhd\!\!\rhd\!\!\rhd s' \\ r \rhd\!\!\rhd\!\!\rhd r' \end{array}\right\} \Longrightarrow r[y := s] \rhd\!\!\rhd\!\!\rhd r'[y := s']$$

to hold. Verifying this implication is not altogether trivial. The problem is that a calculation of a substitution requires some renaming of identifiers. This has to be done in a coherent fashion; if the same occurrence of a bound identifier is renamed twice, then the same renaming identifier should be used. The other possibility is to work up to alphabetic variants.

Suppose a redex removal is matched with something. By unravelling the possible cases we are lead to several instances of the implication above. The target term is then easily constructed.

The case where neither root rule is a redex removal is straightforward. $\square$

C

THE TYPED COMBINATOR CALCULUS

C.1 INTRODUCTION

3.1 The tree to the right shows how the passage from

$$f : R \longrightarrow S \longrightarrow T \quad \text{to} \quad frs : T$$

$$\dfrac{\dfrac{f : (R \longrightarrow (S \longrightarrow T)) \quad r : R}{(fr) : (S \longrightarrow T)} \qquad s : S}{((fr)s) : T}$$

strips off the brackets from the type and puts them on the term. □

C.2 DERIVATION

3.2 The type β is dealt with in Example 3.5. We deal with the other types in a similar fashion.

(γ) The type is

$$\gamma = (\tau \rightarrow \sigma \rightarrow \rho) \rightarrow (\sigma \rightarrow \tau \rightarrow \rho)$$

and the untyped combinator C is given to the right. We carry over the abbreviations $\beta, \delta, \xi, \chi, \eta, \omega, \pi, \epsilon$, and $\kappa_3, \kappa_4, \sigma_3, \sigma_4, \beta_1, \beta_2$ of Solution 1.7(γ). Using these we have the collection of typed combinators with associated housing axioms and derivable statements shown to the right. With $\mathsf{P}_\bullet = \mathsf{B}_1\mathsf{B}_2\mathsf{S}_3$ and $\mathsf{Q}_\bullet = \mathsf{K}_3\mathsf{K}_4$ the derivation below shows that the combinator $\mathsf{C}_\bullet$ inhabits γ.

$\mathsf{C} = \mathsf{S}(\mathsf{BBS})(\mathsf{KK})$

$\mathsf{K}_3 = \mathsf{K}(\beta, \kappa_4)$	$\mathsf{K}_3 : \kappa_3$
$\mathsf{K}_4 = \mathsf{K}(\tau, \sigma)$	$\mathsf{K}_4 : \kappa_4$
$\mathsf{S}_3 = \mathsf{S}(\tau, \sigma, \rho)$	$\mathsf{S}_3 : \sigma_3$
$\mathsf{S}_4 = \mathsf{S}(\beta, \kappa_4, \delta)$	$\mathsf{S}_4 : \sigma_4$
$\mathsf{B}_1 = \mathsf{B}(\beta, \eta, \epsilon)$	$\mathsf{B}_1 : \beta_1$
$\mathsf{B}_2 = \mathsf{B}(\sigma, \xi, \chi)$	$\mathsf{B}_2 : \beta_2$

$\mathsf{C}_\bullet = \mathsf{C}(\tau, \sigma, \rho) = \mathsf{S}_4\mathsf{P}_\bullet\mathsf{Q}_\bullet$

$$\dfrac{\vdash \mathsf{S}_4 : \sigma_4 \qquad \dfrac{\dfrac{\dfrac{\vdots \qquad \vdots}{\vdash \mathsf{B}_1 : \beta_1 \quad \vdash \mathsf{B}_2 : \beta_2}}{\vdash \mathsf{B}_1\mathsf{B}_2 : \sigma_3 \rightarrow \pi} \qquad \vdash \mathsf{S}_3 : \sigma_3}{\vdash \mathsf{P}_\bullet : \pi}}{\dfrac{\vdash \mathsf{S}_4\mathsf{P}_\bullet : \omega \rightarrow \gamma \qquad \dfrac{\vdash \mathsf{K}_3 : \kappa_3 \quad \vdash \mathsf{K}_4 : \kappa_4}{\vdash \mathsf{Q}_\bullet : \omega}}{\vdash \mathsf{C}_\bullet : \gamma}}$$

(δ) The type δ is

$$(\theta \to \psi) \to (\psi \to \phi) \to (\theta \to \phi)$$

and the untyped combinator D is shown to the right. We carry over the abbreviations $\rho, \sigma, \tau, \beta, \xi, \chi, \eta, \zeta$ and $\kappa_5, \kappa_6, \sigma_5, \sigma_6$ of Solution 1.7(δ). With these we have a collection of typed combinators, associated housing axioms, and a derivable statement, as shown. The derivation to the right shows that with $\mathsf{R}_\bullet = \mathsf{K}_5(\mathsf{S}_6\mathsf{B}_\bullet)$ the combinator

$$\mathsf{D}_\bullet = \mathsf{D}(\theta, \psi, \phi) = \mathsf{S}_5\mathsf{R}_\bullet\mathsf{K}_6$$

inhabits δ.

$$\mathsf{D} = \mathsf{S}(\mathsf{K}(\mathsf{SB}))\mathsf{K}$$

$$\begin{array}{ll} \mathsf{K}_5 = \mathsf{K}(\sigma, \eta) & \mathsf{K}_5 : \kappa_5 \\ \mathsf{K}_6 = \mathsf{K}(\tau, \sigma) & \mathsf{K}_6 : \kappa_6 \\ \mathsf{S}_5 = \mathsf{S}(\sigma, \xi, \chi) & \mathsf{S}_5 : \sigma_5 \\ \mathsf{S}_6 = \mathsf{S}(\tau, \sigma, \rho) & \mathsf{S}_6 : \sigma_6 \\ \mathsf{B}_\bullet = \mathsf{B}(\theta, \psi, \phi) & \mathsf{B}_\bullet : \beta \end{array}$$

$$\dfrac{\vdash \mathsf{S}_5 : \sigma_5 \quad \dfrac{\vdash \mathsf{K}_5 : \kappa_5 \quad \dfrac{\vdash \mathsf{S}_6 : \sigma_6 \quad \vdash \mathsf{B}_\bullet : \beta}{\vdash \mathsf{S}_6\mathsf{B}_\bullet : \eta}}{\vdash \mathsf{R}_\bullet : \zeta}}{\dfrac{\vdash \mathsf{S}_5\mathsf{R}_\bullet : \kappa_6 \to \delta \qquad \vdash \mathsf{K}_6 : \kappa_6}{\vdash \mathsf{D}_\bullet : \delta}}$$

(ϵ) The type ϵ is

$$(\psi \to \zeta \to \phi) \to (\theta \to \zeta) \to (\theta \to \psi \to \phi)$$

and the untyped combinator E is to the right. We carry over the abbreviations $\alpha, \xi, \mu, \chi, \eta, \gamma, \beta_1, \beta_2$ of Solution 1.8(ϵ). Using these we have three derivable statement, as shown. The derivation to the right shows that

$$\mathsf{E}(\theta, \psi, \phi, \zeta) = \mathsf{B}_1\mathsf{B}_2\mathsf{C}_\bullet$$

inhabits ϵ.

$$\mathsf{E} = \mathsf{BBC}$$

$$\begin{array}{ll} \mathsf{B}_1 = \mathsf{B}(\eta, \xi, \chi) & \mathsf{B}_1 : \beta_1 \\ \mathsf{B}_2 = \mathsf{B}(\theta, \zeta, \chi) & \mathsf{B}_2 : \beta_2 \\ \mathsf{C}_\bullet = \mathsf{C}(\psi, \zeta, \phi) & \mathsf{C}_\bullet : \gamma \end{array}$$

$$\dfrac{\dfrac{\mathsf{B}_1 : \beta_1 \quad \mathsf{B}_2 : \beta_2}{\vdash \mathsf{B}_1\mathsf{B}_2 : \gamma \to \epsilon} \quad \vdash \mathsf{C}_\bullet : \gamma}{\vdash \mathsf{B}_1\mathsf{B}_2\mathsf{C}_\bullet : \epsilon}$$

(ω) The type is

$$\omega = (\theta \to \theta \to \phi) \to (\theta \to \phi)$$

and $\mathsf{W} = \mathsf{SS}(\mathsf{KI})$ is the untyped combinator. We carry over the abbreviations $\iota, \lambda, \mu, \nu, \kappa, \sigma_1, \sigma_2$ of Solution 1.8(ω). Using these we have a collection of typed combinators and associated housing axioms as shown to the right. The derivation shows that $\mathsf{W}_\bullet$ inhabits ω.

$$\begin{array}{ll} \mathsf{I}_\bullet = \mathsf{I}(\theta) & \mathsf{I}_\bullet : \iota \\ \mathsf{K}_\bullet = \mathsf{K}(\mu, \iota) & \mathsf{K}_\bullet : \kappa \\ \mathsf{S}_1 = \mathsf{S}(\mu, \iota, \lambda) & \mathsf{S}_1 : \sigma_1 \\ \mathsf{S}_2 = \mathsf{S}(\theta, \theta, \phi) & \mathsf{S}_2 : \sigma_2 \end{array}$$

$$\dfrac{\dfrac{\vdash \mathsf{S}_1 : \sigma_1 \quad \vdash \mathsf{S}_2\sigma_2}{\vdash \mathsf{S}_1\mathsf{S}_2 : \nu \to \omega} \quad \dfrac{\vdash \mathsf{K}_\bullet : \kappa \quad \vdash \mathsf{I}_\bullet : \iota}{\vdash \mathsf{K}_\bullet\mathsf{I}_\bullet : \nu}}{\vdash (\mathsf{S}_1\mathsf{S}_2)(\mathsf{K}_\bullet\mathsf{I}_\bullet) : \omega}$$

$$\mathsf{W}(\theta, \phi) = (\mathsf{S}_2\mathsf{S}_1)(\mathsf{K}_\bullet\mathsf{I}_\bullet)$$

(μ) The type μ is quite long

$$\mu = (\zeta \to \rho \to \theta \to \psi \to \phi) \to \zeta \to \rho \to (\theta \to \psi) \to (\theta \to \phi)$$

and $\mathsf{M} = \mathsf{B}(\mathsf{BS})$ is the untyped combinator we must deal with. We carry over the abbreviations $\alpha, \eta, \chi, \omega, \lambda, \xi, \pi, \epsilon, \sigma, \beta_1, \beta_2$ of Solution 1.8(μ). Using these we have a collection of typed combinators and associated derivable statements. The derivation shows that

$$\mathsf{M}(\theta, \psi, \phi, \rho, \zeta) = \mathsf{B}_1(\mathsf{B}_2\mathsf{S}_\bullet)$$

inhabits μ.

$$\begin{array}{ll} \mathsf{S}_\bullet = \mathsf{S}(\theta,\psi,\phi) & \mathsf{S}_\bullet : \sigma \\ \mathsf{B}_1 = \mathsf{B}(\zeta,\chi,\pi) & \mathsf{B}_1 : \beta_1 \\ \mathsf{B}_2 = \mathsf{B}(\rho,\eta,\xi) & \mathsf{B}_2 : \beta_2 \end{array}$$

$$\dfrac{\begin{array}{c}\vdots \\ \vdash \mathsf{B}_1\beta_1\end{array} \quad \dfrac{\begin{array}{c}\vdots \\ \vdash \mathsf{B}_2 : \beta_2\end{array} \quad \vdash \mathsf{S}_\bullet : \sigma}{\vdash \mathsf{B}_2\mathsf{S}_\bullet : \chi}}{\vdash \mathsf{B}_1(\mathsf{B}_2\mathsf{S}_\bullet) : \mu}$$

(ν) Again this type μ is quite long

$$(\rho \to \theta \to \psi \to \phi) \to \rho \to (\theta \to \psi) \to (\theta \to \phi)$$

and the untyped combinator N is given to the right. We carry over the abbreviations $\eta, \chi, \lambda, \xi, \sigma_1, \sigma_2, \kappa$ of Solution 1.8(ν). Using these we have three housing axioms. The derivation shows that

$$\mathsf{N}(\theta, \psi, \phi, \rho) = \mathsf{S}_1(\mathsf{K}_\bullet\mathsf{S}_2)$$

inhabits ν.

$$\mathsf{N} = \mathsf{S}(\mathsf{KS})$$

$$\begin{array}{ll} \mathsf{K}_\bullet = \mathsf{K}(\rho,\sigma_2) & \mathsf{K}_\bullet : \kappa \\ \mathsf{S}_1 = \mathsf{S}(\rho,\eta,\xi) & \mathsf{S}_1 : \sigma_1 \\ \mathsf{S}_2 = \mathsf{S}(\theta,\psi,\phi) & \mathsf{S}_2 : \sigma_2 \end{array}$$

$$\dfrac{\vdash \mathsf{S}_1 : \sigma_1 \quad \dfrac{\vdash \mathsf{K}_\bullet : \kappa \quad \vdash \mathsf{S}_2 : \sigma_2}{\vdash \mathsf{K}_\bullet\mathsf{S}_2 : \chi}}{\vdash \mathsf{S}_1(\mathsf{K}_\bullet\mathsf{S}_2) : \nu}$$

(τ) Once more the type τ is quite long

$$(\theta \to \psi \to \phi) \to (\zeta \to \theta) \to (\zeta \to \psi) \to (\zeta \to \phi)$$

and the untyped combinator $\mathsf{T} = \mathsf{MB}$ is also shown right. We carry over the abbreviations $\alpha, \eta, \omega, \lambda, \epsilon, \sigma, \beta, \mu$ of Solution 1.8(τ). Using these we have a pair of typed combinators and associated derivable statements. The derivation shows that $\mathsf{T}_\bullet$ inhabits τ. □

$$\mathsf{T} = \mathsf{MB}$$

$$\begin{array}{ll} \mathsf{B}_\bullet = \mathsf{B}(\zeta,\theta,\alpha) & \mathsf{B}_\bullet : \beta \\ \mathsf{M}_\bullet = \mathsf{M}(\theta,\psi,\phi,\eta,\omega) & \mathsf{M}_\bullet : \mu \end{array}$$

$$\dfrac{\begin{array}{c}\vdots \\ \vdash \mathsf{M}_\bullet : \mu\end{array} \quad \begin{array}{c}\vdots \\ \vdash \mathsf{B}_\bullet : \beta\end{array}}{\vdash \mathsf{M}_\bullet\mathsf{B}_\bullet : \tau}$$

$$\mathsf{T}(\theta, \psi, \phi, \zeta) = \mathsf{M}_\bullet\mathsf{B}_\bullet$$

3.3 We annotate the derivations of Solution 1.9. Using the same abbreviations

$$\begin{array}{lll} \text{(i)} & \mathsf{W}(1)(\theta,\psi) & = \mathsf{C}(\eta,\theta,\psi)\mathsf{I}(\eta) \\ \text{(ii)} & \mathsf{W}(2)(\theta,\psi,\phi) & = \mathsf{D}(\theta,\nu,\mu)\mathsf{W}(1)(\theta,\psi)\mathsf{W}(1)(\nu,\phi) \\ \text{(iii)} & \mathsf{W}(3)(\theta,\psi,\phi) & = \mathsf{D}(\eta,\rho,\zeta)\mathsf{D}(\theta,\psi,\phi)\mathsf{D}(\xi,\chi,\phi) \\ \text{(iv)} & \mathsf{W}(4)(\theta,\psi,\phi) & = \mathsf{W}(3)(\theta,\nu,\phi)\mathsf{W}(1)(\theta,\psi) \\ \text{(v)} & \mathsf{W}(5)(\theta,\psi,\phi) & = \mathsf{D}(\theta,\lambda,\mu)\mathsf{W}(1)(\theta,\phi)\mathsf{W}(4)(\theta,\psi,\phi) \end{array}$$

are the appropriate combinators. □

3.4 This is a routine induction over the supplied derivation ∇. When ∇ is an Axiom or a Projection, the required conclusions are immediate (by the provisos on the rules). When ∇ arises by Elimination

$$\nabla \quad = \quad \frac{\overset{Q}{\Gamma \vdash q : \pi \to \tau} \quad \overset{P}{\Gamma \vdash p : \pi}}{\Gamma \vdash t : \tau}$$

where $t = qp$, we see by induction (from Q and P) that Γ is legal and

$$\partial(t) = \partial(q) \cup \partial(p) \subseteq \partial(\Gamma)$$

as required (where $\partial(\cdot)$ indicates the occurring identifiers). □

3.5 Informally the algorithm simply takes the supplied derivation

$$(\nabla) \quad \Gamma \vdash t : \tau$$

and replaces each occurrence of Γ by Σ. Formally the algorithm proceeds by recursion on ∇. When ∇ is an Axiom or a Projection we have

$$\Gamma \vdash \mathsf{Z}_\bullet : \zeta \longmapsto \Sigma \vdash \mathsf{Z}_\bullet : \zeta \qquad\qquad \Gamma \vdash x : \sigma \longmapsto \Sigma \vdash x : \sigma$$

respectively, since Σ is legal and in the second $x : \sigma$ occurs in Γ and hence in Σ. When the root rule is an Elimination, $\nabla = QP$, we move to $\nabla^+ = Q^+P^+$ using the algorithm recursively on Q, P.

To handle Contraction we need to use substitution as a structural rule. This could be done here, but we will leave it for later when we look at a more intricate case. □

C.3 Annotation and deletion

$$\cfrac{\cfrac{\overset{\vdots}{\vdash \mathsf{C}_\bullet : \gamma} \quad \vdash \mathsf{S}_\bullet : \sigma}{\vdash \mathsf{C}_\bullet\mathsf{S}_\bullet : \iota \to \omega} \qquad \vdash \mathsf{I}_\bullet : \iota}{\vdash \mathsf{C}_\bullet\mathsf{S}_\bullet\mathsf{I}_\bullet : \omega}$$

3.6 (a) If $(\emptyset, \mathsf{C}_\bullet\mathsf{S}_\bullet\mathsf{I}_\bullet)$ is the code of a derivation then this must have the shape shown to the right for types ι, σ, γ of compatible shapes. We must have $\gamma = \sigma \to \iota \to \omega$ and, because of the nature of $\mathsf{I}_\bullet, \mathsf{S}_\bullet, \mathsf{C}_\bullet$, there must be types $\theta, \phi, \chi, \xi, \lambda, \mu, \nu$ such that

$$\begin{aligned} \iota &= \theta \to \theta \\ \sigma &= (\xi \to \chi \to \phi) \to (\xi \to \chi) \to (\xi \to \phi) \\ \gamma &= (\mu \to \nu \to \lambda) \to (\nu \to \mu \to \lambda) \end{aligned}$$

hold. For these to fit together properly we must have

$$\mu = \xi \to \chi \to \phi \quad \nu = \xi \to \chi \quad \lambda = \xi \to \phi \quad \nu = \iota \quad \omega = \mu \to \lambda$$

so that

$$\xi = \chi = \theta \quad \lambda = \theta \to \phi \quad \mu = \theta \to \theta \to \phi$$

to get

$$\omega = (\theta \to \theta \to \phi) \to (\theta \to \phi)$$

the standard coercion type ω. Thus $\mathsf{C_\bullet S_\bullet I_\bullet}$ encodes a derivation if and only if

$$\mathsf{I_\bullet} = \mathsf{I}(\theta) \quad \mathsf{S_\bullet} = \mathsf{S}(\theta, \theta, \phi) \quad \mathsf{C_\bullet} = \mathsf{C}(\mu, \iota, \lambda)$$

where

$$\mu = \theta \to \theta \to \phi \quad \iota = \theta \to \theta \quad \lambda = \theta \to \phi$$

and θ, ϕ are arbitrary.

(b) If $(\emptyset, \mathsf{S}_1(\mathsf{K_\bullet S}_2))$ is the code of a derivation then this must have shape shown to the right for types $\kappa, \sigma_1, \sigma_2, \alpha, \nu$ of compatible shapes. Because of the nature of $\mathsf{K_\bullet}, \mathsf{S_\bullet}$ we must have

$$\dfrac{\vdash \mathsf{S}_1 : \sigma_1 \qquad \dfrac{\vdash \mathsf{K_\bullet} : \kappa \quad \vdash \mathsf{S}_2 : \sigma_2}{\vdash \mathsf{K_\bullet S}_2 : \alpha}}{\vdash \mathsf{S}_1(\mathsf{K_\bullet S}_2) : \nu}$$

$$\kappa = \sigma_2 \to \rho \to \sigma_2 \quad \alpha = \rho \to \sigma_2 \quad \sigma_1 = \alpha \to \nu$$

for some type ρ. Furthermore, we must have

$$\sigma_2 = (\theta \to \psi \to \phi) \to (\theta \to \psi) \to (\theta \to \phi)$$

for some types θ, ψ, ϕ. Setting

$$\eta = \theta \to \psi \to \phi \qquad \xi = (\theta \to \psi) \to (\theta \to \phi)$$

gives $\alpha = \rho \to \eta \to \xi$ and hence we have $\alpha = (\rho \to \eta) \to (\rho \to \xi)$ using the required shape of σ_1. Thus ν is the type dealt with in Exercise 3.3 and $\mathsf{N_\bullet} = \mathsf{S}_1(\mathsf{K_\bullet S}_2)$ is the derivation given there.

(c) If $(\emptyset, \overline{S}(\overline{S}0))$ is the code of a derivation then this must have shape

$$\dfrac{\dfrac{\vdash \mathsf{S_\bullet} : \sigma \quad \vdash \mathsf{B_\bullet} : \beta}{\vdash \overline{S} : \vartheta} \qquad \dfrac{\dfrac{\vdash \mathsf{S_\bullet} : \sigma \quad \vdash \mathsf{B_\bullet} : \beta}{\vdash \overline{S} : \vartheta} \quad \dfrac{\vdash \mathsf{K_\bullet} : \kappa \quad \vdash \mathsf{I_\bullet} : \iota}{\vdash \overline{0} : \zeta}}{\vdash \overline{S}0 : \omega}}{\vdash \overline{S}(\overline{S}0) : \alpha}$$

for types $\iota, \kappa, \sigma, \beta$ and $\vartheta, \zeta, \omega, \alpha$ of appropriate shapes. For these to fit together properly we must have

$$\sigma = \beta \to \vartheta \qquad \kappa = \iota \to \zeta \qquad \vartheta = \zeta \to \omega \qquad \vartheta = \omega \to \alpha$$

which give

$$\zeta = \omega = \alpha \qquad \vartheta = \alpha' \qquad \sigma = \beta \to \alpha' \qquad \kappa = \iota \to \alpha$$

(where $\tau' = \tau \to \tau$ for each type τ). The usual restrictions on the shapes of ι, κ, σ, and β must be satisfied. There must be types θ, ψ, ϕ with $\iota = \theta'$ and

$\kappa = \psi \to \phi \to \psi$ so that, from above, $\psi = \theta'$ and $\alpha = \phi \to \theta'$ with further restrictions to come from the shapes of σ and β. There must be types λ, μ, ν with

$$\sigma = (\lambda \to \mu \to \nu) \to (\lambda \to \mu) \to (\lambda \to \nu)$$

so that, from above,

$$\beta = (\lambda \to \mu \to \nu) \qquad \alpha = (\phi \to \theta') = (\lambda \to \mu) = (\lambda \to \nu)$$

which gives $\lambda = \phi$ and $\mu = \nu = \theta'$ with further restrictions to come from the shape of β. So far we have

$$\beta = (\lambda \to \mu \to \nu) = (\phi \to \theta' \to \theta')$$

which, recalling the required shape of β, gives $\phi = \theta'$.

Thus we find that θ is arbitrary and

$$\mathsf{I}_\bullet = \mathsf{I}(\theta) : \theta' \qquad \mathsf{K}_\bullet = \mathsf{K}(\theta', \theta') : \theta' \to \theta'' \qquad \mathsf{S}_\bullet = \mathsf{S}(\theta', \theta', \theta') : (\theta' \to \theta'') \to \theta'''$$

and

$$\mathsf{B}_\bullet = \mathsf{B}(\theta', \theta', \theta') : (\theta' \to \theta'')$$

so that $\overline{S} : \theta'''$ and $\overline{0} : \theta''$ to give $\alpha = \theta''$. □

3.7 This is not difficult but it is worth looking at. The proof proceeds by an induction over t with the parameter Γ held rigid. The bases cases, $t = \mathsf{Z}$ and $t = x$, are immediate. For the induction step to $t = qp$, what can a derivation

$$(\nabla) \quad \Gamma \vdash qp : \tau$$

look like? We must have $\nabla = QP$ where

$$(Q) \quad \Gamma \vdash q : \pi \to \tau \qquad (P) \quad \Gamma \vdash p : \pi$$

for some types π, τ. By the induction hypothesis there is just one possible Q and just one possible P. If the generated types match, as shown, then we obtain the only possible ∇. Otherwise there is no derivation ∇. □

3.8 Let Γ be a fixed legal context. We prove each assertion by induction over the derivation ∇. We do the two proofs separately.

($\alpha\,;\delta = id$) Suppose ∇ is an Axiom, i.e. $\nabla = \Gamma^\delta[\zeta]$ for some axiom ζ. Let $\mathsf{Z} : \zeta$ be the associated housing axiom. Then

$$(\Gamma, \nabla)^\alpha = \Gamma[\mathsf{Z}] \qquad (\Gamma, \nabla)^{\alpha\delta} = \Gamma[\mathsf{Z}]^\delta = \Gamma^\delta[\zeta] = \nabla$$

as required. Suppose ∇ is a Projection, i.e. $\nabla = \Gamma^\delta[n]$ for some n. Let $x : \sigma$ be the declaration in the nth position of Γ. Then

$$(\Gamma, \nabla)^\alpha = \Gamma[x] \qquad (\Gamma, \nabla)^{\alpha\delta} = \Gamma[x]^\delta = \Gamma^\delta[n] = \nabla$$

as required. For the induction step to $\nabla = QP$ we have

$$(\Gamma, \nabla)^{\alpha\delta} = ((\Gamma, Q)^{\alpha}(\Gamma, P)^{\alpha})^{\delta} = (\Gamma, Q)^{\alpha\delta}(\Gamma, P)^{\alpha\delta} = QP = \nabla$$

where the penultimate step uses the induction hypothesis.

($\delta\,;\alpha = id$) Suppose ∇ is an Axiom, i.e. $\nabla = \Gamma[\mathsf{Z}]$ for some housing axiom $\mathsf{Z} : \zeta$. Then $\nabla^{\delta} = \Gamma^{\delta}[\zeta]$ so that

$$(\Gamma, \nabla^{\delta})^{\alpha} = (\Gamma, \Gamma^{\delta}[\zeta])^{\alpha} = \Gamma[\mathsf{Z}] = \nabla$$

as required. Suppose ∇ is a Projection, i.e. $\nabla = \Gamma[x]$ where $x : \sigma$ is the declaration in the nth position of Γ. Then $\nabla^{\delta} = \Gamma^{\delta}[n]$ so that

$$(\Gamma, \nabla^{\delta})^{\alpha} = (\Gamma, \Gamma^{\delta}[n])^{\alpha} = \Gamma[x] = \nabla$$

as required. For the induction step to $\nabla = QP$ we have

$$(\Gamma, \nabla^{\delta})^{\alpha} = (\Gamma, Q^{\delta}P^{\delta})^{\alpha} = (\Gamma, Q^{\delta})^{\alpha}(\Gamma, P^{\delta})^{\alpha} = QP = \nabla$$

where the penultimate step uses the induction hypothesis. □

3.9 We require an algorithm which converts each derivation

$$(\nabla) \quad (\Gamma, x : \theta) \vdash t : \tau$$

into a derivation

$$(\nabla^{d}) \quad \Gamma \vdash (-) : (\theta \to \tau)$$

for some term $(-)$. This term is not quite as simple as suggested in the section; it depends on Γ. Once found this term determines ∇^{d}. We set

$$\begin{array}{lll} [x:\theta]x & = x & \\ [x:\theta]y & = \mathsf{K}(\theta,\tau)y & \text{where } y:\tau \text{ is in } \Gamma \\ [x:\theta]\mathsf{Z} & = \mathsf{K}(\theta,\zeta)\mathsf{Z} & \text{where } \mathsf{Z}:\zeta \text{ is a housing axiom} \\ [x:\theta](qp) & = \mathsf{S}(\theta,\pi,\tau)([x:\theta]q)([x:\theta]p) & \text{where } \Gamma, x:\theta \vdash p:\pi \end{array}$$

to generate $[x : \theta]t$. In the second clause y is an identifier declared in Γ. This produces the dependency of $[x : \theta]t$ on Γ. □

3.10 For convenience let

$$\Gamma = \Pi, x : \sigma, \Delta \qquad \Sigma = \Pi, \Delta$$

be the two contexts. Recall that $r * s = r[x := s]$ is generated by

$$x * s = s \quad y * s = y \quad \mathsf{Z} * s = \mathsf{Z} \quad (qp) * s = (q * s)(p * s)$$

in the usual notation. To obtain $R * S$ it suffices to calculate $r * s$ and then use the Generation Algorithm 3.7 to produce the derivation. Using the arboreal code the clauses are

$$\Gamma[x] * S = S \quad \Gamma[y] * S = \Pi[y] \quad \Gamma[\mathsf{Z}] * S = \Pi[\mathsf{Z}] \quad (QP) * S = (Q * S)(P * S)$$

and these match the clauses above. □

C.4 Computation

3.11 From Exercise 3.2(γ) we have

$$\mathsf{C}_\bullet = \mathsf{S}_4\mathsf{P}_\bullet\mathsf{Q}_\bullet \qquad \mathsf{P}_\bullet = \mathsf{B}_1\mathsf{B}_2\mathsf{S}_3 \qquad \mathsf{Q}_\bullet = \mathsf{K}_3\mathsf{K}_4 \qquad \mathsf{B}_2 = \mathsf{S}_1(\mathsf{K}_1\mathsf{S}_2)\mathsf{K}_2$$

for appropriate constants $\mathsf{K}_1, \ldots, \mathsf{K}_4, \mathsf{S}_1, \ldots \mathsf{S}_4$ and compounds $\mathsf{P}_\bullet, \mathsf{Q}_\bullet, \mathsf{B}_1, \mathsf{B}_2$ with the structure of B_1 not shown. Informally

$$\mathsf{C}_\bullet\mathsf{S}_\bullet \rhd\!\!\rhd (\mathsf{P}_\bullet\mathsf{S}_\bullet)(\mathsf{Q}_\bullet\mathsf{S}_\bullet) \rhd\!\!\rhd \mathsf{B}_1(\mathsf{S}_1\mathsf{S}_\bullet)\mathsf{K}_1 \qquad \mathsf{C}_\bullet\mathsf{S}_\bullet\mathsf{I}_\bullet \rhd\!\!\rhd (\mathsf{S}_1\mathsf{S}_\bullet)(\mathsf{K}_1\mathsf{I}_\bullet)$$

to give $\mathsf{S}_* = \mathsf{S}_3$ and $\mathsf{K}_* = \mathsf{K}_4$, as required.

The required computations can be extracted from Solution 2.2. Simplifying the abbreviations let

$$\mathsf{a} = 0 \circ \downharpoonleft 0 \qquad \mathsf{b} = \downharpoonleft\downharpoonleft \mathsf{a} \circ \mathsf{a} \qquad \mathsf{c} = 0 \circ \downharpoonleft \mathsf{b} \circ \downharpoonright 0 \qquad \square = \downharpoonleft \mathsf{c} \circ \mathsf{b}$$

so that

$$\text{(a)}\quad \mathsf{B}_\bullet z \rhd\!\!\rhd \mathsf{S}_\bullet(\mathsf{K}_\bullet z) \qquad \text{(b)}\quad \mathsf{B}_\bullet zyx \rhd\!\!\rhd z(yx)$$

for all x, y, z, and hence

$$\begin{array}{llll} \text{(0)} & \mathsf{C}_\bullet t \rhd\!\!\rhd (\mathsf{P}_\bullet t)(\mathsf{Q}_\bullet t) & \text{(b)} & \mathsf{P}_\bullet t \rhd\!\!\rhd \mathsf{B}_2(\mathsf{S}_3 t) \\ \text{(0)} & \mathsf{Q}_\bullet t \rhd\!\!\rhd \mathsf{K}_4 & \text{(c)} & \mathsf{C}_\bullet t \rhd\!\!\rhd \mathsf{B}_2(\mathsf{S}_3 t)\mathsf{K}_4 \end{array}$$

to give

$$(\square)\quad \mathsf{C}_\bullet ts \rhd\!\!\rhd (\mathsf{S}_3 t)(\mathsf{K}_4 s)$$

for all s, t. There are other computations which give the same result. □

3.12 In some cases there are many possible computations, but each produces the same normal form. We use some of the results of Exercise 2.3. Thus, cleaning up the abbreviations, let

$$\mathsf{a} = 0 \circ \downharpoonleft 0 \quad \mathsf{b} = \downharpoonleft\downharpoonleft \mathsf{a} \circ \mathsf{a} \quad \mathsf{c} = 0 \circ \downharpoonleft \mathsf{b} \circ \downharpoonright 0 \quad \mathsf{d} = \downharpoonleft \mathsf{a} \circ 0 \circ \downharpoonright 0 \quad \mathsf{e} = \downharpoonleft \mathsf{d} \circ \mathsf{b}$$

to organize

$$\begin{array}{llllll} \text{(a)} & \mathsf{B}z \rhd\!\!\rhd \mathsf{S}(\mathsf{K}z) & \text{(a)} & \mathsf{D}z \rhd\!\!\rhd (\mathsf{SB})(\mathsf{K}z) & \text{(b)} & \mathsf{B}zyx \rhd\!\!\rhd z(yx) \\ \text{(c)} & \mathsf{C}z \rhd\!\!\rhd \mathsf{B}(\mathsf{S}z)\mathsf{K} & \text{(d)} & \mathsf{D}zy \rhd\!\!\rhd \mathsf{B}yz & \text{(e)} & \mathsf{D}zyx \rhd\!\!\rhd y(zx) \end{array}$$

for all x, y, z. In the table

i	$\mathsf{W}(i)$	$\mathsf{W}'(i)$	$\mathsf{w}(i)$
1	CI	$\mathsf{S}(\mathsf{S}(\mathsf{SI}))\mathsf{K}$	$\mathsf{c} \circ \downharpoonleft \mathsf{a}$
2	$\mathsf{D}\mathsf{W}'(1)\mathsf{W}'(1)$	$\mathsf{S}(\mathsf{K}\mathsf{W}'(1))\mathsf{W}'(1)$	$\downharpoonleft\downharpoonright \mathsf{w}(1) \circ \downharpoonright \mathsf{w}(1) \circ \mathsf{d} \circ \downharpoonleft \mathsf{a}$
3	DDD	$\mathsf{S}(\mathsf{KD})\mathsf{D}$	$\mathsf{d} \circ \downharpoonleft \mathsf{a}$
4	$\mathsf{W}(3)\mathsf{W}(1)$	$(\mathsf{SB})(\mathsf{KR})$	$\downharpoonright \mathsf{w}(1) \circ \mathsf{e} \circ \mathsf{a} \circ \downharpoonright \mathsf{a}$
5	$\mathsf{D}\mathsf{W}(1)\mathsf{W}(4)$	$\mathsf{S}(\mathsf{K}\mathsf{W}'(4))\mathsf{W}'(1)$	$\downharpoonleft\downharpoonright \mathsf{w}(1) \circ \downharpoonright \mathsf{w}(4) \circ \mathsf{d} \circ \downharpoonleft \mathsf{a}$

for each $1 \leq i \leq 5$ we give the original untyped combinator $\mathsf{W}(i)$, its normal form $\mathsf{W}'(i)$, and a computation $\mathsf{w}(i)$ which organizes the reduction $\mathsf{W}(i) \rhd\!\!\!\rhd \mathsf{W}'(i)$. For $i = 4$ we have $\mathsf{R} = (\mathsf{SB})(\mathsf{KW}'(1))$. The informal reductions

$$\begin{array}{llll}
\mathsf{W}(1) \rhd\!\!\!\rhd \mathsf{B}(\mathsf{SI})\mathsf{K} & \rhd\!\!\!\rhd \mathsf{W}'(1) & & \\
\mathsf{W}(2) \rhd\!\!\!\rhd \mathsf{DW}'(1)\mathsf{W}'(1) & \rhd\!\!\!\rhd \mathsf{W}'(2) & & \\
\mathsf{W}(3) \rhd\!\!\!\rhd \mathsf{BDD} & \rhd\!\!\!\rhd \mathsf{W}'(3) & & \\
\mathsf{W}(4) \rhd\!\!\!\rhd \mathsf{DDDW}'(1) & \rhd\!\!\!\rhd \mathsf{D}(\mathsf{DW}'(1)) & \rhd\!\!\!\rhd \mathsf{W}'(4) \\
\mathsf{W}(5) \rhd\!\!\!\rhd \mathsf{DW}'(1)\mathsf{W}'(4) & \rhd\!\!\!\rhd \mathsf{BW}'(4)\mathsf{W}'(1) & \rhd\!\!\!\rhd \mathsf{W}'(5)
\end{array}$$

can be used to justify the organized versions. □

3.13 Both compounds Y and Z use a combinator $\mathsf{A} = \mathsf{SII}$ which must be typable (if Y and Z are typable). Thus we must have typed constants $\mathsf{S}_1, \mathsf{I}_2, \mathsf{I}_3$ with

$$\mathsf{S}_1 : (\theta \to \psi \to \phi) \to (\theta \to \psi) \to (\theta \to \phi) \qquad \mathsf{I}_2 : (\theta \to \psi \to \phi) \qquad \mathsf{I}_1 : (\theta \to \psi)$$

for some θ, ψ, ϕ. But then, from I_2 and I_1, we must have

$$\theta = \psi \to \phi \qquad \theta = \psi \quad \text{i.e.} \quad \theta = \theta \to \phi$$

which is impossible.

3.14 We are given an untyped computation

$$(\square^*) \quad t^{-\epsilon} \rhd\!\!\!\rhd t^*$$

with a type erased subject, and must show there is a unique typed computation

$$(\square) \quad t^- \rhd\!\!\!\rhd t^+$$

with $\square^\epsilon = \square^*$. In effect we have to show that the arboreal code for $\square^*$ is also the code for some typed computation.

The proof proceeds by induction over $\square^*$. We have the three base cases, so it suffices to look at the three induction steps across LA, RA, and TC.

For the step across LA suppose $\square^* = \downharpoonleft \mathsf{q}^*$ where

$$(\mathsf{q}^*) \quad q^! \rhd\!\!\!\rhd q^* \qquad t^{-\epsilon} = q^! p^! \quad t^* = q^* p^!$$

for some untyped terms $q^!, p^!, q^*$. From the shape of $t^{-\epsilon}$ and the way erasure works we must have $t^- = q^+ p$ for some typed terms q^-, p. Then

$$t^{-\epsilon} = (q^- p)^\epsilon = q^{-\epsilon} p^\epsilon$$

so that $q^! = q^{-\epsilon}$ and $p^! = p^\epsilon$. This gives

$$(\mathsf{q}^*) \quad q^{-\epsilon} \rhd\!\!\!\rhd q^*$$

and hence, by the induction hypothesis, there is a unique typed computation

$$(\mathsf{q}) \quad q^- \rhd\!\!\!\rhd q^+$$

with $\mathsf{q}^\epsilon = \mathsf{q}^*$. With this q^+ let $t^+ = q^+p$, so that

$$t^{+\epsilon} = q^{+\epsilon}p^\epsilon = q^*p^! = t^*$$

and hence $\square = \downharpoonleft\mathsf{q}$ is an example of the required kind of typed computation with $\square^\epsilon = \square^*$. From the shape of t^- this is the only possible example.

The step across RA is similar.

For the step across TC suppose $\square^* = \mathsf{l}^* \circ \mathsf{r}^*$ where

$$(\mathsf{l}^*)\ \ t^{-\epsilon} \rhd\!\!\!\rhd t^! \qquad (\mathsf{r}^*)\ \ t^! \rhd\!\!\!\rhd t^*$$

for some intermediate untyped term $t^!$. By the induction hypothesis there is a unique typed computation

$$(\mathsf{l})\ \ t^- \rhd\!\!\!\rhd t^0$$

with $\mathsf{l}^\epsilon = \mathsf{l}^*$, and then $t^! = t^{0\epsilon}$. But now we have

$$(\mathsf{r}^*)\ \ t^{0\epsilon} \rhd\!\!\!\rhd t^*$$

so, by a second use of the induction hypothesis, there is a unique typed computation

$$(\mathsf{r})\ \ t^0 \rhd\!\!\!\rhd t^+$$

with $\mathsf{r}^\epsilon = \mathsf{r}^*$, and hence $t^* = t^{+\epsilon}$. Let $\square = \mathsf{l} \circ \mathsf{r}$. Then $\square^\epsilon = \square^*$ and, from the shape of $\square^*$, this is the only possible typed computation. $\square$

C.5 SUBJECT REDUCTION

3.15 (a) Carrying over the notation of Example 3.5 we have derivations

$$S_i = \Gamma[\mathsf{S}_i] \quad K_i = \Gamma[\mathsf{K}_i]$$

$$\dfrac{\Gamma \vdash \mathsf{S}_1 : \sigma_1 \qquad \dfrac{\Gamma \vdash \mathsf{K}_1 : \kappa_1 \quad \Gamma \vdash \mathsf{S}_2 : \sigma_2}{\Gamma \vdash \mathsf{K}_1\mathsf{S}_2 : \xi}}{\dfrac{\Gamma \vdash \mathsf{S}_1(\mathsf{K}_1\mathsf{S}_2) : \kappa_2 \to \beta \qquad \Gamma \vdash \mathsf{K}_2 : \kappa_2}{\Gamma \vdash \mathsf{B}_\bullet : \beta}}$$

$$B = S_1(K_1S_2)K_2$$

$$Z = \Gamma[z] \quad Y = \Gamma[y] \quad X = \Gamma[x]$$

$$\dfrac{\dfrac{\dfrac{\overset{B}{\Gamma \vdash \mathsf{B}_\bullet : \beta} \quad \Gamma \vdash z : \tau}{\bullet \qquad \Gamma \vdash y : \sigma}}{\bullet \qquad \Gamma \vdash x : \theta}}{\Gamma \vdash \mathsf{B}_\bullet zyx : \phi}$$

$$\nabla = BZYX$$

where, in both cases, the arboreal code is given below the tree and the appropriate projections are given above. Of course $\mathsf{B}_\bullet = \mathsf{S}_1(\mathsf{K}_1\mathsf{S}_2)\mathsf{K}_2$ is the root term.

(b) From Example 3.13 we see that $\square = \downharpoonleft\downharpoonleft\mathsf{a} \circ \mathsf{a}$ where $\mathsf{a} = \mathsf{0} \circ \downharpoonleft\mathsf{0}$ is a suitable computation.

(c) We have

$$\nabla \cdot \square = (\nabla \cdot \downharpoonleft\downharpoonleft\mathsf{a}) \cdot \mathsf{a} = ((BZ \cdot \mathsf{a})YX) \cdot \mathsf{a}$$

so our first job is to calculate the inner action. We have

$$\begin{aligned} BZ \cdot \mathsf{a} &= ((S_1(K_1S_2)K_2Z) \cdot \mathsf{0}) \cdot \downharpoonleft\mathsf{0} \\ &= ((K_1S_2Z)(K_2Z)) \cdot \downharpoonleft\mathsf{0} \\ &= ((K_1S_2Z) \cdot \mathsf{0})(K_2Z) \qquad = S_2(K_2Z) \end{aligned}$$

and hence

$$\begin{array}{rcll} \nabla\cdot\square & = & (S_2(K_2Z)YX)\cdot \mathsf{a} & \\ & = & ((S_2(K_2Z)YX)\cdot \mathsf{0})\cdot \downharpoonleft\mathsf{0} & \\ & = & ((K_2ZX)(YX))\cdot \downharpoonleft\mathsf{0} & \\ & = & ((K_2ZX)\cdot \mathsf{0})(YX) & = \ Z(YX) \end{array}$$

to produce the expected final result. □

3.16 (a) The derivations are given in Solution 3.2(β, γ, δ). Thus

$$\begin{array}{lll} B = S_1(K_1S_2)K_2 & C = S_4PQ & D = S_5RK_6 \\ P = B_1B_2S_3 & Q = K_3K_4 & R = K_5(S_6B_3) \end{array}$$

where $K_1, \dots, K_6, S_1, \dots, S_6$ are Axioms and B_1, B_2, B_3 are compounds built from Axioms $K_{11}, K_{12}, K_{21}, \dots, S_{11}, S_{12}, \dots, S_{32}$. As in Example 3.16 we find that

$$S_5 = S_{22} \quad K_5 = K_{22} \quad S_6 = SK_3 \quad B_3 = B$$

are the important mediating derivations.

(b) The computations can be extracted from Solution 3.11. With

$$\mathsf{a} = \mathsf{0}\circ\downharpoonleft\mathsf{0} \quad \mathsf{b} = \downharpoonleft\downharpoonleft\mathsf{a}\circ\mathsf{a} \quad \mathsf{p} = \mathsf{b}\circ\mathsf{a}$$

we have

$$(\mathsf{0})\ \ \mathsf{C_\bullet B_\bullet} \rhd\!\!\!\rhd (\mathsf{P_\bullet B_\bullet})(\mathsf{Q_\bullet B_\bullet}) \qquad (\mathsf{p})\ \ \mathsf{P_\bullet B_\bullet} \rhd\!\!\!\rhd \mathsf{S_{22}(K_{22}(S_3B_\bullet))} \qquad (\mathsf{0})\ \ \mathsf{Q_\bullet B_\bullet} \rhd\!\!\!\rhd \mathsf{K_4}$$

and hence $\square = \mathsf{0}\circ\downharpoonleft\mathsf{p}\circ\lfloor\mathsf{0}$ will do. (There are many other ways to organize the reduction $\mathsf{C_\bullet B_\bullet} \rhd\!\!\!\rhd \mathsf{D_\bullet}$.)

(c) We have

$$\begin{array}{rcl} \nabla\cdot\square & = & (S_4PQB\cdot\mathsf{0})\cdot(\downharpoonleft\mathsf{p}\circ\lfloor\mathsf{0}) \\ & = & (PB)(QB)\cdot(\downharpoonleft\mathsf{p}\circ\lfloor\mathsf{0}) \\ & = & (PB\cdot\mathsf{p})(QB\cdot\mathsf{0}) \\ & = & (PB\cdot\mathsf{p})K_4 \end{array} \quad \text{and} \quad \begin{array}{rcl} PB\cdot\mathsf{p} & = & (B_1B_2S_3B\cdot\mathsf{b})\cdot\mathsf{a} \\ & = & B_2(S_3B)\cdot\mathsf{a} \\ & = & S_{22}(K_{22}(S_3B)) \end{array}$$

to give

$$\nabla\cdot\square = S_{22}(K_{22}(S_3B))K_4 = D$$

as required. □

3.17 From Solutions 3.6(a) and 3.2(γ) we have a derivation

$$(\nabla) \quad \vdash \mathsf{C_\bullet S_\bullet I_\bullet} : \omega$$

where

$$\nabla = CSI \quad C = S_4PQ \quad P = B_1B_2S_3 \quad Q = K_3K = 4$$

where I, S, K_3, K_4, S_3, S_4 are Axioms (using constants $\mathsf{I_\bullet}, \mathsf{S_\bullet}, \mathsf{K_3}, \dots, \mathsf{S_4}$) and B_1, B_2 are compound derivations.

As in Solution 3.11 let $\mathsf{a} = 0 \circ \downarrow 0, \mathsf{b} = \downarrow\downarrow \mathsf{a} \circ \mathsf{a}, \mathsf{c} = 0 \circ \downarrow \mathsf{b} \circ \lfloor 0$ and $\square = \downarrow \mathsf{c} \circ \mathsf{b}$ to produce a computation

$$(\square) \quad \mathsf{C_\bullet S_\bullet I_\bullet} \rhd\!\!\!\rhd \mathsf{W_\bullet} \quad \text{where} \quad \mathsf{W_\bullet} = (\mathsf{S_3 S_\bullet})(\mathsf{K_4 I_\bullet})$$

using the combinators from I, S, K_4 and S_3.

Tracking this reduction we have a derivation

$$(W) \quad \vdash \mathsf{W_\bullet} : \omega$$

where $W = \nabla \cdot \square$. This certainly has the same shape and predicate as the derivation of 3.2(ω). We must check that $\mathsf{W_\bullet} = \mathsf{W}(\theta, \phi)$.

We have

$$\begin{array}{lll} \mathsf{C_\bullet} = \mathsf{C}(\mu, \iota, \lambda) & \mathsf{S_\bullet} = \mathsf{S}(\theta, \theta, \phi) & \mathsf{I_\bullet} = \mathsf{I}(\theta) \\ \lambda = \theta \to \phi & \mu = \theta \to \mu & \iota = \theta \to \theta \end{array}$$

and we must pin down $\mathsf{S_3}$ and $\mathsf{K_4}$. Comparing this $\mathsf{C_\bullet}$ with the version in Solution 3.2(γ) we have

$$\mathsf{K_4} = \mathsf{K}(\mu, \iota) \qquad \mathsf{S_3}(\mu, \iota, \lambda)$$

where ι, λ, μ from above are exactly as in 3.2(ω). Thus the comparison between here and 3.2(ω) is

$$\mathsf{S_3} = \mathsf{S_1} \quad \mathsf{S_\bullet} = \mathsf{S_2} \quad \mathsf{K_4} = \mathsf{K_\bullet} \quad \mathsf{I_\bullet} = \mathsf{I_\bullet}$$

to give

$$\mathsf{W_\bullet} = (\mathsf{S_3 S_\bullet})(\mathsf{K_4 I_\bullet}) = (\mathsf{S_1 S_2})(\mathsf{K_\bullet I_\bullet})$$

as required. □

3.18 We gather together, in a uniform notation, the relevant material from Solutions 1.8(μ, ν, τ) and 3.2(μ, ν, τ). (In fact, most of the uniformity is already in those Solutions.) For the given types $\theta, \psi, \phi, \rho, \zeta$ let

$$\eta = \theta \to \psi \to \phi \quad \xi = (\theta \to \psi) \to (\theta \to \phi) \quad \chi = \rho \to \eta \quad \pi = \rho \to \xi$$

so that

$$\mu = (\zeta \to \chi) \to (\zeta \to \pi) \qquad \nu = \chi \to \pi$$

and

$$\tau = \eta \to (\zeta \to \theta) \to (\zeta \to \psi) \to (\zeta \to \phi)$$

are the three target types.

(μ) From 3.2(μ) with

$$\mathsf{B_1} = \mathsf{B}(\zeta, \chi, \pi) \qquad \mathsf{B_2} = \mathsf{B}(\rho, \eta, \xi) \qquad \mathsf{S_\bullet} = \mathsf{S}(\theta, \psi, \phi)$$

we know that

$$\mathsf{M_\bullet} = \mathsf{M}(\theta, \psi, \phi, \rho, \zeta) = \mathsf{B_1}(\mathsf{B_2 S_\bullet})$$

inhabits μ. However, this will reduce quite a lot. From Example 3.13 we have $\mathsf{B_\bullet} = \mathsf{S_1}(\mathsf{K_1 S_*})\mathsf{K_*}$ and we know that $\mathsf{B_\bullet} t \rhd\!\!\!\rhd \mathsf{S_*}(\mathsf{K_*} t)$ for each term t. Thus

$$\mathsf{M_\bullet} \rhd\!\!\!\rhd \mathsf{S_3}(\mathsf{K_2 N_\bullet}) \quad \text{where} \quad \mathsf{N_\bullet} = \mathsf{S_1}(\mathsf{K_\bullet S_\bullet})$$

where $\mathsf{S}_\bullet$ is given and we need to determine the types of $\mathsf{K}_\bullet, \mathsf{S}_1, \mathsf{S}_2, \mathsf{S}_3$. These can be found by tracking the constructions of B_1 and B_2. Doing this gives

$$\begin{array}{lll} \mathsf{S}_\bullet : \sigma & \sigma = \eta \to \xi & = (\theta \to \psi \to \phi) \to (\theta \to \psi) \to (\theta \to \phi) \\ \mathsf{K}_\bullet : \kappa & \kappa = \sigma \to \rho \to \sigma & \\ \mathsf{S}_1 : \sigma_1 & \sigma_1 = (\rho \to \sigma) \to \nu & = (\rho \to \eta \to \xi) \to (\rho \to \eta) \to (\rho \to \xi) \\ \mathsf{K}_2 : \kappa_2 & \kappa_2 = \nu \to \zeta \to \nu & \\ \mathsf{S}_3 : \sigma_3 & \sigma_3 = (\zeta \to \nu) \to \mu & = (\zeta \to \chi \to \pi) \to (\zeta \to \chi) \to (\zeta \to \pi) \end{array}$$

to produce

$$\dfrac{\vdash \mathsf{S}_3 : \sigma_3 \qquad \dfrac{\vdash \mathsf{K}_2 : \kappa_2 \qquad \dfrac{\vdash \mathsf{S}_1 : (\rho \to \sigma) \to \nu \qquad \dfrac{\vdash \mathsf{K}_\bullet : \kappa \qquad \vdash \mathsf{S}_\bullet : \sigma}{\vdash \mathsf{K}_\bullet \mathsf{S}_\bullet : \rho \to \sigma}}{\vdash \mathsf{N}_\bullet : \nu}}{\vdash \mathsf{K}_2\mathsf{N}_\bullet : \rho \to \nu}}{\vdash \mathsf{S}_3(\mathsf{K}_2\mathsf{N}_\bullet) : \mu}$$

and so show that $\mathsf{S}_3(\mathsf{K}_2\mathsf{N}_\bullet)$ is a normal inhabitant of μ.

(ν) The term $\mathsf{N}(\theta, \psi, \phi, \rho)$ of 3.2(ν) is already normal. Notice that this is the term $\mathsf{N}_\bullet$ used in part (μ).

(τ) From Solution 3.2(τ) we have $\mathsf{T}_\bullet = \mathsf{M}_\bullet \mathsf{B}_\bullet$ for suitably typed versions of M and B. Using part (μ) we see that a suitably typed version of $\mathsf{S}_3(\mathsf{K}_2\mathsf{N}_\bullet)\mathsf{B}_\bullet$ inhabits τ. This is not the only normal inhabitant of τ. It can be shown that a typed version of $\mathsf{S}(\mathsf{KR})\mathsf{K}$ where $\mathsf{R} = \mathsf{S}(\mathsf{KN})\mathsf{S}$ inhabits τ. □

3.19 For each $1 \le i \le 5$ we have a derivation and a computation

$$(\nabla(i)) \quad \vdash \mathsf{W}(i) : \omega(i) \qquad (\square(i)) \quad \mathsf{W}(i) \rhd\!\!\!\rhd \mathsf{W}'(i)$$

from Solutions 3.3 and 3.12. The computation

$$(\nabla'(i)) \quad \vdash \mathsf{W}'(i) : \omega(i)$$

can be calculated either from the action $\nabla'(i) = \nabla \cdot \square(i)$ (which merely tracks the reduction $\mathsf{W}(i) \rhd\!\!\!\rhd \mathsf{W}'(i)$), or by using the Generation Algorithm 3.7 to reinstate the types of $\mathsf{W}'(i)$. In practice a mixture of both techniques is the most efficient.

Using the abbreviations $\eta, \xi, \chi, \ldots$ of Solution 1.9 with a few extra ones, we obtain the following results.

(i) Let

$$\eta = \theta \to \psi \qquad \nu = \eta \to \psi = \theta^\psi \qquad \alpha = \eta \to \theta \qquad \epsilon = \alpha \to \nu \qquad \delta = \theta \to \epsilon$$

so that $\omega(1) = \theta \to \nu$. With the combinators

$$\begin{array}{lll} \mathsf{I} : \iota & \iota = \eta \to \eta & \\ \mathsf{K}_1 : \kappa_1 & \kappa_1 = \epsilon \to \delta & = \epsilon \to \theta \to \delta \\ \mathsf{K}_2 : \kappa_2 & \kappa_2 = \theta \to \eta \to \theta & \\ \mathsf{S}_1 : \sigma_1 & \sigma_1 = \delta \to \kappa_2 \to \omega(1) & = (\theta \to \alpha \to \nu) \to (\theta \to \alpha) \to (\eta \to \psi) \\ \mathsf{S}_2 : \sigma_2 & \sigma_2 = \iota \to \epsilon & = (\eta \to \theta \to \psi) \to (\eta \to \theta) \to (\eta \to \psi) \end{array}$$

the compound $\mathsf{S}_1(\mathsf{K}_1(\mathsf{S}_2\mathsf{I}))\mathsf{K}_2$ inhabits $\omega(1)$.

(ii) Let

$$\nu = \theta^{\psi} \quad \mu = \nu^{\phi} \quad \omega(1) = \theta \to \nu \quad \omega = \theta \to \mu \quad \alpha = \theta \to \omega$$

so that $\omega(2) = \theta \to \mu$, $\omega(1)$ is as in part (i), and ω is a variant of $\omega(1)$. With the constants

$$\begin{array}{lll} \mathsf{K}_1 : \kappa_1 & \kappa_1 = \omega \to \alpha & = \omega \to \theta \to \omega \\ \mathsf{S}_1 : \sigma_1 & \sigma_1 = \alpha \to \omega(1) \to \omega(2) & = (\theta \to \nu \to \mu) \to (\theta \to \nu) \to (\theta \to \mu) \end{array}$$

and normal compounds $\mathsf{W} : \omega$ and $\mathsf{W}' : \omega(1)$ as given by part (i), we see that the compound $\mathsf{S}_1(\mathsf{K}_1\mathsf{W})\mathsf{W}'$ inhabits $\omega(2)$.

(iii) With the combinators

$$\begin{array}{lll} \mathsf{D}_1 : \delta_1 & \delta_1 = \rho \to \zeta & = (\xi \to \chi) \to (\chi \to \phi) \to (\xi \to \phi) \\ \mathsf{D}_2 : \delta_2 & \delta_2 = \eta \to \rho & = (\theta \to \psi) \to (\psi \to \phi) \to (\theta \to \phi) \\ \mathsf{K}_1 : \kappa_1 & \kappa_2 = \delta_1 \to \eta \to \delta_1 & \\ \mathsf{S}_1 : \sigma_1 & \sigma_1 = (\eta \to \delta_1) \to \delta_2 \to \omega(3) & = (\eta \to \rho \to \zeta) \to (\eta \to \rho) \to (\eta \to \zeta) \end{array}$$

the compound $\mathsf{S}_1(\mathsf{K}_1\mathsf{D}_1)\mathsf{D}_2$ inhabits $\omega(3)$.

(iv) This is the most interesting of the five parts. We have

$$\mathsf{W}(4) \rhd\!\!\!\!\rhd (\mathsf{S}_1\mathsf{B}_1)(\mathsf{K}_1\mathsf{R}_1) \quad \text{where} \quad \mathsf{R}_1 = (\mathsf{S}_2\mathsf{B}_2)(\mathsf{K}_2\mathsf{R}_2) \quad \text{where} \quad R_2 = \mathsf{W}'(1)$$

for suitably typed versions of the indicated constants and combinators. It is useful to analyse the typing requirements of a compound $(\mathsf{SB})(\mathsf{KR})$.

For arbitrary χ, ϵ, ϕ let $\rho = \epsilon \to \chi$ and suppose we have an inhabitant $\mathsf{R} : \rho$. Let

$$\alpha = \chi \to \phi \quad \tau = \epsilon \to \phi \quad \pi = \alpha \to \rho \quad \zeta = \alpha \to \tau$$

and

$$\begin{array}{l} \kappa = \rho \to \pi \qquad = \rho \to \alpha \to \rho \\ \beta = \alpha \to \rho \to \tau = (\chi \to \phi) \to (\epsilon \to \chi) \to (\epsilon \to \phi) \\ \sigma = \beta \to \pi \to \zeta = (\alpha \to \rho \to \tau) \to (\alpha \to \rho) \to (\alpha \to \tau) \end{array}$$

so there are inhabitants $\mathsf{K} : \kappa$, $\mathsf{B} : \beta$, $\mathsf{S} : \sigma$ to give

$$(\mathsf{SB})(\mathsf{KR}) : \zeta \qquad \zeta = (\chi \to \phi) \to (\epsilon \to \phi)$$

for arbitrary χ, ϵ, ϕ. In short, if

$$\vdash \mathsf{R} : \epsilon \to \chi \quad \text{then} \quad \vdash (\mathsf{SB})(\mathsf{KR}) : (\chi \to \phi) \to (\epsilon \to \phi)$$

for suitable $\mathsf{S}, \mathsf{B}, \mathsf{K}$. (It can be shown that ζ is the most general resulting type.)

Now consider the abbreviations used in Solution 1.9. We have

$$\omega(4) = \lambda \to \mu = (\chi \to \phi) \to (\epsilon \to \phi)$$

where $\epsilon = \nu \to \phi$. It suffices to produce

$$\vdash \mathsf{R}_1 : \epsilon \to \chi$$

and then proceed as above. But

$$\epsilon \to \chi = (\nu \to \phi) \to (\theta \to \phi)$$

so it suffices to produce

$$\vdash \mathsf{R}_2 : \theta \to \nu$$

and then use the above recipe again. Since $\omega(1) = \theta \to \nu$ we may take R_2 to be any normal inhabitant of $\omega(1)$.

(v) Consider

$$\epsilon = \theta \to \omega(4) = \theta \to \lambda \to \mu \qquad \omega = \theta \to \lambda = \theta \to \theta^{\phi}$$

so that ω is a variant of $\omega(1)$. With the constants

$$\begin{array}{lll} \mathsf{K}_1 : \kappa_1 & \kappa_1 = \omega(4) \to \epsilon & = \omega(4) \to \theta \to \omega(4) \\ \mathsf{S}_1 : \sigma_1 & \sigma_1 = \epsilon \to \omega \to \omega(5) = (\theta \to \lambda \to \mu) \to (\theta \to \lambda) \to (\theta \to \mu) & \end{array}$$

and normal compounds $\mathsf{W} : \omega$ and $\mathsf{W}'(4) : \omega(4)$ as given by parts (i,iv), we see that the compound $\mathsf{S}_1(\mathsf{K}_1\mathsf{W}'(4))\mathsf{W}$ inhabits $\omega(5)$. □

3.20 In each case we are given a pair $\square(l), \square(r)$ of computations which organize a common reduction $t^- \rhd\!\!\!\!\rhd t^+$. Both will act on a derivation ∇ with t^- as root subject. In each case we find that $\nabla \cdot \square(l) = \nabla \cdot \square(r)$ so that, in some sense, $\square(l)$ and $\square(r)$ are interchangeable.

(i) We have

$$\begin{array}{l} \nabla \cdot (\mathsf{l} \circ (\mathsf{m} \circ \mathsf{r})) = (\nabla \cdot \mathsf{l}) \cdot (\mathsf{m} \circ \mathsf{r}) = ((\nabla \cdot \mathsf{l}) \cdot \mathsf{m}) \cdot \mathsf{r} \\ \nabla \cdot ((\mathsf{l} \circ \mathsf{m}) \circ \mathsf{r}) = (\nabla \cdot (\mathsf{l} \circ \mathsf{m})) \cdot \mathsf{r} = ((\nabla \cdot \mathsf{l}) \cdot \mathsf{m}) \cdot \mathsf{r} \end{array}$$

to give the equivalence.

(ii) For both computations we must have $\nabla = QP$ for some derivations Q, P. Then

$$\begin{array}{ll} \nabla \cdot \rfloor(\mathsf{l} \circ \mathsf{r}) & = (Q \cdot (\mathsf{l} \circ \mathsf{r}))P = ((Q \cdot \mathsf{l}) \cdot \mathsf{r})P \\ \nabla \cdot (\rfloor\mathsf{l} \circ \rfloor\mathsf{r}) & = (\nabla \cdot \rfloor\mathsf{l}) \cdot \rfloor\mathsf{r} \quad = ((Q \cdot \mathsf{l}) \cdot \mathsf{r})P \end{array}$$

to give the equivalence.

(iii) This is similar to (ii).

(iv) We must have $\nabla = LR$ and then

$$\begin{array}{l} \nabla \cdot (\rfloor\mathsf{l} \circ \lfloor\mathsf{r}) = (\nabla \cdot \rfloor\mathsf{l}) \cdot \lfloor\mathsf{r} = ((L \cdot \mathsf{l})R) \cdot \lfloor\mathsf{r} = (L \cdot \mathsf{l})(R \cdot \mathsf{r}) \\ \nabla \cdot (\lfloor\mathsf{r} \circ \rfloor\mathsf{l}) = (\nabla \cdot \lfloor\mathsf{r}) \cdot \rfloor\mathsf{l} = (L(R \cdot \lfloor) \cdot \rfloor\mathsf{l} = (L \cdot \mathsf{l})(R \cdot \mathsf{r}) \end{array}$$

to give the equivalence.

These examples indicate there is an **algebra of actions** with a notion of 'eventually equivalent' computations. We do not develop this here. □

D
The Typed λ-calculus

D.1 Introduction

4.1 This is easier if we use the version of t given in Solution 2.13.

$$(\lambda w\,.\,((\lambda v\,.\,((\lambda u\,.\,u)v))w))x \qquad (\lambda w : \tau\,.\,((\lambda v : \sigma\,.\,((\lambda u : \rho\,.\,u)v))w))x$$

This version is shown to the left and then a typed version is shown to the right. We can replace u, v, w by x. For this to be well formed we must have u, v of the same type, so that $(\lambda u : \rho\,.\,u)v$ makes sense. Thus we need $\rho = \sigma$, and then this subterm has type σ. The same argument shows that we need $\sigma = \tau$. □

D.2 Derivation

4.2 (β) For the given θ, ψ, ϕ let

$$\rho = \theta \to \phi \qquad \sigma = \theta \to \psi \qquad \tau = \psi \to \phi$$

$$\begin{aligned} \Gamma &= z : \tau, y : \sigma, x : \theta \\ B &= (\Gamma[z](\Gamma[y]\Gamma[x]))\uparrow\uparrow\uparrow \\ \mathsf{B} &= \lambda z : \tau, y : \sigma, x : \theta\,.\,z(yx) \end{aligned}$$

so that $\beta = \tau \to \sigma \to \rho$ is the target type. The derivation B given right shows that the term B inhabits β in the empty context.

$$\dfrac{\dfrac{\Gamma \vdash z : \tau \qquad \dfrac{\Gamma \vdash y : \sigma \quad \Gamma \vdash x : \theta}{\Gamma \vdash yx : \psi}}{\Gamma \vdash z(yx) : \phi}}{\dfrac{\bullet}{\dfrac{\bullet}{\vdash \mathsf{B} : \beta}}}$$

(γ) For the given ρ, σ, τ let

$$\beta = \tau \to \sigma \to \rho \qquad \delta = \sigma \to \tau \to \rho$$

$$\begin{aligned} \Gamma &= w : \beta, y : \sigma, z : \tau \\ C &= (\Gamma[w]\Gamma[z]\Gamma[x])\uparrow\uparrow\uparrow \\ \mathsf{C} &= \lambda w : \beta, y : \sigma, z ; \tau\,.\,wzy \end{aligned}$$

so that $\gamma = \beta \to \delta$ is the target type. The derivation C given right shows that the term C inhabits γ in the empty context.

$$\dfrac{\dfrac{\dfrac{\Gamma \vdash w : \beta \quad \Gamma \vdash z : \tau}{\Gamma \vdash yx : \psi} \qquad \Gamma \vdash y : \sigma}{\Gamma \vdash wzy : \rho}}{\dfrac{\bullet}{\dfrac{\bullet}{\vdash \mathsf{C} : \gamma}}}$$

(δ) For the given θ, ψ, ϕ let

$$\rho = \theta \to \phi \quad \sigma = \theta \to \psi \quad \tau = \psi \to \phi$$

$$\begin{aligned} \Gamma &= y : \sigma, z : \tau, x : \theta \\ D &= (\Gamma[z](\Gamma[y]\Gamma[x]))\uparrow\uparrow\uparrow \\ \mathsf{D} &= \lambda y : \sigma, z : \tau, x; \theta \,.\, z(yx) \end{aligned}$$

so that $\delta = \sigma \to \tau \to \rho$ is the target type. The derivation D given right shows that the term D inhabits δ in the empty context.

$$\dfrac{\dfrac{\dfrac{\Gamma \vdash z : \tau \quad \dfrac{\Gamma \vdash y : \sigma \quad \Gamma \vdash x : \theta}{\Gamma \vdash yx : \psi}}{\Gamma \vdash z(yx) : \phi}}{\bullet \atop {- \atop \bullet}}}{\vdash \mathsf{D} : \delta}$$

□

4.3 In each case we carry over the abbreviates from Solution 1.8 except, for obvious reasons, we replace each use of 'λ' as a type by 'ϑ'. In each case the relationship between the two derivations is dealt with in a later exercise.

(ϵ) Let

$$\Gamma = v : \eta, u : \mu, x : \theta, y : \psi$$

where the identifiers are distinct. From Solution 1.8(ϵ) we have two derivations

$$L = B_1 B_2 C \qquad\qquad R = ((\Gamma[v]\Gamma[y])(\Gamma[u]\Gamma[x]))\uparrow\uparrow\uparrow\uparrow$$

$$\dfrac{\dfrac{\vdots \atop \vdash \mathsf{B}_1 : \beta_1 \quad \vdots \atop \vdash \mathsf{B}_2 : \beta_2}{\vdash \mathsf{B}_1\mathsf{B}_2 : \gamma \to \epsilon} \quad \vdots \atop \vdash \mathsf{C} : \gamma}{\vdash \mathsf{B}_1\mathsf{B}_2\mathsf{C} : \epsilon} \qquad \dfrac{\dfrac{\Gamma \vdash v : \eta \quad \Gamma \vdash y : \psi}{\Gamma \vdash vy : \zeta \to \phi} \quad \dfrac{\Gamma \vdash u : \mu \quad \Gamma \vdash x : \theta}{\Gamma \vdash ux : \zeta}}{\Gamma \vdash (vy)(ux) : \phi}$$

$$\mathsf{B}_1\mathsf{B}_2\mathsf{C} \qquad\qquad \mathsf{E} = \lambda v : \eta, u : \mu, x : \theta, y : \psi \,.\, (vy)(ux)$$

where $\mathsf{B}_1, \mathsf{B}_2, \mathsf{C}$ are suitable terms constructed using Solution 4.2. In each case the arboreal code is given above the tree and the inhabiting term is given below. In the right hand case only the body of the tree is given.

(ω) Let

$$\Gamma = u : \mu, x : \theta$$

where the two identifiers are distinct. From Solution 1.8(ω) we have two derivations

$$L = (S_1 S_2)(KI) \qquad\qquad R = (\Gamma[u]\Gamma[x]\Gamma[x])\uparrow\uparrow\uparrow$$

$$\dfrac{\dfrac{\vdots \atop \vdash \mathsf{S}_1 : \sigma_1 \quad \vdots \atop \vdash \mathsf{S}_2 : \sigma_2}{\vdash \mathsf{S}_1\mathsf{S}_2 : \nu \to \omega} \quad \dfrac{\vdots \atop \vdash \mathsf{K} : \kappa \quad \vdots \atop \vdash \mathsf{I} : \iota}{\vdash \mathsf{KI} : \nu}}{\vdash (\mathsf{S}_1\mathsf{S}_2)(\mathsf{KI}) : \omega} \qquad \dfrac{\dfrac{\Gamma \vdash u : \mu \quad \Gamma \vdash x : \theta}{\Gamma \vdash ux : \vartheta} \quad \Gamma \vdash x : \theta}{\Gamma \vdash uxx : \phi}$$

$$(\mathsf{S}_1\mathsf{S}_2)(\mathsf{KI}) \qquad\qquad \mathsf{W} = \lambda u : \mu, x : \theta \,.\, uxx$$

where $\mathsf{S}_1, \mathsf{S}_2, \mathsf{K}, \mathsf{I}$ are suitable terms as given by Example 4.4.

(μ) Let

$$\Gamma = f : \omega, w : \zeta, v : \rho, u : \vartheta, x : \theta$$

where the identifiers are distinct to give a legal context. From Solution 1.8(μ) we have two derivations

$$L = B_1(B_2C) \qquad R = ((\Gamma[f]\Gamma[w]\Gamma[v]\Gamma[x])(\Gamma[u]\Gamma[x]))\uparrow\uparrow\uparrow\uparrow\uparrow$$

$$\dfrac{\begin{array}{c}\vdots\\ \vdash \mathsf{B}_1 : \beta_1\end{array} \quad \dfrac{\begin{array}{c}\vdots\\ \vdash \mathsf{B}_2 : \beta_2\end{array} \quad \begin{array}{c}\vdots\\ \vdash \mathsf{S} : \sigma\end{array}}{\vdash \mathsf{B}_2\mathsf{S} : \epsilon}}{\vdash \mathsf{B}_1(\mathsf{B}_2\mathsf{S}) : \mu} \qquad \dfrac{\dfrac{\dfrac{\dfrac{\Gamma \vdash f : \omega \quad \Gamma \vdash w : \zeta}{\Gamma \vdash fw : \chi} \quad \Gamma \vdash v : \rho}{\Gamma \vdash fwv : \eta} \quad \Gamma \vdash x : \theta}{\Gamma \vdash fwvx : \alpha} \quad \dfrac{\Gamma \vdash u : \vartheta \quad \Gamma \vdash x : \theta}{\Gamma \vdash ux : \psi}}{\Gamma \vdash (fwvx)(ux) : \phi}$$

$$\mathsf{B}_1(\mathsf{B}_2\mathsf{S}) \qquad \mathsf{M} = \lambda f : \omega, w : \zeta, v : \rho, u : \vartheta, x : \theta\,.\,(fwvx)(ux)$$

where $\mathsf{B}_1, \mathsf{B}_2, \mathsf{S}$ are suitable terms as given by earlier exercises and examples.

(ν) Let

$$\Gamma = w : \chi, v : \rho, u : \vartheta, x : \theta$$

where the identifiers are distinct. From Solution 1.8(ν) we have two derivations

$$L = S_1(KS_1) \qquad R = ((\Gamma[w]\Gamma[v]\Gamma[x])(\Gamma[u]\Gamma[x]))\uparrow\uparrow\uparrow\uparrow$$

$$\dfrac{\begin{array}{c}\vdots\\ \vdash \mathsf{S}_1 : \sigma_1\end{array} \quad \dfrac{\begin{array}{c}\vdots\\ \vdash \mathsf{K} : \kappa\end{array} \quad \begin{array}{c}\vdots\\ \vdash \mathsf{S}_2 : \sigma_2\end{array}}{\vdash \mathsf{KS}_2 : \rho \to \sigma_2}}{\vdash \mathsf{S}_1(\mathsf{KS}_2) : \nu} \qquad \dfrac{\dfrac{\dfrac{\Gamma \vdash w : \chi \; \Gamma \vdash v : \rho}{\Gamma \vdash wv : \eta} \quad \Gamma \vdash x : \theta}{\Gamma \vdash wvx : \psi \to \phi} \quad \dfrac{\Gamma \vdash u : \vartheta \; \Gamma \vdash x : \theta}{\Gamma \vdash ux : \psi}}{\Gamma \vdash (wvx)(ux) : \phi}$$

$$\mathsf{S}_1(\mathsf{KS}_2) \qquad \mathsf{N} = \lambda w : \chi, v : \rho, u : \vartheta, x : \theta\,.\,(wvx)(ux)$$

where $\mathsf{S}_1, \mathsf{S}_2, \mathsf{K}$ are suitable terms given by Example 1.8.

(τ) Let

$$\Gamma = w : \eta, z : \omega, y : \vartheta, x : \zeta$$

where the identifiers are distinct. From Solution 1.8(τ) we have two derivations

$$L = MB \qquad R = (\Gamma[w](\Gamma[z]\Gamma[x])(\Gamma[y]\Gamma[x]))\uparrow\uparrow\uparrow\uparrow$$

$$\dfrac{\begin{array}{c}\vdots\\ \vdash \mathsf{M} : \mu\end{array} \quad \begin{array}{c}\vdots\\ \vdash \mathsf{B} : \beta\end{array}}{\vdash \mathsf{MB} : \tau} \qquad \dfrac{\dfrac{\Gamma \vdash w : \eta \quad \dfrac{\Gamma \vdash z : \omega \quad \Gamma \vdash x : \zeta}{\Gamma \vdash zx : \theta}}{\Gamma \vdash w(zx) : \psi \to \phi} \quad \dfrac{\Gamma \vdash y : \vartheta \quad \Gamma \vdash x : \zeta}{\Gamma \vdash yx : \psi}}{\Gamma \vdash w(zx)(yx) : \phi}$$

$$\mathsf{MB} \qquad \mathsf{T} = \lambda w : \eta, z : \omega, y : \vartheta, x : \zeta\,.\,w(zx)(yx)$$

where M, B are suitable terms from earlier. □

4.4 As in Section 2.4 we set

$$\partial x = \{x\} \quad \partial(qp) = \partial q \cup \partial p \quad \partial(\lambda y : \sigma\,.\,r) = \partial r - \{y\}$$

to generate the support of a λ-term t.

We show

For each derivation $(\nabla)\ \Gamma \vdash t : \tau$ the context Γ is legal and $\partial t \subseteq \partial\Gamma$

by induction over ∇. Only the induction step across a use of I is not immediate.

Consider $\nabla = R\uparrow$ where

$$(R)\quad \Gamma, y : \sigma \vdash r : \rho$$

with $t = \lambda y : \sigma\,.\,r$ and $\tau = \sigma \to \rho$. By the induction hypothesis the extended context $\Gamma, y : \sigma$ is legal and

$$\partial r \subseteq \partial(\Gamma, y : \sigma) = \partial\Gamma \cup \{y\}$$

holds. But then Γ is legal and hence $\partial t \subseteq (\partial\Gamma \cup \{y\}) - \{y\} \subseteq \Gamma$ as required. □

4.5 Clearly we set

$$\Gamma[x]_\lambda = \Gamma[x] \qquad (QP)_\lambda = Q_\lambda P_\lambda \qquad \Gamma[\mathsf{Z}_\bullet] = \text{explicit derivation}$$

using the explicit derivations of Example 4.4. However, there is an annoying problem.

Consider an Axiom

$$(\Gamma[\mathsf{S}_\bullet])\quad \Gamma \vdash \mathsf{S}_\bullet : \sigma$$

where $\mathsf{S}_\bullet = \mathsf{S}(\theta, \psi, \phi)$ and σ is the associated type. As in Example 4.4 let

$$\Sigma = z : \chi, y : \psi, x : \theta$$

and consider the lengthened context (Γ, Σ). Using Projections from this context we produce a $\boldsymbol{\lambda}$-derivation

$$(\Gamma, \Sigma) \vdash (zx)(yx) : \phi$$

and then three uses of I give $\Gamma \vdash \mathsf{S}_{\bullet\lambda} : \sigma$ as required.

Not quite! If one of the identifiers x, y, z is already declared in Γ, then (Γ, σ) is not legal. To get round this we must be prepared to rename the bound identifiers of $\mathsf{S}_\bullet$. In other words the translation $(\cdot)_\lambda$ of derivations is context sensitive. □

D.3 Annotation and deletion

4.6 (i) Any derivation of $\vdash \overline{S}(\overline{S}\,\overline{0}) : \tau$ must end as shown right. Here $\sigma = \mu \to \tau$ and $\sigma = \nu \to \mu$. Thus $\tau = \mu = \nu$ to give $\sigma = \nu'$, and it remains to determine ν and θ, ψ, ϕ.

$$\dfrac{\begin{matrix}\vdots\\ \vdash \overline{S} : \sigma\end{matrix} \qquad \dfrac{\begin{matrix}\vdots\\ \vdash \overline{S} : \sigma\end{matrix} \quad \begin{matrix}\vdots\\ \vdash \overline{0} : \nu\end{matrix}}{\vdash \overline{S}\overline{0} : \mu}\,(E)}{\vdash \overline{S}(\overline{S}\,\overline{0}) : \tau}\,(E)$$

Let

$$\Gamma = u : \phi, y : \psi, x : \theta \qquad \Sigma = y : \psi, x : \theta$$

$$\dfrac{\dfrac{\Sigma \vdash x : \theta}{y : \psi \vdash (\lambda x : \theta . x) : \theta'}}{\vdash \overline{0} : \nu}$$

where it is assumed that u, y, x are distinct (so that Γ and Σ are legal). The derivation $\vdash \overline{0} : \nu$ must be as shown to the right, so that $\nu = \psi \to \theta'$.

The derivation $\vdash \overline{S} : \sigma$ must arise as shown right by three uses of I. This subderivation requires

$$\dfrac{\Gamma \vdash y : \psi \qquad \dfrac{\dfrac{\Gamma \vdash u : \phi \quad \Gamma \vdash y : \psi}{\Gamma \vdash uy : \gamma} \qquad \Gamma \vdash x : \theta}{\vdash uyx : \beta}}{\vdash y(uyx) : \alpha}$$

$$\phi = \psi \to \gamma \quad \gamma = \theta \to \beta \quad \psi = \beta \to \alpha$$

and then

$$\sigma = \phi \to \psi \to \theta \to \alpha$$

is required by the final uses of I.

Since $\sigma = \nu'$ we have $\nu = \phi = \psi \to \theta \to \alpha$, and then $\nu = \psi \to \theta'$ gives $\theta = \alpha$. But now

$$\begin{array}{llll} \phi = \psi \to \gamma & \phi = \psi \to \theta' & \text{requires} & \gamma = \theta' \\ \gamma = \theta \to \beta & \gamma = \theta' & \text{requires} & \beta = \theta \\ \psi = \beta \to \alpha & & & \psi = \theta' \end{array}$$

and hence

$$\psi = \theta' \quad \phi = \theta'' \quad \sigma = \theta''' \quad \tau = \theta''$$

where θ is arbitrary.

(ii) For $t = \overline{A}\,\overline{1}$ let

$$\Gamma = v : \phi, u : \phi, y : \psi, x : \theta \qquad \Sigma = y : \psi, x : \theta$$

where it is assumed that v, u, y, x are distinct (so that Γ and Σ are legal).

A derivation of $\vdash \overline{1} : \alpha$ must pass through an Elimination, as shown, so we require

$$\dfrac{\Sigma \vdash y : \psi \quad \Sigma \vdash x : \theta}{\Sigma \vdash yx : \beta}$$

$$\psi = \theta \to \beta \qquad \alpha = \psi \to \theta \to \beta$$

for some β.

A derivation $\vdash \overline{A} : \gamma$ must go through the compound shown, so we require

$$\dfrac{\dfrac{\Gamma \vdash u : \phi \quad \Gamma \vdash x : \theta}{\Gamma \vdash ux : \delta} \qquad \dfrac{\dfrac{\Gamma \vdash v : \phi \quad \Gamma \vdash y : \psi}{\Gamma \vdash vy : \delta} \qquad \Gamma \vdash x : \theta}{\Gamma \vdash vyx : \epsilon}}{\Gamma \vdash (ux)(vyx) : \xi}$$

$$\begin{array}{ll} \phi = \psi \to \delta & \\ \delta = \theta \to \epsilon & \delta = \epsilon \to \xi \\ \gamma = \ \phi \to \phi \to \psi \to \theta \to \xi & \end{array}$$

for some δ, ϵ, ξ. These two restrictions on δ give $\xi = \epsilon = \theta$, so that $\delta = \theta'$ and $\phi = \psi \to \theta'$.

An Elimination, as shown, needs $\gamma = \alpha \to \tau$ so that $\alpha = \phi$ and $\tau = \phi \to \psi' \to \theta'$ from the previous description of γ.

$$\frac{\vdash \overline{A} : \gamma \quad \vdash \overline{1} : \alpha}{\vdash \overline{A}\,\overline{1} : \tau}$$

But now $\phi = \psi \to \theta'$ and $\phi = \alpha = \psi \to \theta \to \beta$ so that $\beta = \theta$, and hence

$$\phi = \theta'' \quad \psi = \theta' \quad \tau = \theta'''$$

with θ arbitrary.

(iii) For $t = \overline{M}\,\overline{2}$ let

$$\Gamma = v : \phi, u : \phi, y : \psi, x : \theta \qquad \Sigma = y : \psi, x : \theta$$

where it is assumed that v, u, y, x are distinct (so that Γ and Σ are legal).

A derivation $\vdash \overline{2} : \alpha$ must pass through the Elimination shown, so we require β, γ where

$$\frac{\Sigma \vdash y : \psi \quad \dfrac{\Sigma \vdash y : \psi \quad \Sigma \vdash x : \theta}{\Sigma \vdash yx : \beta}}{\Sigma \vdash y^2x : \beta}$$

$$\psi = \theta \to \beta \quad \psi = \beta \to \gamma \quad \alpha = \psi \to \theta \to \gamma$$

hold. These give $\gamma = \beta = \theta$, $\psi = \theta'$, $\alpha = \theta''$.

A derivation of $\vdash \overline{M} : \delta$ must pass through the compound shown so we require

$$\frac{\dfrac{\Gamma \vdash u : \phi \quad \dfrac{\Gamma \vdash v : \phi \quad \Gamma \vdash y : \psi}{\Gamma \vdash vy : \epsilon}}{\Gamma \vdash u(vy) : \xi} \quad \Gamma \vdash x : \theta}{\Gamma \vdash u(vy)x : \chi}$$

$$\phi = \psi \to \epsilon \quad \phi = \epsilon \to \xi \quad \xi = \theta \to \chi$$

and hence $\xi = \epsilon = \psi = \theta'$ to give $\chi = \theta$ and $\phi = \theta''$.

This gives

$$\delta = \phi \to \phi \to \psi \to \theta \to \chi = \theta'' \to \theta'' \to \theta''$$

and hence $\tau = \theta'''$ is the housing type of $t = \overline{M}\,\overline{2}$. □

D.4 Substitution

4.7 The algorithm proceeds by recursion on ∇.

For the base case we have a Projection

$$(\nabla) \quad \Pi, x : \sigma, \Delta \vdash y : \tau$$

where $y \neq x$ and $y : \tau$ is declared in Π, Δ. Thus the Projection

$$(\nabla^-) \quad \Pi, \Delta \vdash y : \tau$$

is the required output.

There are two recursion steps where the root rule of ∇ is either I or E.

For the step across I we have $\nabla = R\!\uparrow$ where

$$(R) \quad \Pi, x : \sigma, \Delta, y : \xi \vdash r : \rho$$

with $t = \lambda y : \xi . r$ and $\tau = \xi \to \rho$. Note that $y \neq x$ (for the context of R must be legal). But $\partial r \subseteq \partial t \cup \{y\}$ so that $x \notin \partial r$, and hence we obtain

$$(R^-) \quad \Pi, \Delta, y : \xi \vdash r : \rho$$

by a recursion call on the algorithm. Thus $\nabla^- = R^-\uparrow$ is the required output.

The step across E is immediate. □

4.8 The algorithm proceeds by recursion on ∇. Only the recursion step across a use of I is not immediate.

For this case we have $\nabla = R\uparrow$ where

$$(R) \quad \Pi, x : \sigma, \Delta, y : \xi \vdash r : \rho$$

with $t = \lambda y : \xi . r$ and $\tau = \xi \to \rho$. The identifier u does not appear in R (otherwise it appears in ∇), so a recursion call on the algorithm gives

$$(R') \quad \Pi, u : \sigma, \Delta, y : \xi \vdash r' : \rho$$

where r' is obtained from r by replacing each occurrence of x by u. But then $t' = \lambda y : \xi . r'$ and we may set $\nabla' = R'\uparrow$ for the required output. □

4.9 A first attempt at this algorithm will probably fail. We need a more powerful algorithm.

Let $\Gamma \sqsubseteq \Sigma$ mean that the two contexts are legal and every declaration of Γ appears in Σ. We describe an algorithm which, when supplied with

$$(\nabla) \quad \Gamma \vdash t : \tau \qquad \Gamma \sqsubseteq \Sigma$$

where each $x \in \partial\Sigma - \partial\Gamma$ does not appear in ∇, will return a derivation

$$(\nabla^+) \quad \Sigma \vdash t : \tau$$

in the longer context. A particular case will give the required algorithm.

The algorithm proceeds by recursion on ∇. Only the recursion step across a use of I is not immediate.

For this case we have $\nabla = R\uparrow$ where

$$(R) \quad \Gamma, y : \xi \vdash r : \rho$$

with $t = \lambda y : \xi . r$ and $\tau = \xi \to \rho$. Let $\Gamma^+ = \Gamma, y : \xi$ and $\Sigma^+ = \Sigma, y : \xi$. We show that Σ^+ is legal, and hence $\Gamma^+ \sqsubseteq \Sigma^+$.

Suppose Σ^+ is not legal. Then, since Σ is legal, we see that y is already declared in Σ. We know that $y \notin \partial\Gamma$ (since Γ^+ is legal) so that $y \in \partial\Sigma - \partial\Gamma$. But such an identifier can not appear in ∇, which is a direct contradiction.

We have $\Gamma^+ \sqsubseteq \Sigma^+$ and

$$\partial\Sigma^+ - \partial\Gamma^+ = \partial\Sigma - \partial\Gamma$$

so a recursion call on the algorithm produces

$$(R^+) \quad \Sigma, y : \xi \vdash r : \rho$$

and hence $\nabla^+ = R^+\uparrow$ is the required output. □

4.10 We make use of the algorithm $(\cdot)^d$ described at the end of Section 3.3 and in Exercise 3.9. This acts on **C**-derivations

$$(\nabla) \quad \Gamma, x : \sigma \vdash r : \rho \quad \longmapsto \quad (\nabla^d) \quad \Gamma \vdash [x : \sigma]r : \sigma \to \rho$$

where $[x : \sigma]r$ is a suitably constructed combinator term (which is Γ-sensitive). We produce a translation

$$(\nabla) \quad \Gamma \vdash t : \tau \quad \longmapsto \quad (\nabla^\Gamma) \quad \Gamma \vdash t_\Gamma : \tau$$

from **C**-derivations to **λ**-derivations. As indicated this translation is context sensitive.

The algorithm proceeds by recursion over ∇. The routine clauses are

$$\Gamma[x]^\Gamma = \Gamma[x] \qquad x_\Gamma = x$$
$$(QP)^\Gamma = (Q^\Gamma)(P^\Gamma) \qquad (qp)_\Gamma = (q_\Gamma)(p_\Gamma)$$

but the step across a use of I is more complicated.

We have

$$\nabla = R\uparrow \quad t = \lambda y : \sigma . r \quad \tau = \sigma \to \rho \qquad (R) \quad \Gamma, y : \sigma \vdash r : \rho$$

where the controlling **λ**-derivation is shown on the right. For convenience let $\Sigma = \Gamma, y : \sigma$. By recursion we obtain a **C**-derivation

$$(R^\Sigma) \quad \Sigma \vdash r_\Sigma : \rho$$

and then $(\cdot)^d$ produces

$$(R^{\Sigma d}) \quad \Gamma \vdash [x : \sigma](r_\Sigma) : \tau$$

so we set

$$(R\uparrow)^\Gamma = R^{\Sigma d} \qquad (\lambda y : \sigma . r)_\Gamma = [y : \sigma](r_\Sigma)$$

for this clause. □

D.5 Computation

4.11 We carry over the abbreviations used in Exercise 4.2 and 4.3.

(ϵ) From Solutions 4.3(ϵ) and 4.2 we have the λ-combinators shown (where the identifiers of E have been used to choose the identifiers of $\mathsf{B}_1, \mathsf{B}_2, \mathsf{C}$). We must organize $\mathsf{B}_1\mathsf{B}_2\mathsf{C} \trianglerighteq \mathsf{E}$.

$$\begin{array}{ll} \mathsf{E} & = \lambda v : \eta, u : \mu, x : \theta, y : \psi . (vy)(ux) \\ \mathsf{B}_1 = & \lambda b : \beta_2, w : \gamma, v : \eta . b(wv) \\ \mathsf{B}_2 = & \lambda a : \xi, u : \mu, x : \theta . a(ux) \\ \mathsf{C} = & \lambda v : \eta, z : \zeta, y : \psi . vyz \end{array}$$

The first step must be

$$(1)\quad \mathsf{B}_1\mathsf{B}_2 \rhd\!\!> \lambda w:\gamma, v:\eta\,.\,\mathsf{B}_2(wv)$$

and then we can organize

$$(1)\quad \mathsf{B}_1\mathsf{B}_2\mathsf{C} \rhd\!\!> \lambda v:\eta, u:\mu, x:\theta\,.\,(\mathsf{C}v)(ux)$$

by either of

$$\downharpoonright(1\circ\uparrow\uparrow 1)\circ 1 \qquad 2\circ\uparrow 1$$

where $2 = \downharpoonright 1\circ 1$. In other words either we can reduce $\mathsf{B}_2(wv)$, introduce C, and then remove a redex; or we can introduce C, remove a redex, and then attack the body. After this we have

$$(2)\quad (\mathsf{C}v)(ux) \rhd\!\!> (vy)(ux)$$

so that

$$\downharpoonright(1\circ\uparrow\uparrow 1)\circ 1\circ\uparrow\uparrow\uparrow 2 \qquad 2\circ\uparrow 1\circ\uparrow\uparrow\uparrow 2$$

are two possible computations. It is also possible to attack C in the middle of the computation rather than the end.

(ω) From Solution 4.3(ω) and Example 4.4 we have the λ-combinators shown (where the identifiers of W have been used to choose some of the identifiers of $\mathsf{S}_1, \mathsf{S}_2, \mathsf{K}, \mathsf{I}$). We must organize $(\mathsf{S}_1\mathsf{S}_2)(\mathsf{K}\mathsf{I}) \rhd\!\!> \mathsf{W}$.

$$\begin{array}{ll}
\mathsf{W} = & \lambda u:\mu, x:\theta\,.\,uxx\\
\mathsf{S}_1 = & \lambda s:\sigma_2, v:\nu, u:\mu\,.\,(su)(vu)\\
\mathsf{S}_2 = & \lambda u:\mu, y:\iota, x:\theta\,.\,(ux)(yx)\\
\mathsf{K} = & \lambda y:\iota, u:\mu\,.\,y\\
\mathsf{I} = & \lambda x:\theta\,.\,x
\end{array}$$

With $2 = \downharpoonright 1\circ 1$ we have

$$(1\circ\uparrow\uparrow 2)\quad \mathsf{S}_1\mathsf{S}_2 \rhd\!\!> \lambda v:\nu, u:\mu, x:\theta\,.\,(ux)(vux) \qquad (1)\quad \mathsf{K}\mathsf{I} \rhd\!\!> V \qquad (2)\quad \mathsf{K}\mathsf{I}u \rhd\!\!> I$$

where $V = \lambda u:\mu\,.\,\mathsf{I}$. In the compound $(\mathsf{S}_1\mathsf{S}_2)(\mathsf{K}\mathsf{I})$ we can attack either the left hand component or the right hand one

$$\downharpoonright(1\circ\uparrow\uparrow 2)\circ 1 \qquad \downharpoonleft 1\circ\downharpoonright(1\circ\uparrow\uparrow 2)\circ 1$$

to get

$$(\mathsf{S}_1\mathsf{S}_2)(\mathsf{K}\mathsf{I}) \rhd\!\!> \lambda u:\mu, x:\theta\,.\,(ux)(\mathsf{K}\mathsf{I}ux) \qquad (\mathsf{S}_1\mathsf{S}_2)(\mathsf{K}\mathsf{I}) \rhd\!\!> \lambda u:\mu, x:\theta\,.\,(ux)(Vux)$$

and hence

$$\downharpoonright(1\circ\uparrow\uparrow 2)\circ 1\circ\uparrow\uparrow\downharpoonleft(\downharpoonright 2\circ 1) \qquad \downharpoonleft 1\circ\downharpoonright(1\circ\uparrow\uparrow 2)\circ 1\circ\uparrow\uparrow\downharpoonleft 2$$

are two possible computations.

(μ) Solution 4.3(μ) and elsewhere give us the λ-combinators shown (where the identifiers of M have been used to choose some of the identifiers of $\mathsf{B}_1, \mathsf{B}_2, \mathsf{S}$). We must organize $\mathsf{B}_1(\mathsf{B}_2\mathsf{S}) \rhd\!\!> \mathsf{M}$.

$$\begin{array}{ll}
\mathsf{M} = & \lambda f:\omega, w:\zeta, v:\rho, u\vartheta, x:\theta\,.\,(fwvx)(ux)\\
\mathsf{B}_1 = & \lambda e:\epsilon, f:\omega, w:\zeta\,.\,e(fw)\\
\mathsf{B}_2 = & \lambda s:\sigma, z:\chi, v:\rho\,.\,s(zv)\\
\mathsf{S} = & \lambda z:\eta, u:\vartheta, x:\theta\,.\,(zx)(ux)
\end{array}$$

Let

$$Z = \lambda v : \rho, u : \vartheta, x : \theta \,.\, (fwvx)(ux)$$

so that $\mathsf{M} = \lambda f : \omega, w : \zeta \,.\, Z$. We have

$$(2 \circ \uparrow 1) \quad (\mathsf{B}_2\mathsf{S})(fw) \rhd\!\!\!\rhd Z$$

and hence

$$1 \circ \uparrow\uparrow(2 \circ \uparrow 1) \qquad \downharpoonleft(2 \circ \uparrow 1) \circ 1$$

are two possible computations.

(ν) Solution 4.3(ν) and elsewhere give us the λ-combinators shown (where the identifiers of N suggest some of the identifiers of $\mathsf{S}_1, \mathsf{S}_2, \mathsf{K}$). We must organize $\mathsf{S}_1(\mathsf{K}\mathsf{S}_2) \rhd\!\!\!\rhd \mathsf{N}$.

$$\begin{array}{ll} \mathsf{N} = & \lambda w : \chi, v : \rho, u\vartheta, x : \theta \,.\, (wvx)(ux) \\ \mathsf{S}_1 = & \lambda t : -, x : \chi, v : \rho \,.\, (tv)(wv) \\ \mathsf{S}_2 = & \lambda z : \eta, u : \vartheta, x : \theta \,.\, (zx)(ux) \\ \mathsf{K} = & \lambda s : \sigma_2, v : \rho \,.\, s \end{array}$$

Let

$$T = \lambda u\vartheta, x : \theta \,.\, (wvx)(ux)$$

so that $\mathsf{N} = \lambda w : \chi, v : \rho \,.\, T$. We have

$$(\downarrow 2 \circ 1) \quad (\mathsf{K}\mathsf{S}_2)v(wv) \rhd\!\!\!\rhd T$$

and hence

$$1 \circ \uparrow\uparrow(\downarrow 2 \circ 1) \qquad \downharpoonleft(\downarrow 2 \circ 1) \circ 1$$

are two possible computations.

(τ) Solution 4.3(τ) and above give us the λ-combinators shown (where the identifiers of T suggest some of the identifiers of M, B). We must organize $\mathsf{MB} \rhd\!\!\!\rhd \mathsf{T}$.

$$\begin{array}{ll} \mathsf{T} = & \lambda w : \eta, z : \omega, y : \vartheta, x : \theta \,.\, w(zx)(yx) \\ \mathsf{M} = & \lambda b : \beta, w : \eta, z : \omega, y : \vartheta, x : \theta \,.\, (bwzx)(yx) \\ \mathsf{B} = & \lambda w : \eta, z : \omega, , x : \theta \,.\, w(zy) \end{array}$$

We have

$$(1) \quad \mathsf{MB} \rhd\!\!\!\rhd \lambda z : \omega, y : \vartheta, x : \theta \,.\, (\mathsf{B}wzx)(yx)$$

so that

$$1 \circ \uparrow\uparrow\uparrow\downarrow 3 \quad \text{where} \quad 3 = \uparrow(\uparrow 1 \circ 1) \circ 1$$

is the only reasonable computation. □

4.12 The proof proceeds by an induction over t. Only the step across an abstraction

$$t = (\lambda y : \sigma \,.\, r)$$

is not immediate. For this we have

$$t[x : s] = \lambda v : \sigma \,.\, r'[x := s]$$

where $r' = r[y := v]$ for some suitable v. Thus

$$(t[x : s])^\epsilon = \lambda v : \sigma \,.\, (r'[x := s])^\epsilon = \lambda v : \sigma \,.\, r'^\epsilon[x := s^\epsilon]$$

where

$$(r')^\epsilon = (r[y := v])^\epsilon = r^\epsilon[y := v^\epsilon] r^\epsilon[y := v]$$

by two uses of the induction hypothesis. (!) But

$$t^\epsilon[x := s^\epsilon] = (\lambda y \,.\, r^\epsilon)[x := s^\epsilon] = \lambda v \,.\, r^{\epsilon\prime}[x := s^\epsilon]$$

where

$$r^{\epsilon\prime} = r^\epsilon[y := v] = r'^\epsilon$$

provided the renaming of y is done in the same way on both occasions. Thus

$$(t[x := s])^\epsilon = \lambda v \,.\, r'^\epsilon[x := s^\epsilon] = \lambda v \,.\, r^{\epsilon\prime}[x := s^\epsilon] = t^\epsilon[x := s^\epsilon]$$

as required. □

4.13 Only the proof of the second part is not immediate. This proceeds by induction over the given untyped computation $\square^*$. There is one base case and four induction steps.

For the base case suppose $\square^* = \mathsf{1}$ is an untyped redex removal. Thus

$$t^{-\epsilon} = (\lambda y \,.\, r^!)s^! \qquad t^* = r^![y := s^!]$$

for some untyped terms $r^!, s^!$. From the way erasure works we must have

$$t^- = (\lambda y : \sigma \,.\, r)s$$

for some typed terms r, s. Then $r^! = r^\epsilon$ and $s^! = s^\epsilon$. Let

$$t^+ = r[y := s]$$

so that

$$(\mathsf{1}) \quad t^- \rhd\!\!\!\rhd t^+$$

is the only possible redex removal with t^- as subject. Also

$$t^{+\epsilon} = (r[y := s])^\epsilon = r^\epsilon[y := s^\epsilon] = t^*$$

so we have a typed computation $\square$ with $\square^\epsilon = \square^*$.

The steps across LA, RA, and TC are similar to the corresponding steps for the proof for combinator computations.

For the step across Ab we have $\square^* = \uparrow\mathsf{r}^*$ where

$$(\mathsf{r}^*) \quad r^! \rhd\!\!\!\rhd r^* \qquad t^{-\epsilon} = \lambda y \,.\, r^! \qquad t^* = \lambda y \,.\, r^*$$

for some untyped terms $r^!, r^*$. From the way erasure works we must have $t^- = \lambda y : \sigma \,.\, r^-$ for some typed term r^-. Then $r^! = r^{-\epsilon}$ and hence the untyped computation

$$(\mathsf{r}^*) \quad r^{-\epsilon} \rhd\!\!\!\rhd r^*$$

has a type erased subject. By the induction hypothesis there is a unique typed computation

$$(\mathsf{r}) \quad r^- \rhd\!\!\!\rhd r^+$$

with $\mathsf{r}^\epsilon = \mathsf{r}^*$, and then $r^* = r^{+\epsilon}$. Let $t^+ = \lambda y : \sigma \,.\, r^+$ and $\square = \uparrow\mathsf{r}$, to obtain the required typed computation. □

D.6 Subject reduction

4.14 By way of example consider the triple

$$(L) \quad \vdash \mathsf{L} : \epsilon \qquad (R) \quad \vdash \mathsf{R} : \epsilon \qquad (\Box) \quad \mathsf{L} \rhd\!\!\!\rhd \mathsf{R}$$

where

$$\mathsf{L} = \mathsf{B}_1\mathsf{B}_2\mathsf{C}_2 \qquad \mathsf{R} = \mathsf{E} = \lambda v : \eta, u : \mu, x : \theta, y : \psi \,.\, (vy)(ux)$$

as given by part (ϵ) of Exercises 4.3 and 4.11.

Using the context and body derivations shown in the top two lines

$$\begin{array}{lll}
\Delta \ = b : \beta_2, w : \gamma, v : \eta & \Sigma \ = z : \xi, u : \mu, x : \theta & \Pi = v : \eta, z : \zeta, y : \psi \\
D \ = \Delta[b](\Delta[w]\Delta[v]) & S \ = \Sigma[a](\Sigma[u]\Sigma[x]) & P = \Pi[v]\Pi[y]\Pi[z] \\
B_1 = D\uparrow\uparrow\uparrow & B_2 = S\uparrow\uparrow\uparrow & C = P\uparrow\uparrow\uparrow \\
\mathsf{B}_1 = \lambda\Delta \,.\, b(wv) & \mathsf{B}_2 = \lambda\Sigma \,.\, a(ux) & \mathsf{C} = \lambda\Pi \,.\, vyx
\end{array}$$

we have derivations and subject terms shown in the bottom two lines, where the prefix of each term has been abbreviated in an obvious way. We have

$$L = B_1B_2C \quad \mathsf{L} = \mathsf{B}_1\mathsf{B}_2\mathsf{C} \quad \Box = \mathsf{2} \circ \uparrow\mathsf{1} \circ \uparrow\uparrow\uparrow\mathsf{2}$$

where $\mathsf{2} = \downharpoonleft\mathsf{1} \circ \mathsf{1}$, and we must calculate $L \cdot \Box$.

With

$$\begin{aligned}
\mathsf{L}' \ &= (\lambda w : \gamma, v : \eta \,.\, \mathsf{B}_2(wv))\mathsf{C} \\
\mathsf{L}'' \ &= \lambda v : \eta \,.\, \mathsf{B}_2(\mathsf{C}v) \\
\mathsf{L}''' &= \lambda v : \eta, u : \mu, x : \theta \,.\, (\mathsf{C}v)(ux)
\end{aligned}$$

we have

$$\mathsf{L} \rhd\!\!\!\rhd \mathsf{L}' \rhd\!\!\!\rhd \mathsf{L}'' \rhd\!\!\!\rhd \mathsf{L}''' \rhd\!\!\!\rhd \mathsf{E}$$

where these steps are coded by $\downharpoonleft\mathsf{1}, \mathsf{1}, \uparrow\mathsf{1}$ and $\uparrow\uparrow\uparrow\mathsf{2}$, respectively. The calculation of $L \cdot \Box$ will track this reduction but will require a renaming of the v from L'' onwards.

For the first phase we have

$$L \cdot \downharpoonleft\mathsf{1} = ((B_1B_2) \cdot \mathsf{1})C$$

and

$$(B_1B_2) \cdot \mathsf{1} = (B_1B_2) \bullet \mathsf{1} = D\uparrow\uparrow * B_2 = (D' * B_2')\uparrow\uparrow$$

where, for the second equality, b is the nominated identifier. The third equality requires two calls on the crucial clause of the grafting algorithm. We need to determine the modified versions D', B_2' of D, B_2.

With $\Lambda = w : \gamma, v ; \eta$ (so that $\Delta = b : \beta_2, \Lambda$) we have

$$(D) \quad b : \beta_2, \Lambda \vdash b(wv) \qquad (B_2) \quad \vdash \mathsf{B}_2 : \beta_2$$

and we require

$$(D') \quad b : \beta_2, \Lambda' \vdash b(w'v') \qquad (B_2') \quad \Lambda' \vdash \mathsf{B}_2 : \beta_2$$

for some alphabetic variant Λ' of Λ. In fact, B_2 is a combinator, so there is no danger of identifier capture, and w, v do not appear in B_2, so we may leave $\Lambda' = \Lambda$. For later, observe that

$$T = (\Lambda, \Sigma)[a]((\Lambda, \Sigma)[u](\Lambda, \Sigma)[x]) \qquad B_2' = T\uparrow\uparrow\uparrow$$

gives the modified version of B_2.

Since $D' = D$ and

$$\Delta[b] * B_2' = B_2' \quad \Delta[w] * B_2' = \Lambda[w] \quad \Delta[v] * B_2' = \Lambda[v]$$

with

$$M = B_2'(\Lambda[w]\Lambda[v]) \qquad L' = (M\uparrow\uparrow)C$$

we have $D' * B_2' = M$ and

$$L \cdot \downarrow\mathbf{1} = (B_1 B_2 \cdot \mathbf{1})C = (M\uparrow\uparrow)C = L'$$

to complete the first phase of the full calculation. Observe that

$$(L') \quad \vdash \mathsf{L}' : \epsilon$$

holds.

For the second phase we have

$$L' \cdot \mathbf{1} = L' \bullet \mathbf{1} = M\uparrow * C = (M' * C')\uparrow$$

where w is the nominated identifier in this call on the leaf recipe. We have

$$(M) \quad w : \gamma, v : \eta \vdash \mathsf{B}_2(wv) : \chi \qquad (C) \quad \vdash \mathsf{C} : \gamma$$

and we require

$$(M') \quad w : \gamma, v' : \eta \vdash \mathsf{B}_2(wv') : \chi \qquad (C) \quad v' : \eta \vdash \mathsf{C} : \gamma$$

for some suitable v'. Since v occurs in C we must rename. (This could have been avoided by changing the v in C, but it's too late for that now.)

With a fresh identifier h let

$$\begin{aligned} \Lambda' &= w : \gamma, h : \eta & H &= (h : \eta \vdash h : \eta) \\ \Xi &= h : \eta, \Pi & X &= \Xi[v]\Xi[y]\Xi[z] \\ \Theta &= \Lambda', \Sigma & Q &= \Theta[a](\Theta[u]\Theta[x]) \\ \Psi &= h : \eta, \Sigma & R &= \Psi[a](\Psi[u]\Psi[x]) \end{aligned}$$

so that

$$B_2'' = Q\uparrow\uparrow\uparrow \quad M' = B_2''(\Lambda'[w]\Lambda'[h]) \quad C' = X\uparrow\uparrow\uparrow$$

are the modified derivations. We have

$$\Lambda'[w] * C' = C' \qquad \Lambda'[h] * C' = H$$

and by another call on the grafting algorithm

$$B_2'' * C' = (Q' * C'')\uparrow\uparrow\uparrow = R\uparrow\uparrow\uparrow$$

where this last step holds since the nominated identifier w does not occur in Q. These give

$$M' * C' = (R\uparrow\uparrow\uparrow)F$$

where $F = C'H$, and hence

$$L \cdot 2 = L' \cdot 1 = ((R\uparrow\uparrow\uparrow)F)\uparrow$$

completes the second phase of the whole calculation. We may check that

$$(L \cdot 2) \quad \vdash (\lambda h : \eta \,.\, \mathsf{B}_2(\mathsf{C}h)) : \epsilon$$

and this subject is an alphabetic variant of L'' (obtained by replacing v by h).

For the third phase we have

$$(L \cdot 2) \cdot \uparrow 1 = (((R\uparrow\uparrow\uparrow)F) \cdot 1)\uparrow = ((R\uparrow\uparrow * F))\uparrow = (R' * F')\uparrow\uparrow\uparrow$$

with a as the nominated identifier. By inspection we see there is no danger of clashing symbols. Thus with

$$\Phi = h : \eta, u : \mu, x; \theta \qquad \Omega = \Phi, \Pi = h : \eta, u : \mu, x; \theta, v : \eta, z : \zeta, y : \psi$$

and

$$Y = \Omega[v]\Omega[y]\Omega[z] \qquad F' = (Y\uparrow\uparrow\uparrow)\Phi[h]$$

we find that

$$R * F' = F'(\Phi[u]\Phi[x])$$

and hence

$$L \cdot (2 \circ \uparrow 1) = (F'(\Phi[u]\Phi[x]))\uparrow\uparrow\uparrow$$

completes the third phase. We may check that

$$(L \cdot (2 \circ \uparrow 1)) \quad \vdash (\lambda h : \eta, u : \mu, x : \theta \,.\, (\mathsf{C}h)(ux)) : \epsilon$$

and this subject is an alphabetic variant of L'''.

Finally let

$$\Upsilon = h : \eta, u : \mu, x; \theta, z : \zeta, y : \psi \qquad \Gamma = h : \eta, u : \mu, x; \theta, y : \psi$$

and

$$Z = \Phi[u]\Phi[x] \qquad G = (\Gamma[h]\Gamma[y])(\Gamma[u]\Gamma[x])$$

so that G is an alphabetic variant of the body of the derivation R. We find that

$$F' \cdot 1 = Y\uparrow\uparrow * \Phi[h] = (\Upsilon[h]\Upsilon[y]\Upsilon[z])\uparrow\uparrow$$

and then

$$F'Z \cdot 2 = ((F \cdot 1)Z) \cdot 1 = (\Upsilon[h]\Upsilon[y]\Upsilon[z])\uparrow * Z = G$$

to give

$$L \cdot \square = (F'Z)\uparrow\uparrow\uparrow \cdot \uparrow\uparrow\uparrow 2 = ((F'Z) \cdot 2)\uparrow\uparrow\uparrow = G\uparrow\uparrow\uparrow$$

to complete the calculation. □

As you can see, a calculation of $\nabla \cdot \square$ can be quite long even for comparatively small ∇ and $\square$. This is because in the middle of the calculation the reduction algorithm 4.17 makes repeated calls of the leaf recipe 4.16 which, in essence, recalls itself. Many parts of the calculation must be suspended until a subsidiary calculation has been completed (which itself may be suspended at some point to wait for a sub-subsidiary calculation). This kind of suspended nesting produces an inefficient algorithm. The cause of all this is the nesting in the substitution algorithm of Definition 4.10. In Chapter 6 we reformulate this to produce a more efficient subject reduction algorithm.

E
Substitution Algorithms

E.1 Introduction

5.1 For convenience let $s = (\lambda x : \rho \,.\, yx)$ so that $t = \lambda y : \sigma \,.\, zsx$ and the supports are $\partial s = \{y\}$, $\partial t = \{z, x\}$, $\partial \alpha = \{x, y, z\}$.

Since $\partial t \cap \partial \alpha = \{x, z\}$ we have

$$M(t, \alpha) = \partial(x\alpha) \cup \partial(z\alpha) = \{y, x\} \qquad U(t, \alpha) = \{z, y, x\}$$

so that w is the next available renaming identifier. With $\beta = [y := w]$ we have

$$\begin{aligned} t \cdot \alpha &= \lambda w : \sigma \,.\, (((zsx) \cdot \beta) \cdot \alpha) \\ &= \lambda w : \sigma \,.\, (z \cdot \beta \cdot \alpha) s''(x \cdot \beta \cdot \alpha) \\ &= \lambda w : \sigma \,.\, (z \cdot \alpha) s''(x \cdot \alpha) \qquad = \lambda w : \sigma \,.\, x s'' y \end{aligned}$$

where $s'' = (s \cdot \beta) \cdot \alpha$ is the central component.

We have $\partial s = \{y\}$ and $\partial \beta = \{y\}$ so that

$$M(s, \beta) = \partial(y\beta) = \{w\} \qquad U(s, \beta) = \{y, w\}$$

and hence z is the next available renaming identifier. With $\gamma = [x := z]$ we have

$$\begin{aligned} s' &= s \cdot \beta \\ &= \lambda z : \rho \,.\, ((((yx) \cdot \gamma)) \cdot \beta) \\ &= \lambda z : \rho \,.\, (y \cdot \gamma \cdot \beta)(x \cdot \gamma \cdot \beta) \\ &= \lambda z : \rho \,.\, (y \cdot \beta)(z \cdot \beta) \qquad = \lambda z : \rho \,.\, wz \end{aligned}$$

and we now require $s' \cdot \alpha$.

We have $\partial s' = \{w\}$ and $\partial \alpha = \{x, y, z\}$ with $\partial s' \cap \partial \alpha = \emptyset$ so that $M(s', \alpha) = \emptyset$ and

$$U(s', \alpha) = \{w, x, y, z\}$$

and hence v is the next available renaming identifier. With $\delta = [z := v]$ we have

$$\begin{aligned} s'' &= (s \cdot \beta) \cdot \alpha \\ &= s' \cdot \alpha \\ &= \lambda v : \rho \,.\, ((((wz) \cdot \delta)) \cdot \alpha) \\ &= \lambda v : \rho \,.\, (w \cdot \delta \cdot \alpha)(z \cdot \delta \cdot \alpha) \\ &= \lambda v : \rho \,.\, (w \cdot \alpha)(v \cdot \alpha) \qquad = \lambda v : \rho \,.\, wv \end{aligned}$$

to complete the calculation.

Thus we have

$$t \cdot \alpha = \lambda w : \sigma \,.\, x(\lambda v : \rho \,.\, wv)y$$

as the final result. □

5.2 For convenience let $s = (\lambda y : \sigma \,.\, zy)$ and $r = (\lambda x : \rho \,.\, yx)$ so that the given term is $t = \lambda z : \tau \,.\, sr$ and

$$\partial r = \{y\} \qquad \partial s = \{z\} \qquad \partial t = (\partial s \cup \partial r) - \{z\} = \{y\} \qquad \partial\alpha = \{y, x\}$$

are the supports.

Since $\partial t \cap \partial\alpha = \{y\}$ we have

$$M(t, \alpha) = \partial(y\alpha) = \{x\} \qquad U(t, \alpha) = \{y, x\}$$

so that z is the next available renaming identifier. This is the outermost bound identifier in t so no renaming is necessary. We have

$$t \cdot \alpha = \lambda z : \tau \,.\, ((sr) \cdot \alpha) = \lambda z : \tau \,.\, (s \cdot \alpha)(r \cdot \alpha)$$

and we must now calculate $s \cdot \alpha$ and $r \cdot \alpha$.

Since $\partial s \cap \partial\alpha = \emptyset$ we have $M(s, \alpha) = \emptyset$ and

$$U(s, \alpha) = \{z\} \cup \{y, x\} = \{z, y, x\}$$

so that w is the next available renaming identifier. With $\beta = [y := w]$ we have

$$\begin{aligned} s \cdot \alpha &= \lambda w : \sigma \,.\, (((zy) \cdot \beta) \cdot \alpha) \\ &= \lambda w : \sigma \,.\, (z \cdot \beta \cdot \alpha)(y \cdot \beta \cdot \alpha) \\ &= \lambda w : \sigma \,.\, (z \cdot \alpha)(w \cdot \alpha) \qquad = \lambda w : \sigma \,.\, zw \end{aligned}$$

for the first body component.

Since $\partial r \cap \partial\alpha = \{y\}$ we have

$$M(r, \alpha) = \partial(y\alpha) = \{x\} \qquad U(r, \alpha) = \{y, x\}$$

so that z is the next available renaming identifier. With $\gamma = [x := z]$ we have

$$\begin{aligned} r \cdot \alpha &= \lambda z : \rho \,.\, (((yx) \cdot \gamma) \cdot \alpha) \\ &= \lambda z : \rho \,.\, (y \cdot \gamma \cdot \alpha)(x \cdot \gamma \cdot \alpha) \\ &= \lambda z : \rho \,.\, (y \cdot \alpha)(z \cdot \alpha) \qquad = \lambda z : \rho \,.\, xz \end{aligned}$$

for the second component.

Thus we obtain

$$t \cdot \alpha = \lambda z : \tau \,.\, (\lambda w : \sigma \,.\, zw)(\lambda z : \rho \,.\, xz)$$

as the final result. □

E.2 Formal replacements

5.3 We have

$$\partial([z \mapsto y][x \mapsto w]\mathfrak{i}) = \begin{cases} \partial(x \mapsto w]\mathfrak{i}) \cup \{z\} & z \neq y \\ \partial(x \mapsto w]\mathfrak{i}) - \{z\} & z = y \end{cases}$$

$$= \begin{cases} \partial\mathfrak{i} \cup \{x\} \cup \{z\} & z \neq y,\ x \neq w \\ (\partial\mathfrak{i} - \{x\}) \cup \{z\} & z \neq y,\ x = w \\ (\partial\mathfrak{i} \cup \{x\}) - \{z\} & z = y,\ x \neq w \\ (\partial\mathfrak{i} - \{x\}) - \{z\} & z = y,\ x = w \end{cases}$$

$$= \begin{cases} \{x, z\} & z \neq y,\ x \neq w \\ \{z\} & z \neq y,\ x = w \\ \{x\} - \{z\} & z = y,\ x \neq w \\ \emptyset & z = y,\ x = w \end{cases}$$

$$= \begin{cases} \{x, z\} & z \neq y, x \neq w \\ \{z\} & z \neq y, x = w \\ \{x\} & z = y \neq x \neq w \\ \emptyset & z = y = x \neq w \\ \emptyset & z = y, x = w \end{cases}$$

$$= \begin{cases} \{x, z\} & z \neq y, x \neq w \\ \{z\} & z \neq y, x = w \\ \{x\} & z = y \neq x \neq w \\ \emptyset & z = y, x \in \{y, w\} \end{cases}$$

to show that the support can be one of four sets. □

5.4 For each pair $\mathcal{V} \subseteq \mathcal{W}$ of sets of identifiers we have $\mathcal{V} \cup \mathcal{W} = \mathcal{W}$, so that $\mathcal{V}^{\langle\mathfrak{a}\rangle} \cup \mathcal{W}^{\langle\mathfrak{a}\rangle} = \mathcal{W}^{\langle\mathfrak{a}\rangle}$ to give $\mathcal{V}^{\langle\mathfrak{a}\rangle} \subseteq \mathcal{W}^{\langle\mathfrak{a}\rangle}$, as required to show that $\cdot^{\langle\mathfrak{a}\rangle}$ is montone.

Consider $\mathfrak{a} = [x \mapsto y][y \mapsto z]\mathfrak{i}$ and $\mathcal{W} = \{x\}$ where x, y, z are distinct identifiers. Then

$$\{x\}^{\langle\mathfrak{a}\rangle} = \partial(x\mathfrak{a}) = \{y\} \qquad \{y\}^{\langle\mathfrak{a}\rangle} = \partial(y\mathfrak{a}) = \{z\}$$

to show that

$$\mathcal{W} \nsubseteq \mathcal{W}^{\langle\mathfrak{a}\rangle} \qquad \mathcal{W}^{\langle\mathfrak{a}\rangle} \nsubseteq \mathcal{W} \qquad (\mathcal{W}^{\langle\mathfrak{a}\rangle})^{\langle\mathfrak{a}\rangle} \neq \mathcal{W}^{\langle\mathfrak{a}\rangle}$$

can happen. □

E.3 The generic algorithm

5.5 The replacement

$$\mathfrak{a} = [x \mapsto y][y \mapsto z][z \mapsto x]\mathfrak{i}$$

names α.

As in Solution 5.1 let $s = \lambda x : \rho \,.\, yx$ so that $t = \lambda y : \sigma \,.\, zsx$ is the term we must hit with $\mathfrak{a}$.

We have $\partial t = \{z, x\}$ so that

$$(\partial t)^{\langle \mathfrak{a} \rangle} = \partial(z\mathfrak{a}) \cup \partial(x\mathfrak{a}) = \{x, y\}$$

and hence z is the next available renaming identifier. With $'\mathfrak{a} = [y \mapsto z]\mathfrak{a}$ we have

$$t \cdot \mathfrak{a} = \lambda z : \sigma \,.\, ((zsx) \cdot {'\mathfrak{a}}) = \lambda z : \sigma \,.\, (z \cdot {'\mathfrak{a}})s''(x \cdot {'\mathfrak{a}}) = \lambda z : \sigma \,.\, xs''y$$

where $s'' = s \cdot {'\mathfrak{a}}$.

We have $\partial s = \{y\}$ so that

$$(\partial s)^{\langle '\mathfrak{a} \rangle} = \partial(y\,'\mathfrak{a}) = \{z\}$$

and hence y is the next available renaming identifier. With $''\mathfrak{a} = [x \mapsto y]\,'\mathfrak{a}$ we have

$$s'' = s \cdot {'\mathfrak{a}} = \lambda y : \rho \,.\, ((yx) \cdot {''\mathfrak{a}}) = \lambda y : \rho \,.\, (y \cdot {''\mathfrak{a}})(x \cdot {''\mathfrak{a}}) = \lambda y : \rho \,.\, zy$$

to give

$$t \cdot \mathfrak{a} = \lambda z : \sigma \,.\, x(\lambda y : \rho \,.\, zy)y$$

as the final result. □

5.6 The replacement

$$\mathfrak{a} = [y \mapsto x][x \mapsto y]\mathfrak{i}$$

names α.

As in Solution 5.2 let $s = \lambda y : \sigma \,.\, zy$ and $r = \lambda x : \rho \,.\, yx$ so that $t = \lambda z : \tau \,.\, sr$ is the term we must hit with $\mathfrak{a}$.

We have $\partial t = \{y\}$ so that

$$(\partial t)^{\langle \mathfrak{a} \rangle} = \partial(y\mathfrak{a}) = \{x\}$$

and hence z is the next available renaming identifier. With $'\mathfrak{a} = [z \mapsto z]\mathfrak{a}$ we have

$$t \cdot \mathfrak{a} = \lambda z : \tau \,.\, ((sr) \cdot {'\mathfrak{a}}) = \lambda z : \sigma \,.\, (s \cdot {'\mathfrak{a}})(r \cdot {'\mathfrak{a}})$$

and we must now calculate $s \cdot {'\mathfrak{a}}$ and $r \cdot {'\mathfrak{a}}$.

We have $\partial s = \{z\}$ so that

$$(\partial s)^{\langle '\mathfrak{a} \rangle} = \partial(z\,'\mathfrak{a}) = \{z\}$$

and hence y is the next available renaming identifier. With $''\mathfrak{a} = [y \mapsto y]\,'\mathfrak{a}$ we have

$$s \cdot {'\mathfrak{a}} = \lambda y : \sigma \,.\, ((zy) \cdot {''\mathfrak{a}}) = \lambda y : \sigma \,.\, zy = s$$

i.e. the first body component doesn't change.

We have $\partial r = \{y\}$ so that

$$(\partial r)^{\langle '\mathfrak{a} \rangle} = \partial(y\,'\mathfrak{a}) = \{x\}$$

and hence z is the next available renaming identifier. With $'''\mathfrak{a} = [x \mapsto z]'\mathfrak{a}$ we have

$$r \cdot '\mathfrak{a} = \lambda z : \rho \,.\, ((yx) \cdot '''\mathfrak{a}) = \lambda z : \rho \,.\, (y \cdot '''\mathfrak{a})(x \cdot '''\mathfrak{a}) = \lambda z : \rho \,.\, xz$$

for the second body component.

Thus

$$t \cdot \mathfrak{a} = \lambda z : \tau \,.\, (\lambda y : \sigma \,.\, zy)(\lambda z : \rho \,.\, xz)$$

is the final result. □

5.7 Since

$$w\,'\mathfrak{a} = \begin{cases} v & \text{if } w = y \\ w\mathfrak{a} & \text{if } w \neq y \end{cases} \qquad \text{we have} \qquad \partial(w\,'\mathfrak{a}) = \begin{cases} \{v\} & \text{if } w = y \\ \partial(w\mathfrak{a}) & \text{if } w \neq y \end{cases}$$

so that

$$w \overset{'\mathfrak{a}}{\longleftarrow} u \iff \begin{cases} u = v & \text{if } w = y \\ w \overset{\mathfrak{a}}{\longleftarrow} u & \text{if } w \neq y \end{cases}$$

is the required description. □

5.8 We proceed by induction over t.

When $t = \mathsf{k}$ we have $t \cdot \mathfrak{a} = \mathsf{k}$ and $\partial\mathsf{k} = \emptyset$. The property $\emptyset^{\langle\mathfrak{a}\rangle} = \emptyset$ gives the required result.

When $t = x$ we have $\partial t = \{x\}$ so that

$$v \in (\partial t)^{\langle\mathfrak{a}\rangle} \iff (\exists w \in \partial t)[v \in \partial(w\mathfrak{a})] \iff v \in \partial(x\mathfrak{a}) = \partial(t \cdot \mathfrak{a})$$

to give $(\partial t)^{\langle\mathfrak{a}\rangle} = \partial(t \cdot \mathfrak{a})$ as required.

When $t = qp$ we have

$$\partial t = \partial q \cup \partial p \qquad t \cdot \mathfrak{a} = (q \cdot \mathfrak{a})(p \cdot \mathfrak{a}) \qquad \partial(t \cdot \mathfrak{a}) = \partial(q \cdot \mathfrak{a}) \cup \partial(p \cdot \mathfrak{a})$$

and hence

$$\partial(t \cdot \mathfrak{a}) = (\partial q)^{\langle\mathfrak{a}\rangle} \cup (\partial p)^{\langle\mathfrak{a}\rangle} = (\partial q \cup \partial p)^{\langle\mathfrak{a}\rangle} = (\partial t)^{\langle\mathfrak{a}\rangle}$$

using the induction hypothesis and the additive property of $\cdot^{\langle\mathfrak{a}\rangle}$.

The induction step across an abstraction is given in the proof of Lemma 5.9. This step requires variation in the parameter $\mathfrak{a}$. □

E.4 THE MECHANISTIC ALGORITHM

5.9 As before let $s = \lambda x : \rho \,.\, yx$ so that $t = \lambda y : \sigma \,.\, zsx$ is the term we must hit with $\mathfrak{a}$ with $\mathcal{U} = \{z\}$ as the current set of untouchables.

As before we have $(\partial t)^{\langle\mathfrak{a}\rangle} = \{x, y\}$, so that $(\partial t)^{\langle\mathfrak{a}\rangle} \cup \mathcal{U} = \{x, y, z\}$ is the current set of unsafe identifiers, and hence w is the next safe identifier.

With

$$'\mathfrak{a} = [y \mapsto w]\mathfrak{a} \qquad \mathcal{U}' = \{z, w\}$$

we have

$$t \cdot \mathfrak{a} = \lambda w : \sigma \,.\, ((zsx) \cdot '\mathfrak{a}) = \lambda w : \sigma \,.\, xs''y$$

where $s'' = s \cdot {}'\mathfrak{a}$.

Since $\partial s = \{y\}$ we have $(\partial s)^{\langle '\mathfrak{a}\rangle} = \partial(y\,'\mathfrak{a}) = \{w\}$ so that $(\partial s)^{\langle '\mathfrak{a}\rangle} \cup \mathcal{U}' = \{z, w\}$ is the current set of unsafe identifiers, and hence y is the next safe identifier.

With

$$''\mathfrak{a} = [x \mapsto y]\,'\mathfrak{a} \qquad \mathcal{U}'' = \{z, y, w\}$$

we have

$$s'' = s \cdot {}'\mathfrak{a} = \lambda y : \rho \,.\, ((yx) \cdot {}''\mathfrak{a}) = \lambda y : \rho \,.\, wy$$

to give

$$t \cdot \mathfrak{a} = \lambda w : \sigma \,.\, x(\lambda y : \rho \,.\, wy)y$$

as the final result. □

5.10 As before let $s = \lambda y : \sigma \,.\, zy$ and $r = \lambda x : \rho \,.\, yx$ so that $t = \lambda z : \tau \,.\, sr$ is the term we must hit with $\mathfrak{a}$ with $\mathcal{U} = \{z\}$ as the current set of untouchables.

As before $(\partial t)^{\langle \mathfrak{a}\rangle} = \{x\}$ so that $(\partial t)^{\langle \mathfrak{a}\rangle} \cup \mathcal{U} = \{x, z\}$ is the current set of unsafe identifiers, and hence y is the next safe identifier.

With

$$'\mathfrak{a} = [z \mapsto y]\mathfrak{a} \qquad \mathcal{U}' = \{z, y\}$$

we have

$$t \cdot \mathfrak{a} = \lambda y : \tau \,.\, ((sr) \cdot {}'\mathfrak{a}) = \lambda y : \tau \,.\, (s \cdot {}'\mathfrak{a})(r \cdot {}'\mathfrak{a})$$

and we must now calculate $s \cdot {}'\mathfrak{a}$ and $r \cdot {}'\mathfrak{a}$.

We have $\partial s = \{z\}$, so that $(\partial s)^{\langle '\mathfrak{a}\rangle} = \partial(z\,'\mathfrak{a}) = \{y\}$, and hence the current set of unsafe identifiers is $(\partial s)^{\langle '\mathfrak{a}\rangle} \cup \mathcal{U}' = \{z, y\}$, to give x as the next safe identifier.

With

$$''\mathfrak{a} = [y \mapsto x]\,'\mathfrak{a} \qquad \mathcal{U}'' = \{z, y, x\}$$

we have

$$s \cdot {}'\mathfrak{a} = \lambda x : \sigma \,.\, ((zy) \cdot {}''\mathfrak{a}) = \lambda x : \sigma \,.\, (z \cdot {}''\mathfrak{a})(y \cdot {}''\mathfrak{a}) = \lambda x : \sigma \,.\, yx$$

as the first body component.

We have $\partial r = \{y\}$ so that $(\partial r)^{\langle '\mathfrak{a}\rangle} = \partial(y\,'\mathfrak{a}) = \{y\}$, and hence the current set of unsafe identifiers is $(\partial r)^{\langle '\mathfrak{a}\rangle} \cup \mathcal{U}'' = \{z, y, x\}$, to give w as the next safe identifier.

With

$$'''\mathfrak{a} = [x \mapsto w]\,'\mathfrak{a} \qquad \mathcal{U}''' = \{z, y, x, w\}$$

we have

$$r \cdot {}'\mathfrak{a} = \lambda w : \rho \,.\, ((yx) \cdot {}'''\mathfrak{a}) = \lambda w : \rho \,.\, (y \cdot {}'''\mathfrak{a})(x \cdot {}'''\mathfrak{a}) = \lambda w : \rho \,.\, xw$$

for the second body component.

Thus

$$t \cdot \mathfrak{a} = \lambda z : \tau \,.\, (\lambda x : \sigma \,.\, yx)(\lambda w : \rho \,.\, xw)$$

is the final result.

Notice that the algorithm doesn't tell us in which order $s \cdot {}'\mathfrak{a}$ and $r \cdot {}'\mathfrak{a}$ should be calculated. In some circumstances this could make a difference to the bound identifiers in the final result. □

5.11 Let

$$q = (\lambda y : \rho \,.\, xy) \qquad r = (\lambda x : \sigma \,.\, qz) \qquad s = (\lambda y : \tau \,.\, yr)$$

so that

$$t = \lambda x : \tau \,.\, sx$$

is the term we must standardize. Note that

$$\partial q = \{x\} \qquad \partial r = \{z\} \qquad \partial s = \{z\} \qquad \partial t = \{z\}$$

are the supports. In particular $\mathcal{U} = \{z\}$ is the current set of untouchables.

Since $\partial t = \{z\}$ we have $(\partial t)^{\langle \mathfrak{i} \rangle} = \partial(z\mathfrak{i}) = \{z\}$, so that $(\partial t)^{\langle \mathfrak{i} \rangle} \cup \mathcal{U} = \{z\}$ is the current set of unsafe identifiers, and hence y is the next safe identifier. With

$$\mathfrak{a} = [x \mapsto y]\mathfrak{i} \qquad \mathcal{U}' = \{z, y\}$$

we have

$$t \cdot \mathfrak{i} = \lambda y : \tau \,.\, ((sx) \cdot \mathfrak{a}) = \lambda y : \tau \,.\, ((s \cdot \mathfrak{a})(x \cdot \mathfrak{a})) = \lambda y : \tau \,.\, ((s \cdot \mathfrak{a})y)$$

so we must now calculate $s \cdot \mathfrak{a}$.

Since $\partial s = \{z\}$ we have $(\partial s)^{\langle \mathfrak{a} \rangle} = \partial(z\mathfrak{a}) = \{z\}$, so that $(\partial s)^{\langle \mathfrak{a} \rangle} \cup \mathcal{U}' = \{z, y\}$ is the current set of unsafe identifiers, and hence x is the next safe identifier. With

$$\mathfrak{b} = [y \mapsto x]\mathfrak{a} \qquad \mathcal{U}'' = \{z, y, x\}$$

we have

$$s \cdot \mathfrak{a} = \lambda x : \tau \,.\, ((yr) \cdot \mathfrak{b}) = \lambda x : \tau \,.\, ((y \cdot \mathfrak{b})(r \cdot \mathfrak{b})) = \lambda x : \tau \,.\, (x(r \cdot \mathfrak{b}))$$

so we must now calculate $r \cdot \mathfrak{b}$.

Since $\partial r = \{z\}$ we have $(\partial r)^{\langle \mathfrak{b} \rangle} = \partial(z\mathfrak{b}) = \{z\}$, so that $(\partial r)^{\langle \mathfrak{a} \rangle} \cup \mathcal{U}'' = \{z, y, x\}$ is the current set of unsafe identifiers, and hence w is the next safe identifier. With

$$\mathfrak{c} = [x \mapsto w]\mathfrak{b} \qquad \mathcal{U}''' = \{z, y, x, w\}$$

we have

$$r \cdot \mathfrak{b} = \lambda w : \sigma \,.\, ((qz) \cdot \mathfrak{c}) = \lambda w : \sigma \,.\, ((q \cdot \mathfrak{c})(z \cdot \mathfrak{c})) = \lambda w : \sigma \,.\, ((q \cdot \mathfrak{c})z)$$

so we must now calculate $q \cdot \mathfrak{c}$.

Since $\partial q = \{x\}$ we have $(\partial q)^{\langle \mathfrak{c} \rangle} = \partial(x\mathfrak{c}) = \{w\}$, so that the current set of unsafe identifiers is $(\partial r)^{\langle \mathfrak{a} \rangle} \cup \mathcal{U}''' = \{z, y, x, w\}$, and hence v is the next safe identifier. With

$$\mathfrak{d} = [y \mapsto v]\mathfrak{c} \qquad \mathcal{U}'''' = \{z, y, x, w, v\}$$

we have

$$q \cdot \mathfrak{c} = \lambda v : \rho \,.\, ((xy) \cdot \mathfrak{d}) = \lambda v : \rho \,.\, ((x \cdot \mathfrak{d})(y \cdot \mathfrak{d})) = \lambda v : \rho \,.\, wv$$

to complete the calculation.

The comparison

$$\begin{aligned} t &= \lambda x : \tau . ((\lambda y : \tau . (y(\lambda x : \sigma . ((\lambda y : \rho . xy)z))))x) \\ t \cdot \mathfrak{i} &= \lambda y : \tau . ((\lambda x : \tau . (x(\lambda w : \sigma . ((\lambda v : \rho . wv)z))))y) \end{aligned}$$

shows that $t \cdot \mathfrak{i}$ is slightly easier to read. □

5.12 For each i let

$$\mathcal{U}^{(i)} = \mathcal{U} \cup \{u_1, \ldots, u_i\}$$

so that $\mathcal{U}_i = (\partial t)^{\langle \mathfrak{a} \rangle} \cup \mathcal{U}^{(i)}$. We show by induction on m that

$$t \cdot \mathfrak{a} = \lambda u_1 : \rho_1, \ldots, u_m : \rho_m . (s \cdot {}^*\mathfrak{a})$$

with $\mathcal{U}^{(m)}$ as output untouchables.

We look at the induction step, $m \mapsto m'$.

Let

$$s = \lambda v : \sigma . r \qquad v = \mathit{fresh}(\mathcal{U}_m) \qquad \mathfrak{b} = [y \mapsto v]{}^*\mathfrak{a}$$

so that the desired result is

$$t \cdot \mathfrak{a} = \lambda u_1 : \rho_1, \ldots, u_m : \rho_m, v : \sigma . (r \cdot \mathfrak{b})$$

with $\mathcal{U}^{(m)} \cup \{v\}$ as output untouchables.

The first phase in the calculation of $t \cdot \mathfrak{a}$ concludes as above with $s \cdot {}^*\mathfrak{a}$ left to calculate. At that stage

$$(\partial s)^{\langle {}^*\mathfrak{a} \rangle} \cup \mathcal{U}^{(m)} = \partial(s \cdot {}^*\mathfrak{a}) \cup \mathcal{U}^{(m)}$$

is the current set of unsafe identifiers. Using the induction hypothesis we have

$$(\partial t)^{\langle \mathfrak{a} \rangle} = \partial(t \cdot \mathfrak{a}) \subseteq \partial(s \cdot {}^*\mathfrak{a}) \subseteq \partial(t \cdot \mathfrak{a}) \cup \mathcal{U}^{(m)} = (\partial t)^{\langle \mathfrak{a} \rangle} \cup \mathcal{U}^{(m)}$$

and hence

$$(\partial s)^{\langle {}^*\mathfrak{a} \rangle} \cup \mathcal{U}^{(m)} = (\partial t)^{\langle \mathfrak{a} \rangle} \cup \mathcal{U}^{(m)} = \mathcal{U}_m$$

is the current set of unsafe identifiers. Hence we obtain the desired result. □

5.13 This is not easy, is it? One problem is that we do not have a formal definition of 'alphabetic variant'. Another problem is that we need to prove more than the stated result. As $t \cdot \mathfrak{i}$ is calculated the updating will convert $\mathfrak{i}$ into a renaming replacement. Thus, to proceed by induction across t, we need to prove a result about the action of all renaming replacements. It is not clear what this is.

The easiest way out of this is to treat an equality $t_1 \cdot \mathfrak{i} = t_2 \cdot \mathfrak{i}$ as the definition of alphabetic variants. □

E.5 Some properties of substitution

5.14 We show

$$x \cdot (\mathfrak{b} ; \mathfrak{a}) = (x \cdot \mathfrak{b}) \cdot \mathfrak{a}$$

by induction on $\mathfrak{b}$. The base case, $\mathfrak{b} = \mathfrak{i}$, is straight forward. For the induction step, $\mathfrak{b} \mapsto {}'\mathfrak{b} = [y \mapsto s]\mathfrak{b}$, we have

$$'\mathfrak{b} ; \mathfrak{a} = [y \mapsto s \cdot \mathfrak{a}](\mathfrak{b} ; \mathfrak{a})$$

so that

$$x \cdot ({}'\mathfrak{b} ; \mathfrak{a}) = \begin{cases} s \cdot \mathfrak{a} & \text{if } x = y \\ x \cdot (\mathfrak{b} ; \mathfrak{a}) & \text{if } x \neq y \end{cases}$$

$$(x \cdot {}'\mathfrak{b}) \cdot \mathfrak{a} = \begin{cases} s \cdot \mathfrak{a} & \text{if } x = y \\ (x \cdot \mathfrak{b}) \cdot \mathfrak{a} & \text{if } x \neq y \end{cases}$$

and the induction hypothesis gives the required result. □

5.15 Note first that

$$\partial(w(\mathfrak{b} ; \mathfrak{a})) = \partial((w\mathfrak{b}) \cdot \mathfrak{a}) = \partial(w\mathfrak{b})^{\langle \mathfrak{a} \rangle}$$

using Exercise 5.14 and Lemma 5.9. Thus

$$\begin{aligned} w \xleftarrow{\mathfrak{b};\mathfrak{a}} u &\iff u \in \partial(w(\mathfrak{b} ; \mathfrak{a})) = \partial(w\mathfrak{b})^{\langle \mathfrak{a} \rangle} \\ &\iff (\exists v \in \partial(w\mathfrak{b}))[v \xleftarrow{\mathfrak{a}} u] \qquad \iff (\exists v)[w \xleftarrow{\mathfrak{b}} v \xleftarrow{\mathfrak{a}} u] \end{aligned}$$

as required.

With this we have

$$\begin{aligned} u \in (\mathcal{T}^{\langle \mathfrak{b} \rangle})^{\langle \mathfrak{a} \rangle} &\iff (\exists v \in \mathcal{T}^{\langle \mathfrak{b} \rangle})[v \xleftarrow{\mathfrak{a}} u] \\ &\iff (\exists v \in \mathcal{T}^{\langle \mathfrak{b} \rangle}, w \in \mathcal{T})[w \xleftarrow{\mathfrak{b}} v \xleftarrow{\mathfrak{a}} u] \\ &\iff (\exists w \in \mathcal{T}, v \in \mathcal{T}^{\langle \mathfrak{b} \rangle})[w \xleftarrow{\mathfrak{b}} v \xleftarrow{\mathfrak{a}} u] \\ &\iff (\exists w \in \mathcal{T})[w \xleftarrow{\mathfrak{b};\mathfrak{a}} u] \qquad \iff u \in \mathcal{T}^{\langle \mathfrak{b};\mathfrak{a} \rangle} \end{aligned}$$

as required. □

5.16 We investigate a proof of

$$\mathfrak{c} ; (\mathfrak{b} ; \mathfrak{a}) = (\mathfrak{c} ; \mathfrak{b}) ; \mathfrak{a}$$

by induction on $\mathfrak{c}$. The base case, $\mathfrak{c} = \mathfrak{i}$, is immediate. For the induction step, $\mathfrak{c} \mapsto {}'\mathfrak{c} = [y \mapsto s]\mathfrak{c}$, we have

$$\begin{aligned} {}'\mathfrak{c} ; (\mathfrak{b} ; \mathfrak{a}) &= [y \mapsto s]\mathfrak{c} ; (\mathfrak{b} ; \mathfrak{a}) &&= [y \mapsto s \cdot (\mathfrak{b} ; \mathfrak{a})](\mathfrak{c} ; (\mathfrak{b} ; \mathfrak{a})) \\ ({}'\mathfrak{c} ; \mathfrak{b}) ; \mathfrak{a} &= ([y \mapsto s \cdot \mathfrak{b}](\mathfrak{c} ; \mathfrak{b})) ; \mathfrak{a} &&= [y \mapsto (s \cdot \mathfrak{b}) ; \mathfrak{a}]((\mathfrak{c} ; \mathfrak{b}) ; \mathfrak{a}) \end{aligned}$$

and hence the equalities

$$s \cdot (\mathfrak{b} ; \mathfrak{a}) = (s \cdot \mathfrak{b}) ; \mathfrak{a} \qquad \mathfrak{c} ; (\mathfrak{b} ; \mathfrak{a}) = (\mathfrak{c} ; \mathfrak{b}) ; \mathfrak{a}$$

will suffice. The second is the induction hypothesis. The first is Theorem 5.16 but, as we saw there, this requires a parallel computation. □

5.17 (a) Consider $'\mathfrak{a} = [y \mapsto s]\mathfrak{a}$. Then

$$'\mathfrak{a} ; \mathfrak{i} = [y \mapsto s \cdot \mathfrak{i}](\mathfrak{a} ; \mathfrak{i})$$

so that

$$'\mathfrak{a} ; \mathfrak{i} = {}'\mathfrak{a} \iff s \cdot \mathfrak{i} = s \text{ and } \mathfrak{a} ; \mathfrak{i} = \mathfrak{a}$$

i.e.

$$'\mathfrak{a} ; \mathfrak{i} \text{ is standard } \iff s \text{ is standard and } \mathfrak{a} \text{ is standard}$$

(for replacements are syntactic objects so can be compared component by component). Since $\mathfrak{i}$ is standard, an obvious induction gives the required result.

(b) If $\mathfrak{a}$ is standard then

$$(t \cdot \mathfrak{a}) \cdot \mathfrak{i} = t \cdot (\mathfrak{a} ; \mathfrak{i}) = t \cdot \mathfrak{a}$$

so that $t \cdot \mathfrak{a}$ is standard for each term t.

(c) Suppose $\mathfrak{a}$ is such that $t \cdot \mathfrak{a}$ is standard for all terms t. In particular, for each identifier x we have

$$x \cdot (\mathfrak{a} ; \mathfrak{i}) = (x \cdot \mathfrak{a}) \cdot \mathfrak{i} = x \cdot \mathfrak{a}$$

i.e.

$$\mathfrak{a} ; \mathfrak{i} \text{ and } \mathfrak{a}$$

agree on all identifiers. When $\mathfrak{a} ; \mathfrak{i}$ and $\mathfrak{a}$ agree in this way, Lemma 5.14 gives

$$t \cdot (\mathfrak{a} ; \mathfrak{i}) = t \cdot \mathfrak{a}$$

and hence $t \cdot \mathfrak{a}$ is standard for all terms t. Thus the converse of (b) can be rephrased as

If $\mathfrak{a} ; \mathfrak{i}$ and $\mathfrak{a}$ agree on all identifiers, then $\mathfrak{a}$ is standard.

(and (b) itself has a similar rephrasing).

Now consider

$$\mathfrak{a} = [x \mapsto s][x \mapsto r]\mathfrak{i}$$

where s is standard. Then

$$\mathfrak{a} ; \mathfrak{i} = [x \mapsto s \cdot \mathfrak{i}]([x \mapsto r]\mathfrak{i} ; \mathfrak{i}) = [x \mapsto s][x \mapsto r \cdot \mathfrak{i}]\mathfrak{i}$$

and hence $\mathfrak{a} ; \mathfrak{i}$ and $\mathfrak{a}$ agree on all identifiers. (The inner updates are never activated.) But $\mathfrak{a}$ need not be standard (for r is arbitrary.)

The converse of (b) is false. For a replacement, being standard is not a property of its behaviour, but of its syntactic form. □

5.18 Consider a term

$$t = \lambda y : \sigma . r$$

and suppose $\mathcal{U}$ is the current set of untouchables. Then

$$t \cdot \mathsf{i} = (\lambda v : \sigma \, . \, r \cdot {}'\mathsf{i}) \qquad \text{where } {}'\mathsf{i} = [y \mapsto v]\mathsf{i} \qquad \text{with } v = \mathit{fresh}(\partial t \cup \mathcal{U})$$

(since $\mathcal{T}^{\langle \mathsf{i} \rangle} = \mathcal{T}$ for all sets $\mathcal{T}$). If t is standard then $y = v$, i.e.

$$y = \mathit{fresh}(\partial t \cup \mathcal{U})$$

holds.

A different environment may have a different set $\mathcal{V}$ of untouchables, and then

$$y \neq \mathit{fresh}(\partial t \cup \mathcal{V})$$

may hold so that t is not standard in this environment! □

5.19 Consider the replacements

$$\begin{aligned} \mathfrak{b} &= [y \mapsto s]\mathsf{i} \, ; \mathfrak{a} && = [y \mapsto s \cdot \mathfrak{a}]\mathfrak{a} \\ \mathfrak{c} &= {}'\mathfrak{a} \, ; [v \mapsto s \cdot \mathfrak{a}]\mathsf{i} && \\ &= [y \mapsto v \cdot [v \mapsto s \cdot \mathfrak{a}]\mathsf{i}](\mathfrak{a} \, ; [v \mapsto s \cdot \mathfrak{a}]\mathsf{i}) && = [y \mapsto s \cdot \mathfrak{a}](\mathfrak{a} \, ; [v \mapsto s \cdot \mathfrak{a}]\mathsf{i}) \end{aligned}$$

so we want $\mathfrak{b}, \mathfrak{c}$ to agree on ∂t.

They certainly agree at y, so it suffices to consider $x \in \partial t - \{y\}$. For such an x we have

$$\begin{aligned} x \cdot \mathfrak{b} &= x \cdot \mathfrak{a} && = (x \cdot \mathfrak{a}) \cdot \mathsf{i} \\ x \cdot \mathfrak{b} &= x \cdot (\mathfrak{a} \, ; [v \mapsto s \cdot \mathfrak{a}]\mathsf{i}) && = (x \cdot \mathfrak{a}) \cdot [v \mapsto s \cdot \mathfrak{a}]\mathsf{i} \end{aligned}$$

since $\mathfrak{a}$ is standard (i.e. $\mathfrak{a} \, ; \mathsf{i} = \mathfrak{a}$). Thus it suffices to show that i and $[v \mapsto s \cdot \mathfrak{a}]\mathsf{i}$ agree on $\partial(x \cdot \mathfrak{a})$. The only possible discrepancy is at v, but

$$v \in \partial(x \cdot \mathfrak{a}) \Rightarrow v \in (\partial t - \{y\})^{\langle \mathfrak{a} \rangle}$$

(since $x \in (\partial t - \{y\})$), so the choice of v ensures the required result. □

F
APPLIED λ-CALCULI

F.1 INTRODUCTION

6.1 We certainly want to name each natural number. This is achieved using two constants $0, \mathsf{S}$ to name zero and the successor function. These have $0 : \mathcal{N}$ and $\mathsf{S} : \mathcal{N}'$ as housing axioms (as described in Section 6.2). With these the term $\ulcorner m \urcorner = \mathsf{S}^m 0$ is a name for $m \in \mathbb{N}$. To name other functions on $\mathbb{N}$ we need more constants which name other attributes of $\mathbb{N}$. This is the topic of Chapters 7 and 9.

We certainly want to name boolean values. Thus we have two constants with

$$\mathsf{true} : \mathcal{B} \qquad \mathsf{false} : \mathcal{B}$$

as housing axioms. We also want to name the boolean connectives. More generally, for a type σ we use a constant

$$\mathsf{Cnd}_\sigma : \sigma \to \sigma \to \mathcal{B} \to \sigma$$

where $\mathsf{Cnd}_\sigma sru$ is read as 'if u then r else s' for $u : \mathcal{B}$ and $r, s : \sigma$. With this

$$\mathsf{Cnd}_\sigma sr\mathsf{true} \rhd r \qquad \mathsf{Cnd}_\sigma sr\mathsf{false} \rhd s$$

are the reduction axioms (as described in Section 6.5). The constant $\mathsf{Cnd}_\mathcal{B}$ gives all the boolean connectives. □

F.2 DERIVATION

6.2 (B) We are interested in derivations $B = (Z(YX))\uparrow\uparrow\uparrow$ where

$$(Z)\ \ \Pi \vdash z : \tau \qquad (Y)\ \ \Pi \vdash y : \sigma \qquad (X)\ \ \Pi \vdash x : \theta$$

are the three component extractions. From the order of the declarations in Π,

$$(X)\ \ \Pi[x] \qquad (Y)\ \ \Pi[y], \Pi(y) \qquad (Z)\ \ \Pi[z], \Pi^\uparrow[z]\downarrow, \Pi(z)$$

are the only possibilities. This gives us $1 \times 2 \times 3$ different derivations.

(C) We have $C = (WZY)\uparrow\uparrow\uparrow$ where

$$(Y)\ \ \Gamma[y], \Gamma(y) \qquad (Z)\ \ \Gamma[z] \qquad (W)\ \ \Gamma[w], \Gamma^\uparrow[w]\downarrow, \Gamma(w)$$

are the only possible components.

(D) We have $D = (Z(YX))\uparrow\uparrow\uparrow$ where

$$(X)\ \ \Delta[x] \qquad (Y)\ \ \Delta[y],\ \Delta^{\uparrow}[y]\downarrow,\ \Delta(y) \qquad (Z)\ \ \Delta[z],\ \Delta(z)$$

are the only possible components.

Later we see that there are no other derivations of these judgements. □

6.3 With this triple use of the identifier z there is only one possible derivation. With the information on the left we generate the derivation T on the right.

$$\begin{array}{lll} Z = \Sigma[z] & \Sigma = z : \sigma \\ R = Z\uparrow & \vdash r : \sigma' \\ R' = (R, z, \sigma)\downarrow & \Sigma \vdash r : \sigma' \\ \\ S = (R'Z)\uparrow & \vdash s : \sigma' \\ S' = (S, z, \sigma)\downarrow & \Sigma \vdash s : \sigma' \\ \\ T = (S'Z)\uparrow & \vdash t : \sigma' \end{array}
\qquad
\cfrac{\cfrac{\cfrac{\cfrac{\cfrac{\cfrac{\cfrac{Z}{R}\,(\uparrow)}{R'}\,(\downarrow) \quad Z}{\bullet}}{S}\,(\uparrow)}{S'}\,(\downarrow) \quad Z}{\bullet}}{T}\,(\uparrow)$$

With

$$r = \lambda x : \sigma \,.\, x \qquad s = \lambda y : \sigma \,.\, ry \qquad t = \lambda z : \sigma \,.\, sz$$

where x, y, z are distinct, and with the information on the left, we generate a new derivation T^* on the right with distinguished bindings.

$$\begin{array}{ll} X = \Gamma[x] & \Gamma = z : \sigma, y : \sigma, x : \sigma \\ Y = \Pi[y] & \Pi = z : \sigma, y : \sigma \\ R^* = X\uparrow & \Pi \vdash r : \sigma' \\ \\ Z = \Sigma[z] & \Sigma = x : \sigma \\ S^* = (R^*Y)\uparrow & \Sigma \vdash s : \sigma' \\ \\ T^* = (S^*Z)\uparrow & \vdash t : \sigma' \end{array}
\qquad
\cfrac{\cfrac{\cfrac{\cfrac{\cfrac{X}{R^*}\,(\uparrow) \quad Y}{\bullet}}{S^*}\,(\uparrow) \quad Z}{\bullet}}{T^*}\,(\uparrow)$$

This derivation has all the uses of W as near to the leaves as possible (i.e. none at all). It is also possible to delay some of these uses (and not use the distinctness of x, y, z). □

6.4 On the whole the derivations we have produced have root subjects of the form

$$\lambda \text{ declaration} \,.\, \text{body}$$

where 'body' is built up by application. For such a derivation the only choices are the kind of projections used at each leaf: whether an identifier should be declared as early as possible or as late as possible, or somewhere in between. For instance, a variant of the derivation of Solution 4.2(ϵ) has the form

$$(((V\Downarrow)Y)((U\Downarrow)(X\Downarrow)\Downarrow))\uparrow\uparrow\uparrow\uparrow$$

where V, U, X, Y are projections and each $\Downarrow$ is an appropriate number of Weakenings. There are, respectively, 4,3,2,1 choices for V, U, X, Y, so there are $4\times3\times2\times1 = 24$ possible derivations. □

F.3 Type synthesis

6.5 The successful part of the search tree gives us several possible solutions. Reading that tree from root upwards and taking the alternatives in turn (left first then right) gives

$$(\Pi[z]\Pi[x])\uparrow \quad (\Pi[z]\Pi^{\uparrow}[x]\downarrow)\uparrow \quad (\Pi[z]\Pi\langle x\rangle)\uparrow \quad (\Lambda[z]\Lambda[x])\uparrow\downarrow \quad (\Lambda[z]\Lambda\langle x\rangle)\uparrow\downarrow$$

in arboreal code.

When $z = y$ the context Π is not legal, so only the last two are derivations. □

6.6 In each case we give the full generation tree (the successful part of the search tree) and the subproblem solved at some nodes. Notice how the various extractions are created.

(B) With $\Pi = z:\tau, y:\sigma, x:\theta$ we obtain the tree shown.

$$\dfrac{\dfrac{\dfrac{\dfrac{\dfrac{\begin{matrix}\Pi[z]\\ \cdot\, l\uparrow\uparrow\uparrow\end{matrix}\text{ or }\dfrac{\begin{matrix}\Pi^{\uparrow}[z]\\ \cdot\downarrow l\uparrow\uparrow\uparrow\end{matrix}\text{ or }\dfrac{\begin{matrix}\Pi^{\uparrow\uparrow}[z]\\ \cdot\downarrow\downarrow l\uparrow\uparrow\uparrow\end{matrix}}{\downarrow\downarrow l\uparrow\uparrow\uparrow}}{\downarrow l\uparrow\uparrow\uparrow}}{l\uparrow\uparrow\uparrow}\text{ and }\dfrac{\dfrac{\begin{matrix}\Pi[y]\\ \cdot\, lr\uparrow\uparrow\uparrow\end{matrix}\text{ or }\dfrac{\begin{matrix}\Pi^{\uparrow}[y]\\ \cdot\downarrow lr\uparrow\uparrow\uparrow\end{matrix}}{\downarrow lr\uparrow\uparrow\uparrow}}{lr\uparrow\uparrow\uparrow}\text{ and }\dfrac{\begin{matrix}\Pi[x]\\ \cdot\, rr\uparrow\uparrow\uparrow\end{matrix}}{rr\uparrow\uparrow\uparrow}}{r\uparrow\uparrow\uparrow}}{\uparrow\uparrow\uparrow}}{\uparrow\uparrow}}{\uparrow}}{\perp}$$

$$\begin{array}{rl}(lr\uparrow\uparrow\uparrow) & (\Pi, y)\\ (rr\uparrow\uparrow\uparrow) & (\Pi, yx)\\ (l\uparrow\uparrow\uparrow) & (\Pi, z)\\ (r\uparrow\uparrow\uparrow) & (\Pi, yx)\\ (\uparrow\uparrow\uparrow) & (\Pi, z(yx))\end{array}$$

The crucial nodes are given on the right.

(C) With $\Gamma = w:\beta, y:\sigma, z:\tau$ we obtain the tree shown.

$$\dfrac{\dfrac{\dfrac{\dfrac{\dfrac{\dfrac{\begin{matrix}\Gamma[w]\\ \cdot\, ll\uparrow\uparrow\uparrow\end{matrix}\text{ or }\dfrac{\begin{matrix}\Gamma^{\uparrow}[w]\\ \cdot\downarrow ll\uparrow\uparrow\uparrow\end{matrix}\text{ or }\dfrac{\begin{matrix}\Gamma^{\uparrow\uparrow}[w]\\ \cdot\downarrow\downarrow ll\uparrow\uparrow\uparrow\end{matrix}}{\downarrow\downarrow ll\uparrow\uparrow\uparrow}}{\downarrow ll\uparrow\uparrow\uparrow}}{ll\uparrow\uparrow\uparrow}\text{ and }\dfrac{\begin{matrix}\Gamma[z]\\ \cdot\, rl\uparrow\uparrow\uparrow\end{matrix}}{rl\uparrow\uparrow\uparrow}}{l\uparrow\uparrow\uparrow}\text{ and }\dfrac{\begin{matrix}\Gamma[y]\\ \cdot\, r\uparrow\uparrow\uparrow\end{matrix}\text{ or }\dfrac{\begin{matrix}\Gamma^{\uparrow}[y]\\ \cdot\downarrow r\uparrow\uparrow\uparrow\end{matrix}}{\downarrow r\uparrow\uparrow\uparrow}}{r\uparrow\uparrow\uparrow}}{\uparrow\uparrow\uparrow}}{\uparrow\uparrow}}{\uparrow}}{\perp}$$

$$\begin{array}{rl}(ll\uparrow\uparrow\uparrow) & (\Gamma, w)\\ (rl\uparrow\uparrow\uparrow) & (\Gamma, z)\\ (l\uparrow\uparrow\uparrow) & (\Gamma, wz)\\ (r\uparrow\uparrow\uparrow) & (\Gamma, y)\\ (\uparrow\uparrow\uparrow) & (\Gamma, wzy)\end{array}$$

The crucial nodes are given on the right.

(D) With $\Delta = y : \sigma, z : \tau, x : \theta$ we obtain the tree shown.

$$
\dfrac{\dfrac{\dfrac{\dfrac{\dfrac{\Delta[z] \atop \cdot\, l\uparrow\uparrow\uparrow \quad \text{or} \quad \dfrac{\Delta^{\uparrow}[z] \atop \cdot \downarrow l\uparrow\uparrow\uparrow}{\downarrow l\uparrow\uparrow\uparrow}}{l\uparrow\uparrow\uparrow} \quad \text{and} \quad \dfrac{\dfrac{\Delta[y] \atop \cdot\, lr\uparrow\uparrow\uparrow \quad \text{or} \quad \dfrac{\Delta^{\uparrow}[y] \atop \cdot \downarrow lr\uparrow\uparrow\uparrow \quad \text{or} \quad \dfrac{\Delta^{\uparrow\uparrow}[y] \atop \cdot \downarrow\downarrow lr\uparrow\uparrow\uparrow}{\downarrow\downarrow lr\uparrow\uparrow\uparrow}}{\downarrow lr\uparrow\uparrow\uparrow}}{lr\uparrow\uparrow\uparrow} \quad \text{and} \quad \dfrac{\Delta[x] \atop \cdot\, rr\uparrow\uparrow\uparrow}{rr\uparrow\uparrow\uparrow}}{r\uparrow\uparrow\uparrow}}{\uparrow\uparrow\uparrow}}{\uparrow\uparrow}}{\uparrow}}{\perp}
\qquad
\begin{array}{rl}
(lr\uparrow\uparrow\uparrow) & (\Delta, y) \\
(rr\uparrow\uparrow\uparrow) & (\Delta, x) \\
(l\uparrow\uparrow\uparrow) & (\Delta, z) \\
(r\uparrow\uparrow\uparrow) & (\Delta, yx) \\
(\uparrow\uparrow\uparrow) & (\Delta, z(yx))
\end{array}
$$

The crucial nodes are given on the right. □

6.7 Consider $r = \lambda x : \sigma \,.\, x$, $s = \lambda y : \sigma \,.\, ry$, $t = \lambda z : \sigma \,.\, sz$ where x, y, z need not be distinct. Consider

$$
\begin{array}{lll}
\Sigma = z : \sigma & \Pi = z : \sigma, y : \sigma & \Gamma = z : \sigma, y : \sigma, x : \sigma \\
\Lambda = y : \sigma & \Delta = y : \sigma, x : \sigma & \\
\Xi = x : \sigma & \Psi = z : \sigma, x : \sigma &
\end{array}
$$

where some of these will be illegal and others will be equal if there are equalities amongst the identifiers.

For all cases the successful part of the search tree is contained in the following.

$$
\dfrac{\dfrac{\dfrac{\dfrac{\dfrac{\Gamma[x]}{(\Gamma, x)} \quad \text{or} \quad \dfrac{\dfrac{\Psi[x]}{(\Psi, x)} \quad \dfrac{\dfrac{\Xi[x]}{(\Xi, x)}}{(\emptyset, r)}}{(\Sigma, r)}}{(\Pi, r)} \quad \text{and} \quad \dfrac{\Pi[y]}{(\Pi, y)}}{(\Pi, ry)} \quad \text{or} \quad \dfrac{\dfrac{\dfrac{\dfrac{\Delta[x]}{(\Delta, x)} \quad \text{or} \quad \dfrac{\dfrac{\Xi[x]}{(\Xi, x)}}{(\emptyset, r)}}{(\Lambda, r)} \quad \text{and} \quad \dfrac{\Lambda[y]}{(\Lambda, y)}}{(\Lambda, ry)}}{(\emptyset, s)}}{(\Sigma, s)} \quad \text{and} \quad \dfrac{\Sigma[z]}{(\Sigma, z)}}{\dfrac{(\Sigma, sz)}{(\emptyset, t)}}
$$

If x, y, z are distinct then this is the full search tree, and it produces five different derivations. If there are equalities amongst x, y, z then some contexts are illegal, and these abort certain branches. In particular, if $x = y = z$ then $\Xi = \Lambda = \Sigma$ are the only legal contexts, and the search gives just one derivation T (as in Solution 6.3). □

6.8 (a) Let

$$|l, a| = l + 2a$$

for all $l, a \in \mathbb{N}$. Then

$$|l+1, a| = l + 1 + 2a < l + 2a + 2 = l + 2(a+1) = |l, a+1|$$

and

$$|l, a| = l + 2a < l + 1 + 2a = |l+1, a|$$

to verify (w, i).

(b) This is proved by a progressive induction over $h(\nabla)$.

When $h(\nabla) = 0$, we know that ∇ is a Leaf and $|t| = 0$ (since t is a constant or an identifier).

When $h(\nabla) \neq 0$, we know that ∇ has a root rule W, I, or E, which we look at in turn.

(W) Here $\nabla = T\downarrow$ where the context of ∇ has length l (say) and that of T has length $l - 1$. Both ∇ and T have the same subject t. Also $h(\nabla) = h(T) + 1$. Thus

$$|t| \leq h(T) < h(\nabla)$$

using the induction hypothesis. Also

$$h(\nabla) = h(T) + 1 \leq |l-1, |t|| < |l, |t||$$

again using the induction hypothesis.

(I) Here $\nabla = R\uparrow$ with $t = \lambda y : \sigma \,.\, r$ where the context of ∇ has length l (say) and that of R has length $l + 1$. Also $h(\nabla) = h(R) + 1$. Thus

$$|t| = |r| + 1 \leq h(R) + 1 = h(\nabla)$$

where the central comparison follows by the induction hypothesis. Also

$$|l, |t|| = l + 2|t| = l + 2|r| + 2 = (l+1) + 2|r| + 1 = |l+1, |r|| + 1$$

and hence

$$h(\nabla) = h(R) + 1 \leq |l+1, |r|| < |l, |t||$$

using the induction hypothesis.

(E) here $\nabla = QP$ with $t = qp$ and where P, Q, ∇ have the same context with length l, say. We have

$$|t| = \max(|q| + 1, |p| + 1) \leq \max(h(Q)| + 1, h(P)| + 1) = h(\nabla)$$

where the comparison follows by the induction hypothesis. Similarly

$$h(Q) \leq l + 2|q| \qquad h(P) \leq l + 2|p|$$

(by the induction hypothesis) so that

$$h(\nabla) \leq \max(l+2|q|+1, l+2|p|+1) < l+2\max(|q|+1, |p|+1) = l+2|qp| = |l, |t||$$

as required.

(c) Any solution ∇ to a synthesis problem (Γ, t) must satisfy

$$h(\nabla) \leq |l(\Gamma), |t||$$

which bounds the *size* of ∇, but doesn't restrict the *number* of such ∇.

We need to prove this result by a progressive induction over $|\Gamma, t|$.

Each solution of (Γ, t) is either a Leaf or arises from a solution of a cheaper problem by W, I, or E.

There is at most one Leaf solution to (Γ, t).

By the induction hypothesis each recursion step produces just finitely many derivations which must be combined in certain ways to produce the required solution. There are only finitely many possible combinations, and hence only finitely many possible solutions. $\square$

6.9 We prove this by induction on the cost

$$|\Gamma, t| = l(\Gamma) + 2|t|$$

of the synthesis problem. However, we can not use the obvious induction hypothesis.

Let us say context Σ is a **part** of a context Γ if Σ can be obtained from Γ by omitting some declarations. These omissions may occur throughout Γ; they need not be consecutive.

We say a problem (Γ, t) is **good** if for each pair of derivations

$$(\nabla_i) \quad \Sigma_i \vdash t : \tau_i \qquad (i = 1, 2)$$

where both Σ_1 and Σ_2 are parts of Γ, we have $\tau_1 = \tau_2$. In particular, if a problem is good then it can synthesize at most one type.

We show that each problem with a legal context is good. (It is *not* true that every problem is good. Look at $(x : \rho, x : \sigma, x)$ where $\rho \neq \sigma$.) To this end, for each $x \in \mathbb{N}$ let $\langle x \rangle$ abbreviate

For each problem (Γ, t), if $|\Gamma, t| \leq x$ and Γ is legal, then (Γ, t) is good

so we require $\forall x : \mathbb{N} . \langle x \rangle$. We prove this by a progressive induction on x.

Thus, for an arbitrary $x \in \mathbb{N}$, consider a problem (Γ, t) with $|\Gamma, t| \leq x$ and Γ legal. To show (Γ, t) is good consider a pair of derivations ∇_1, ∇_2 (as above). We must show that $\tau_1 = \tau_2$. Note that $l(\Sigma_i) \leq l(\Gamma)$.

These derivations need not be principal, but they do arise by W from principal derivations

$$(\nabla_i) \quad \Pi_i \vdash t : \tau_i \qquad (i = 1, 2)$$

where Π_i is an initial part of Σ_i, and hence $l(\Pi_i) \leq l(\Sigma_i) \leq l(\Gamma)$. We look at the various possible shapes of t.

(A) When $t = \mathsf{k}$ (a constant) the two principal derivations must be Axioms, so $\tau_1 = \kappa = \tau_2$ where $k : \kappa$ is the axiom statement for k.

(P) When $t = x$ (an identifier) the two principal derivations must be Projections. Thus, for $i = 1, 2$, we see that $x : \tau_i$ occurs in Π_i and hence $x : \tau_i$ occurs in Γ (since Π_i is a part of Γ). But Γ is legal, so $\tau_1 = \tau_2$.

(Neither of these cases calls on the induction hypothesis.)

(I) When $t = (\lambda y : \sigma . r)$ (an abstraction) we have derivations

$$\Pi_i, y : \sigma \vdash r : \rho_i$$

where $\tau_i = \sigma \to \rho_i$ (for $i = 1, 2$). Let Π be the amalgam of Π_1 and Π_2 as occurring in Π. Note that

$$|(\Pi, y : \sigma), r| < |\Pi, t| \leq x$$

so we may apply the induction hypothesis to the simpler problem $((\Pi, y : \sigma), r)$.

We need to check that $(\Pi, y : \sigma)$ is legal. There can be no clash in Π (for Π is a part of the legal context Γ). If y clashes with some z in Π, then z occurs in some Π_i, and this clash contradicts the legality of $(\Pi_i, y : \sigma)$.

This shows that $(\Pi, y : \sigma)$ is legal and hence the problem $((\Pi, y : \sigma), r)$ is good (by the induction hypothesis). Thus, looking at the derivations in $(\Pi_i, y : \sigma)$ we have $\rho_1 = \rho_2$ and hence $\tau_1 = \tau_2$ as required.

(E) When $t = qp$ (an application) we have derivations

$$\Pi_i \vdash q : \pi_i \to \tau_i$$

where $|\Gamma, q| < |\Gamma, t| \leq x$ (since $|q| < |t|$). Since Γ is legal the induction hypothesis ensures that (Γ, q) is good, and hence $\pi_1 \to \tau_1 = \pi_2 \to \tau_2$ to give $\tau_1 = \tau_2$, as required.

This completes the induction. □

6.10 When presented with a problem (Γ, qp) the original algorithm looks for alternative solutions

$$(\Gamma, q) \text{ and } (\Gamma, p) \quad \text{or} \quad (\Gamma^{\uparrow}, qp)$$

(where $\Gamma^{\uparrow}$ is a shortened version of Γ). If $(\Gamma^{\uparrow}, qp)$ succeeds, then both (Γ, q) and (Γ, p) will succeed. So the second alternative need not be investigated.

Here is a procedure which pursues only the first alternative.

Determine the legality of Γ.

- If Γ is illegal then abort (in failure).
- If Γ is legal then follow the path determined by t.

$(t = \mathsf{k})$ Return $\Gamma[\mathsf{k}]$.

$(t = x)$ Determine whether or not x is declared in Γ.

- If x is not declared then abort (in failure).
- If x is declared then return $\Gamma[x]$.

$(t = \lambda y : \sigma . r)$ Test both

$$((\Gamma, y : \sigma), r) \quad (\Gamma^{\uparrow}, t)$$

and if either succeeds in

$$R \text{ or } T \quad \text{then return} \quad R\uparrow \text{ or } T\downarrow$$

as appropriate.

$(t = qp)$ Test both

$$(\Gamma, q) \quad (\Gamma, p)$$

and require both to succeed with compatible types $\pi \to \tau$, τ (for some π, τ). Return all derivations

$$QP$$

for all possible solutions Q of (Γ, t) and P of (Γ, p) with compatible types.

Note that the case $t = \lambda y : \sigma \,.\, r$ cannot be simplified, for either alternative may succeed and the other fail. The two problems

$$(x : \rho,\ \lambda y : \sigma \,.\, yx) \quad (x : \sigma,\ \lambda y : \sigma \,.\, y)$$

illustrate these possibilities.

We may define a canonical derivation as one with as few uses of W as possible. To produce a canonical solution, first use the above algorithm to generate a search tree. Then pass through the tree from leaves to root. At each abstraction attempt always select $R\uparrow$ over $T\downarrow$ whenever there is a choice.

There is, of course, no good reason why this notion of a canonical derivation has any intrinsic value. In practice it is probably sensible to keep contexts as short as possible, i.e. any use of W should be as near to the root as possible. To find such a sensible derivation we must be prepared to generate the full search tree, and at each alternative always take the rightmost path and then backtrack if this aborts. This algorithm is quite costly. □

F.4 Mutation

6.11 Consider any $y \in (\partial\Sigma)^{\langle \mathfrak{a} \rangle}$. There is some $x \in \partial\Sigma$ with $y \in \partial s$ where $s = x \cdot \mathfrak{a}$. With this x and s we have

$$(x \cdot \mathfrak{A}) \quad \Gamma \vdash s : \sigma$$

(where $x : \sigma$ is in Σ), and hence $y \in \partial s \subseteq \partial\Gamma$, as required. □

6.12 We have $\nabla = \Sigma^{\Uparrow}[x]\Downarrow$ for some matching sequence $\Uparrow$ of shortenings and $\Downarrow$ of Weakenings. Then with the corresponding restriction $\Uparrow\mathfrak{A}$ of $\mathfrak{A}$ we have

$$\nabla \cdot \mathfrak{A} = \Sigma^{\Uparrow}[x] \cdot \Uparrow\mathfrak{A} = x \cdot \Uparrow\mathfrak{A} = x \cdot \mathfrak{A}$$

which depends only on $\mathfrak{A}$ and x. □

6.13 (i) We will need extractions

$$\begin{array}{lll} (Z)\ \Gamma \vdash z : \sigma'' & (Y)\ \Gamma \vdash y : \sigma' & (X)\ \Gamma \vdash x : \sigma \\ (W)\ \Gamma^{\uparrow} \vdash z : \sigma'' & (V)\ \Gamma^{\uparrow} \vdash y : \sigma' & \end{array}$$

where

$$X = \Gamma[x] = \Gamma(x) \qquad V = \Gamma^{\uparrow}[y] = \Gamma^{\uparrow}(y)$$

are unique but there are 2,2,3 possibilities for W, Y, Z.

With these we have

$$\nabla = ZYX \quad \text{or} \quad \nabla = ((WV)\downarrow)X$$

to give eight different solutions ∇.

(ii) Each relevant mutation

$$(\mathfrak{A})\quad \Gamma \xrightarrow{\ \mathfrak{i}\ } \Gamma \quad \text{has} \quad z \cdot \mathfrak{A} = Z \quad y \cdot \mathfrak{A} = Y \quad x \cdot \mathfrak{A} = X$$

where X, Y, Z are extractions. This gives six different mutations. We also need the lengthening

$$({}'\mathfrak{A})\quad \Gamma^{\uparrow} \xrightarrow{\ \mathfrak{i}\ } \Gamma \quad \text{with} \quad z \cdot {}'\mathfrak{A} = z \cdot \mathfrak{A} = Z \quad y \cdot {}'\mathfrak{A} = y \cdot \mathfrak{A} = Y$$

as nominated derivations.

(iii) For each extraction Z we have $Z \cdot \mathfrak{A} = z \cdot \mathfrak{A}$, which is also an extraction *but need not be the same as* Z. There are similar remarks for the extraction Y, and for ${}'\mathfrak{A}$.

When $\nabla = ZYX$ we have

$$\nabla \cdot \mathfrak{A} = (z \cdot \mathfrak{A})(y \cdot \mathfrak{A})(x \cdot \mathfrak{A})$$

which, as just observed, need not be ∇.

Also

$$WV \cdot {}'\mathfrak{A} = (z \cdot {}'\mathfrak{A})(y \cdot {}'\mathfrak{A}) = (z \cdot \mathfrak{A})(y \cdot \mathfrak{A})$$

so that, when $\nabla = ((WV)\downarrow)X$, we have

$$\nabla \cdot \mathfrak{A} = (WV \cdot {}'\mathfrak{A})(x \cdot \mathfrak{A}) = (z \cdot \mathfrak{A})(y \cdot \mathfrak{A})(x \cdot \mathfrak{A})$$

as before.

(iv) In all cases

$$\nabla \cdot \mathfrak{A} = (z \cdot \mathfrak{A})(y \cdot \mathfrak{A})(x \cdot \mathfrak{A})$$

i.e. the output does not depend on ∇ but is completely determined by $\mathfrak{A}$. In particular, a solution of (Γ, zyx) of the form $((WV)\downarrow)X$ can not arise as $\nabla \cdot \mathfrak{A}$. Any other solution can arise by a suitable choice of $\mathfrak{A}$. □

6.14 Observe that $\Pi^{\downarrow\uparrow} = \Pi$ for any context Π. Thus we get no more than 15 different mutations, as follows.

$$\begin{array}{llll}
(\mathfrak{A})\ \Sigma \xrightarrow{\ \mathfrak{a}\ } \Gamma & (\uparrow\mathfrak{A})\ \Sigma^{\uparrow} \xrightarrow{\ \mathfrak{a}\ } \Gamma & ({}'\mathfrak{A})\ \Sigma^{\downarrow} \xrightarrow{\ {}'\mathfrak{a}\ } \Gamma^{\downarrow} & \\
(\uparrow\uparrow\mathfrak{A})\ \Sigma^{\uparrow\uparrow} \xrightarrow{\ \mathfrak{a}\ } \Gamma & ({}'\uparrow\mathfrak{A})\ \Sigma^{\uparrow\downarrow} \xrightarrow{\ {}'\mathfrak{a}\ } \Gamma^{\downarrow} & (\uparrow{}'\mathfrak{A})\ \Sigma \xrightarrow{\ {}'\mathfrak{a}\ } \Gamma^{\downarrow} & ({}''\mathfrak{A})\ \Sigma^{\downarrow\downarrow} \xrightarrow{\ {}''\mathfrak{a}\ } \Gamma^{\downarrow\downarrow} \\
(\uparrow\uparrow\uparrow\mathfrak{A})\ \Sigma^{\uparrow\uparrow\uparrow} \xrightarrow{\ \mathfrak{a}\ } \Gamma & ({}'\uparrow\uparrow\mathfrak{A})\ \Sigma^{\uparrow\uparrow\downarrow} \xrightarrow{\ {}'\mathfrak{a}\ } \Gamma^{\downarrow} & (\uparrow{}'\uparrow\mathfrak{A})\ \Sigma^{\uparrow} \xrightarrow{\ {}'\mathfrak{a}\ } \Gamma^{\downarrow} & ({}''\uparrow\mathfrak{A})\ \Sigma^{\uparrow\downarrow\downarrow} \xrightarrow{\ {}''\mathfrak{a}\ } \Gamma^{\downarrow\downarrow} \\
(\uparrow\uparrow{}'\mathfrak{A})\ \Sigma^{\uparrow} \xrightarrow{\ {}'\mathfrak{a}\ } \Gamma^{\downarrow} & ({}'\uparrow{}'\mathfrak{A})\ \Sigma^{\downarrow} \xrightarrow{\ {}^{\bullet}\mathfrak{a}\ } \Gamma^{\bullet} & (\uparrow{}''\mathfrak{A})\ \Sigma^{\downarrow} \xrightarrow{\ {}''\mathfrak{a}\ } \Gamma^{\downarrow\downarrow} & ({}'''\mathfrak{A})\ \Sigma^{\downarrow\downarrow\downarrow} \xrightarrow{\ {}'''\mathfrak{a}\ } \Gamma^{\downarrow\downarrow\downarrow}
\end{array}$$

Here

$$'\mathfrak{a} = [u \mapsto u]\mathfrak{a} \qquad ''\mathfrak{a} = [v \mapsto v]\,'\mathfrak{a} \qquad '''\mathfrak{a} = [w \mapsto w]\,''\mathfrak{a}$$

and $^{\bullet}\mathfrak{a}$ and $\Gamma^{\bullet}$ will be described shortly.

At first sight it seems that

$$\uparrow'\!\uparrow\!\mathfrak{A} = \uparrow\uparrow'\!\mathfrak{A} \qquad '\!\uparrow'\!\mathfrak{A} = \uparrow''\!\mathfrak{A}$$

are possible. Let's look at the details of these.

For the left hand equality let

$$\Pi = \Sigma^{\uparrow} \quad \Sigma^{\downarrow} = \Sigma, u : \theta \quad \Gamma^{\downarrow} = \Gamma, u : \theta \quad \Pi^{\downarrow} = \Pi, u : \theta$$

so that

$$\begin{array}{rlrl}
(\mathfrak{A}) & \Sigma \xrightarrow{\mathfrak{a}} \Gamma & (\mathfrak{A}) & \Sigma \xrightarrow{\mathfrak{a}} \Gamma \\
(\uparrow\mathfrak{A}) & \Pi \xrightarrow{\mathfrak{a}} \Gamma & ('\mathfrak{A}) & \Sigma^{\downarrow} \xrightarrow{'\mathfrak{a}} \Gamma^{\downarrow} \\
('\!\uparrow\!\mathfrak{A}) & \Pi^{\downarrow} \xrightarrow{'\mathfrak{a}} \Gamma^{\downarrow} & (\uparrow'\!\mathfrak{A}) & \Sigma \xrightarrow{'\mathfrak{a}} \Gamma^{\downarrow} \\
(\uparrow'\!\uparrow\!\mathfrak{A}) & \Pi \xrightarrow{'\mathfrak{a}} \Gamma^{\downarrow} & (\uparrow\uparrow'\!\mathfrak{A}) & \Pi \xrightarrow{'\mathfrak{a}} \Gamma^{\downarrow}
\end{array}$$

gives the construction of the two left hand mutations to verify that equality.

For the right hand equality we have

$$\Sigma^{\downarrow\downarrow} = \Sigma^{\downarrow}, v : \psi \quad \Gamma^{\downarrow\downarrow} = \Gamma^{\downarrow}, v : \psi$$

in $'\!\uparrow'\!\mathfrak{A}$ and $\uparrow''\!\mathfrak{A}$. But then in $'\!\uparrow'\!\mathfrak{A}$ we have

$$\Sigma^{\downarrow} = \Sigma, u : \theta \quad \Gamma^{\bullet} = \Gamma^{\downarrow}, u' : \theta \quad {}^{\bullet}\mathfrak{a} = [u \mapsto u']\,'\mathfrak{a}$$

where $u' \notin \partial(\Gamma^{\downarrow})$. In particular, $^{\bullet}\mathfrak{a} \neq {}''\mathfrak{a}$ and, more importantly,

$$\Gamma^{\bullet} = \Gamma, u : \theta, u' : \theta \neq \Gamma, u : \theta, v : \psi = \Gamma^{\downarrow\downarrow}$$

and the right hand equality does not hold. □

6.15 To hit T with the base mutation $\mathfrak{T} = \mathfrak{I}$ we will need various lengthenings and restrictions. Let

$$\Sigma = z : \sigma \quad Z = \Sigma[z] \quad \Pi = z : \sigma, y : \sigma \quad Y = \Pi[y] \quad \Gamma = z : \sigma, y : \sigma, x : \sigma \quad X = \Gamma[x]$$

and consider

$$\begin{array}{rlll}
(\mathfrak{T}) & \emptyset \xrightarrow{\mathfrak{i}} \emptyset & & \\
('\mathfrak{T}) & \Sigma \xrightarrow{\mathfrak{t}} \Sigma & \mathfrak{t} = [z \mapsto z]\mathfrak{i} & z \cdot '\mathfrak{T} = Z \\
(\mathfrak{S} = \uparrow'\mathfrak{T}) & \emptyset \xrightarrow{\mathfrak{t}} \Sigma & & \\
('\mathfrak{S}) & \Sigma \xrightarrow{\mathfrak{s}} \Pi & \mathfrak{s} = [z \mapsto y]\mathfrak{t} & z \cdot '\mathfrak{S} = Y \\
(\mathfrak{R} = \uparrow'\mathfrak{S}) & \emptyset \xrightarrow{\mathfrak{s}} \Pi & & \\
('\mathfrak{R}) & \Sigma \xrightarrow{\mathfrak{r}} \Gamma & \mathfrak{r} = [z \mapsto x]\mathfrak{s} & z \cdot '\mathfrak{R} = X
\end{array}$$

(where these contexts and extractions occurred in Solution 6.3). Remembering the construction of T given in that solution we have

$$\begin{array}{lllll} T \cdot \mathfrak{T} & = (S'Z \cdot '\mathfrak{T})\uparrow & = ((S' \cdot '\mathfrak{T})(Z \cdot '\mathfrak{T}))\uparrow & = ((S' \cdot '\mathfrak{T})Z)\uparrow \\ S \cdot \mathfrak{T} & = (R'Z \cdot '\mathfrak{S})\uparrow & = ((R' \cdot '\mathfrak{S})(Z \cdot '\mathfrak{S}))\uparrow & = ((R' \cdot '\mathfrak{S})Y)\uparrow \\ R \cdot \mathfrak{R} & = (Z \cdot '\mathfrak{R})\uparrow & = X\uparrow \end{array}$$

so that

$$T \cdot \mathfrak{I} = (((X\uparrow)Y)\uparrow Z)\uparrow = T^*$$

where T^* is as in the second part of Solution 6.3.

In a similar way we find

$$S \cdot \mathfrak{I}((Y\uparrow)Z)\uparrow = S^* \qquad R \cdot \mathfrak{I} = S^* = R$$

(and hence $\mathfrak{I}$ doesn't change R). □

6.16 Carrying over the notation of Example 6.5 and Exercise 6.5, with the context $\Gamma = x : \theta, y : \psi$ we have $\Theta = \Gamma^{\uparrow}$, $\Pi = \Gamma^{\downarrow}$, $\Lambda = \Gamma^{\uparrow\downarrow}$ where the two extensions (to form Π and Λ) are by $z : \phi$. We assume that x, y, z are distinct.

For the mutation

$$(\mathfrak{A}) \;\; \Gamma \xrightarrow{\mathfrak{i}} \Gamma \quad \text{we have} \quad (x \cdot \mathfrak{A}) \;\; \Gamma \vdash x : \theta \qquad (y \cdot \mathfrak{A}) \;\; \Gamma \vdash y : \psi$$

so these nominated derivations must be extractions. There are two choices $\Gamma[x]$ or $\Gamma(x)$ for $x \cdot \mathfrak{A}$, but just one choice for $y \cdot \mathfrak{A}$. This gives the two versions of $\mathfrak{A}$. Notice that $\mathfrak{A} = \mathfrak{I}_\Gamma$ when $x \cdot \mathfrak{A} = \Gamma(x)$.

For convenience let $X = x \cdot \mathfrak{A}$ and $Z = \Pi[z]$ (so that X is one of the two extractions of x from Γ). We need to generate several mutations from $\mathfrak{A}$. With $\mathfrak{z} = [z \mapsto z]\mathfrak{i}$ these are

$$(\uparrow\mathfrak{A}) \;\; \Theta \xrightarrow{\mathfrak{i}} \Gamma \qquad ('\uparrow\mathfrak{A}) \;\; \Lambda \xrightarrow{\mathfrak{z}} \Pi \qquad ('\mathfrak{A}) \;\; \Pi \xrightarrow{\mathfrak{z}} \Pi$$

with

$$x \cdot \uparrow\mathfrak{A} = X \qquad x \cdot '\uparrow\mathfrak{A} = X\downarrow \;\; z \cdot '\uparrow\mathfrak{A} = Z \qquad x \cdot '\mathfrak{A} = X\downarrow \;\; z \cdot '\mathfrak{A} = Z$$

as the relevant nominated derivations. (The nominated derivations for y are not needed.)

There are two forms of ∇

$$(\Pi(z)P)\uparrow \qquad (\Lambda(z)L)\uparrow\downarrow$$

where P is one of the three extractions of x from Π and L is one of the two extractions of x from Λ. In all cases we have

$$P \cdot '\mathfrak{A} = x \cdot '\mathfrak{A} = X\downarrow \qquad L \cdot '\uparrow\mathfrak{A} = x \cdot '\uparrow\mathfrak{A} = X\downarrow$$

so that

$$(\Pi(z)P)\uparrow\cdot\mathfrak{A} = ((\Pi(z)P)\cdot{}'\mathfrak{A})\uparrow = (Z(X\downarrow))\uparrow \quad (\Lambda(z)L)\uparrow\downarrow\cdot\mathfrak{A} = ((\Lambda(z)L)\cdot{}'\uparrow\mathfrak{A})\uparrow = (Z(X\downarrow))\uparrow$$

in other words

$$\nabla\cdot\mathfrak{A} = (Z(X\downarrow))\uparrow$$

in all cases. This result has the form

$$(\Pi(z)P)\uparrow$$

but only $P = \Gamma[x]\downarrow$ and $P = \Gamma(x)\downarrow = \Pi(x)$ can occur. The extra possibility $P = \Pi[x]$ never occurs in $\nabla\cdot\mathfrak{A}$. □

6.17 As in Solutions 6.2 and 6.6, using the context $\Gamma = w:\beta, y:\sigma, z:\tau$ we have $C = \Gamma(C)\uparrow\uparrow\uparrow$ where $\Gamma(C) = WZY$ and where W, Y, Z are extractions of w, y, z from Γ, respectively. There are 3,2,1 of these.

The redex removal algorithm gives

$$CB\bullet 1 = \Gamma(C)\uparrow\uparrow\cdot\mathcal{B} = \Gamma(C)\cdot{}''\mathfrak{B}$$

for a certain mutation $\mathfrak{B}$ and extensions $'\mathfrak{B}$, $''\mathfrak{B}$. Since there is no danger of identifiers clashing these mutations are

$$(\mathfrak{B})\ \Gamma^{\uparrow\uparrow} \xrightarrow{\mathfrak{b}} \emptyset \qquad ('\mathfrak{B})\ \Gamma^{\uparrow} \xrightarrow{'\mathfrak{b}} y:\sigma \qquad (''\mathfrak{B})\ \Gamma \xrightarrow{''\mathfrak{b}} \Lambda$$

where

$$\mathfrak{b} = [w\mapsto \mathsf{B}]\mathfrak{i} \qquad {}'\mathfrak{b} = [y\mapsto y]\mathfrak{b} \qquad {}''\mathfrak{b} = [z\mapsto z]{}'\mathfrak{b} \qquad \Lambda = y:\sigma, z:\tau$$

are the shafts and the final target, and with

$$\begin{array}{rclrclrcl}
w\cdot\mathfrak{B} &=& B \\
w\cdot{}'\mathfrak{B} &=& B\downarrow & y\cdot{}'\mathfrak{B} &=& \Lambda^{\uparrow}[y] \\
w\cdot{}''\mathfrak{B} &=& B\downarrow\downarrow & y\cdot{}''\mathfrak{B} &=& \Lambda(y) & z\cdot{}''\mathfrak{B} &=& \Lambda(z)
\end{array}$$

as nominated derivations.

Now

$$\begin{aligned}
W\cdot{}''\mathfrak{B} &= w\cdot{}''\mathfrak{B} = B\downarrow\downarrow\\
Z\cdot{}''\mathfrak{B} &= z\cdot{}''\mathfrak{B} = \Lambda(z)\\
Y\cdot{}''\mathfrak{B} &= y\cdot{}''\mathfrak{B} = \Lambda(y)
\end{aligned}$$

for all versions of W, Z, Y are used. Thus

$$D^* = CB\bullet 1 = ((B\downarrow\downarrow)\Lambda(z)\Lambda(y))\uparrow\uparrow$$

is the generated derivation. This has the shape shown right, perhaps not what you expected.

$$\dfrac{\dfrac{\dfrac{\dfrac{\dfrac{\dfrac{B}{\bullet}(\downarrow)}{\bullet}(\downarrow)\quad \Lambda(z)}{\bullet}\quad \dfrac{y:\sigma\vdash y:\sigma}{\bullet}(\downarrow)}{\bullet}}{\bullet}(\uparrow)}{\vdash \mathsf{D}^*:\delta}(\uparrow)$$

Hitting D^* with $\mathfrak{I}$ gives

$$D^* \cdot \mathfrak{I} = ((B{\downarrow}{\downarrow})\Lambda[z]\Lambda(y) \cdot {}''\mathfrak{I}){\uparrow}{\uparrow} = ((B \cdot \mathfrak{I})(z \cdot {}''\mathfrak{I})(y \cdot {}''\mathfrak{I})){\uparrow}{\uparrow}$$

for certain mutations $'\mathfrak{I}, {}''\mathfrak{I}, \uparrow{}''\mathfrak{I}$, and $\mathfrak{I} = {\uparrow}{\uparrow}{}''\mathfrak{I}$ generated from $\mathfrak{I}$.

We have

$$\begin{array}{rllll}
(\mathfrak{I}) & \emptyset \xrightarrow{\mathfrak{i}} \emptyset & & & \\
('\mathfrak{I}) & y:\sigma \xrightarrow{'\mathfrak{i}} y:\sigma & '\mathfrak{i} = [y \mapsto y]\mathfrak{i} & y \cdot {}'\mathfrak{I} = \Lambda^{\uparrow}[y] & \\
(''\mathfrak{I}) & \Lambda \xrightarrow{''\mathfrak{i}} \Lambda & ''\mathfrak{i} = [z \mapsto z]'\mathfrak{i} & y \cdot {}''\mathfrak{I} = \Lambda(y) & z \cdot {}''\mathfrak{I} = \Lambda[z] \\
(\uparrow{}''\mathfrak{I}) & y:\sigma \xrightarrow{\mathfrak{j}} \Lambda & \mathfrak{j} = {}''\mathfrak{i} & y \cdot \uparrow{}''\mathfrak{I} = \Lambda(y) & \\
(\mathfrak{I} = {\uparrow}{\uparrow}{}''\mathfrak{I}) & \emptyset \xrightarrow{\mathfrak{j}} \Lambda & & &
\end{array}$$

to give

$$D^* \cdot \mathfrak{I} = (B \cdot \mathfrak{I})\Lambda[z]\Lambda(y)$$

and it remains to calculate $B \cdot \mathfrak{I}$.

So far no renaming identifiers have been necessary, but now we will require three lengthenings of $\mathfrak{I}$, and since the target is Λ we must choose the new target identifiers carefully.

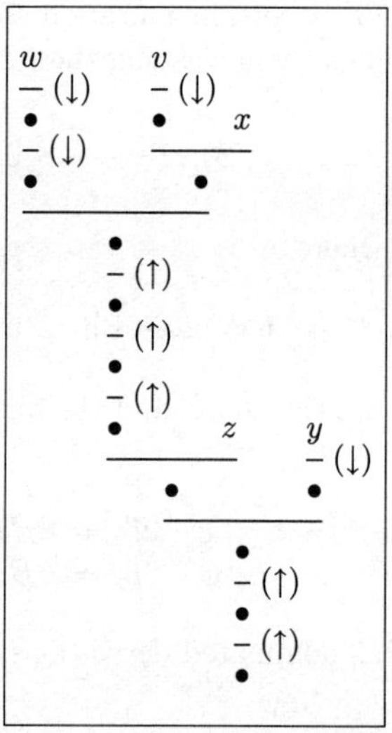

With $\Pi = z:\tau, y:\sigma, x:\theta$ we have $B = \Pi(B){\uparrow}{\uparrow}{\uparrow}$ where $\Pi(B) = Z(YX)$ to give

$$B \cdot \mathfrak{I} = (\Pi(B) \cdot \mathfrak{K}){\uparrow}{\uparrow}{\uparrow}$$

for a lengthening $\mathfrak{K}$ of $\mathfrak{I}$. With $\Xi = \Lambda, w:\tau, v:\sigma, x:\theta$ and $\mathfrak{K} = {}'''\mathfrak{I}$ we use

$$\begin{array}{rlll}
(\mathfrak{I}) & \emptyset \xrightarrow{\mathfrak{j}} \Lambda & & \\
('\mathfrak{I}) & \Pi^{\uparrow\uparrow} \xrightarrow{'\mathfrak{j}} \Xi^{\uparrow\uparrow} & '\mathfrak{j} = [z \mapsto w]\mathfrak{j} & z \cdot {}'\mathfrak{I} = \Xi^{\uparrow\uparrow}[w] \\
(''\mathfrak{I}) & \Pi^{\uparrow} \xrightarrow{''\mathfrak{j}} \Xi^{\uparrow} & ''\mathfrak{j} = [y \mapsto v]'\mathfrak{j} & z \cdot {}''\mathfrak{I} = \Xi^{\uparrow\uparrow}[w]{\downarrow} \\
 & & & y \cdot {}''\mathfrak{I} = \Xi^{\uparrow}[v] \\
(\mathfrak{K}) & \Pi \xrightarrow{\mathfrak{k}} \Xi & \mathfrak{k} = [x \mapsto x]''\mathfrak{j} & z \cdot \mathfrak{K} = \Xi(w) \\
 & & & y \cdot \mathfrak{K} = \Xi(v) \\
 & & & x \cdot \mathfrak{K} = \Xi[x]
\end{array}$$

w — ($\downarrow$) • — ($\downarrow$) • ; v — ($\downarrow$) • ; x • ; • — ($\uparrow$) • — ($\uparrow$) • — ($\uparrow$) • ; z • ; y — ($\downarrow$) • ; • — ($\uparrow$) • — ($\uparrow$) •

to get

$$\Pi(B) \cdot \mathfrak{K} = \Xi(w)(\Xi(v)\Xi[x]) = \Xi(B) \quad \text{(say)}$$

so that

$$D^* \cdot \mathcal{I} = ((\Xi(B){\uparrow}{\uparrow}{\uparrow})\Lambda[z]\Lambda(y)){\uparrow}{\uparrow}$$

is the final result. All the leaves are gate Projections of the indicated identifier. □

6.18 This is proved by induction over ∇ with allowable variation of $\mathfrak{A}$.

For the two base cases

$$\nabla = \Sigma[\mathsf{k}] \qquad \nabla = \Sigma[x]$$

(where k is a constant and x is an identifier) we have

$$\nabla \cdot \mathfrak{A} = \Gamma(\mathsf{k}) \qquad \nabla \cdot \mathfrak{A} = x \cdot \mathfrak{A} = \Gamma(x \cdot \mathfrak{a})$$

both of which are standard extractions.

For the induction step across W we must consider $\nabla = \nabla^{\uparrow}\downarrow$ where

$$(\nabla^{\uparrow}) \quad \Sigma^{\uparrow} \vdash t : \tau$$

with a shortened context. Observe that with the mutation

$$(\uparrow\mathfrak{A}) \quad \Sigma^{\uparrow} \xrightarrow{\mathfrak{a}} \Gamma$$

is renaming. Thus, using the induction hypothesis on $\nabla^{\uparrow}$ with the parameter $\uparrow\mathfrak{A}$, we see that

$$\nabla \cdot \mathfrak{A} = \nabla^{\uparrow} \cdot \uparrow\mathfrak{A}$$

is standard, as required.

For the induction step across I we must consider $\nabla = \nabla'\uparrow$ where

$$(\nabla') \quad \Sigma, y : \sigma \vdash r : \rho$$

with $t = (\lambda y : \sigma . r)$. Observe that with $'\mathfrak{a} = [y \mapsto v]\mathfrak{a}$ where $v \notin \partial\Gamma$ the lengthening

$$('\mathfrak{A}) \quad \Sigma, y : \sigma \xrightarrow{'\mathfrak{a}} \Gamma, v : \sigma$$

is renaming. Thus, using the induction hypothesis on ∇' with the parameter $'\mathfrak{A}$, we see that

$$\nabla \cdot \mathfrak{A} = (\nabla' \cdot {'\mathfrak{A}})\uparrow$$

is standard, as required.

The induction step across E is immediate. □

6.19 This is false. In fact, for such a mutation $\mathfrak{A}$ the result $\nabla \cdot \mathfrak{A}$ need not be (fully) standard even when ∇ is already.

For an example of this consider the derivation ∇ of Examples 6.3 and 6.10. We know that ∇ is fully standard. Still using Example 6.10 consider the mutation

$$(\mathfrak{B}) \quad \Xi \xrightarrow{\mathfrak{a}} \Xi \quad \text{where} \quad z \cdot \mathfrak{B} = \nabla$$

which is fully standard. In the Example we obtained

$$\nabla \cdot \mathfrak{B} = \nabla^{\eta}$$

where ∇^{η} is not fully standard. This gives a counterexample. □

6.20 (a) We must check that

$$x \cdot (\mathfrak{B}\,;\mathfrak{A}) \quad \text{has the form} \quad \Gamma \vdash (x \cdot (\mathfrak{b}\,;\mathfrak{a})) : \xi$$

for all declarations $x : \xi$ of Π. But

$$x \cdot \mathfrak{B} \quad \text{has the form} \quad \Sigma \vdash (x \cdot \mathfrak{b}) : \xi$$

so that

$$(x \cdot \mathfrak{B}) \cdot \mathfrak{A} \quad \text{has the form} \quad \Gamma \vdash ((x \cdot \mathfrak{b}) \cdot \mathfrak{a}) : \xi$$

and hence the known identity

$$x \cdot (\mathfrak{b} \,; \mathfrak{a}) = (x \cdot \mathfrak{b}) \cdot \mathfrak{a}$$

gives the required result.

(b) This is false. Consider the derivations and mutations of Example 6.10. There we showed that

$$(\nabla \cdot \mathfrak{B}) \cdot \mathfrak{A} = \nabla^*$$

holds. Also

$$z \cdot (\mathfrak{B} \,; \mathfrak{A}) = (z \cdot \mathfrak{B}) \cdot \mathfrak{A} = \nabla \cdot \mathfrak{A} = \nabla$$

and hence $\mathfrak{B} \,; \mathfrak{A} = \mathfrak{B}$. But then

$$\nabla \cdot (\mathfrak{B} \,; \mathfrak{A}) = \nabla \cdot \mathfrak{B} = \nabla^{\eta}$$

which, since $\nabla^* \neq \nabla^{\eta}$, gives the required counterexample.

It can be shown that for any composable $\mathfrak{B}, \mathfrak{A}$, if there is some $\mathfrak{C}$ such that

$$(\nabla \cdot \mathfrak{B}) \cdot \mathfrak{A} = \nabla \cdot \mathfrak{C}$$

for all compatible ∇, then $\mathfrak{C} = \mathfrak{B} \,; \mathfrak{A}$ has this property. Thus, in general, there is no acceptable composition of mutations.

(c) Again this is false. Given three mutations $\mathfrak{A}, \mathfrak{B}, \mathfrak{C}$ with the appropriate compatibility we have

$$x \cdot ((\mathfrak{C} \,; \mathfrak{B}) \,; \mathfrak{A}) = (x \cdot (\mathfrak{C} \,; \mathfrak{B})) \cdot \mathfrak{A} = ((x \cdot \mathfrak{C}) \cdot \mathfrak{B}) \cdot \mathfrak{A} \qquad x \cdot (\mathfrak{C} \,; (\mathfrak{B} \,; \mathfrak{A})) = (x \cdot \mathfrak{C}) \cdot (\mathfrak{B} \,; \mathfrak{A})$$

for each relevant identifier x. As we have just seen, with $x \cdot \mathfrak{C} = \nabla$

$$(\nabla \cdot \mathfrak{B}) \cdot \mathfrak{A} \neq \nabla \cdot (\mathfrak{B} \,; \mathfrak{A})$$

can happen and hence

$$(\mathfrak{C} \,; \mathfrak{B}) \,; \mathfrak{A} \neq \mathfrak{C} \,; (\mathfrak{B} \,; \mathfrak{A})$$

for the required counterexample. □

6.21 This is rather tricky. For each term t let $\langle t \rangle$ abbreviate the following.

For each compatible pair

$$(\nabla) \;\; \Sigma \vdash t : \tau \qquad (\mathfrak{A}) \;\; \Sigma \xrightarrow{\mathfrak{a}} \Gamma$$

where $\mathfrak{A}$ is renaming, the algorithm succeeds and

$$\nabla \cdot \mathfrak{A} = \nabla[\Gamma, t, \mathfrak{a}]$$

holds.

We show that $\langle t \rangle$ holds for all terms t by induction over t with variations of the parameters ∇ and $\mathfrak{A}$.

Notice that in all cases we have

$$\nabla = \nabla^{\Uparrow}\Downarrow \quad \text{where} \quad (\nabla^{\Uparrow}) \;\; \Sigma^{\Uparrow} \vdash t : \tau$$

where $\nabla^{\Uparrow}$ is principal. The restriction

$$(\Uparrow\mathfrak{A}) \;\; \Sigma^{\Uparrow} \xrightarrow{\;\mathfrak{a}\;} \Gamma$$

is renaming and

$$\nabla \cdot \mathfrak{A} = \nabla^{\Uparrow} \cdot \Uparrow\mathfrak{A}$$

holds.

We look at the two base cases $(t = \mathsf{k})$ and $(t = x)$ and the two induction steps $(t = qp)$ and $(t = \lambda y : \sigma \,.\, r)$ in turn.

$(t = \mathsf{k})$ The existence of $\mathfrak{A}$ ensures that Γ is legal so

$$\nabla[\Gamma, t, \mathfrak{a}] = \Gamma(\mathsf{k})$$

exists. Also

$$\nabla^{\Uparrow} = \Sigma^{\Uparrow}[\mathsf{k}] \quad \text{so} \quad \nabla \cdot \mathfrak{A} = \Sigma^{\Uparrow}[\mathsf{k}] \cdot \Uparrow\mathfrak{A} = \Gamma(\mathsf{k})$$

as required.

$(t = x)$ The existence of $\mathfrak{A}$ ensures that Γ is legal and gives

$$x \cdot \mathfrak{A} = \Gamma(x \cdot \mathfrak{a}) = \nabla[\Gamma, t, \mathfrak{a}]$$

i.e. the algorithm does terminate. Also

$$\nabla^{\Uparrow} = \Sigma^{\Uparrow}[x] \quad \text{so} \quad \nabla \cdot \mathfrak{A} = \Sigma^{\Uparrow}[x] \cdot \Uparrow\mathfrak{A} = x \cdot \Uparrow\mathfrak{A} = x \cdot \mathfrak{A}$$

as required.

$(t = qp)$ With

$$(Q) \;\; \Sigma^{\Uparrow} \vdash q : \pi \to \tau \qquad (P) \;\; \Sigma^{\Uparrow} \vdash p : \pi$$

we have $\nabla^{\Uparrow} = QP$ and the induction hypothesis provides

$$Q \cdot \Uparrow\mathfrak{A} = \nabla[\Gamma, q, \mathfrak{a}] \qquad P \cdot \Uparrow\mathfrak{A} = \nabla[\Gamma, p, \mathfrak{a}]$$

(where both do exist). Thus

$$\nabla \cdot \mathfrak{A} = (QP) \cdot \Uparrow\mathfrak{A} = (Q \cdot \Uparrow\mathfrak{A})(P \cdot \Uparrow\mathfrak{A}) = \nabla[\Gamma, q, \mathfrak{a}]\nabla[\Gamma, p, \mathfrak{a}] = \nabla[\Gamma, t, \mathfrak{a}]$$

as required.

$(t = \lambda y : \sigma \,.\, r)$ With

$$(\nabla^{\Uparrow\downarrow}) \;\; \Sigma^{\Uparrow}, y : \sigma \vdash r : \rho$$

(for some type ρ) we have $\nabla^{\Uparrow} = \nabla^{\Uparrow\downarrow}\uparrow$. With $\Sigma^{\Uparrow\downarrow} = \Sigma^{\Uparrow}, y : \sigma$ and

$$v = \mathit{fresh}(\partial\Gamma) \quad \Gamma^{\downarrow} = \Gamma, v : \sigma \quad {}'\mathfrak{a} = [y \mapsto v]\mathfrak{a}$$

the lengthened mutation

$$('\Uparrow\mathfrak{A}) \quad \Sigma^{\Uparrow\downarrow} \xrightarrow{'\mathfrak{a}} \Gamma^{\downarrow}$$

is renaming. (This is because the nominated derivations are

$$x \cdot {'\Uparrow\mathfrak{A}} = (x \cdot \Uparrow\mathfrak{A})\downarrow = (x \cdot \mathfrak{A})\downarrow \qquad y \cdot {'\Uparrow\mathfrak{A}} = \Gamma^{\downarrow}(v)$$

for each identifier x declared in $\Sigma^{\Uparrow}$ and any Weakening of a standard extraction is a standard extraction). The induction hypothesis ensures that

$$\nabla^{\Uparrow\downarrow} \cdot {'\Uparrow\mathfrak{A}} = \nabla[\Gamma^{\downarrow}, r, {'\mathfrak{a}}]$$

exists. Thus

$$\nabla \cdot \mathfrak{A} = \nabla^{\Uparrow} \cdot \Uparrow\mathfrak{A} = (\nabla^{\Uparrow\downarrow} \cdot {'\Uparrow\mathfrak{A}})\uparrow = \nabla[\Gamma^{\downarrow}, r, {'\mathfrak{a}}]\uparrow = \nabla[\Gamma, t, \mathfrak{a}]$$

as required. □

F.5 Computation

6.22 Recall that the housing axioms are $\mathsf{I} : \iota$, $\mathsf{K} : \kappa$, $\mathsf{S} : \sigma$ where $\mathsf{I} = \mathsf{I}(\theta)$, $\mathsf{K} = \mathsf{K}(\theta, \psi)$, $\mathsf{S} = \mathsf{S}(\theta, \psi, \phi)$ and

$$\iota = \theta \to \theta \qquad \kappa = \psi \to \theta \to \psi \qquad \sigma = (\theta \to \psi \to \phi) \to (\theta \to \psi) \to (\theta \to \phi)$$

for arbitrary types θ, ψ, ϕ. The associated reduction axioms are

$$\mathsf{I}u \rhd u \qquad \mathsf{K}vu \rhd v \qquad \mathsf{S}wvu \rhd (wu)(vu)$$

for arbitrary terms u, v, w. We need the three associated recipes $\nabla \longmapsto \nabla \bullet \mathsf{0}$. Remember these apply only to principal derivations.

(I) Any principal derivation

$$(\nabla) \quad \Gamma \vdash \mathsf{I}u : \theta$$

has the form $\nabla = IU$ where

$$(I) \quad \Gamma \vdash \mathsf{I} : \iota \qquad (U) \quad \Gamma \vdash u : \theta$$

so we set $\nabla \bullet \mathsf{0} = U$ for the required modified derivation.

(K) Any principal derivation

$$(\nabla) \quad \Gamma \vdash \mathsf{K}vu : \psi$$

has the form $\nabla = ((KV)\Downarrow)U$ where

$$(K) \quad \Gamma^{v} \vdash \mathsf{K} : \kappa \quad (V) \quad \Gamma^{v} \vdash V : \psi \quad (U) \quad \Gamma \vdash u : \theta$$

where Γ^{v} is an initial part of Γ. We set $\nabla \bullet \mathsf{0} = V\Downarrow$ using enough Weakenings to build up Γ from Γ^{v} (as in ∇).

(S) Any principal derivation

$$(\nabla) \quad \Gamma \vdash \mathsf{S}wvu : \phi$$

has the form $\nabla = ((((SW)\Downarrow)V)\Downarrow)U$ where $(S) \quad \Gamma^w \vdash \mathsf{S} : \sigma$ and

$$(W) \quad \Gamma^w \vdash w : (\theta \to \psi \to \phi) \qquad (V) \quad \Gamma^v \vdash V : (\theta \to \psi) \qquad (U) \quad \Gamma \vdash u : \theta$$

where Γ^w is an initial part of Γ^v and this is an initial part of Γ. We set

$$\nabla \bullet \mathsf{0} = ((W\Downarrow)U)((V\Downarrow)U)$$

using enough Weakenings to build up Γ^w and Γ^v to Γ. □

6.23 Recall that

$$r = \lambda z : \sigma . z \qquad s = \lambda z : \sigma . rz \qquad t = \lambda z : \sigma . sz$$

are the component terms. There are two computations

$$(\square) \quad t \mathrel{\rhd\!\!\rhd} r \quad \text{where} \quad \square = \uparrow(\mathsf{I} \circ \mathsf{1}) \quad \text{where} \quad \mathsf{I} = \downarrow\uparrow\mathsf{1} \text{ or } \mathsf{I} = \mathsf{1}$$

depending on how we choose to normalize t.

The obvious strategy is to normalize r, s, and t as they are built, i.e. we attack the innermost redex of t. Since r is already normal, the first job is to normalize s. With $\mathfrak{z} = [z \mapsto z]\mathfrak{i}$ we obtain the computation shown to the left below. We then repeat to get the full computation □ shown in the centre below.

$$\dfrac{\dfrac{\dfrac{rz \rhd z \cdot \mathfrak{z}}{rz \mathrel{\rhd\!\!\rhd} z}(1)}{s \mathrel{\rhd\!\!\rhd} r}(\uparrow)}{sz \mathrel{\rhd\!\!\rhd} rz}(\downarrow)$$

$$\dfrac{\dfrac{\dfrac{\dfrac{\dfrac{rz \rhd z \cdot \mathfrak{z}}{rz \mathrel{\rhd\!\!\rhd} z}(1)}{s \mathrel{\rhd\!\!\rhd} r}(\uparrow)}{sz \mathrel{\rhd\!\!\rhd} rz}(\downarrow) \qquad \dfrac{rz \rhd z \cdot \mathfrak{z}}{rz \mathrel{\rhd\!\!\rhd} z}(1)}{sz \mathrel{\rhd\!\!\rhd} z}(\circ)}{t \mathrel{\rhd\!\!\rhd} r}(\uparrow)$$

$$\dfrac{\dfrac{\dfrac{sz \rhd (-) \cdot \mathfrak{z}}{sz \mathrel{\rhd\!\!\rhd} (\lambda w : \sigma . w)z}(1) \qquad \dfrac{(-)z \rhd w \cdot \mathfrak{z}}{(-)z \mathrel{\rhd\!\!\rhd} z}(1)}{sz \mathrel{\rhd\!\!\rhd} z}(\circ)}{t \mathrel{\rhd\!\!\rhd} r}(\uparrow)$$

A different strategy is to attack the outermost redex sz of t. We have $sz \rhd (rz) \cdot \mathfrak{z}$ where $\mathfrak{z} = [z \mapsto z]\mathfrak{i}$ and then

$$rz \cdot \mathfrak{z} = (r \cdot \mathfrak{z})z = (\lambda w : \sigma . (z \cdot {}'\mathfrak{z}))z = (\lambda w : \sigma . w)z \quad \text{where} \quad {}'\mathfrak{z} = [z \mapsto w]\mathfrak{z}$$

for a suitable w. In fact we could take $w = z$, but this calculation will illustrate a point.

With this we have obtained the full computation shown to the right above. Notice how the obvious reduction strategy leads to a less efficient computation. □

6.24 We will need various terms and replacements. Let

$$\begin{array}{lll}
\mathfrak{b} \; = [w \mapsto \mathsf{B}]\mathfrak{i} & \mathfrak{z} \; = [z \mapsto z]\mathfrak{i} & \mathfrak{y} = [y \mapsto y]\mathfrak{i} \\
B'' = \lambda x : \theta . z(yx) & B' = \lambda y : \sigma . B'' & \mathsf{B} = \lambda z : \tau . B' \\
C'' = \lambda z : \tau . wzy & C' = \lambda y : \sigma . C'' & \mathsf{C} = \lambda w : \beta . C' \\
D'' = \lambda x : \theta . z(yx) & D' = \lambda z : \tau . D'' & \mathsf{D} = \lambda y : \sigma . D' \\
 & E' = \lambda z : \tau . \mathsf{B}zy & E = \lambda y : \sigma . E'
\end{array}$$

so that B, C, and D are the standard terms, but the others are merely for current use. Note that $B'' = D''$ and

$$C' \cdot \mathfrak{b} = E \quad B' \cdot \mathfrak{z} = B' \quad B'' \cdot \mathfrak{y} = B''$$

hold. With these we see that

$$\square = \mathbf{1} \circ \uparrow\uparrow(\downarrow\mathbf{1} \circ \mathbf{1})$$

shown right organizes the reduction.

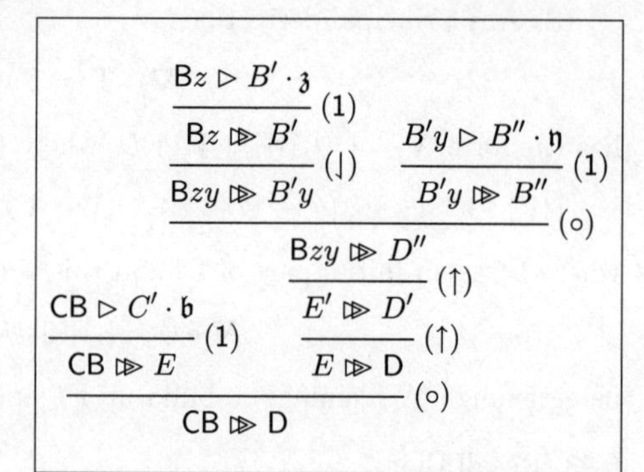

We could also use

$$\square = \mathbf{1} \circ \uparrow\uparrow\downarrow\mathbf{1} \circ \uparrow\uparrow\mathbf{1}$$

which pushes the Abstractions towards the leaves (and thereby duplicates the use of that rule). Of the two computations, the first is probably more efficient. □

6.25 We have

H		T	
$h0 = 1$	$w0 = 1$	$h0 = 1$	$w0 = 1$
$hm' = hm + 2$	$wm' = wm + 1$	$hm' = hm + 1$	$wm' = wm + 1$

where the step rules come from the computations $\square(H, m')$ and $\square(T, m')$ of Example 6.17. These give

$$hm = 2m + 1 \quad wm = m + 1 \qquad hm = m + 1 \quad wm = m + 1$$

for the head and tail version, respectively. □

F.6 Type inhabitation

6.26 (a) The normal inhabitants are

(i) None	(ii) $\lambda x : X . x$	(iii) $\lambda x : X, y : X . v$
(iv) None	(v) $\lambda x : X, y : X, z : X . w$	(vi) $\lambda x : X, y : X' . y^m x$

where

$$\text{(iii)} \ v \in \{x, y\} \qquad \text{(v)} \ w \in \{x, y, z\} \qquad \text{(vi)} \ m \in \mathbb{N}$$

to give

(i) 0 (ii) 1 (iii) 2 (iv) 0 (v) 3 (vi) ∞

solutions, respectively.

There are several general arguments which lead to these results. We describe these for $\boldsymbol{\lambda\emptyset}$ (and indicate where the presence of constants would require more work).

First of all for any normal derivation one of

$$\begin{array}{lllll} \text{(docked)} & t = \lambda y : \sigma \,.\, r & \tau = \sigma \to \rho & \text{with} & \Gamma, y : \sigma \vdash r : \rho \\ \text{(anchored)} & t = hp_1 \cdots p_m & & \text{with} & \Gamma \vdash h : \pi_1 \to \cdots \to \pi_m \to \tau \\ & & & & \Gamma \vdash p_i : \pi_i \end{array}$$

must hold. Here r, h and each p_i is normal, and h is an identifier declared in Γ (or a constant). The length m is unknown, but the context (or housing axioms) gives us the possible values.

When $\Gamma = \emptyset$, the anchored case can't arise. This means that τ must be an arrow type. Thus we reduce each of the original problems to a problem

$$\text{(i)}\ \ \text{None} \qquad \text{(ii, iii, v, vi)}\ \ x : X \vdash t : \tau \qquad \text{(iv)}\ \ y : X' \vdash t : \tau$$

respectively.

Consider any anchored solution to

$$x : X \vdash t : \tau$$

for some given τ. Then $t = x$ or $t = xp$ or $t = xpq$ or $\ldots$. If $t = xp \cdots$ then a derivation

$$\frac{x : X \vdash x : X \quad x : X \vdash p : \pi}{x : X \vdash xp :?}$$

must be possible, and it isn't (because the housing type X of x is not an arrow type). Thus only $t = x$ can arise in this way (and this means $\tau = X$).

There may also be docked solutions, but then τ must be an arrow type.

This argument leads to the given solution for (ii) and leaves us with problems

$$\begin{array}{llll} \text{(iii)} & x : X, y : X \vdash t : X & \text{(iv)} & y : X' \vdash t : X \\ \text{(v)} & x : X, y : X \vdash t : X' & \text{(vi)} & x : X, y : X' \vdash t : X \end{array}$$

for the other four cases.

Consider a problem

$$x : X, y : X \vdash t : \tau$$

for a given τ. If t is anchored then $t = x$ or $t = y$ with $\tau = X$. This leads to the given solutions for (iii), and to a problem

$$x : X, y : X, z : X \vdash t : \tau$$

for (v). For this a similar argument leads to the given solution.

Finally consider a problem

$$\Gamma \vdash t : X$$

where

$$\text{(iv)}\ \ \Gamma = (y : X') \qquad \text{(vi)}\ \ \Gamma = (x : X, y : X')$$

for the two remaining problems. Since X is not an arrow type, only a docked solution can arise.

Consider first the possible docked solutions

$$t = y \quad t = yp \quad t = ypq \quad \ldots$$

with y as head. (This is the only possibility for (iv).) By a type comparison we see that only the case

$$t = yp \quad \text{where} \quad \Gamma \vdash p : X$$

is possible. We now argue by induction.

What is the smallest possible solution (of the whole problem)? Any docked solution with a body of the form $t = yp$ requires a smaller solution p. Thus

(iv) there is no smallest solution, and hence no solution at all,

(vi) the smallest solution is $t = x$

for the two cases. This leads to the indicated solutions.

(b) It is true that the given terms are normal inhabitants of the given types, but there are also other normal inhabitants.

For the type X'' we see that

$$\lambda y : X'' \,.\, y$$

is a normal inhabitant. This and the listed terms are all the normal inhabitants.

For the other type the terms

$$\lambda x : X, y : X', z : X'' \,.\, (z^n y)^k (y^m x)$$

are normal inhabitants for all $m, n, k \in \mathbb{N}$ (and there are other normal inhabitants as well).

6.27 The arguments for the three types β, γ, δ are similar, with the γ case slightly more complicated (and not producing the expected result). We look at that first.

(γ) The same kind of argument is used several times, so let's go through a general version of that first.

With

$$\Xi = w : \beta, y : \sigma, z : \tau, x : X$$

consider the problem

$$\Xi \vdash t : \xi$$

where ξ is some given type. There may be docked solutions, but these can occur only when ξ is an arrow type. What anchored solutions can arise?

If t is anchored then its head must be one of the declared identifiers w, y, z, x, and this head must be combined with certain terms $p, q, r, \ldots$ to produce an inhabitant of ξ. This puts some restrictions on the possible types of $p, q, r, \ldots$, and on ξ. Because of the shapes of β, σ, τ, X, no more than nine cases need be considered.

t	ξ	t	ξ	t	ξ	t	ξ
w	β	y	σ	z	τ	x	X
wp	$\sigma \to \rho$	yp	Y	zp	Z		
wpq	ρ						
$wpqr$	Z						

If ξ is not one of the nine listed types, then no anchored solution occurs. If ξ is one of the nine types then there is one possible shape for t and this shape must be investigated further.

Let's look at the cases $\xi = X, Y, Z$.

($\xi = X$) Here we must have $t = x$.
($\xi = Y$) Here we must have $t = yp$ where $\Xi \vdash p : X$. Hence $t = yx$.
($\xi = Z$) Here we must have $t = zp$ where $\Xi \vdash p : Y$. Hence $t = z(yx)$.

Similar arguments can be carried through for subcontexts of Ξ; there are fewer cases to consider.

With this preliminary we can look at the full problem.

Because the context is empty, any solution c of

$$\vdash c : \gamma$$

can not be anchored, and so must be docked. Thus

$$w : \beta \vdash d : \delta$$

must be solved.

Looking at the first column of the table we see there are no anchored solutions to this (since δ does not occur as a possible type ξ). Thus d must be docked and

$$w : \beta, y : \sigma \vdash e : (\tau \to \rho)$$

must be solved.

The first and second columns of the table show that e must be docked, and hence

$$w : \beta, y : \sigma, z : \tau \vdash f : \rho$$

must be solved.

Let Γ be this context. From the first column we see that

$$f = wpq \quad \text{where} \quad \Gamma \vdash p : \tau \quad \Gamma \vdash q : \sigma$$

for a docked solution is a possibility we must pursue. We must also consider

$$\Xi \vdash g : Z$$

for an anchored solution. From above we see that

$$\lambda w : \beta, y : \sigma, z ; \tau, x : X \,.\, z(yx) \quad \text{i.e.} \quad \lambda w : \beta \,.\, \mathsf{D}$$

is the fully anchored solution.

What can

$$(P) \quad \Gamma \vdash p : \tau \qquad (Q) \quad \Gamma \vdash q : \sigma$$

be to provide a partly docked solution? Looking at the first three columns of the table we see that

$$p = z \qquad q = y$$

are possible solutions. These give

$$\lambda w:\beta, y:\sigma, z;\tau\,.\,wzy \quad \text{i.e.} \quad \mathsf{C}$$

the expected solution. But either of p, q may be docked, i.e. arise from

$$\Xi \vdash p' : Z \qquad \Xi \vdash q' : Y$$

respectively. Using the table we see that

$$p = \lambda x : X\,.\,z(yx) \qquad q = \lambda x : X\,.\,yx$$

lead to possible solutions.

Putting all these together we get five solutions to the original problem, C as expected and

$$\lambda w:\beta\,.\,\mathsf{D} \quad \lambda w:\beta, y:\sigma, z:\tau\,.\,wzq \quad \lambda w:\beta, y:\sigma, z:\tau\,.\,wpy \quad \lambda w:\beta, y:\sigma, z:\tau\,.\,pq$$

where p, q are as above.

(β, δ) Using a similar argument we find that only the expected solutions B, D occur.

When X, Y, Z are not distinct variables, the type comparison arguments break down and other possibilities can arise. □

6.28 For each $m \in \mathbb{N}$ let $\langle m \rangle$ abbreviate the following.

For each derivation

$$(\nabla^?) \quad \Gamma \vdash t : \tau$$

where t is anchored at depth m in Γ, there are

- a type $\eta = \pi_1 \to \cdots \to \pi_m \to \tau$
- a statement $h : \eta$ which is either a housing axiom or a declaration of Γ, and which has an associated normal derivation $\nabla_0 = \Gamma[h]$
- a list

$$(\nabla_i) \quad \Gamma \vdash p_i : \pi_i \quad (i \leq i \leq m)$$

of normal derivations

such that

$$t = hp_1 \cdots p_m \quad \text{where} \quad \nabla = \nabla_0\nabla_1 \cdots \nabla_m$$

is a derivation of the supplied judgement (but need not be the supplied derivation $\nabla^?$).

We prove $(\forall m : \mathbb{N})\langle m \rangle$ by induction over m.

For the base case, $\langle 0 \rangle$, we are given a derivation

$$(\nabla^?) \quad \Gamma \vdash t : \tau$$

where t is either a constant or an identifier declared in Γ. Thus we have a statement $t : \eta$ which is either a housing axiom or a declaration in Γ. This gives a Leaf

$$(\nabla_0) \quad \Gamma \vdash t : \eta$$

which we compare with $\nabla^?$. By (a simple case of) unicity of types we have $\tau = \eta$ to verify $\langle 0 \rangle$. Note that we can *not* conclude that $\nabla^? = \nabla_0$, for $\nabla^?$ may be a Leaf followed by a certain number of Weakenings, whereas we have chosen ∇_0 as a Leaf.

For the induction step, $\langle m \rangle \Rightarrow \langle m' \rangle$, the hypothesis of $\langle m' \rangle$ provides derivations

$$(\nabla^?) \quad \Gamma \vdash t : \tau \qquad (\nabla_q) \quad \Gamma \vdash q : \pi \to \tau \qquad (\nabla_p) \quad \Gamma \vdash p : \pi$$

where $t = qp$, q is anchored at depth m in Γ, and p is normal. Applying the induction hypothesis $\langle m \rangle$ to

$$(\nabla_q) \quad \Gamma \vdash q : \pi \to \tau \qquad q \text{ is anchored at depth } m \text{ in } \Gamma$$

we obtain normal derivations

$$(\nabla_0) \quad \Gamma \vdash h : \eta \qquad (\nabla_i) \quad \Gamma \vdash p_i : \pi_i \quad 1 \leq i \leq m$$

where

- $\eta = \pi_1 \to \cdots \to \pi_m \to (\pi \to \tau)$
- $q = hp_1 \cdots p_m$
- $\nabla_0 \nabla_1 \cdots \nabla_m$ derives $\Gamma \vdash q : \pi \to \tau$

with h a constant or an identifier declared in Γ. Note that the generated derivation $\nabla_0 \nabla_1 \cdots \nabla_m$ need not be the one supplied $\nabla^?$.

With $\pi_{m'} = \pi$, $p_{m'} = p$, $\nabla_{m'} = \nabla_p$ we have normal derivations

$$(\nabla_0) \quad \Gamma \vdash h : \eta \qquad (\nabla_i) \quad \Gamma \vdash p_i : \pi_i \quad 1 \leq i \leq m'$$

where

- $\eta = \pi_1 \to \cdots \to \pi_m \to \pi_{m'} \to \tau$
- $t = hp_1 \cdots p_m p_{m'}$
- $\nabla_0 \nabla_1 \cdots \nabla_m \nabla_{m'}$ derives $\Gamma \vdash t : \tau$

as required.

Notice that the recursive procedure which can be extracted from this induction need not return the supplied derivation $\nabla^?$. The returned derivation does not use Weakening at the generated leaf ∇_0 which produces the head. Also, we do not know how the other generated derivations ∇_i (for $1 \leq i \leq m$) relate to the supplied derivation.

This analysis of normal type inhabitation is designed to return all normal inhabitants of a supplied type (in a supplied context). It doesn't attempt to return all the associated derivations. If required, these derivations may be generated by running the synthesis algorithm on the known term (with the given context). □

F.7 SUBJECT REDUCTION

6.29 Recall that

$$R = Z\uparrow \quad R' = R\downarrow \quad S = (R'Z)\uparrow \quad S' = S\downarrow \quad T = (S'Z)\uparrow$$

gives the source derivation (where $Z = \Sigma[z]$), and

$$\square = \uparrow(\mathsf{I} \cdot \mathbf{1}) \quad \text{where} \quad \mathsf{I} = \begin{cases} \downarrow\uparrow\mathbf{1} \\ \mathbf{1} \end{cases}$$

are the two possible computations. We have

$$T \cdot \square = S'Z\uparrow \cdot \uparrow(\mathsf{I} \cdot \mathbf{1}) = (S'Z \cdot (\mathsf{I} \cdot \mathbf{1}))\uparrow = ((S'Z \cdot \mathsf{I}) \cdot \mathbf{1})\uparrow$$

so our first job is to calculate the inner action. We look at the two cases for I.

($\mathsf{I} = \downarrow\uparrow\mathbf{1}$) Since $S' = (S, z, \sigma)\downarrow$ we have

$$S'Z \cdot \mathsf{I} = (S' \cdot \uparrow\mathbf{1})Z = (S \cdot \uparrow\mathbf{1}, z, \sigma)\downarrow Z$$

where

$$S \cdot \uparrow\mathbf{1} = (R'Z)\uparrow \cdot \uparrow\mathbf{1} = (R'Z \bullet \mathbf{1})\uparrow = (((Z\uparrow\downarrow)Z) \bullet \mathbf{1})\uparrow$$

and we must now call on the redex removal algorithm. With $\Sigma = z : \sigma$ we have

$$(Z\uparrow\downarrow) \;\; \Sigma \vdash (\lambda z : \sigma \,.\, z) : \sigma' \qquad (Z) \;\; \Sigma \vdash z : \sigma$$

so that Z is both the applicator and the applicant derivation, and $\mathfrak{z} = [z \mapsto z]\mathrm{i}$ is the shaft of the applicant mutation

$$(\mathfrak{Z}) \;\; \Sigma \xrightarrow{\;\mathfrak{z}\;} \Sigma$$

which has $z \cdot \mathfrak{Z} = Z$ as its sole nominated derivation. Notice how this description disguises the two uses of z.

With this we have

$$((Z\uparrow\downarrow)Z) \cdot \mathbf{1} = Z \cdot \mathfrak{Z} = Z$$

to give

$$S \cdot \uparrow\mathbf{1} = Z\uparrow = R \qquad S'Z \cdot \mathsf{I} = (Z\uparrow\downarrow)Z$$

and hence

$$T \cdot \square = ((Z\uparrow\downarrow)Z \cdot \mathbf{1})\uparrow$$

and we must make a second call on the redex removal algorithm. In fact, this call is a repeat of the first, so we get

$$T \cdot \square = Z\uparrow = R$$

as the final result (as expected).

($\mathsf{I} = \mathbf{1}$) Here we have

$$S'Z \cdot \mathsf{I} = ((R'Z)\uparrow\downarrow)Z \cdot \mathbf{1}$$

so we must call on the redex removal algorithm. With the same mutation $\mathfrak{z}$ as above we have

$$S'Z \cdot \mathsf{I} = R'Z \cdot \mathfrak{z} = (R' \cdot \mathfrak{z})(Z \cdot \mathfrak{z}) = (R \cdot \uparrow\mathfrak{z})Z = (R \cdot {}'\uparrow\mathfrak{z})\uparrow Z$$

where

$$(\uparrow\mathfrak{z}) \quad \emptyset \longrightarrow \Sigma \qquad ('\uparrow\mathfrak{z}) \quad \Sigma \xrightarrow{'\mathfrak{z}} \Gamma$$

where $\Gamma = \Sigma, y : \sigma$ and $'\mathfrak{z} = [z \mapsto y]\mathfrak{z}$ with $z \cdot {}'\uparrow\mathfrak{z} = \Gamma(y)$. Thus

$$S'Z \cdot \mathsf{I} = (\Gamma(y)\uparrow)Z \quad \text{to give} \quad T \cdot \Box = ((\Gamma(y)\uparrow)Z \cdot \mathbf{1})\uparrow$$

and we must now make a second call on the redex removal algorithm.

We have

$$T \cdot \Box = (\Gamma(y) \cdot \mathfrak{y})\uparrow = (y \cdot \mathfrak{y})\uparrow$$

where

$$(\uparrow\mathfrak{y}) \quad \Gamma \xrightarrow{\mathfrak{y}} \Sigma$$

with $\mathfrak{y} = [y \mapsto z]\mathfrak{i}$ and $y \cdot \mathfrak{y} = Z$ as the relevant nominated derivation. Hence we have

$$T \cdot \Box = Z\uparrow = R$$

as before. Notice how this calculation uses a dummy identifier y. □

6.30 (a) Making use of Solutions 6.2 and 6.6 we have a table of contexts and extractions

		w	x	y	z
Π	$z : \tau, y : \sigma, x : \theta$	–	P	Q	R
Γ	$w : \beta, y : \sigma, z : \tau$	W	–	U	V
Δ	$y : \sigma, z : \tau, x : \theta$	–	L	M	N

where there are different possibilities for most of these. Setting

$$\Pi(B) = R(QP) \quad \Gamma(C) = WVU \quad \Delta(D) = N(ML)$$

we have

$$B = \Pi(B)\uparrow\uparrow\uparrow \quad C = \Gamma(C)\uparrow\uparrow\uparrow \quad D = \Delta(D)\uparrow\uparrow\uparrow$$

as the required derivations.

(b) As in Solution 6.24

$$\Box = \mathbf{1} \circ \mathsf{r} \quad \text{where} \quad \mathsf{r} = \begin{cases} \uparrow\uparrow(\downarrow\mathbf{1} \circ \mathbf{1}) \\ \uparrow\uparrow\downarrow\mathbf{1} \circ \uparrow\uparrow\mathbf{1} \end{cases}$$

where there are two possibilities for r as indicated.

(c) We have

$$CB \cdot \Box = (CB \cdot \mathbf{1}) \cdot \mathsf{r} = (CB \bullet \mathbf{1}) \cdot \mathsf{r}$$

where $CB \bullet \mathbf{1}$ is calculated in Solution 6.17. Thus with

$$\Lambda = y : \sigma, z : \tau = \Delta^{\uparrow}$$

we have $CB \bullet \mathbf{1} = \Lambda(D)\uparrow\uparrow$ where $\Lambda(D) = (B\downarrow\downarrow)\Lambda\langle z\rangle\Lambda y$ and where now both y and z are used in two different ways (inside B and outside).

Now, in the upper case,

$$\Lambda(D)\uparrow\uparrow \cdot \mathsf{r} = (\Lambda(D) \cdot (\downharpoonleft\mathbf{1} \circ \mathbf{1}))\uparrow\uparrow = ((\Lambda(D) \cdot \downharpoonleft\mathbf{1}) \cdot \mathbf{1})\uparrow\uparrow$$

and, in the lower case,

$$\Lambda(D)\uparrow\uparrow \cdot \mathsf{r} = (\Lambda(D)\uparrow\uparrow \cdot \uparrow\uparrow\downharpoonleft\mathbf{1}) \cdot \uparrow\uparrow\mathbf{1} = (\Lambda(D) \cdot \downharpoonleft\mathbf{1})\uparrow\uparrow \cdot \uparrow\uparrow\mathbf{1} = ((\Lambda(D) \cdot \downharpoonleft\mathbf{1}) \cdot \mathbf{1})\uparrow\uparrow$$

and we need to calculate

$$(\Lambda(D) \cdot \downharpoonleft\mathbf{1}) \cdot \mathbf{1} = ((E \cdot \mathbf{1})\Lambda(y)) \cdot \mathbf{1} \quad \text{where} \quad E = (B\downarrow\downarrow)\Lambda(z)$$

in both cases. This is done in two steps.

For the inner one a call on the redex removal algorithm gives

$$E \bullet \mathbf{1} = \Pi(B)\uparrow\uparrow \cdot \mathfrak{z} = (\Pi(B) \cdot {''}\mathfrak{z})\uparrow\uparrow$$

for certain mutations $\mathfrak{z}, {'}\mathfrak{z}, {''}\mathfrak{z}$. With

$$\mathfrak{z} = [z \mapsto z]\mathfrak{i} \quad {'}\mathfrak{z} = [y \mapsto v]\mathfrak{z} \quad {''}\mathfrak{z} = [x \mapsto x]{'}\mathfrak{z} \qquad \Xi = \Lambda, v : \sigma, x : \theta$$

(with v suitably fresh) these are

$$(\mathfrak{z}) \ \ \Pi^{\uparrow\uparrow} \xrightarrow{\mathfrak{z}} \Lambda \qquad ({'}\mathfrak{z}) \ \ \Pi^{\uparrow} \xrightarrow{{'}\mathfrak{z}} \Xi^{\uparrow} \qquad ({''}\mathfrak{z}) \ \ \Pi \xrightarrow{{''}\mathfrak{z}} \Xi$$

where

$$\begin{array}{rcl@{\qquad}rcl@{\qquad}rcl}
z \cdot \mathfrak{z} & = & \Lambda(z) & & & & & & \\
z \cdot {'}\mathfrak{z} & = & \Xi^{\uparrow}(z) & y \cdot {'}\mathfrak{z} & = & \Xi^{\uparrow}(v) & & & \\
z \cdot {''}\mathfrak{z} & = & \Xi(z) & y \cdot {''}\mathfrak{z} & = & \Xi(v) & x \cdot {''}\mathfrak{z} & = & \Xi(x)
\end{array}$$

are the nominated derivations. (All of these are standard extensions.)

We know that

$$P \cdot {''}\mathfrak{z} = \Xi(x) \qquad Q \cdot {''}\mathfrak{z} = \Xi(y) \qquad R \cdot {''}\mathfrak{z} = \Xi(z)$$

no matter which versions of P, Q, R are used. Thus with

$$\Xi(E) = \Xi(z)(\Xi(v)\Xi(x))$$

we have

$$\Pi(B) \cdot {''}\mathfrak{z} = \Xi(E) \quad \text{and hence} \quad E \bullet \mathbf{1} = \Xi(E)\uparrow\uparrow$$

and we have completed the removal of the inner redex.

We need a second call on the redex removal algorithm. We have

$$CB \cdot \square = ((\Xi(E)\uparrow\uparrow)\Lambda(y) \bullet \mathbf{1})\uparrow\uparrow = (\Xi(E)\uparrow \cdot \mathfrak{y})\uparrow\uparrow = (\Xi(E) \cdot {'}\mathfrak{y})\uparrow\uparrow\uparrow$$

for a mutation $\mathfrak{y}$ and an lengthening $'\mathfrak{y}$. With

$$\mathfrak{y} = [v \mapsto y]\mathfrak{i} \qquad '\mathfrak{y} = [x \mapsto x]\mathfrak{y}$$

these are

$$(\mathfrak{y}) \;\; \Xi^{\uparrow} \xrightarrow{\mathfrak{y}} \Lambda \qquad ('\mathfrak{y}) \;\; \Xi \xrightarrow{'\mathfrak{y}} \Delta$$

where

$$\begin{array}{cccc} y \cdot \mathfrak{y} = \Lambda(y) & z \cdot \mathfrak{y} = \Lambda(z) & v \cdot \mathfrak{y} = \Lambda(y) & \\ y \cdot '\mathfrak{y} = \Delta(y) & z \cdot '\mathfrak{y} = \Delta(z) & v \cdot '\mathfrak{y} = \Delta(y) & x \cdot '\mathfrak{y} = \Delta(x) \end{array}$$

as nominated derivations. (Again, all of these are standard extractions.)

These give

$$\Xi(E) \cdot '\mathfrak{y} = \Delta(z)(\Delta(y)\Delta(x))$$

and hence

$$CB \cdot \square = (\Delta(z)(\Delta(y)\Delta(x)))\uparrow\uparrow\uparrow$$

is the final result. This is D where the body $\Delta(D) = N(ML)$ uses standard extractions. Notice that the final result is independent of the choice of B and C, and indicates that the extractions L, M, N (in D) should be chosen as standard.

Notice how a dummy identifier v is used in the middle of the calculation. $\square$

G
Multi-recursive Arithmetic

G.1 Introduction

7.1 The shape of the derivation for $\nabla(m)$ is shown at the right. There are two kinds of leaves

$$(S)\quad \vdash \mathsf{S} : \mathcal{N}' \qquad (Z)\quad \vdash 0 : \mathcal{N}$$

where (S) occurs m times and (Z) just once. With these two leaves we have $\nabla(m) = S^m Z$, in other words

$$\nabla(0) = Z \qquad \nabla(m') = S\nabla(m)$$

$$\dfrac{\vdash \mathsf{S} : \mathcal{N}' \qquad \dfrac{\begin{matrix} \vdots \\ \vdash \ulcorner m-1 \urcorner : \mathcal{N} \end{matrix}}{}}{\vdash \ulcorner m \urcorner : \mathcal{N}} \qquad \dfrac{\vdash \mathsf{S} : \mathcal{N}' \qquad \dfrac{\vdash \mathsf{S} : \mathcal{N}' \quad \vdash 0 : \mathcal{N}}{\vdash \ulcorner 1 \urcorner : \mathcal{N}}}{\ulcorner 2 \urcorner}$$

for each $m \in \mathbb{N}$. □

7.2 Both the constant S and the compound term $\lambda x : \mathcal{N} \,.\, \mathsf{S}x$ are normal representations of the successor function. The representation S uses the reflexive reduction relation. □

7.3 We have terms $\vdash \ulcorner \psi \urcorner : \mathcal{N}[k']$ and $\vdash \ulcorner \theta_i \urcorner : \mathcal{N}[l']$ for $1 \le i \le k$ which represent the given functions. Using the context $\Gamma = x_l : \mathcal{N}, \dots, x_1 : \mathcal{N}$ we can Weaken to get

$$\Gamma \vdash \ulcorner \psi \urcorner : \mathcal{N}[k'] \qquad \Gamma \vdash \ulcorner \theta_i \urcorner : \mathcal{N}[l] \quad (1 \le i \le k)$$

and then

$$\Gamma \vdash \ulcorner \theta_i \urcorner x_l \cdots x_1 : \mathcal{N} \quad (1 \le i \le k) \quad \text{followed by} \quad \Gamma \vdash \ulcorner \psi \urcorner (\ulcorner \theta_1 \urcorner \mathsf{x}) \cdots (\ulcorner \theta_k \urcorner \mathsf{x})$$

and concluded with an abstraction

$$\vdash \ulcorner \phi \urcorner : \mathcal{N}[l'] \quad \text{where} \quad \ulcorner \phi \urcorner = \lambda \mathsf{x} : \mathcal{N} \,.\, \ulcorner \psi \urcorner (\ulcorner \theta_1 \urcorner \mathsf{x}) \cdots (\ulcorner \theta_k \urcorner \mathsf{x})$$

gives the required term $\ulcorner \phi \urcorner$. (Here x is the list of declared identifiers.) □

7.4 Consider first the normal terms t such that

$$x : \mathcal{N} \vdash t : \mathcal{N}$$

holds. What can t be? By the Type Inhabitation Lemma such a t is either $(x : \mathcal{N})$-docked or $(x : \mathcal{N})$-anchored. Since the predicate $\mathcal{N}$ of the judgement is an atom, t can not be an abstraction, so t is $(x : \mathcal{N})$-docked. Thus

$$t = hp_1 \cdots p_m \quad \text{where} \quad h = x \quad \text{or} \quad h = \mathsf{0} \quad \text{or} \quad h = \mathsf{S}$$

with ps to match. We look at the three possibilities.

If $h = x$ then $m = 0$ and $t = x$.

If $h = \mathsf{0}$ then $m = 0$ and $t = \mathsf{0}$.

If $h = \mathsf{S}$ then $m = 1$ and $t = hp$ where $x : \mathcal{N} \vdash p : \mathcal{N}$.

Thus, by induction on the complexity of t, we have $t = \mathsf{S}^r x$ or $t = \mathsf{S}^r\mathsf{0} = \ulcorner r \urcorner$ for some r.

Now consider the problem $\vdash \ulcorner f \urcorner : \mathcal{N}$ where $\ulcorner f \urcorner$ is normal. If $\ulcorner f \urcorner$ is $\emptyset$-anchored then $f = \mathsf{S}$. If $\ulcorner f \urcorner$ is $\emptyset$-docked then, using the previous argument

$$\ulcorner f \urcorner = \lambda x : \mathcal{N} \,.\, \mathsf{S}^r x \qquad \text{or} \qquad \ulcorner f \urcorner = \lambda x : \mathcal{N} \,.\, \mathsf{S}^r\mathsf{0} = \ulcorner r \urcorner$$

for some r. Thus f is either linear or constant, i.e. $fx = x + r$ or $fx = r$. □

G.2 THE SPECIFICS OF $\boldsymbol{\lambda G}$

7.5 Consider the computations $(\boxed{m} \mid m \in \mathbb{N})$ generated by

$$\boxed{0} = \mathsf{0} \qquad \boxed{m'} = \mathsf{0} \circ \lfloor \boxed{m}$$

(for $m \in \mathbb{N}$). For instance $\boxed{3} = \mathsf{0} \circ \lfloor(\mathsf{0} \circ \lfloor(\mathsf{0} \circ \lfloor \mathsf{0}))$ and, in general, $\boxed{m}$ is constructed entirely from $\mathsf{0}$. Using induction on m we check that

$$(\boxed{m}) \quad \mathsf{I}_\sigma ts \ulcorner m \urcorner \mathrel{\triangleright\!\!\!\triangleright} t^m s$$

and hence $\square(\mathsf{I}_\sigma, m) = \boxed{m}$ will do. Notice that $\square(\mathsf{I}_\sigma, m)$ does not depend on σ.

For the base case, $m = 0$, we have the small computation shown on the left

$$(\boxed{0}) \ \dfrac{\mathsf{I}_\sigma ts\mathsf{0} \triangleright s}{\mathsf{I}_\sigma ts\mathsf{0} \mathrel{\triangleright\!\!\!\triangleright} s} \qquad (\boxed{m'}) \ \dfrac{\dfrac{\mathsf{I}_\sigma ts\ulcorner m' \urcorner \triangleright t(\mathsf{I}_\sigma ts \ulcorner m \urcorner)}{\mathsf{I}_\sigma ts\ulcorner m' \urcorner \mathrel{\triangleright\!\!\!\triangleright} t(\mathsf{I}_\sigma ts \ulcorner m \urcorner)} \qquad \dfrac{\begin{matrix}\vdots \\ \mathsf{I}_\sigma ts \ulcorner m \urcorner \mathrel{\triangleright\!\!\!\triangleright} t^m s\end{matrix}}{t(\mathsf{I}_\sigma ts\ulcorner m \urcorner) \mathrel{\triangleright\!\!\!\triangleright} t^{m'} s}}{\mathsf{I}_\sigma ts\ulcorner m' \urcorner \mathrel{\triangleright\!\!\!\triangleright} t^{m'} s}$$

using the appropriate reduction axiom. For the induction step, $m \mapsto m'$, the template on the right generates the required result. Here the left hand branch is a reduction axiom and the right hand branch comes from the induction hypothesis.

It is possible that other computations give the same reduction. □

7.6 (a) We use the obvious derivations

$$(R) \ \vdash r : \sigma' \quad (I^+) \ \vdash \mathsf{I}^+ : \sigma^{+\prime} \to \sigma^{++} \quad (I) \ \vdash \mathsf{I} : \sigma' \to \sigma^+ \quad (U) \ \vdash \mathsf{1} : \mathcal{N}$$

where $R = X{\uparrow}$ with X the projection $x : \sigma \vdash x : \sigma$. With $\Gamma = y : \sigma^+, x : \sigma$ let

$$Y' = \Gamma[y] \quad X' = \Gamma[x] \quad S = ((I{\downarrow})((Y'X'(U{\downarrow\downarrow})){\uparrow})){\uparrow}$$

to get

$$(S) \quad \vdash s : \sigma : \sigma^{+\prime}$$

as a crucial component. With this

$$\nabla = I^{+}S(IR)U \quad \text{gives} \quad (\nabla) \quad \vdash t : \sigma^{+}$$

to show that $\tau = \sigma^{+}$ is the required type. The shape of the derivation is quite complicated, as shown to the right.

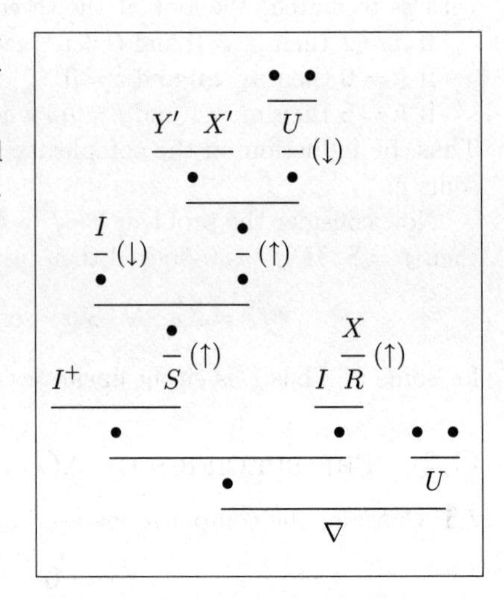

(b) We find that $t^{*} = \mathsf{I}r$ is the normal reduct but where 'r' is not the original one. As in Exercise 7.5 let $\boxed{1} = \mathsf{0} \circ \lfloor\mathsf{0}$ so both

$$(\boxed{1}) \;\; t \rhd\!\!\rhd s(\mathsf{I}r) \qquad (\boxed{1}) \;\; \mathsf{I}rx\ulcorner 1\urcorner \rhd\!\!\rhd rx$$

hold. Also

$$(\mathsf{1}) \quad s(\mathsf{I}r) \rhd\!\!\rhd \mathsf{I}(\lambda x : \sigma\,.\,\mathsf{I}rx\ulcorner 1\urcorner)$$

and rx is a redex, so that

$$\begin{array}{rr} ((\boxed{1} \circ \mathsf{1})) & (\mathsf{I}rx\ulcorner 1\urcorner) \rhd\!\!\rhd x \\ (\uparrow(\boxed{1} \circ \mathsf{1})) & (\lambda x : \sigma\,.\,\mathsf{I}rx\ulcorner 1\urcorner) \rhd\!\!\rhd r \\ (\lfloor\uparrow(\boxed{1} \circ \mathsf{1})) & \mathsf{I}(\lambda x : \sigma\,.\,\mathsf{I}rx\ulcorner 1\urcorner) \rhd\!\!\rhd \mathsf{I}r \end{array}$$

and hence

$$\square = \boxed{1} \circ \mathsf{1} \circ \lfloor\uparrow(\boxed{1} \circ \mathsf{1}) \quad \text{with} \quad (\square) \quad t \rhd\!\!\rhd t^{*}$$

is the required computation.

(c) The derivation ∇^{*} is shown right and is much simpler than ∇.

$$\dfrac{I \qquad \dfrac{X}{R}}{\bullet}$$

(d) We find that

$$\nabla \cdot \boxed{1} = S(IR) \qquad \nabla \cdot (\boxed{1} \circ \mathsf{1}) = S(IR) \cdot \mathsf{1} = I(Z\uparrow)$$

where

$$Z = ((IR)\downarrow)X(U\downarrow)$$

so that, with $\boxed{z} = \boxed{1} \circ \mathsf{1}$, we have

$$\nabla \cdot \square = I(Z\uparrow) \cdot \lfloor\uparrow\boxed{z} = I(Z\uparrow \cdot \uparrow\boxed{z}) = I((Z \cdot \boxed{z})\uparrow)$$

so it remains to calculate $Z \cdot \boxed{z}$. With a little care we find that

$$Z \cdot \boxed{1} = (R\downarrow)X \qquad Z \cdot \boxed{z} = (R\downarrow)X \cdot \mathsf{1} = X$$

and hence

$$\nabla \cdot \square = I(X\uparrow) = IR$$

as required. □

7.7 As in Exercise 7.5 let $\boxed{0} = \mathsf{0}$ and $\boxed{m'} = \mathsf{0} \circ \lfloor\boxed{m}$ (for $m \in \mathbb{N}$), so that

$$(\boxed{m}) \quad \mathsf{I}_\sigma ts\ulcorner m\urcorner \rhd t^m s$$

for all appropriate σ, t, s, m. In particular, since $\mathsf{Add}\ulcorner n\urcorner\ulcorner m\urcorner = \mathsf{IS}\ulcorner n\urcorner\ulcorner m\urcorner$, we see that the computation

$$\square(Add, n, m) = \boxed{m}$$

will do for addition. This computation is independent of n.

For multiplication we have

$$(\downharpoonleft\mathbf{1} \circ \mathsf{0}) \quad \mathsf{Mlt}\ulcorner n\urcorner\ulcorner m\urcorner \rhd \mathsf{I}(\mathsf{Add}\ulcorner n\urcorner)\mathsf{0}\ulcorner m\urcorner \rhd (\mathsf{Add}\ulcorner n\urcorner)^m\mathsf{0}$$

so it suffices to evaluate the iteration. To do this set

$$M^0(n) = \text{`do nothing'} \qquad M^{m'}(n) = \lfloor M^m(n) \circ \boxed{k} \quad \text{where } k = nm$$

for each $m \in \mathbb{N}$. Then

$$(M^m(n)) \quad (\mathsf{Add}\ulcorner n\urcorner)^m\mathsf{0} \rhd\!\!\rhd \ulcorner mn\urcorner \quad \text{and hence} \quad \square(Mlt, m) = \downharpoonleft\mathbf{1} \circ \mathsf{0} \circ M^m(n)$$

will do for multiplication. $\square$

7.8 We show that the computations are

$$\begin{aligned} \square(Grz, F, n, m) &= \downharpoonleft(\downharpoonleft\mathbf{1} \circ \mathbf{1}) \circ \boxed{m} \circ \square^m(F, n) \\ \square(GRZ, F, i, n, m) &= \downharpoonleft\downharpoonleft(\downharpoonleft\mathbf{1} \circ \mathbf{1} \circ \boxed{i}) \circ \square^i(Grz, F, n, m) \end{aligned}$$

where the $\boxed{\cdot}$ are as in Exercise 7.5, and $\square^m(F, n), \square^i(Grz, F, n, m)$ are computations to be generated.

(Grz) First of all we have

$$\begin{array}{rl} (\downharpoonleft(\downharpoonleft\mathbf{1} \circ \mathbf{1})) & \ulcorner Grz\urcorner\ulcorner F\urcorner\ulcorner n\urcorner\ulcorner m\urcorner \rhd \mathsf{I}(\ulcorner F\urcorner\ulcorner n\urcorner)1\ulcorner m\urcorner \\ (\boxed{m}) & \mathsf{I}(\ulcorner F\urcorner\ulcorner n\urcorner)1\ulcorner m\urcorner \rhd (\ulcorner F\urcorner\ulcorner n\urcorner)^m 1 \end{array}$$

so we require a computation

$$(\square^m(F, n)) \quad (\ulcorner F\urcorner\ulcorner n\urcorner)^m 1 \rhd\!\!\rhd \ulcorner Fnm\urcorner$$

for the third component. We generate this by recursion on m. Thus

$$\square^0(F, n) = \text{`do nothing'} \quad \square^{m'}(F, n) = \lfloor\square^m(F, n) \circ \square(F, n, k) \quad \text{where } k = F'nm$$

give the required computations.

(GRZ) The indicated first component organizes

$$\mathsf{GRZ}\ulcorner F\urcorner\ulcorner i\urcorner\ulcorner n\urcorner\ulcorner m\urcorner \rhd \ulcorner Grz\urcorner^i\ulcorner F\urcorner\ulcorner n\urcorner\ulcorner m\urcorner$$

so that

$$\begin{aligned} \square^0(Grz, F, N, M) &= \square(F, n, m) \\ \square^{i'}(Grz, F, N, M) &= \lfloor\square^i(Grz, F, N, M) \circ \square(F, n, k) \quad \text{where } k = F^{(i)}nm \end{aligned}$$

generates the required second component. $\square$

7.9 Suppose we have a derivation $\vdash n : \mathcal{N}$ where n is normal. By the Type Inhabitation Lemma 6.19 each such term is either $\emptyset$-anchored at $\mathcal{N}$ or $\emptyset$-docked at $\mathcal{N}$, so we investigate both possibilities.

If n is $\emptyset$-docked at $\mathcal{N}$ then

$$n = \lambda x : \sigma . r \quad \text{where} \quad x : \sigma \vdash r : \rho$$

for some term r and types σ, ρ. But then $\mathcal{N} = \sigma \rightarrow \rho$, which is not so. Thus this case does not arise. (This is the crucial observation which enables the whole argument to be carried through.)

We know that n must be $\emptyset$-anchored. (Now comes the second most important observation.) There are no identifiers declared in $\emptyset$, so $n = hp_1 \cdots p_m$ where h is a constant and $p_1, \ldots, p_m$ are normal with

$$\vdash h : \pi_1 \rightarrow \cdots \rightarrow \pi_m \rightarrow \mathcal{N} \qquad \vdash p_i : \pi_i \quad 1 \leq i \leq m$$

for some types $\pi_1, \ldots, \pi_m$. This constant is either $\mathsf{0}, \mathsf{S}$ or I_σ for some type σ. Let's look at these in turn.

($h = \mathsf{0}$) There are no derivations $\vdash \mathsf{0}p : \tau$, so $m = 0$ and $h = \mathsf{0}$.

($h = \mathsf{S}$) Since $S : \mathcal{N}'$ we must have $n = \mathsf{S}p$ where $\vdash p : \mathcal{N}$ with p normal. We suspend this case for a moment.

($h = \mathsf{I}_\sigma$) Since $\mathsf{I}_\sigma : \sigma' \rightarrow \sigma^+$ where $\sigma^+ = \sigma \rightarrow \mathcal{N} \rightarrow \sigma$, we have

$$n = \mathsf{I}_\sigma tspp_1 \cdots p_k \quad \text{where} \quad \vdash t : \sigma' \quad \vdash s : \sigma \quad \vdash p : \mathcal{N} \quad \cdots$$

with $t, s, p, p_1, \ldots, p_k$ normal. The nature of $p_1, \ldots, p_k$ depends on the form of σ.

This preliminary analysis gives the following.

> If $\vdash n : \mathcal{N}$ where n is normal then either
>
> $$n = \mathsf{0} \quad \text{or} \quad n = \mathsf{S}p \quad \text{or} \quad n = \mathsf{I}_\sigma tsp \cdots$$
>
> where $\vdash p : \mathcal{N}$ with p normal (and other possible restrictions).

We can now proceed by an induction on the syntactic complexity of n.

If $n = \mathsf{0}$, then we are done. Otherwise $n = \mathsf{S}p$ or $n = \mathsf{I}_\sigma tsp \cdots$ for some normal p with $\vdash p : \mathcal{N}$. This p is a subterm of n and hence, by the induction hypothesis, there is some $m \in \mathbb{N}$ with $p = \ulcorner m \urcorner$. The first alternative gives $n = \ulcorner m' \urcorner$. The second alternative gives $n = \mathsf{I}_\sigma ts\ulcorner m \urcorner \cdots$ which is not normal, and so does not occur.

What happens if we try to modify this argument to classify the normal terms f with $\vdash f : \mathcal{N}'$? We can not dismiss the $\emptyset$-docked case, for we could have

$$f = \lambda x : \mathcal{N} . r \quad \text{where} \quad x : \mathcal{N} \vdash r : \mathcal{N}$$

for some normal r. Thus we must re-do the above analysis with a non-empty context. In the anchored case the head h could be an identifier. This feeds back into the induction and then p need not be a numeral. Very soon this whole attempted classification blows up and is impossible to control.

In this chapter we see that very many 1-placed functions are represented in $\boldsymbol{\lambda G}$. These cannot be classified using these simple minded syntactic methods. □

7.10 Let B_σ and A_σ be, respectively, the following λ-terms.

$$\lambda z : \sigma^+ \to \sigma' \,,\, y : \sigma' \to \sigma^+ \,,\, x : \sigma' \,.\, z(yx) \qquad \lambda z : \sigma' \,,\, y : \sigma \,,\, x : \mathcal{N} \,.\, zy$$

(Thus B_σ is a version of the usual B-combinator). With $\mathsf{B} = \mathsf{B}_\sigma, \mathsf{R} = \mathsf{R}_\sigma, \mathsf{A} = \mathsf{A}_\sigma$ we have

$$\mathsf{I}tsr = \mathsf{BRA}tsr \mathrel{\rhd\!\!\!\rhd} \mathsf{R}(\mathsf{A}t)sr$$

for all terms r, s, t. In particular

$$\mathsf{I}ts\mathsf{0} \mathrel{\rhd\!\!\!\rhd} \mathsf{R}(\mathsf{A}t)s\mathsf{0} \mathrel{\rhd\!\!\!\rhd} s$$

and

$$\mathsf{I}ts(\mathsf{S}r) \mathrel{\rhd\!\!\!\rhd} \mathsf{R}(\mathsf{A}t)s(\mathsf{S}r) \mathrel{\rhd\!\!\!\rhd} \mathsf{A}t(\mathsf{R}(\mathsf{A}t)sr)r \rhd t(\mathsf{R}(\mathsf{A}t)sr)$$

to show that the terms $\mathsf{I}ts(\mathsf{S}r)$ and $t(\mathsf{I}tsr)$ have a common reduct. Thus anything computable with I_σ is also computable with $\mathsf{B}_\sigma\mathsf{R}_\sigma\mathsf{A}_\sigma$ but the shapes of the computations will be different. □

7.11 For arbitrary $w : \sigma^+, s : \sigma$ consider $f = \mathsf{R}_\sigma ws$. Thus

$$f : \mathbb{N} \longrightarrow \sigma \quad \text{is given by} \quad f0 = s \qquad f(r+1) = w(fr)r$$

(for $r \in \mathbb{N}$). With $\tau = \sigma \times \mathbb{N}$ consider

$$W : \tau' \quad \text{given by} \quad W\langle x, r\rangle = \langle wxr, r+1\rangle$$

where $\langle \cdot, \cdot \rangle$ is the pair forming operation. Now set $F = \mathsf{I}_\tau W\langle s, 0\rangle$ so that F is given recursively by

$$F0 = \langle s, 0\rangle \qquad F(r+1) = W(Fr)$$

(for $r \in \mathbb{N}$). This allows us to check that $Fr = \langle fr, r\rangle$ and hence $f = \mathsf{left} \circ F$. □

G.3 Forms of recursion and induction

7.12 For the given data functions θ, ψ, κ (as in Definition 7.8) let $\Psi : \mathbb{N} \longrightarrow \mathbb{F}'$ be given by

$$\Psi rfp = \psi r(fp^+)p \quad \text{where} \quad p^+ = \kappa rp$$

for $r : \mathbb{N}$, $f : \mathbb{F}$, $p : \mathbb{P}$. Consider $\Phi : \mathbb{N} \longrightarrow \mathbb{F}$ obtained by a G-recursion from $\theta : \mathbb{F}$ and Ψ, i.e.

$$\Phi 0 = \theta \qquad \Phi r' = \Psi r(\Phi r)$$

for each $r \in \mathbb{N}$. A routine induction over r shows that $\phi r = \Phi r$ and hence $\phi = \Phi$.

The 'problem' with this simulation is that it uses types higher than the original recursion. For instance, when $\mathbb{P} = \mathbb{S} = \mathbb{N}$ we have $\theta : \mathbb{N}[1]$, $\psi : \mathbb{N}[2]$, $\kappa : \mathbb{N}[2]$ all of which are first order, but $\Psi : \mathbb{N} \longrightarrow \mathbb{N}''$ which is second order. The received dogma is that this is not acceptable. □

7.13 (i,ii) The easiest way to verify these is to proceed by induction on s.

Since $\mathbb{M}_0 = \mathbb{N}$, the base case is immediate.

For the induction step $s \mapsto s'$ observe that each member of $\mathbb{M}_{s'}$ has the form (i, i) where $\mathsf{i} \in \mathbb{M}_s$. We lift the required properties from $\mathbb{M}_s$ to $\mathbb{M}_{s'}$.

(Reflexive) This is immediate (and doesn't need the induction hypothesis).

(Transitive) Suppose

$$(i,\mathsf{i}) \le_{s'} (j,\mathsf{j}) \le_{s'} (k,\mathsf{k}) \quad \text{i.e.} \quad \begin{array}{c} i < j \\ \text{or} \\ i = j \text{ and } \mathsf{i} \le_s \mathsf{j} \end{array} \quad \text{and} \quad \begin{array}{c} j < k \\ \text{or} \\ j = k \text{ and } \mathsf{j} \le_s \mathsf{k} \end{array}$$

hold. In all cases we have $i \le j \le k$. If $i < k$ then $(i,\mathsf{i}) \le_{s'} (k,\mathsf{k})$. If $i = k$ then $i = j = k$, so $\mathsf{i} \le_s \mathsf{j} \le_s \mathsf{k}$ must hold. By the induction hypothesis this gives $\mathsf{i} \le_s \mathsf{k}$, and hence $(i,\mathsf{i}) \le_{s'} (k,\mathsf{k})$, as required.

(Antisymmetric) Given

$$(i,\mathsf{i}) \le_{s'} (j,\mathsf{j}) \le_{s'} (i,\mathsf{i})$$

we deduce $i \le j \le i$ and $\mathsf{i} \le_s \mathsf{j} \le_s \mathsf{i}$ so that $i = j$ and $\mathsf{i} = \mathsf{j}$ (by the induction hypothesis), as required.

(Linear) Consider any pair $(i,\mathsf{i}), (j,\mathsf{j})$ in $\mathbb{M}_{s'}$. One of

$$i < j \qquad i = j \qquad j < i$$

must hold and, by the induction hypothesis, one of

$$\mathsf{i} \le_s \mathsf{j} \qquad \mathsf{j} \le_s \mathsf{i}$$

must hold. If $i < j$ or $j < i$ then $(i,\mathsf{i}) \le_{s'} (j,\mathsf{j})$ or $(j,\mathsf{j}) \le_{s'} (i,\mathsf{i})$, accordingly. If $i = j$ then $(i,\mathsf{i}) \le_{s'} (j,\mathsf{j})$ or $(j,\mathsf{j}) \le_{s'} (i,\mathsf{i})$ according as the comparison between i and j.

(Well ordered) Consider any non-empty subset $X \subseteq \mathbb{M}_{s'}$. We use

$$j \in Y \iff (\exists \mathsf{j} \in \mathbb{M}_s)[(j,\mathsf{j}) \in X]$$

to produce a non-empty subset $Y \subseteq \mathbb{N}$. Let i be the least member of Y. With this i we use

$$\mathsf{j} \in Z \iff (i,\mathsf{j}) \in X$$

to produce a non-empty subset $Z \subseteq \mathbb{M}_s$. By the induction hypothesis Z has a least member i. It is now easy to check that (i,i) is the least member of X.

(iii) The successor of i is obtained by replacing i_0 by $i_0{}'$.

(iv) Each i with $i_0 = 0$ can not be an immediate successor, and so does not have an immediate predecessor. When $s > 0$, there are infinitely many such i. □

7.14 Assume that $F : N^4 \longrightarrow \mathbb{A}$ is specified by an $\mathbb{M}_3$-recursion and the five conditions $[\perp, 0, 1, 2, 3]$ are known. To verify $\langle\rangle$ we proceed by four nested inductions with some intermediate implications.

In turn we show the following (for $k, j, i \in \mathbb{N}$).

[1.0] $\langle k,j,i,0\rangle \Rightarrow \langle k,j,i\rangle$	[2.1] $\langle k,j,0\rangle \Rightarrow \langle k,j\rangle$	[3.2] $\langle k,0\rangle \Rightarrow \langle k\rangle$	[4.3] $\langle 0\rangle \Rightarrow \langle\rangle$
[1.⊥] $\langle 0,0,0\rangle$	[2.⊥] $\langle 0,0\rangle$	[3.⊥] $\langle 0\rangle$	[4.⊥] $\langle\rangle$

Using the given [0] an induction over r gives [1.0]. This with [⊥] gives [1.⊥].
Using the given [1] an induction over i gives [2.1]. This with [1.⊥] gives [2.⊥].
Using the given [2] an induction over j gives [3.2]. This with [2.⊥] gives [3.⊥].
Using the given [3] an induction over k gives [4.3]. This with [3.⊥] gives [4.⊥].
The condition [4.⊥] is the required conclusion. □

Before the next solution we repeat the information used in that and many later solutions. Thus we have some property $\langle k,j,i,r\rangle$ of 4-tuples of natural numbers (as given by the local conditions) and we introduce several abbreviations, as shown on the left.

$\langle k,j,i,r\rangle$	abbrv	as given	$(0.\bot)$	$\langle 0,0,0,0\rangle$		
$\langle k,j,i\rangle$	abbrv	$(\forall r:\mathbb{N})\langle k,j,i,r\rangle$	(0.0)	$\langle k,j,i,r\rangle$	$\Rightarrow$	$\langle k,j,i,r'\rangle$
$\langle k,j\rangle$	abbrv	$(\forall i:\mathbb{N})\langle k,j,i\rangle$	(0.1)	$\langle k,j,i\rangle$	$\Rightarrow$	$\langle k,j,i',0\rangle$
$\langle k\rangle$	abbrv	$(\forall j:\mathbb{N})\langle k,j\rangle$	(0.2)	$\langle k,j\rangle$	$\Rightarrow$	$\langle k,j',0,0\rangle$
$\langle\rangle$	abbrv	$(\forall k:\mathbb{N})\langle k\rangle$	(0.3)	$\langle k\rangle$	$\Rightarrow$	$\langle k',0,0,0\rangle$

Thus $\langle\rangle$ is the target condition. We prove a base condition and four step conditions, as shown on the right. Solution 7.15 justifies this method. We refer to these as the usual abbreviations and the usual conditions.

7.15 The given condition $(0.\bot)$ is the required [⊥], and the given implication (0.0) is stronger than the required [0]. Thus it suffices to verify [1,2,3]. In fact, we show

$$[1]^+\ \langle k,j,i\rangle \Rightarrow \langle k,j,i'\rangle \qquad [2]^+\ \langle k,j\rangle \Rightarrow \langle k,j'\rangle \qquad [3]^+\ \langle k\rangle \Rightarrow \langle k'\rangle$$

which are stronger than the required [1,2,3]. Along the way we also verify the implications [1.0, 2.1, 3.2, 4.3] of Solution 7.14.

From the given (0.0) an induction over r gives [1.0].

Taking $i := i'$ in [1.0] and using the given (0.1) produces $[1]^+$.

From $[1]^+$ an induction over i gives [2.1].

Using (0.2), [1.0], [2.1] gives

$$\langle k,j\rangle \Rightarrow \langle k,j',0,0\rangle \Rightarrow \langle k,j',0\rangle \Rightarrow \langle k,j'\rangle$$

to verify $[2]^+$.

From $[2]^+$ an induction over j gives [3.2].

Using (0.3), [1.0], [2.1], [3.2] gives

$$\langle k\rangle \Rightarrow \langle k,0,0,0\rangle \Rightarrow \langle k,0,0\rangle \Rightarrow \langle k,0\rangle \Rightarrow \langle k\rangle$$

to verify $[3]^+$.

This completes the required proof, but notice that [4.3] follows from $[3]^+$ by an induction over k. □

7.16 Let

$$\langle k,j,i,r\rangle \text{ abbreviate } F(k,j,i,r) = 8k+4j+2j+r$$

and with this consider the usual abbreviations so that $\langle\rangle$ is the target assertion. We verify the usual conditions and then appeal to the method of Exercise 7.15.

By the spec clause $(\bot)$ the base case $(0.\bot)$ is immediate.

Using the spec clause (0) and assuming $\langle k,j,i,r\rangle$ we have

$$F(k,j,i,r') = F(k,j,i,r)+1 = 8k+4j+2i+r+1$$

to obtain $\langle k,j,i,r'\rangle$. This verifies (0.0).

Using the spec clause (1) and assuming $\langle k,j,i\rangle$ we have

$$F(k,j,i',0) = F(k,j,i,1)+1 = 8k+4j+2i+1+1 = 8k+4j+2i'$$

to obtain $\langle k,j,i',0\rangle$. This verifies (0.1).

Using the spec clause (2) and assuming $\langle k,j\rangle$ we have

$$F(k,j',0,0) = F(k,j,1,0)+2 = 8k+4j+2+2 = 8k+4j'$$

to obtain $\langle k,j',0,0\rangle$. This verifies (0.2).

Using the spec clause (3) and assuming $\langle k\rangle$ we have

$$F(k',0,0,0) = F(k,1,2,0) = 8k+4+4 = 8k'$$

to obtain $\langle k',0,0,0\rangle$. This verifies (0.3), and so completes the inductions. □

7.17 Let

$$\langle k,j,i,r\rangle \text{ abbreviate } (\forall x:\mathbb{N})[F(k,j,i,r)x = x^3k+x^2j+xi+r]$$

and with this consider the usual abbreviations so that $\langle\rangle$ is the target assertion. We verify the usual conditions and then appeal to the method of Exercise 7.15.

For instance, using the spec clause (1) and assuming $\langle k,j,i\rangle$ (in the form $\langle k,j,i,x\rangle$) we have

$$F(k,j,i',0)x = F(k,j,i,x)x = x^3k+x^2j+xi+x = x^3k+x^2j+xi'$$

to deduce $\langle k,j,i',0\rangle$. This verifies (0.1). □

7.18 For convenience we write

$$G = [f,g,h]$$

to indicate the construction of G. Using the identically zero function *Zero* consider a particular case H of G as shown to the right. Let

$$H = [Suc, Zero, h]$$

$(H.\bot)$	$H(0,0,0,0)x = 0$
$(H.0)$	$H(k,j,i,r')x = 1 + H(k,j,i,r)(hx)$
$(H.1)$	$H(k,j,i',0)x = H(k,j,i,1)x$
$(H.2)$	$H(k,j',0,0)x = H(k,j,x,1)x$
$(H.3)$	$H(k',0,0,0)x = H(k,x^2,x,2)x$

$$\begin{array}{lll}\langle k,j,i,r,x\rangle & \text{abbreviate} & G(k,j,i,r)x = (f^e\circ g\circ h^e)x \quad\text{where}\quad e = H(k,j,i,r)x\\ \langle k,j,i,r\rangle & \text{abbreviate} & (\forall x:\mathbb{N})\langle k,j,i,r,x\rangle\end{array}$$

where we have used an extra parameter x not in the official $\mathbb{M}_3$-induction scheme. This will help in the first phase of the nested inductions. Consider the usual abbreviations so that $\langle\rangle$ is the target assertion. In this first phase we show that the implications to the right are simple consequences of the specs of G and H.

$(*.\bot)$	$\langle 0,0,0,0,x\rangle$		
$(*.0)$	$\langle k,j,i,r,hx\rangle$	$\Rightarrow$	$\langle k,j,i,r',x\rangle$
$(*.1)$	$\langle k,j,i,1,x\rangle$	$\Rightarrow$	$\langle k,j,i',0,x\rangle$
$(*.2)$	$\langle k,j,x,1,x\rangle$	$\Rightarrow$	$\langle k,j',0,0,x\rangle$
$(*.3)$	$\langle k,x^2,x,1,x\rangle$	$\Rightarrow$	$\langle k',0,0,0,x\rangle$

The base $(*.\bot)$ is an immediate consequence of the spec clause $(\bot)$ and $(H.\bot)$. For $(*.0)$ let

$$y = hx \quad a = H(k,j,i,r')y \quad e = 1+y \quad \text{so that} \quad e = H(k,j,i,r')x$$

by $(H.0)$. Then, assuming $\langle k,j,i,r,y\rangle$, we have

$$\begin{aligned} G(k,j,i,r')x &= (f \circ G(k,j,i,r') \circ h)x \\ &= f(G(k,j,i,r')y) \\ &= f((f^a \circ g \circ h^a)y) \\ &= f^{1+a} \circ g \circ h^{a+1}x \qquad = (f^e \circ g \circ h^e)x \end{aligned}$$

to deduce $\langle k,j,i,r',x\rangle$. Here the first step uses the spec clause (0), and the third uses the assumed $\langle k,j,i,r,y\rangle$.

For $(*.1)$ let

$$e = H(k,j,i,1)x = H(k,j,i',0)x$$

using $(H.1)$. Then, assuming $\langle k,j,i,1,x\rangle$, we have

$$G(k,j,i',0)x = G(k,j,i,1)x = (f^e \circ g \circ h^e)x$$

to deduce $\langle k,j,i',0,x\rangle$. Here the steps follow by the spec clause (1) and the assumed $\langle k,j,i,1,x\rangle$, respectively.

For $(*.2)$ let

$$e = H(k,j,x,1)x = H(k,j',0,0)x$$

using $(H.2)$. Then, assuming $\langle k,j,x,1,x\rangle$, we have

$$G(k,j',0,0)x = G(k,j,x,1)x = (f^e \circ g \circ h^e)x$$

to deduce $\langle k,j',0,0,x\rangle$. Here the steps follow by the spec clause (2) and the assumed $\langle k,j,x,1,x\rangle$, respectively.

For $(*.3)$ let

$$e = H(k,x^2,x,1)x = H(k',0,0,0)x$$

using $(H.3)$. Then, assuming $\langle k,x^2,x,1,x\rangle$, we have

$$G(k',0,0,0)x = G(k,x^2,x,1)x = (f^e \circ g \circ h^e)x$$

to deduce $\langle k',0,0,0,x\rangle$. Here the steps follow by the spec clause (3) and the assumed $\langle k,x^2,x,1,x\rangle$, respectively. This completes the first phase.

For the second phase we observe that the usual implications $(0.\bot,\dots,0.3)$ follow from $(*.\bot,\dots,*.3)$ by simple rules of quantification. For instance, for each x we have

$$\langle k\rangle \Rightarrow \langle k,x^2\rangle \Rightarrow \langle k,x^2,x\rangle \Rightarrow \langle k,x^2,x,1\rangle \Rightarrow \langle k,x^2,1,x\rangle \Rightarrow \langle k',0,0,0,x\rangle$$

where the first four steps are instantiations and the last is $(*.3)$. Thus

$$\langle k\rangle \Rightarrow \langle k',0,0,0,x\rangle$$

and then we may quantify out the x to obtain (1.3).

Finally, a nested induction, as in Exercise 7.15, gives $\langle\rangle$, as required. □

7.19 The function $\ell : \mathbb{N}^4 \longrightarrow \mathbb{N}$ is specified to the right (for $k,j,i,r \in \mathbb{N}$). It can be checked that

$$\ell(k,j,i,r) = K(k) + J(j) + I(i) + r$$

$(\ell.\bot)$	$\ell(0,0,0,0) = 0$
$(\ell.0)$	$\ell(k,j,i,r') = \ell(k,j,i,r) + 1$
$(\ell.1)$	$\ell(k,j,i',0) = \ell(k,j,i,2i) + 1$
$(\ell.2)$	$\ell(k,j',0,0) = \ell(k,j,j,0)$
$(\ell.3)$	$\ell(k',0,0,0) = \ell(k,0,k,k)$

where K,J,I are polynomials. In fact I is quadratic and J,K are cubic. The specs are

$$I(i') = I(i) + 2i \qquad J(j') = J(j) + I(j) \qquad K(k') = K(j) + I(k) + k$$

with $I(0) = J(0) = K(0) = 0$. We show that

$$E(k,j,i,r) = 2^l \quad \text{where } l = \ell(k,j,i,r)$$

is the required exponent function. To this end let

$$\langle k,j,i,r\rangle \quad \text{abbreviate} \quad (\forall f : \mathbb{N}')[L(k,j,i,r)f = f^l \text{ where } l = \ell(k,j,i,r)]$$

and then obtain the usual abbreviations with $\langle\rangle$ as the target condition. We verify the usual conditions and then appeal to the standard method (of Exercise 7.15).

The base $(0.\bot)$ is an immediate consequence of the spec clause $(\bot)$ and $(\ell.\bot)$.

To verify (0.0) consider $f : \mathbb{N}'$. Assuming $\langle k,j,i,r\rangle$ we have

$$L(k,j,i,r')f = L(k,j,i,r)f^2 = (f^2)^{2^{l(k,j,i,r)}} = f^{2^{1+l(k,j,i,r)}} = f^{2^{l(k,j,i,r')}}$$

to deduce $\langle k,j,i,r'\rangle$. Here the first step uses the spec clause (0), the second step uses an instantiation of $\langle k,j,i,r\rangle$, and the last step uses the spec clause $(\ell.0)$.

To verify (0.1) consider $f : \mathbb{N}'$. Assuming $\langle k,j,i\rangle$ in the form $\langle k,j,i,2i\rangle$ we have

$$L(k,j,i',0)f = L(k,j,i,2i)f^2 = (f^2)^{2^{l(k,j,i,2i)}} = f^{2^{1+l(k,j,i,2i)}} = f^{2^{l(k,j,i',0)}}$$

to deduce $\langle k,j,i',0\rangle$. Here the first step uses the spec clause (1), the second step uses $\langle k,j,i\rangle$, and the last step uses the spec clause $(\ell.1)$.

To verify (0.2) consider $f : \mathbb{N}'$. Assuming $\langle k,j\rangle$ in the form $\langle k,j,j,0\rangle$ we have

$$L(k,i',0,0)f = L(k,j,j,0)f^2 = f^{2^{l(k,j,j,0)}} = f^{2^{l(k,j',0,0)}}$$

to deduce $\langle k, j', 0, 0\rangle$. Here the first step uses the spec clause (2), the second step uses $\langle k, j\rangle$, and the last step uses the spec clause $(\ell.2)$.

To verify (0.3) consider $f : \mathbb{N}'$. Assuming $\langle k\rangle$ in the form $\langle k, 0, k, k\rangle$ we have

$$L(k', 0, 0, 0)f = L(k, 0, k, k)f^2 = f^{2^{l(k,0,k,k)}} = f^{2^{l(k',0,0,0)}}$$

to deduce $\langle k', 0, 0, 0\rangle$. Here the first step uses the spec clause (3), the second step uses $\langle k\rangle$, and the last step uses the spec clause $(\ell.3)$. □

7.20 Given $F : \mathbb{N}''$ let F_0, F_1, F_2, F_3 (all of type $\mathbb{N}''$) be defined by $F_0 = F$ and

$$F_1 fx = F_0^x fx \quad F_2 fx = (F_0^x \circ F_1^x)fx = F_1^{x+1}fx \quad F_3 fx = (F_0^x \circ F_1^x \circ F_2^x)fx = F_2^{x+1}fx$$

(for $f : \mathbb{N}', x : \mathbb{N}$). Let

$$\langle k, j, i, r\rangle \quad \text{abbreviate} \quad (\forall F : \mathbb{N}'')[\Phi(k, j, i, r)f = F_0^r \circ F_1^i \circ F_2^j \circ F_3^k]$$

and with this consider the usual abbreviations so that $\langle\rangle$ is the target assertion. We verify the usual conditions and then appeal to the method of Exercise 7.15.

The base $(0.\bot)$ is an immediate consequence of the spec clauses $(\bot)$.

To verify (0.0), assuming $\langle k, j, i, r\rangle$, we have for each $F : \mathbb{N}''$

$$\Phi(k, j, i, r')F = F \circ \Phi(k, j, i, r)F = F \circ F_0^r \circ F_1^i \circ F_2^j \circ F_3^k = F_0^{r'} \circ F_1^i \circ F_2^j \circ F_3^k$$

to deduce $\langle k, j, i, r'\rangle$.

To verify (0.1), assuming $\langle k, j, i\rangle$, we have for each $F : \mathbb{N}'', f : \mathbb{N}', x : \mathbb{N}$

$$\begin{aligned}
\Phi(k, j, i', 0)Ffx &= \Phi(k, j, i, x)Ffx \\
&= (F_0^x \circ F_1^i \circ F_2^j \circ F_3^k)fx \\
&= F_0^x((F_1^i \circ F_2^j \circ F_3^k)f)x \\
&= F_1((F_1^i \circ F_2^j \circ F_3^k)f)x \quad = \quad (F_1^{i'} \circ F_2^j \circ F_3^k)fx
\end{aligned}$$

to deduce $\langle k, j, i', 0\rangle$.

To verify (0.2), assuming $\langle k, j\rangle$, we have for each $F : \mathbb{N}'', f : \mathbb{N}', x : \mathbb{N}$

$$\begin{aligned}
\Phi(k, j', 0, 0)Ffx &= \Phi(k, j, x, x)Ffx \\
&= (F_0^x \circ F_1^x \circ F_2^j \circ F_3^k)fx \\
&= (F_0^x \circ F_1^x)((F_2^j \circ F_3^k)f)x \\
&= F_2((F_2^j \circ F_3^k)f)x \quad = \quad (F_2^{j'} \circ F_3^k)fx
\end{aligned}$$

to deduce $\langle k, j', 0, 0\rangle$.

To verify (0.3), assuming $\langle k\rangle$, we have for each $F : \mathbb{N}'', f : \mathbb{N}', x : \mathbb{N}$

$$\begin{aligned}
\Phi(k', 0, 0, 0)Ffx &= \Phi(k, x, x, x)Ffx \\
&= (F_0^x \circ F_1^x \circ F_2^x \circ F_3^k)fx \\
&= (F_0^x \circ F_1^x \circ F_2^x)((F_3^k)f)x \\
&= F_3((F_3^k)f)x \quad = \quad F_3^{k'}fx
\end{aligned}$$

to deduce $\langle k', 0, 0, 0\rangle$. □

7.21 (a) Let's look at Exercises 7.16–7.20 in turn.

For 7.16 consider $\mathbb{A} = \mathbb{N}$, $a = 0$, $A = Suc$, and

$$\begin{array}{llll} \mathcal{A}_0 : (\mathbb{N} \longrightarrow \mathbb{N}) \longrightarrow \mathbb{N} & \text{where} & \mathcal{A}_0 p = p1 + 1 & \text{for} \quad p : \mathbb{N} \longrightarrow \mathbb{N} \\ \mathcal{A}_1 : (\mathbb{N}^2 \longrightarrow \mathbb{N}) \longrightarrow \mathbb{N} & \text{where} & \mathcal{A}_0 p = p(1,0) + 2 & \text{for} \quad p : \mathbb{N}^2 \longrightarrow \mathbb{N} \\ \mathcal{A}_2 : (\mathbb{N}^3 \longrightarrow \mathbb{N}) \longrightarrow \mathbb{N} & \text{where} & \mathcal{A}_0 p = p(1,0,0) & \text{for} \quad p : \mathbb{N}^3 \longrightarrow \mathbb{N} \end{array}$$

for the three remaining attributes. Then $F = \mathfrak{A}(-)$.

For 7.17 consider $\mathbb{A} = \mathbb{N}'$, $a = Zero$, $A = Suc \circ -$, and

$$\begin{array}{llll} \mathcal{A}_0 : (\mathbb{N} \longrightarrow \mathbb{N}') \longrightarrow \mathbb{N}' & \text{where} & \mathcal{A}_0 px = pxx & \text{for} \quad p : \mathbb{N} \longrightarrow \mathbb{N}' \\ \mathcal{A}_1 : (\mathbb{N}^2 \longrightarrow \mathbb{N}') \longrightarrow \mathbb{N}' & \text{where} & \mathcal{A}_0 px = p(x,0)x & \text{for} \quad p : \mathbb{N}^2 \longrightarrow \mathbb{N}' \\ \mathcal{A}_2 : (\mathbb{N}^3 \longrightarrow \mathbb{N}') \longrightarrow \mathbb{N}' & \text{where} & \mathcal{A}_0 px = p(x,0,0)x & \text{for} \quad p : \mathbb{N}^3 \longrightarrow \mathbb{N}' \end{array}$$

for each $x \in \mathbb{N}$. Then $F = \mathfrak{A}(-)$.

For 7.18 consider $\mathbb{A} = \mathbb{N}'$, $a = g$, $A = f \circ - \circ h$, and

$$\begin{array}{llll} \mathcal{A}_0 : (\mathbb{N} \longrightarrow \mathbb{N}') \longrightarrow \mathbb{N}' & \text{where} & \mathcal{A}_0 p = p1 & \text{for} \quad p : \mathbb{N} \longrightarrow \mathbb{N}' \\ \mathcal{A}_1 : (\mathbb{N}^2 \longrightarrow \mathbb{N}') \longrightarrow \mathbb{N}' & \text{where} & \mathcal{A}_0 px = p(x,1)x & \text{for} \quad p : \mathbb{N}^2 \longrightarrow \mathbb{N}' \\ \mathcal{A}_2 : (\mathbb{N}^3 \longrightarrow \mathbb{N}') \longrightarrow \mathbb{N}' & \text{where} & \mathcal{A}_0 px = p(x^2,x,2)x & \text{for} \quad p : \mathbb{N}^3 \longrightarrow \mathbb{N}' \end{array}$$

for each $x \in \mathbb{N}$. Then $G = \mathfrak{A}(-)$.

The L of 7.19 can not be given in the form. The spec clauses (1,2,3) are too index sensitive.

For 7.20 consider $\mathbb{A} = \mathbb{N}'''$, $a = id_{\mathbb{N}''}$, $A\phi F = F \circ \phi F$, for $\phi : \mathbb{N}''', F : \mathbb{N}''$, and

$$\begin{array}{llll} \mathcal{A}_0 : (\mathbb{N} \longrightarrow \mathbb{N}''') \longrightarrow \mathbb{N}''' & \text{where} & \mathcal{A}_0 pFfx = pxFfx & \text{for} \quad p : \mathbb{N} \longrightarrow \mathbb{N}''' \\ \mathcal{A}_1 : (\mathbb{N}^2 \longrightarrow \mathbb{N}''') \longrightarrow \mathbb{N}''' & \text{where} & \mathcal{A}_0 pFfx = p(x,x)Ffx & \text{for} \quad p : \mathbb{N}^2 \longrightarrow \mathbb{N}''' \\ \mathcal{A}_2 : (\mathbb{N}^3 \longrightarrow \mathbb{N}''') \longrightarrow \mathbb{N}''' & \text{where} & \mathcal{A}_0 pFfx = p(x,x,x)Ffx & \text{for} \quad p : \mathbb{N}^3 \longrightarrow \mathbb{N}''' \end{array}$$

for $F : \mathbb{N}''$, $f : \mathbb{N}'$, $x : \mathbb{N}$. Then $G = \mathfrak{A}(-)$.

(b) Given two 3-structures $\mathfrak{B}, \mathfrak{A}$ and a morphism ϕ, as in the question, let

$$\langle k,j,i,r \rangle \quad \text{abbreviate} \quad \phi(\mathfrak{B}(k,j,i,r)) = \mathfrak{A}(k,j,i,r)$$

and with this consider the usual abbreviations so that $\langle \rangle$ is the target assertion. We verify the usual conditions and then appeal to the method of Exercise 7.15.

The properties $\phi b = a$ and $\phi \circ B = A$ give $(0.\bot)$ and (0.0), respectively. Then

$$\phi \circ \mathcal{B}_l = \mathcal{A}_l \bullet \phi$$

for $l = 0,1,2$ give (0.1, 0.2, 0.3), respectively. Let us check (0.2).

Given k,j let $p = \mathfrak{B}(k,j,\cdot,\cdot)$ so that $p : \mathbb{M}_1 \longrightarrow \mathbb{B}$ and

$$\phi \circ p = \mathcal{A}(k,j,\cdot,\cdot)$$

is a restatement of the induction hypothesis $\langle k,j \rangle$. Then, using the spec clause (2) for both $\mathfrak{B}$ and $\mathfrak{A}$, we have

$$(\phi \circ \mathfrak{B})(k,j',0,0) = \phi(\mathfrak{B}(k,j',0,0)) = \phi(\mathcal{B}_1 p) = \mathcal{A}_1(\phi \circ p) = \mathcal{A}_1 \mathfrak{A}(k,j,\cdot,\cdot) = \mathfrak{A}(k,j',0,0)$$

to verify $\langle k,j',0,0 \rangle$, as required. □

7.22 Suppose F, G are two functions both satisfying the $\mathbb{M}_3$-recursion scheme (where in both cases the clauses $(\bot, \dots, 3)$ use the same recipes). Let

$$\langle k, j, i, r \rangle \quad \text{abbreviate} \quad F(k, j, i, r) = G(k, j, i, r)$$

and obtain $\langle k, j, i \rangle, \langle k, j \rangle, \langle k \rangle, \langle \rangle$ in the usual way. Each recursion clause immediately gives the corresponding clause of the $\mathbb{M}_3$-induction scheme. Hence $\langle \rangle$. □

G.4 Small jump operators

7.23 Property (i) follows by induction on r. For (ii) the comparisons between r, s, x, y give

$$f^r x \leq f^s x \leq f^s y$$

since each iterate of f is monotone and inflationary. □

7.24 Assuming $\phi \sqsubseteq f$ we have some $a \in \mathbb{N}$ such that

$$\{a \leq x,\ \mathsf{x} \leq x\} \Rightarrow \phi\mathsf{x} \leq fx$$

holds for all $x \in \mathbb{N}$ and argument sequences x of ϕ. Here '$\mathsf{x} \leq x$' means $x_i \leq x$ for each component x_i of x.

Consider any r such that $a \leq f^r 0$ (so $r = a$ will do). We show that

$$\mathsf{x} \leq x \Rightarrow \phi\mathsf{x} \leq f^{r+1} x$$

holds. Thus consider any pair $\mathsf{x} \leq x$. If $a \leq x$ then $\phi\mathsf{x} \leq fx \leq f^{r+1}x$ as required. If $x < a$ then $\mathsf{x} \leq a$ so $\phi\mathsf{x} \leq fa \leq f^{r+1}0 \leq f^{r+1}x$ as required.

It is always possible to eliminate 'eventual domination' in favour of 'dominated by an iterate', but this would be inconvenient. In practice, we often need to know the start of an eventual domination, hence we use $\leq_a$ or even $\sqsubseteq$ when possible. □

7.25 (a) For all $r \leq s$ and $x \leq y$ we have

$$f_r x \leq f_s x \leq f_s y \qquad f_r x \leq f_r y \leq f_s y$$

so that $gx = f_x x \leq f_y y = gy$ to show that g is monotone. Trivially we have $x + 1 \leq f_x x = gx$ and hence the diagonal limit g is a snake.

(b) Consider the functions f_r where

$$f_r x = \begin{cases} x + 2 & \text{if } r < x \\ x + 1 & \text{if } x \leq r \end{cases}$$

so that each f_r is a snake. Notice also that $f_r \sqsubseteq f_{r'}$ but $f_r \not\leq f_{r'}$ (since $f_r r' = r + 3$ but $f_{r'} r' = r + 2$). The diagonal limit g of this chain is

$$gx = f_x x = x + 1$$

which does not dominate or even eventually dominate any f_r! □

7.26 (a) Let $h = g \circ f$. By a couple of easy arguments, we have $f \leq h$ and $g \leq h$.

(b) If $g, h \in \mathcal{B}(f)$ then $g \leq f^{r+1}$, $h \leq f^{s+1}$ for some r, s, and hence $h \circ g \leq f^{t+1}$ for $t = s + r + 1$. Trivially $\mathcal{B}(f)$ is downward closed, so is a basket containing f.

If $g \in \mathcal{B}(f)$ where $f \in \mathcal{B}$ for some basket $\mathcal{B}$, then $g \leq f^{r+1} \in \mathcal{B}$ for some r and hence $g \in \mathcal{B}$, to give $\mathcal{B}(f) \subseteq \mathcal{B}$.

(c) The intersection of any set of baskets is a basket. □

7.27 (a) We show that

$$f \leq pol f \leq_2 brw f \leq rob f \leq_1 ack f \leq exp f$$

holds for each snake. Most of these are simple consequences of the snake properties of f. The second comparison is the most complicated. For this we check that $r+2 \leq f^r 2$ (by induction on r) and hence $f^2(r+2) \leq f^{r+2}2$ to give the comparison.

(b) For $F = brw$ property (1) of Definition 7.18 fails, but $f \leq_1 Ff$ does hold. This is not a serious problem. For $F = rob, f = Suc$ we see that property (3) fails badly. In spite of this rob is a useful jump operator. The properties (0,1,2,3,4) have been chosen specifically for this book. As yet there is no generally accepted notion of a 'useful operator'. □

7.28 Using $\mathsf{I} = \mathsf{I}_{\mathcal{N}}$ and $\mathsf{J} = \mathsf{I}_{\mathcal{N}'}$, the terms

$$\begin{aligned}
\mathsf{pol} &= \lambda y : \mathcal{N}', x : \mathcal{N} \,.\, \mathsf{I} y x \ulcorner 2 \urcorner \\
\mathsf{brw} &= \lambda y : \mathcal{N}', x : \mathcal{N} \,.\, \mathsf{I} y \ulcorner 2 \urcorner x \\
\mathsf{rob} &= \lambda y : \mathcal{N}', x : \mathcal{N} \,.\, \mathsf{I} y \ulcorner 1 \urcorner (\mathsf{S} x) \\
\mathsf{ack} &= \lambda y : \mathcal{N}', x : \mathcal{N} \,.\, \mathsf{I} y x (\mathsf{S} x) \\
\mathsf{exp} &= \lambda y : \mathcal{N}', x : \mathcal{N} \,.\, \mathsf{J} \mathsf{pol} y x x
\end{aligned}$$

will do. □

7.29 Given $f : \mathbb{N}'$ consider the functions F specified, respectively, by

$$\begin{array}{lll}
\mathbb{N}^2 \longrightarrow \mathbb{N}' & \mathbb{N}^2 \longrightarrow \mathbb{N} & \mathbb{N}^2 \longrightarrow \mathbb{N} \\
(\bot)\ F(0,0)x = fx & (\bot)\ F(0,x) = fx & (\bot)\ F(0,x) = fx \\
(0)\ F(i',0)x = F(i,x)x & (0)\ F(i',0) = F(i,1) & (0)\ F(i',0) = 2 \\
(1)\ F(i,r')x = F(i,0)y & (1)\ F(i',x') = F(i,y) & (1)\ F(i',x') = F(i,y) \\
\quad \text{where } y = F(i,r)x & \quad \text{where } y = F(i',x) & \quad \text{where } y = F(i',x)
\end{array}$$

for all $i, r, x \in \mathbb{N}$. The type of each function is given above its spec. In the left hand spec the x is a parameter, but in the other two specs it is a recursion argument. Three easy inductions show that these functions are

$$(i,r,x) \longmapsto (ack^i f)^{r+1}x \qquad (i,x) \longmapsto rob^i fx \qquad (i,x) \longmapsto brw^i fx$$

respectively. In particular, the left hand function is Ack_f. □

7.30 We show that $F = Ack_f$. To this end let

$$\langle i, r \rangle \text{ abbreviate } (\forall x)[F(i,r)x = (ack^i f)^{r+1}x]$$

and obtain $\langle i \rangle$ and $\langle \rangle$ by first quantifying out r and then i. Thus $\langle \rangle$ is the target assertion. We proceed by a double induction. The three spec clauses give

$$[\perp] \quad \langle 0, 0 \rangle \qquad [0] \quad \langle i, a \rangle, \langle i, b \rangle \Rightarrow \langle i, r' \rangle \qquad [1] \quad \langle i \rangle \Rightarrow \langle i', 0 \rangle$$

respectively. Only [0] is not immediate.

Using the suggested $a = \lfloor r/2 \rfloor$ and $b = \lceil r/2 \rceil$ we have $b + a = r$ but any pair $0 \leq a, b \leq r$ with this property will do. (The usual spec of Ack_f uses $a = r, b = 0$).

Let $g = ack^i f$. Thus $\langle i, b \rangle, \langle i, a \rangle$ give

$$F(i, b)y = g^{b+1}y \qquad F(i, a)x = g^{a+1}x$$

for all $x, y \in \mathbb{N}$. With $y = g^{a+1}x$ the spec clause (0) gives

$$F(i, r')x = F(i, b)(F(i, a)x) = F(i, b)y = g^{b+1}y = g^{b+a+2}x = g^{r'+1}x$$

to deduce $\langle i, r' \rangle$, and so verify [0].

To use $[\perp, 0, 1]$ let

$$\langle\langle i, r \rangle\rangle \quad \text{abbreviate} \quad (\forall s \leq r)\langle i, r \rangle \qquad \text{so that} \qquad \langle\langle i, r' \rangle\rangle \iff \langle\langle i, r \rangle\rangle \text{ and } \langle i, r' \rangle$$

and hence

$$[[0]] \quad \langle\langle i, r \rangle\rangle \Rightarrow \langle\langle i, r' \rangle\rangle$$

is a simple consequence of [0]. An induction over r gives

$$\langle\langle i, 0 \rangle\rangle \Rightarrow (\forall r)\langle\langle i, r \rangle\rangle \qquad \text{and hence} \qquad \langle i, 0 \rangle \Rightarrow \langle i \rangle$$

(by some simple quantifier manipulation). This with $[\perp]$ and [1] gives $\langle 0 \rangle$ and $\langle i \rangle \Rightarrow \langle i' \rangle$, and hence the required $\langle \rangle$ follows by an induction over i. □

7.31 For each sequence x of natural numbers and each natural number x, we write $\mathsf{x} \leq x$ to indicate $x_i \leq x$ holds for each component x_i of x.

(a) We have

$$\left.\begin{array}{r} \mathsf{x} \leq x \\ a \leq x \end{array}\right\} \Rightarrow \theta_j \mathsf{x} \leq gx \qquad \left.\begin{array}{r} \mathsf{y} \leq y \\ a \leq y \end{array}\right\} \Rightarrow \psi \mathsf{y} \leq hy$$

for all sequences x, y of appropriate lengths, all $x, y \in \mathbb{N}$, and all $1 \leq j \leq n$ (the number of θ_j and the length of y). Thus with $\mathsf{y} = (\theta_1 \mathsf{x}, \ldots, \theta_n \mathsf{x})$ and $y = gx$ we have

$$\left.\begin{array}{r} \mathsf{x} \leq x \\ a \leq x \end{array}\right\} \Rightarrow \left\{\begin{array}{c} \mathsf{y} \leq gx = y \\ a \leq x \leq y \end{array}\right\} \Rightarrow \phi \mathsf{x} \leq hy = (h \circ g)\mathsf{x}$$

as required.

(b) We have

$$\phi 0 \mathsf{x} = \theta \mathsf{x} \qquad \phi r' \mathsf{x} = \psi r s \mathsf{x} \quad \text{where } s = \phi r \mathsf{x}$$

and $\theta, \psi \leq_a g$, and we wish to show that

$$(\forall r)[\phi r \leq_a g^{r+1}]$$

holds. Let $\langle r \rangle$ be the body of the assertion. We proceed by induction over r.

The base case $\langle 0 \rangle$ is immediate since $\theta \leq_a g$.

For the induction step, $r \mapsto r'$, we are given $\psi \leq_a g$, i.e.

$$\{r \leq y,\ s \leq y,\ \mathsf{x} \leq y,\ a \leq y\} \Rightarrow \psi r s \mathsf{x} \leq gy$$

for all s, a, y and x of the appropriate length. Given x let $y = g^{r+1}x$, so that $z = r + 1 + x \leq y$. Thus, using $\langle r \rangle$, we have

$$\left.\begin{matrix} \mathsf{x} \leq x \\ a \leq x \end{matrix}\right\} \Rightarrow \left\{\begin{matrix} r \leq z \leq y \\ s = \phi r \mathsf{x} \leq y \\ \mathsf{x} \leq x \leq y \\ a \leq x \leq y \end{matrix}\right\} \Rightarrow \phi r' \mathsf{x} = \psi r s \mathsf{x} \leq gy = g^{r+2}x$$

to verify $\langle r' \rangle$, as required.

Finally

$$\{r \leq x,\ \mathsf{x} \leq x,\ a \leq x\} \Rightarrow \phi r \mathsf{x} \leq g^{r+1}x \leq g^{x+1}x = ack\, g\, x$$

as required.

(c) Using parts (a,b), a comparison $\phi \leq_a ack^i f$ follows by induction over the construction of ϕ from f. The index i measures the number of steps in the construction. This comparison gives $\phi \leq (ack^i f)^{r+1}$ for some large r, to prove Theorem 7.22. □

G.5 THE MULTI-RECURSIVE HIERARCHIES

7.32 Rephrasing Solution 7.29 we obtain a spec of F_0 as given on the right. This is a double recursion over $i \in \mathbb{N}$ and $r \in \mathbb{N}$ with x as a parameter.

$$\begin{array}{rrcl} (\bot 0) & F_0 00x & = & fx \\ (/) & F_0 i r' x & = & F_0 i 0 y \\ & \text{where } y & = & F_1(j,i) r x \\ (0) & F_0 i' 0 x & = & F_0 i x x \end{array}$$

(a) To obtain a spec of F_1 we add an extra clause to the spec of F_0 above. This is a double recursion over $(j, i) \in \mathbb{M}_1$ and $r \in \mathbb{N}$ with x as a parameter. Note that $F_0 i = F_1(0, i)$. The reason for the apparently eccentric labelling, $(\bot 0, /, 00, 1)$, will become clear soon.

$$\begin{array}{rrcl} (\bot 0) & F_1(0,0)0x & = & fx \\ (/) & F_1(j,i)r'x & = & F_1(j,i)0y \\ & \text{where } y & = & F_1(j,i)rx \\ (00) & F_1(j,i')0x & = & F_1(j,i)xx \\ (1) & F_1(j',0)0x & = & F_1(j,x)0x \end{array}$$

(b) To obtain a spec of F_2 we add an extra clause to the spec of F_1. This is a double recursion over $(k, j, i) \in \mathbb{M}_2$ and $r \in \mathbb{N}$ with x as a parameter. Note that $F_1(j, i) = F_2(0, j, i)$.

$$\begin{array}{rrcl} (\bot 0) & F_2(0,0,0)0x & = & fx \\ (/) & F_2(k,j,i)r'x & = & F_2(k,j,i)0y \\ & \text{where } y & = & F_2(k,j,i)rx \\ (00) & F_2(k,j,i')0x & = & F_2(k,j,i)xx \\ (1) & F_2(k,j',0)0x & = & F_2(k,j,x)0x \\ (2) & F_2(k',0,0)0x & = & F_2(k,x,0)0x \end{array}$$

To obtain a spec of F_3 we add an extra clause to the spec of F_2 above. This is a double recursion over $(l,k,j,i) \in \mathbb{M}_2$ and $r \in \mathbb{N}$ with x as a parameter. Note that $F_2(k,j,i) = F_3(0,l,j,i)$.

(In Exercise 9.3 we will see the spec of F_s for an arbitrary $s < \omega$.)

$$\begin{array}{llcl}
(\bot 0) & F_3(0,0,0,0)0x & = & fx \\
(/) & F_3(l,k,j,i)r'x & = & F_3(l,k,j,i)0y \\
& \text{where } y & = & F_3(l,k,j,i)rx \\
(00) & F_3(l,k,j,i')0x & = & F_3(l,k,j,i)xx \\
(1) & F_3(l,k,j',0)0x & = & F_3(l,k,j,x)0x \\
(2) & F_3(l,k',0,0)0x & = & F_3(l,k,x,0)0x \\
(3) & F_3(l',0,0,0)0x & = & F_3(l,x,0,0)0x
\end{array}$$

(c) The spec of (b) is a nested recursion over the *five* arguments l,k,j,i,r with x as a parameter. This is an instance of $\mathbb{M}_4$-recursion. However, it doesn't use the full facilities of $\mathbb{M}_4$-recursion, and is best seen as a composite of an $\mathbb{N}$-recursion (over r) and an $\mathbb{M}_3$-recursion (over (l,k,j,i)). The induction proof of part (d) will explain this.

(d) Recalling the condition $(*)$ let

$$\begin{array}{lll}
\langle l,k,j,i\rangle[r,x] & \text{abbreviate} & (*) \\
\langle l,k,j,i\rangle[r] & \text{abbreviate} & (\forall x:\mathbb{N})\langle l,k,j,i\rangle[r,x] \\
\langle l,k,j,i\rangle & \text{abbreviate} & (\forall r:\mathbb{N})\langle l,k,j,i\rangle[r]
\end{array}$$

and then obtain $\langle l,k,j\rangle, \langle l,k\rangle, \langle l\rangle, \langle\rangle$ in the usual way so that $\langle\rangle$ is the target assertion. The clauses of the spec give

$$\begin{array}{lllll}
[\bot 0] & \langle 0,0,0,0\rangle[0] & & & \\
[/] & \left.\begin{array}{l}\langle l,k,j,i\rangle[0]\\ \langle l,k,j,i\rangle[r]\end{array}\right\} & \Rightarrow & \langle l,k,j,i\rangle[r'] & \\
[00] & \langle l,k,j,i\rangle & \Rightarrow & \langle l,k,j,i'\rangle[0] &
\end{array}
\qquad
\begin{array}{llll}
[1] & \langle l,k,j\rangle & \Rightarrow & \langle l,k,j',0\rangle \\
[2] & \langle l,k\rangle & \Rightarrow & \langle l,k',0,0\rangle \\
[3] & \langle l\rangle & \Rightarrow & \langle l',0,0,0\rangle
\end{array}$$

where the hypothesis of [/] is a conjunction. Using [/] an induction over r gives

$$\langle l,k,j,i\rangle[0] \Rightarrow \langle l,k,j,i\rangle$$

and this with $[\bot 0]$ and [00] gives

$$[\bot] \quad \langle 0,0,0,0\rangle \qquad [0] \quad \langle l,k,j,i\rangle \Rightarrow \langle l,k,j,i'\rangle$$

which we can add to [1, 2, 3]. This gives a standard form of $\mathbb{M}_3$-induction (as in Exercise 7.15) which leads to $\langle\rangle$, as required. □

7.33 Rephrasing Solution 7.29 we obtain a spec of both cases of G_0 as given to the left in Table G.1. This has three clauses ($\bot$, 00, 0+) and is a double recursion over $i \in \mathbb{N}$ and $x \in \mathbb{N}$. In (00) the upper alternative gives *rob* and the lower one gives *brw*.

(a) For a spec of G_1 we add an extra clause (1) to the spec of G_0, as in the table. This is a double recursion over $(j,i) \in \mathbb{M}_1$ and $x \in \mathbb{N}$. Note that $G_0 i = G_1(0,i)$.

(b) For G_3 we add extra clauses (2,3) to the spec of G_1, as in the table. This is a double recursion over $(l,k,j,i) \in \mathbb{M}_2$ and $x \in \mathbb{N}$. Note that $G_1(j,i) = G_3(0,0,j,i)$.

$(\bot)$	$G_0 0x = fx$	$G_1(0,0)x = fx$	$G_3(0,0,0,0)x = fx$
(00)	$G_0 i'0 = \begin{cases} G_0 i1 \\ 2 \end{cases}$	$G_1(j,i')0 = \begin{cases} G_1(j,i)1 \\ 2 \end{cases}$	$G_3(l,k,j,i')0 = \begin{cases} G_3(l,k,j,i)1 \\ 2 \end{cases}$
$(0+)$	$G_0 i'x' = G_0 iy$ where $y = G_0 i'x$	$G_1(j,i')x' = G_1(j,i)y$ where $y = G_1(j,i')x$	$G_3(l,k,j,i')x' = G_3(l,k,j,i)y$ where $y = G_3(l,k,j,i')x$
(1)		$G_1(j',0)x = G_1(j,x)x$	$G_3(l,k,j',0)x = G_3(l,k,j,x)x$
(2)			$G_3(l,k',0,0)x = G_3(l,k,x,0)x$
(3)			$G_3(l',0,0,0)x = G_3(l,x,0,0)x$

Table G.1: Specs for Solution 7.33

(c) For the time being let $g = G_3(l,k,j,i)$ and $h = G_3(l,k,j,i')$ where l,k,j,i are fixed. Then clauses $(00, 0+)$ rephrase as

$$(00) \quad h0 = \begin{cases} g1 & \text{if } jmp = rob \\ 2 & \text{if } jmp = brw \end{cases} \qquad (0+) \quad hx' = g(hx)$$

from which we obtain $hx = jmp\, g\, x$ by a simple induction over x. Thus the whole spec can be rephrased as

$$\begin{array}{lrcl}
(\bot) & G_3(0,0,0,0)x & = & fx \\
(0) & G_3(l,k,j,i')x & = & jmp\, G_3(l,k,j,i)x \\
(1) & G_3(l,k,j',0)x & = & G_3(l,k,j,x)x \\
(2) & G_3(l,k',0,0)x & = & G_3(l,k,x,0)x \\
(3) & G_3(l',0,0,0)x & = & G_3(l,x,0,0)x
\end{array}$$

and this is a standard $\mathbb{M}_3$-recursion.

(d) Let

$$\langle l,k,j,i\rangle \quad \text{abbreviate} \quad G_3(l,k,j,i) = (jmp_0^i \circ jmp_1^j \circ jmp_2^k \circ jmp_3^l)f$$

to obtain $\langle l,k,j\rangle, \langle l,k\rangle, \langle l\rangle, \langle\rangle$ so that $\langle\rangle$ is the target assertion. The rephrased spec gives the usual conditions (with a slightly different use of the index symbols) and then a standard $\mathbb{M}_3$-induction gives $\langle\rangle$, as required. □

7.34 (a) The required spec is

$$\begin{array}{lrcl}
(\bot) & \mathbb{G}_3(0,0,0,0)jmp\, f & = & f \\
(0) & \mathbb{G}_3(l,k,j,i')jmp\, f & = & jmp \circ (\mathbb{G}_3(l,k,j,i')jmp) \\
(1) & \mathbb{G}_3(l,k,j',0)jmp\, f\, x & = & \mathbb{G}_3(l,k,j,x)jmp\, f\, x \\
(2) & \mathbb{G}_3(l,k',0,0)jmp\, f\, x & = & \mathbb{G}_3(l,k,x,0)jmp\, f\, x \\
(3) & \mathbb{G}_3(l',0,0,0)jmp\, f\, x & = & \mathbb{G}_3(l,x,0,0)jmp\, f\, x
\end{array}$$

where $jmp : \mathbb{N}'', f : \mathbb{N}', x : \mathbb{N}$ are the non-recursion parameters.

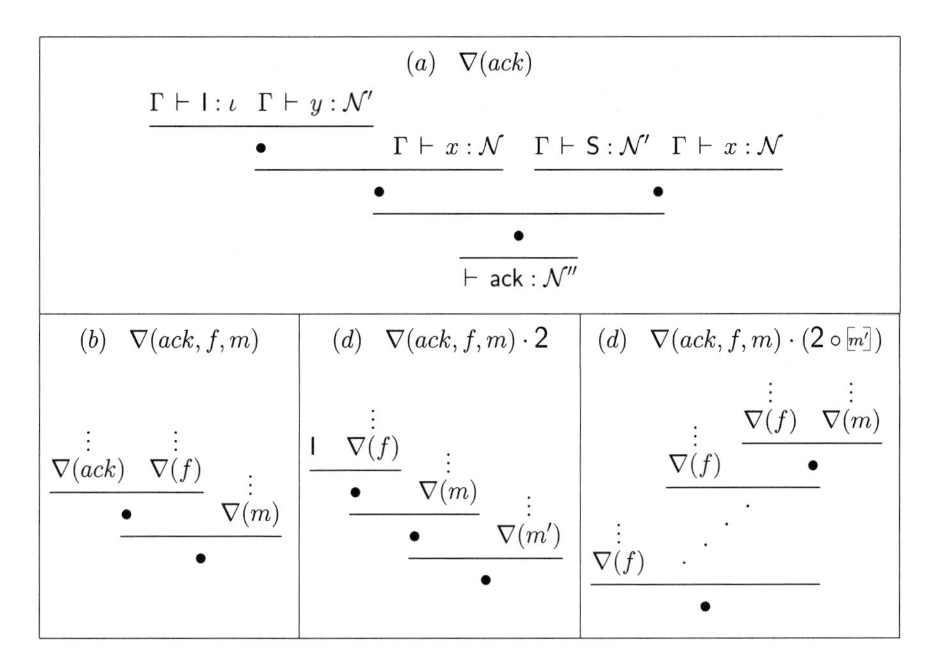

Table G.2: For Solution 7.35

(b) There are two points here.

Firstly, there is a difference between a first order spec and a higher order spec. The spec of (a) is higher order because of the use of the parameters jmp and f. This is true even when jmp and f are fixed; clause (0) is not first order. The specs of F_3 and G_3 are first order because the particular nature of the jmp used allows an expansion of clause (0). However, these specs are not an instance of $\mathbb{M}_3$-recursion precisely because of this expanded version of (0).

Secondly, there is a difference between a spec of a function F being an instance of $\mathbb{M}_3$-recursion and a function F being $\mathbb{M}_3$-recursive (in given functions). The second of these means that F can be specified by a battery of no more than $\mathbb{M}_3$-recursions (and compositions). The functions F_3 and G_3 are $\mathbb{M}_3$-recursive in the given function f, but the specs used are composites of $\mathbb{M}_3$-recursion and $\mathbb{N}$-recursion. $\square$

G.6 The extent of $\boldsymbol{\lambda}\mathbf{G}$

7.35 With $\Gamma = y : \mathcal{N}', x : \mathcal{N}$ the derivations of Table G.2 occur in this solution.

(a) The derivation $\nabla(ack)$ of the table will do.

(b) The derivation $\nabla(ack, f, m) = \nabla(ack)\nabla(f)\nabla(m)$ is given in the table. Here $\nabla(ack)$ is from (a), $\nabla(f)$ depends on f, and $\nabla(m)$ is an ascending staircase of height m.

(c) Using the given template $\square(f,\cdot)$ we set

$$\square^0(f,n) = \text{'do nothing'} \qquad \square^{m'}(f,n) = \downharpoonleft\square^m(f,n)\circ\square(f,k) \quad \text{where } k = f^m n$$

to produce

$$(\square^m(f,n)) \quad \ulcorner f\urcorner^m\ulcorner n\urcorner \rhd\!\!\!\rhd \ulcorner f^m n\urcorner$$

for all $m,n \in \mathbb{N}$. With $2 = \downharpoonleft 1 \circ 1$ and $\boxed{m'}$ as in Solution 7.5, we have

$$(2) \quad \mathsf{ack}\ulcorner f\urcorner\ulcorner m\urcorner \rhd\!\!\!\rhd \mathsf{I}\ulcorner f\urcorner\ulcorner m\urcorner\ulcorner m'\urcorner \qquad (\boxed{m'}) \quad \mathsf{I}\ulcorner f\urcorner\ulcorner m\urcorner\ulcorner m'\urcorner \rhd\!\!\!\rhd \ulcorner f\urcorner^{m'}\ulcorner m\urcorner$$

so that

$$(\square^{m'}(f,m)) \ \ulcorner f\urcorner^{m'}\ulcorner m\urcorner \rhd\!\!\!\rhd \ulcorner ack\, f\, m\urcorner \quad \text{and hence} \quad \square(ack,f,m) = 2\circ\boxed{m'}\circ\square^{m'}(f,m)$$

is the required computation.

(d) Derivations $\nabla(ack,f,m)\cdot 2$ and $\nabla(ack,f,m)\cdot(2\circ\boxed{m'})$ are in the table and then $\nabla(ack,f,m)\cdot\square(ack,f,m)$ is an ascending staircase of height $ack\, f\, m$. □

7.36 Let

$$\mathsf{Ack} = \lambda y : \mathcal{N}' \,.\, \lambda i, r, x : \mathcal{N} \,.\, \mathsf{I}(\mathsf{Jack} y i) x (\mathsf{S} r)$$

where $\mathsf{I} = \mathsf{I}_{\mathcal{N}}$, $\mathsf{J} = \mathsf{I}_{\mathcal{N}'}$, and ack is the term of Solution 7.28. With

$$\Gamma = y : \mathcal{N}', i : \mathcal{N}, r : \mathcal{N}, x : \mathcal{N}$$

the following derivation $\nabla(Ack)$ shows that Ack is well typed.

$$\dfrac{\dfrac{\dfrac{\dfrac{\dfrac{\Gamma \vdash \mathsf{I} : \mathcal{N}' \to \mathcal{N}^+ \qquad \dfrac{\dfrac{\dfrac{\dfrac{\vdash \mathsf{J} : \mathcal{N}'' \to \mathcal{N}'^+ \qquad \overset{\nabla(ack)}{\vdash \mathsf{ack} : \mathcal{N}''}}{\vdash \mathsf{Jack} : \mathcal{N}'^+}}{\Gamma \vdash \mathsf{Jack} : \mathcal{N}'^+} \qquad \Gamma \vdash y : \mathcal{N}'}{\Gamma \vdash \mathsf{Jack}y : \mathcal{N} \to \mathcal{N}'} \qquad \Gamma \vdash i : \mathcal{N}'}{\Gamma \vdash \mathsf{Jack}yi : \mathcal{N}'}}{\Gamma \vdash \mathsf{I}(--) : \mathcal{N}^+} \qquad \Gamma \vdash x : \mathcal{N}}{\Gamma \vdash \mathsf{I}(--)x : \mathcal{N}'} \qquad \dfrac{\Gamma \vdash \mathsf{S} : \mathcal{N}' \quad \Gamma \vdash r : \mathcal{N}}{\Gamma \vdash (\mathsf{S}r :)\mathcal{N}}}{\Gamma \vdash \mathsf{I}(\mathsf{Jack}yi)x(\mathsf{S}r) : \mathcal{N}}}{\dfrac{\dfrac{\bullet}{\bullet}}{\dfrac{\bullet}{\vdash \mathsf{Ack} : \mathcal{N}' \to \mathcal{N} \to \mathcal{N}^+}}}$$

Suppose that $\ulcorner f\urcorner$ represents f. Then for each $i, r, m \in \mathbb{N}$ we have

$$\begin{array}{lll} \mathsf{Ack}\ulcorner f\urcorner\ulcorner i\urcorner\ulcorner r\urcorner\ulcorner m\urcorner & \rhd\!\!\!\rhd & \mathsf{I}(\mathsf{Jack}\ulcorner f\urcorner\ulcorner i\urcorner)\ulcorner m\urcorner\ulcorner r'\urcorner \\ & \rhd\!\!\!\rhd & \mathsf{I}(\mathsf{ack}^i\ulcorner f\urcorner)\ulcorner m\urcorner\ulcorner r'\urcorner \\ & \rhd\!\!\!\rhd & (\mathsf{ack}^i\ulcorner f\urcorner)^{r'}\ulcorner m\urcorner \\ & \rhd\!\!\!\rhd & \ulcorner(ack^i f)^{r'} m\urcorner \qquad = \ulcorner Ack_f irm\urcorner \end{array}$$

where the last reduction step is proved by a double induction over i, r. □

7.37 Suppose the term $\vdash \ulcorner jmp \urcorner : \mathcal{N}''$ and operator $jmp : \mathbb{N}''$ are such that

> (1) For each term $\vdash \ulcorner f \urcorner : \mathcal{N}'$, if $\ulcorner f \urcorner$ represents a function $f : \mathbb{N}'$, then $\ulcorner jmp \urcorner \ulcorner f \urcorner$ represents $jmp\, f$.

holds. Then for each such pair $\vdash \ulcorner f \urcorner : \mathcal{N}'$ and $\ulcorner f \urcorner$, and $r \in \mathbb{N}$

$$\ulcorner jmp \urcorner^r \ulcorner f \urcorner \text{ represents } jmp^r f$$

follows by a simple induction over r.

Thus, for each term $\ulcorner jmp \urcorner$ which satisfies (1), and each term $\ulcorner f \urcorner$ which represents a function f, and each $m \in \mathbb{N}$, we have

$$\begin{array}{rll} \mathsf{Pet}\ulcorner jmp \urcorner \ulcorner f \urcorner \ulcorner m \urcorner & \rhd\!\!\!\rhd & \mathsf{J}\ulcorner jmp \urcorner \ulcorner f \urcorner \ulcorner m \urcorner \ulcorner m \urcorner \\ & \rhd\!\!\!\rhd & \ulcorner jmp \urcorner^m \ulcorner f \urcorner \ulcorner m \urcorner \\ & \rhd\!\!\!\rhd & \ulcorner jmp^m f\, m \urcorner \qquad = \ulcorner Pet\, jmp\, f\, m \urcorner \end{array}$$

where the crucial third step uses the representing property of $\ulcorner jmp \urcorner^m \ulcorner f \urcorner$. Thus we see that

> (2) If $\ulcorner jmp \urcorner$ represents an operator jmp, and if $\ulcorner f \urcorner$ represents a function f, then $\mathsf{Pet}\ulcorner jmp \urcorner \ulcorner f \urcorner$ represents $Pet\, jmp\, f$

holds. This is a rather complicated idea of 'representation'. See Section 7.7 for a discussion of this. □

7.38 Using the term Pet of Exercise 7.37 let $z_r = \mathsf{Pet}^r z$ for each identifier z and $r \in \mathbb{N}$. Note that $z : \mathcal{N} \vdash z_r : \mathcal{N}$ is derivable, and z_r contains the iterator J nested to a depth r. Given a declaration

$$\Gamma = i_s : \mathcal{N}, \ldots, i_0 : \mathcal{N}, z : \mathcal{N}'', y : \mathcal{N}'$$

let

$$\begin{array}{rcl} y_s & = & \mathsf{J} z_s y i_s \\ y_{s-1} & = & \mathsf{J} z_{s-1} y i_{s-1} \\ & \vdots & \\ y_1 & = & \mathsf{J} z_1 y i_1 \\ y_0 & = & \mathsf{J} z_0 y i_0 \end{array}$$

to get $\Gamma \vdash y_r : \mathcal{N}'$ for all $s \geq r \geq 0$. Finally set

$$\mathsf{M}_s = \lambda i_s : \mathcal{N}, \ldots, i_0 : \mathcal{N}, z : \mathcal{N}'', y : \mathcal{N}' \,.\, y_0 \qquad \mu_s = \mathcal{M} \to \cdots \to \mathcal{N} \to \mathcal{N}''$$

to get

$$\vdash \mathsf{M}_s : \mu_s \qquad \mathsf{M}_s \ulcorner i_s \urcorner \cdots \ulcorner i_0 \urcorner \ulcorner jmp \urcorner \ulcorner f \urcorner \ulcorner m \urcorner \rhd\!\!\!\rhd \ulcorner M_s \mathsf{i}\, jmp\, f\, m \urcorner$$

for all suitable terms $\ulcorner jmp \urcorner, \ulcorner f \urcorner, \ulcorner m \urcorner$ and $\mathsf{i} \in \mathbb{M}_s$. □

G.7 Naming in $\boldsymbol{\lambda G}$

7.39 Suppose the derivation

$$(F) \quad \vdash \ulcorner f \urcorner : \mathcal{N}'$$

names a function f. We must show that

$$\ulcorner f \urcorner \ulcorner m \urcorner \rhd\!\!\!\rhd \ulcorner fm \urcorner$$

holds for each $m \in \mathbb{N}$. Using the standard derivation

$$(M) \quad \vdash \ulcorner m \urcorner : \mathcal{N}$$

we obtain a derivation

$$(FM) \quad \vdash \ulcorner f \urcorner \ulcorner m \urcorner : \mathcal{N}$$

to show that the subject of the required reduction is well formed. By Normalization there is a computation

$$(\square) \quad \ulcorner f \urcorner \ulcorner m \urcorner \rhd\!\!\!\rhd t$$

to a normal term t. By Influence we have

$$(FM \cdot \square) \quad \vdash t : \mathcal{N}$$

and then by Lemma 7.5 (proved in Exercise 7.9) the term t is a numeral $\ulcorner n \urcorner$ for some $n \in \mathbb{N}$. Coherence gives $[\![FM \cdot \square]\!] = [\![FM]\!]$ so that $n = fm$, as required. (This last step depends on the recipe for the semantics of a derivation). $\square$

H

Ordinals and Ordinal Notations

H.1 Introduction

8.1 I hope you are now feeling refreshed. □

H.2 Ordinal arithmetic

8.2 (a) There are three parts.

For the first part we show

$$0 + \alpha = \alpha$$

by induction on α.

The base case, $\alpha = 0$, is immediate.

For the induction step, $\alpha \mapsto \alpha'$, we have

$$0 + \alpha' = (0 + \alpha)' = \alpha'$$

where the first equality uses the recursive construction of addition and the second uses the induction hypothesis.

For the induction leap to a limit ordinal λ we have

$$0 + \lambda = \bigvee\{0 + \alpha \mid \alpha < \lambda\} = \bigvee\{\alpha \mid \alpha < \lambda\} = \lambda$$

where the first equality uses the recursive construction of addition, the second uses the induction hypothesis, and the third uses the limiting property of λ.

For the second part we show

$$(\gamma + \beta) + \alpha = \gamma + (\beta + \alpha)$$

by induction on α.

The base case, $\alpha = 0$, is immediate.

For the induction step, $\alpha \mapsto \alpha'$, we have

$$(\gamma + \beta) + \alpha' = ((\gamma + \beta) + \alpha)' = (\gamma + (\beta + \alpha))' = \gamma + (\beta + \alpha)' = \gamma + (\beta + \alpha')$$

where the second equality uses the induction hypothesis and the rest use the recursive construction of addition.

For the induction leap to a limit ordinal λ we have

$$\begin{aligned}(\gamma+\beta)+\lambda &= \bigvee\{(\gamma+\beta)+\alpha \mid \alpha<\lambda\}\\ &= \bigvee\{\gamma+(\beta+\alpha) \mid \alpha<\lambda\}\\ &= \gamma+\bigvee\{\beta+\alpha \mid \alpha<\lambda\} \quad = \gamma+(\beta+\lambda)\end{aligned}$$

where the second equality uses the induction hypothesis. Notice also that the third equality should have some prior justification (concerning the continuity of addition).

For the third part note that

$$1+\omega=\bigvee\{1+r \mid r<\omega\}=\omega$$

so that

$$1+\omega=\omega\neq\omega+1$$

to show that addition is not commutative.

(b) An attempt to prove this by induction over α runs into difficulties. We need to prove a stronger result

$$\beta+\alpha\leq\beta+\gamma\Rightarrow\alpha\leq\gamma$$

by induction over α. This will require some information about the comparison $\leq$.

The base case, $\alpha=0$, is immediate.

For the induction step, $\alpha\mapsto\alpha'$, assume $\beta+\alpha'\leq\beta+\gamma$ holds. Then

$$\beta+\alpha\leq(\beta+\alpha)'=\beta+\alpha'\leq\beta+\gamma$$

so that $\alpha\leq\gamma$ (by the induction hypothesis). If $\alpha=\gamma$ then

$$(\beta+\alpha)'=\beta+\alpha'\leq\beta+\gamma=\beta+\alpha$$

which is not true (since there are no ordinals δ with $\delta'\leq\delta$). Thus $\alpha<\gamma$ and hence $\alpha'\leq\gamma$ as required.

For the induction leap to a limit ordinal λ assume $\beta+\lambda\leq\beta+\gamma$ holds. Then for each $\alpha<\lambda$ we have

$$\beta+\alpha\leq\beta+\lambda\leq\beta+\gamma$$

so that $\alpha\leq\gamma$ by the induction hypothesis. Thus

$$\lambda=\bigvee\{\alpha \mid \alpha<\lambda\}\leq\gamma$$

as required. □

8.3 (a) There are three parts

For the first part we have

$$\beta\times 1=\beta\times 0'=\beta\times 0+\beta=0+\beta=\beta$$

to give the first identity. The other two are proved by induction on α.

Thus

$$0 \times \alpha' = 0 \times \alpha + 0 = 0 + 0 = 0 \qquad 1 \times \alpha' = 1 \times \alpha + 1 = \alpha + 1 = \alpha'$$

give the two induction steps, and

$$\begin{aligned} 0 \times \lambda &= \bigvee\{0 \times \alpha \,|\, \alpha < \lambda\} \\ &= \bigvee\{0 \,|\, \alpha < \lambda\} \quad = 0 \end{aligned} \qquad \begin{aligned} 1 \times \lambda &= \bigvee\{1 \times \alpha \,|\, \alpha < \lambda\} \\ &= \bigvee\{\alpha \,|\, \alpha < \lambda\} \quad = \lambda \end{aligned}$$

give the two induction leaps.

For the second part we show

$$(\gamma \times \beta) \times \alpha = \gamma \times (\beta \times \alpha)$$

by induction on α.

The base case, $\alpha = 0$, is straight forward.

For the induction step, $\alpha \mapsto \alpha'$, we have

$$(\gamma\times\beta)\times\alpha' = (\gamma\times\beta)\times\alpha+\gamma\times\beta = \gamma\times(\beta\times\alpha)+\gamma\times\beta = \gamma\times(\beta\times\alpha+\beta) = \gamma\times(\beta\times\alpha')$$

where the second equality uses the induction hypothesis and the third uses a distributive law (which will be proved in (c) but should have been proved earlier).

For the induction leap to a limit ordinal λ we have

$$\begin{aligned} (\gamma \times \beta) \times \lambda &= \bigvee\{(\gamma \times \beta) \times \alpha \,|\, \alpha < \lambda\} \\ &= \bigvee\{\gamma \times (\beta \times \alpha) \,|\, \alpha < \lambda\} \\ &= \gamma \times \bigvee\{\beta \times \alpha \,|\, \alpha < \lambda\} \quad = \gamma \times (\beta \times \lambda) \end{aligned}$$

where the second equality uses the induction hypothesis. Notice also that the third equality should have some prior justification (concerning the continuity properties of addition).

For the third part a simple calculation shows that

$$2 \times \omega = \omega \qquad \omega \times 2 = \omega + \omega$$

so that $2 \times \omega \neq \omega \times 2$.

(b) We show

$$\beta \times \alpha \leq \beta \times \gamma \Rightarrow \alpha \leq \gamma$$

by induction over α. This will require some information about the comparison $\leq$.

The base case, $\alpha = 0$, is immediate.

For the induction step, $\alpha \mapsto \alpha'$, assume $\beta \times \alpha' \leq \beta \times \gamma$ holds. Then

$$\beta \times \alpha \leq \beta \times \alpha + \beta = \beta \times \alpha' \leq \beta \times \gamma$$

so that $\alpha \leq \gamma$ (by the induction hypothesis). If $\alpha = \gamma$ then, since $\beta \neq 0$,

$$(\beta \times \alpha)' \leq \beta \times \alpha = \beta = \beta \times \alpha' \leq \beta \times \gamma = \beta \times \alpha$$

which is not true. Thus $\alpha < \gamma$ and hence $\alpha' \leq \gamma$ as required.

For the induction leap to a limit ordinal λ assume $\beta \times \lambda \leq \beta \times \gamma$ holds. Then for each $\alpha < \lambda$ we have

$$\beta \times \alpha \leq \beta \times \lambda \leq \beta \times \gamma$$

so that $\alpha \leq \gamma$ by the induction hypothesis. Thus

$$\lambda = \bigvee \{\alpha \,|\, \alpha < \lambda\} \leq \gamma$$

as required.

(c) We show

$$\gamma \times (\beta + \alpha) = \gamma \times \beta + \gamma \times \alpha$$

by induction on α.

The base case, $\alpha = 0$, is immediate. For the induction step, $\alpha \mapsto \alpha'$, using the recursive construction of $+$, the recursive construction of $\times$, the induction hypothesis, the associativity of $+$, and the recursive construction of $+$, in that order, we have

$$\begin{aligned} \gamma \times (\beta + \alpha') &= \gamma \times (\beta + \alpha)' \\ &= \gamma \times (\beta + \alpha) + \gamma \\ &= (\gamma \times \beta + \gamma \times \alpha) + \gamma \\ &= \gamma \times \beta + (\gamma \times \alpha + \gamma) \quad = \quad \gamma \times \beta + \gamma \times \alpha' \end{aligned}$$

to give the required result.

For the induction leap to a limit ordinal λ, using the recursive construction of $+$, a continuity property of $\times$, the induction hypothesis, a continuity property of $+$, the recursive construction of $+$, in that order, we have

$$\begin{aligned} \gamma \times (\beta + \lambda) &= \gamma \times \bigvee \{\beta + \alpha \,|\, \alpha < \lambda\} \\ &= \bigvee \{\gamma \times (\beta + \alpha) \,|\, \alpha < \lambda\} \\ &= \bigvee \{\gamma \times \beta + \gamma \times \alpha \,|\, \alpha < \lambda\} \\ &= \gamma \times \beta + \bigvee \{\gamma \times \alpha \,|\, \alpha < \lambda\} \quad = \quad \gamma \times \beta + \gamma \times \lambda \end{aligned}$$

to give the required result.

Finally we have

$$(1 + 1) \times \omega = 2 \times \omega = \omega \neq \omega + \omega = 1 \times \omega + 1 \times \omega$$

so the other distributive law doesn't hold. □

8.4 (i) We have

$$\gamma^1 = \gamma^{0'} = \gamma^0 \times \gamma = 1 \times \gamma = \gamma$$

as required.

(ii) We prove this by induction on α. The base case is straight forward. Also

$$\begin{aligned} \gamma^{(\beta+\alpha')} &= \gamma^{(\beta+\alpha)'} \\ &= \gamma^{(\beta+\alpha)} \times \gamma \\ &= (\gamma^\beta \times \gamma^\alpha) \times \gamma \\ &= \gamma^\beta \times (\gamma^\alpha \times \gamma) = \gamma^\beta \times \gamma^{\alpha'} \end{aligned} \qquad \begin{aligned} \gamma^{(\beta+\lambda)} &= \gamma^{\bigvee\{\beta+\alpha \,|\, \alpha<\lambda\}} \\ &= \bigvee \{\gamma^{\beta+\alpha} \,|\, \alpha < \lambda\} \\ &= \bigvee \{\gamma^\beta \times \gamma^\alpha \,|\, \alpha < \lambda\} \\ &= \gamma^\beta \times \bigvee \{\gamma^\alpha \,|\, \alpha < \lambda\} = \gamma^\beta \times \gamma^\lambda \end{aligned}$$

give, respectively, the induction step and the induction leap. (You should work out why each equality of these arguments is valid.)

(iii) We prove this by induction on α. The base case is straight forward. Also

$$\begin{aligned}(\gamma^\beta)^{\alpha'} &= (\gamma^\beta)^\alpha \times \gamma^\beta \\ &= \gamma^{\beta\times\alpha} \times \gamma^\beta \\ &= \gamma^{\beta\times\alpha+\beta} \\ &= \gamma^{(\beta\times\alpha)'} \quad = \gamma^{\beta\times\alpha'}\end{aligned} \qquad \begin{aligned}(\gamma^\beta)^\lambda &= \bigvee\{(\gamma^\beta)^\alpha \,|\, \alpha < \lambda\} \\ &= \bigvee\{\gamma^{\beta\times\alpha} \,|\, \alpha < \lambda\} \\ &= \gamma^{\bigvee\{\beta\times\alpha \,|\, \alpha<\lambda\}} \quad = \gamma^{\beta\times\lambda}\end{aligned}$$

give, respectively, the induction step and the induction leap. (Again you should work out why each equality of these arguments is valid.) □

H.3 FUNDAMENTAL SEQUENCES

8.5 Since

$$\omega^\alpha \leq \omega^\beta + \omega^\alpha$$

it suffices to show

$$\beta < \alpha \Rightarrow \omega^\beta + \omega^\alpha \leq \omega^\alpha$$

and we proceed by induction on α.

The base case, $\alpha = 0$, holds vacuously (since there is no possible $\beta < 0$).

For the induction step, $\alpha \mapsto \alpha'$, we have

$$\beta < \alpha' \Rightarrow \beta \leq \alpha \Rightarrow \omega^\beta + \omega^{\alpha'} \leq \omega^\alpha + \omega^{\alpha'}$$

and

$$\omega^\alpha + \omega^{\alpha'} = \omega^\alpha \times 1 + \omega^\alpha \times \alpha = \omega^\alpha \times (1+\omega) = \omega^\alpha \times \omega = \omega^{\alpha'}$$

(since $1 + \omega = \omega$ is known), to give the required result. This argument does not use the induction hypothesis.

For the induction leap to a limit ordinal μ let

$$A = \{\alpha \,|\, \beta < \alpha < \mu\}$$

where $\beta < \mu$ is assumed. We have $\beta' \in A$ and $\bigvee A = \mu$. Thus

$$\begin{aligned}\omega^\beta + \omega^\mu &= \omega^\beta + \bigvee\{\omega^\alpha \,|\, \alpha \in A\} \\ &= \bigvee\{\omega^\beta + \omega^\alpha \,|\, \alpha \in A\} \\ &= \bigvee\{\omega^\alpha \,|\, \alpha \in A\} \quad = \omega^\mu\end{aligned}$$

where the penultimate step uses the induction hypothesis. □

8.6 Both addition and multiplication are associative, and the distributive law gives

$$\beta\alpha = (\beta\omega^{\omega+1}2) + (\beta\omega^\omega) + \beta 3$$

so it suffices to calculate each of these three components (and absorb if necessary).

We have

$$\begin{aligned}\beta 2 &= \beta + \beta \\ &= \omega^{\omega+2}3 + \omega^{\omega} + \omega + 7 + \omega^{\omega+2}3 + \omega^{\omega} + \omega + 7 \\ &= \omega^{\omega+2}3 + \omega^{\omega+2}3 + \omega^{\omega} + \omega + 7 \qquad = \omega^{\omega+2}6 + \omega^{\omega} + \omega + 7\end{aligned}$$

since the tail of β is absorbed by $\omega^{\omega+2}$. Similarly

$$\beta 3 = \omega^{\omega+2}9 + \omega^{\omega} + \omega + 7$$

and then

$$\beta r' = \omega^{\omega+2}(3r) + \beta$$

for all $r < \omega$.

We say two ascending chains

$$\Theta = (\theta_r \mid r < \omega) \qquad \Psi = (\psi_s \mid s < \omega)$$

of ordinals interlace if both

$$(\forall r)(\exists s)[\theta_r \leq \psi_s] \qquad (\forall s)(\exists r)[\psi_s \leq \theta_r]$$

hold. When this happens we have $\bigvee\Theta = \bigvee\Psi$ by the construction of suprema.

From above we see that

$$\Theta = (\beta r \mid r < \omega) \qquad \Psi = (\omega^{\omega+2} r \mid r < \omega)$$

interlace. Hence

$$\beta\omega = \bigvee\Theta = \bigvee\Psi = \omega^{\omega+2}\omega = \omega^{\omega+3}$$

by the interlacing principle.

We now have

$$\beta\omega^2 = (\beta\omega)\omega = \omega^{\omega+3}\omega = \omega^{\omega+4}$$

and then

$$\beta\omega^r = \omega^{\omega+3+r}$$

for all $r < \omega$. This shows that

$$\Theta = (\beta\omega^r \mid r < \omega) \qquad \Psi = (\omega^{\omega+r} \mid r < \omega)$$

interlace, hence

$$\beta\omega^{\omega} = \bigvee\Theta = \bigvee\Psi = \omega^{\omega+\omega} = \omega^{\omega 2}$$

by another use of the interlacing principle.

Finally

$$\beta\omega^{\omega+1} = \beta\omega^{\omega}\omega = \omega^{\omega 2+1}$$

and hence we obtain

$$\beta\alpha = \omega^{\omega 2+1}2 + \omega^{\omega 2} + \omega^{\omega+2}9 + \omega^{\omega} + \omega + 7$$

as the required result. □

8.7 (a) From the spec of exponentiation we have $0^0 = 1$ and then $0^{r'} = 0^r \times 0 = 0$ for all $r < \omega$. But now

$$0^\omega = \bigvee\{0^r \mid r < \omega\} = 1$$

(since 0^0 occurs in the supremum). After that $0^{\omega+r'} = 0^{\omega+r} \times 0 = 0$ and then

$$0^{\omega+\omega} = \bigvee\{0^{\omega+r} \mid r < \omega\} = 1$$

(since $0^\omega = 1$). In this way see see that

$$0^\alpha = \begin{cases} 1 & \text{if } \alpha \text{ is not a successor} \\ 0 & \text{if } \alpha \text{ is a successor} \end{cases}$$

is the required description.

(b) Using part (a) we see that

$$\ell\alpha = \alpha \times 0^\alpha$$

is a suitable definition. □

H.4 Some particular ordinals

8.8 (a) For

$$\mathsf{i} = (\mathsf{l}, i', 0, 0)$$

with i' in position t', we have

$$\text{ind}\,\mathsf{i} = \text{ind}\,(\mathsf{l}, 0, 0, 0) + \omega^{t+1}i' = \zeta + \omega^{t+1}i' = \zeta + \omega^{t+1}i + \omega^{t+1} = \eta + \omega^{t+1}$$

where $\zeta = \text{ind}\,(\mathsf{l}, 0, 0, 0)$ and $\eta = \zeta + \omega^{t+1}i$ so that $\zeta \gg \omega^{t+1}$ and $\eta \gg \omega^{t+1}$ hold. This gives

$$(\text{ind}\,\mathsf{i})[x] = (\eta + \omega^{t+1})[x] = \eta + \omega^{t+1}[x] = \eta + \omega^t x = \text{ind}\,(\mathsf{l}, i, x, 0)$$

as required.

(b) The spec

$$\text{ind}\,0 = 0 \qquad \text{ind}\,\mathsf{i}' = (\text{ind}\,\mathsf{i})' \qquad \text{ind}\,(\mathsf{i}, i', 0, 0)[\cdot] = \text{ind}\,(\mathsf{i}, i, \cdot, 0)$$

tells us all we need to know about ind. □

8.9 (a) We use the canonical expansions

$$\mu = \omega^{\mu(m)} + \cdots + \omega^{\mu(1)} \qquad \nu = \omega^{\nu(n)} + \cdots + \omega^{\nu(1)} \qquad \alpha = \omega^{\alpha(m)} + \cdots + \omega^{\alpha(1)}$$

throughout this part.

Suppose $\mu \gg \nu \geq \alpha$. We have $\mu(1) \geq \nu(n)$ and we require $\mu(1) \geq \alpha(a)$. If $\mu(1) < \alpha(a)$ then

$$\nu(1) \leq \cdots \leq \nu(n) \leq \mu(1) < \alpha(a)$$

so that

$$\nu \leq \omega^{\mu(n)} + \cdots + \omega^{\mu(n)} = \omega^{\mu(n)}n < \omega^{\mu(n)+1} \leq \omega^\alpha$$

i.e. $\nu < \alpha$, which is not so.

Suppose $\mu \gg \nu$. Since $\nu \geq \nu[x]$, the first part gives $\mu \gg \nu[x]$.

Suppose $\mu \gg \nu \gg \alpha$. We have

$$\mu(1) \geq \nu(n) \geq \nu(1) \geq \alpha(a)$$

so that $\mu(1) \geq \alpha(a)$ and hence $\mu \gg \alpha$.

Suppose $\mu \gg \nu$. Then

$$\mu + \nu = \omega^{\mu(m)} + \cdots + \omega^{\mu(1)} + \omega^{\nu(n)} + \cdots + \omega^{\nu(1)}$$

and this is a meshing sum. The standard way of producing $(\mu + \nu)[x]$ and $\nu[x]$ gives the required equality.

(b) Consider

$$\mu = \omega^2 + \omega \qquad \nu = \omega^2 \qquad \alpha = \omega^2$$

so that $\mu \geq \nu \gg \alpha$ and $\mu[x] = \omega^2 + x$. Then $\mu \not\gg \alpha$ and $\mu \not\gg \mu[x]$.

With the same μ, ν we have

$$\mu + \nu = \omega^2 + \omega^2 \qquad (\mu + \nu)[x] = \omega^2 + \omega x \qquad \nu[x] = \omega x \qquad \mu + \nu[x] = \omega^2 + \omega(x + 1)$$

to give the required inequality.

(c) All exponents in the canonical expansion of μ must be the same. Thus $\mu = \omega^\alpha m$ for some $\alpha \neq 0 \neq m$.

(d) All exponents in the canonical expansion of both μ and ν must be the same. Thus $\mu = \omega^\alpha m$ and $\nu = \omega^\alpha n$ for some $\alpha \neq 0$ and non-zero $m, n \in \mathbb{N}$. □

H.5 Ordinal notations

8.10 The natural numbers form the canonical expansions for the finite ordinals, so we can concentrate on the ordinals α with $\omega \leq \alpha$.

The function $\omega^\bullet$ is monotone, and $\omega^0 < \alpha < \omega^\alpha$ (since $\alpha < \epsilon_0$), so

$$\beta \in B \iff \omega^\beta \leq \alpha$$

defines an initial section B of $\mathbb{O}$. Also

$$\omega^{\bigvee B} = \bigvee\{\omega^\beta \mid \beta \in B\} \leq \alpha$$

so B has a maximum member $\bigvee B$. Let β be this largest member. Thus the sandwich $\omega^\beta \leq \alpha < \omega^{\beta+1}$ determines β precisely.

The function $\omega^\beta + \bullet$ is monotone, and (since $\omega^\beta + \omega^{\beta+1} = \omega^{\beta+1}$) we have

$$\omega^\beta + 0 \leq \alpha < \omega^\beta + \omega^{\beta+1}$$

so that

$$\gamma \in C \iff \omega^\beta + \gamma \leq \alpha$$

defines an initial section C of $\mathbb{O}$. Also

$$\omega^\beta + \bigvee C = \bigvee\{\omega^\beta + \gamma \mid \gamma \in C\} \leq \alpha$$

so C has a maximum member $\bigvee C$. Let γ be this largest member. Thus the sandwich

$$\omega^\beta + \gamma \leq \alpha < \omega^\beta + \gamma'$$

determines γ precisely. In fact $\alpha = \omega^\beta + \gamma$ since $\omega^\beta + \gamma'$ is the successor of $\omega^\beta + \gamma$.

We now check that both $\beta < \alpha$ and $\gamma < \alpha$ hold.

Since $\alpha < \omega^\alpha$ we have $\alpha \notin B$, and hence $\beta < \alpha$.

Suppose $\alpha \leq \gamma$. Then

$$\omega^\beta + \alpha \leq \omega^\beta + \gamma \leq \alpha$$

and, assuming $\omega^\beta r \leq \alpha$, we have

$$\omega^\beta r' = \omega^\beta + \omega^\beta r \leq \omega^\beta + \alpha \leq \alpha$$

to give the induction step, which eventually shows

$$\omega^{\beta+1} = \omega^\beta \omega = \bigvee\{\omega^\beta r \,|\, r < \omega\} \leq \alpha$$

which is false. Thus $\gamma < \alpha$, by contradiction.

Finally $\gamma < \alpha < \omega^{\beta+1}$ so that $\omega^\beta \gg \gamma$ and hence $\alpha = \omega^\beta + \gamma$ is a meshing sum with $\gamma, \beta < \alpha$. Thus we may proceed by recursion. □

8.11 (a) The exponential polynomial $\{\alpha\}x$ is generated by recursion on the *canonical notation* for α, not on the size of α. Thus

$$\{0\}x = 0 \quad \{\omega\}x = x \quad \{\alpha'\} = 1 + \{\alpha\}x \quad \{\omega^\alpha\}x = x^{\{\alpha\}x} \quad \{\zeta + \eta\}x = \{\zeta\}x + \{\eta\}x$$

are the appropriate clauses. In the fifth clause ζ and η are limit ordinals with $\zeta \gg \eta$.

(b) Each limit ordinal μ has one of the forms

$$\omega^{\alpha+1} \qquad \omega^\nu \qquad \zeta + \eta$$

where ν, ζ, η are limit ordinals with $\zeta \gg \eta$ (and η is a power of ω). We prove

$$\{\mu[x]\}x = \{\mu\}x$$

by induction on the notation for μ.

The base case is $\mu = \omega$. Observe that for each $m \in \mathbb{N}$

$$\{m\}x = m + \{0\}x = m$$

and hence

$$\{\omega[x]\}x = \{x\}x = x\{\omega\}x$$

as required. This also covers the case $\mu = \omega^1$.

For the induction step we consider the three possible shapes for μ, as above.

For each limit ordinal $\zeta = \omega^\alpha$ where $\alpha > 0$ and $m \in \mathbb{N}$ we have

$$\{\zeta m\}x = \{\zeta + \cdots + \zeta\}x = \{\zeta\}x + \cdots + \{\zeta\}x = (\{\zeta\}x)m$$

by an iterated use of the last spec clause. Thus, with $y = \{\alpha\}x$, we have

$$\{\omega^{\alpha'}[x]\}x = \{\omega^{\alpha}x\}x = (\{\omega^{\alpha}\}x)x = (x^y)x = x^{1+y} = x^{\{\alpha'\}x} = \{\omega^{\alpha'}\}x$$

as required. The first step uses the definition of $\omega^{\alpha'}[\cdot]$, the second uses the spec of $\{\cdot\}\cdot$, the third uses the induction hypothesis, and the fifth and sixth use the spec of $\{\cdot\}\cdot$.

For $\mu = \omega^{\nu}$ we have

$$\{\omega^{\nu}[x]\}x = \{\omega^{\nu[x]}\}x = x^{\{\nu[x]\}x} = x^{\{\nu\}x} = \{\omega^{\nu}\}x$$

as required. The third step uses the induction hypothesis, and the others use various specs.

For $\mu = \zeta + \eta$ with $\zeta \gg \eta$ we have

$$\{(\zeta+\eta)[x]\}x = \{\zeta + \eta[x]\}x = \{\zeta\}x + \{\eta[x]\}x = \{\zeta\}x + \{\eta\}x = \{\zeta+\eta\}x$$

as required. The second step holds since $\zeta \gg \eta[x]$ and the third uses the induction hypothesis.

(c) Using part (b) we have

$$\{0\} = \mathsf{zero} \qquad \{\alpha'\} = suc \circ \{\alpha\} \qquad \{\mu\}x = \{\mu[x]\}x$$

for all ordinals α, limit ordinals μ, and $x \in \mathbb{N}$. This is a reorganized version of the spec of (a).

(d) Since

$$\epsilon[0] = \omega \qquad \epsilon[r'] = \omega^{\epsilon[r]}$$

we have

$$\{\epsilon[0]\}x = x \qquad \{\epsilon[r']\}x = x^{\{\epsilon[r]\}x}$$

and hence

$$\{\epsilon[r]\}x = \beth(r, x, x)$$

by a simple induction on r.

The function $x \longmapsto \beth(x, x, x)$ is a reasonable limit of $(\{\epsilon[r]\} \mid r < \omega)$. □

8.12 (a) We show first that ω^{α} is additively critical by induction on α.

The base case, $\alpha = 0$, is straight forward (since $\omega^0 = 1$).

For the induction step, $\alpha \mapsto \alpha'$, we have

$$\omega^{\alpha'} = \omega^{\alpha}\omega = \bigvee\{\omega^{\alpha}r \mid r < \omega\}$$

so that if $\beta < \omega^{\alpha'}$ then $\beta < \omega^{\alpha}r$ for some $r < \omega$. But then

$$\beta + \omega^{\alpha'} \le \omega^{\alpha}r + \omega^{\alpha}\omega = \omega^{\alpha}(r+\omega) = \omega^{\alpha}\omega = \omega^{\alpha'}$$

as required. (In fact this argument is not quite complete. It uses the case $\alpha = 1$, i.e. the additive criticality of ω, which should be verified directly.)

For the induction leap to a limit ordinal λ we have

$$\omega^{\lambda} = \bigvee\{\omega^{\alpha} \mid \alpha < \lambda\}$$

so that if $\beta < \omega^\lambda$ then $\beta < \omega^\alpha$ for some $\alpha < \lambda$. But then (!) $\lambda = \alpha + \gamma$ for some $0 < \gamma \le \lambda$, and hence

$$\beta + \omega^\lambda \le \omega^\alpha + \omega^{\alpha+\gamma} = \omega^\alpha 1 + \omega^\alpha \omega^\gamma = \omega^\alpha(1 + \omega^\gamma) = \omega^\alpha \omega^\gamma = \omega^{\alpha+\gamma} = \omega^\lambda$$

as required.

Conversely, suppose that θ is additively critical and consider the ordinal α with

$$\omega^\alpha \le \theta < \omega^{\alpha'}$$

(i.e. the least ordinal α such that $\theta < \omega^{\alpha+1}$). If $\omega^\alpha < \theta$ then

$$\omega^\alpha 2 = \omega^\alpha + \omega^\alpha \le \omega^\alpha + \theta = \theta$$

(since θ is additively critical). We then get $\omega^\alpha r < \theta$ by induction on r, and hence

$$\omega^{\alpha'} = \omega^\alpha \omega = \bigvee\{\omega^\alpha r \,|\, r < \omega\} \le \theta$$

which contradicts the choice of α.

Thus, with this α, we have $\theta = \omega^\alpha$, as required.

(b) We show first that ω^{ω^α} is multiplicatively critical by induction on α.
The base case, $\alpha = 0$, is straight forward (since $\omega^{\omega^0} = \omega$).
For the induction step, $\alpha \mapsto \alpha'$, we have

$$\omega^{\omega^{\alpha'}} = \omega^{\omega^\alpha \omega} = \bigvee\{\omega^{\omega^\alpha r} \,|\, r < \omega\}$$

so that if $\beta < \omega^{\omega^{\alpha'}}$ then $\beta < \omega^{\omega^\alpha r}$ for some $r < \omega$. But then

$$\beta\omega^{\omega^{\alpha'}} \le \omega^{\omega^\alpha r}\omega^{\omega^\alpha \omega} = \omega^{\omega^\alpha r + \omega^\alpha \omega} = \omega^{\omega^\alpha(r+\omega)} = \omega^{\omega^\alpha \omega} = \omega^{\omega^{\alpha'}}$$

as required.

For the induction leap to a limit ordinal λ we have

$$\omega^{\omega^\lambda} = \bigvee\{\omega^{\omega^\alpha} \,|\, \alpha < \lambda\}$$

so that if $\beta < \omega^{\omega^\lambda}$ then $\beta < \omega^{\omega^\alpha}$ for some $\alpha < \lambda$. But then $\omega^\alpha < \omega^\lambda$ and ω^λ is additively critical, so that

$$\beta\omega^{\omega^\lambda} \le \omega^{\omega^\alpha}\omega^{\omega^\lambda} = \omega^{\omega^\alpha + \omega^\lambda} = \omega^{\omega^\lambda}$$

as required.

Conversely, suppose θ is multiplicatively critical and consider the ordinal α with

$$\omega^{\omega^\alpha} \le \theta < \omega^{\omega^{\alpha'}}$$

(i.e. the least ordinal α such that $\theta < \omega^{\omega^{\alpha+1}}$). If $\omega^{\omega^\alpha} < \theta$ then

$$\omega^{\omega^\alpha 2} = (\omega^{\omega^\alpha})^2 \le \omega^{\omega^\alpha}\theta = \theta$$

(since θ is multiplicatively critical). We then get $\omega^{\omega^\alpha r} < \theta$ by induction on r, and hence

$$\omega^{\omega^{\alpha'}} = \omega^{\omega^\alpha \omega} = \bigvee\{\omega^{\omega^\alpha r} \,|\, r < \omega\} \le \theta$$

which contradicts the choice of α.

Thus, with this α, we have $\theta = \omega^{\omega^\alpha}$, as required. $\square$

8.13 Since $\theta = \omega^\theta$ we know that θ is multiplicatively critical. Consider $2 \leq \alpha < \theta$. Then $\alpha\theta = \theta$ so that

$$\theta \leq \alpha^\theta \leq (\omega^\alpha)^\theta = \omega^{\alpha\theta} = \omega^\theta = \theta$$

as required. □

8.14 A simple induction gives

$$\beth(\omega, \omega, r) = \epsilon[r]$$

(for $r < \omega$) and hence

$$\beth(\omega, \omega, \omega) = \bigvee\{\beth(\omega, \omega, r) \mid r < \omega\} = \epsilon_0$$

which is the smallest critical ordinal.

We show

$$\beth(\alpha, \beta, \omega) = \begin{cases} \epsilon_0 & \text{if } \omega \leq \beta \\ \epsilon_0 & \text{if } 2 \leq \beta < \omega,\ \omega < \alpha \\ \omega & \text{if } 2 \leq \beta < \omega,\ \alpha \leq \omega \\ \max(1, \alpha) & \text{if } \beta \leq 1 \end{cases}$$

(where $\alpha, \beta < \epsilon_0$ in all cases).

We show first that

$$\beth(\alpha, \beta, r) < \epsilon_0$$

for all $r < \omega$ and $\alpha, \beta < \epsilon_0$. In fact

$$\beth(\alpha, \beta, 0) = \alpha < \epsilon_0$$

gives the base case $r = 0$. Then, since ϵ_0 is critical,

$$\beth(\alpha, \beta, r') = \beta^{\beth(\alpha,\beta,r)} \leq \beta^{\epsilon_0} = \epsilon_0$$

gives the induction step.

We now deal with the various cases from the bottom clause up. Remembering how to calculate 0^α, we have

$$\begin{aligned} \beth(\alpha, 0, 0) &= \alpha \\ \beth(\alpha, 0, 1) &= \begin{cases} 1 & \text{if } \alpha \text{ is not a successor} \\ 0 & \text{if } \alpha \text{ is a successor} \end{cases} \\ \beth(\alpha, 0, 2) &= \begin{cases} 0 & \text{if } \alpha \text{ is not a successor} \\ 1 & \text{if } \alpha \text{ is a successor} \end{cases} \\ \beth(\alpha, 0, 3) &= \begin{cases} 1 & \text{if } \alpha \text{ is not a successor} \\ 0 & \text{if } \alpha \text{ is a successor} \end{cases} \\ &\vdots \end{aligned}$$

and hence, for $r < \omega$

$$\beth(\alpha, 0, r) \in \{0, 1, \alpha\}$$

and all three possible values do occur for appropriate r. Thus

$$\beth(\alpha, 0, \omega) = \max(1, \alpha)$$

and

$$\beth(\alpha, 1, \omega) = \max(1, \alpha)$$

follows by a similar (and simpler) calculation.

From now on we may assume $2 \leq \beta$ and appeal to the monotone properties of $\beth(\cdot, \cdot, \cdot)$.

Consider any $2 \leq \beta < \omega$. Then $(\beta^r \mid r < \omega)$ is an ascending chain of finite ordinals, and hence $\beta^\omega = \omega$ holds. In particular,

$$\beth(\omega, \beta, r) = \omega$$

for all $r < \omega$, to give

$$\beth(\omega, \beta, \omega) = \omega$$

and so deal with the range $2 \leq \beta < \omega$, $\alpha = \omega$.

For any $2 \leq \beta < \omega$ and $\alpha < \omega$ we know that $(\beth(\alpha, \beta, r) \mid r < \omega)$ is an ascending chain of finite ordinals with supremum ω. This, with the previous calculation, deals with the range $2 \leq \beta < \omega$, $\alpha \leq \omega$: the second clause up.

For the next case, as a preliminary we first show

$$\beth(\omega + 1, 2, \omega) = \epsilon_0$$

and then use a monotony argument.

We first show $2^{\omega^{1+\gamma}} = \omega^{\omega^\gamma}$ for all $\gamma < \epsilon_0$. We know $2^\omega = \omega$, which gives the $\gamma = 0$ case. Then, assuming $2^{\omega^{1+\gamma}} = \omega^{\omega^\gamma}$, we have

$$2^{\omega^{1+\gamma'}} = 2^{(\omega^{1+\gamma} \times \omega)} = (2^{\omega^{1+\gamma}})^\omega = (\omega^{\omega^{1+\gamma}})^\omega = \omega^{(\omega^{1+\gamma} \times \omega)} = \omega^{\omega^{1+\gamma'}}$$

to produce an induction step. Using the finite cases we have

$$2^{\omega^\omega} = \bigvee\{2^{\omega^{r'}} \mid r < \omega\} = \bigvee\{\omega^{\omega^r} \mid r < \omega\} = \omega^{\omega^\omega}$$

to verify the first limit case. (You should make sure you understand this.) For a limit ordinal $\mu > \omega$ we have

$$2^{\omega^\mu} = \bigvee\{2^{\omega^\gamma} \mid \omega \leq \gamma < \mu\} = \bigvee\{\omega^{\omega^\gamma} \mid \omega \leq \gamma < \mu\} = \omega^{\omega^\mu}$$

to make the induction leaps.

Using this result we have

$$\begin{array}{lllllll}
\beth(\omega+1, 2, 0) & = & \omega + 1 & & & & \\
\beth(\omega+1, 2, 1) & = & 2^{\omega+1} & = & (2^\omega)2 & = & \omega 2 \\
\beth(\omega+1, 2, 2) & = & 2^{\omega 2} & = & (2^\omega)^2 & = & \omega^2 \\
\beth(\omega+1, 2, 3) & = & 2^{\omega^2} & = & \omega^\omega & = & \epsilon[1]
\end{array}$$

and then

$$\beth(\omega + 1, 2, r + 3) = \epsilon[r + 1]$$

will follow by induction on r. Thus

$$\beth(\omega+1,2,r'+3) = 2^{\beth(\omega+1,2,r+3)} = 2^{\epsilon[r+1]} = 2^{\omega^{\epsilon[r]}} = \omega^{\omega^{\epsilon[r]}} = \epsilon[r+2]$$

gives the induction step. The penultimate equality uses the previous observation. Finally

$$\beth(\omega+1,2,\omega) = \bigvee\{\beth(\omega+1,2,r)\,|\,r<\omega\} = \epsilon_0$$

as required to complete the preliminary.

Consider $\omega < \alpha$ and $2 \leq \beta < \omega$. We have

$$\beth(\omega+1,2,r) \leq \beth(\alpha,\beta,r) \leq \epsilon_0$$

for all $r < \omega$. Thus, taking a supremum over r, we have

$$\epsilon_0 = \beth(\omega+1,2,\omega) = \beth(\alpha,\beta,\omega) = \epsilon_0$$

to complete the third clause up.

Turning to the top clause we have $\omega \leq \beta < \epsilon_0$, $\alpha < \epsilon_0$. Then

$$\beth(0,\omega,0) = 0 \qquad \beth(0,\omega,1) = \omega^0 = 1 \qquad \beth(0,\omega,2) = \omega^1 = \omega$$

and a simple induction gives

$$\beth(0,\omega,r+2) = \epsilon[r]$$

for all $r < \omega$. Thus

$$\beth(0,\omega,\omega) = \epsilon_0$$

and then

$$\beth(\alpha,\beta,\omega) = \epsilon_0$$

for $\omega \leq \beta$ follows by a monotony argument. □

I
Higher Order Recursion

I.1 Introduction

9.1 We must show

$$\langle \mathsf{i} \rangle \quad \mathfrak{P}\mathsf{i}F = F_0^{i_0} \circ \cdots \circ F_s^{i_s}$$

for each $F : \mathbb{N}''$ and multi-index $\mathsf{i} = (i_s, \ldots, i_0)$. We may fix F and proceed by induction on i using the three suggested phases.

(base) For $\mathsf{i} = \mathsf{0}$ this identity is immediate.

(step) Recall that $\mathsf{i}' = (i_s, \ldots, i_1, i_0{}')$, hence using the (step) clause of the spec of $\mathfrak{P}$, the induction hypothesis $\langle \mathsf{i} \rangle$, and the identity $F_0 = F$ we have

$$\mathfrak{P}\mathsf{i}'F = F \circ \mathfrak{P}\mathsf{i}F = F \circ F_0^{i_0} \circ \cdots \circ F_s^{i_s} = F_0^{i_0{}'} \circ \cdots \circ F_s^{i_s}$$

as required.

(leap) Consider any multi-index $\mathsf{i} = (\mathsf{l}, i', 0, 0)$ where the i' and 0 occur in positions $t + 1$ and t. For each $r \in \mathbb{N}$ the multi-index $(\mathsf{l}, i, r, \mathsf{0})$ is smaller than i, so the equality

$$\mathfrak{P}(\mathsf{l}, i, r, \mathsf{0})F = F_t^r \circ F_{t+1}^i \circ G \quad \text{where } G = F_{t+2}^{i_{t+2}} \circ \cdots \circ F_s^{i_s}$$

is an instance of the induction hypothesis. Consider any $f : \mathbb{N}'$ and for convenience let $g = Gf$ and $h = F_{t+1}^i g$. For each $x \in \mathbb{N}$, using the (leap) clause of the spec of $\mathfrak{P}$, the induction hypothesis in the indicated form with $r = x$, the construction of F_{t+1}, and some simple calculations, we have

$$\mathfrak{P}\mathsf{i}Ffx = \mathfrak{P}(\mathsf{l}, i, x, \mathsf{0})Ffx = (F_t^x \circ F_{t+1}^i \circ G)fx = F_t^x hx = F_{t+1}hx = F_{t+1}^{i'}gx$$

so that

$$\mathfrak{P}\mathsf{i}Ffx = (F_{t+1}^{i'} \circ F_{t+2}^{i_{t+2}} \circ \cdots \circ F_s^{i_s})fx$$

to give the required result. □

9.2 (a) We mimic the spec of the Péter function. Thus

$$\begin{array}{llcl} \text{(base)} & \mathfrak{P}_r\mathsf{0} & = & Id_{\mathbb{N}(r)'} \\ \text{(step)} & \mathfrak{P}_r\mathsf{i}'ts & = & t \circ \mathfrak{P}_r\mathsf{i}t \\ \text{(leap)} & \mathfrak{P}_r(\mathsf{i}, i', 0, 0)ts\phi_r \cdots \phi_1 x & = & \mathfrak{P}_r(\mathsf{i}, i, x, \mathsf{0})ts\phi_r \cdots \phi_1 x \end{array}$$

for each $t : \mathbb{N}(r)', s : \mathbb{N}(r), \phi_r : \mathbb{N}^{(r)}, \ldots, \phi_1 : \mathbb{N}', x : \mathbb{N}$ and multi-indexes as indicated.

(b) Almost trivially $\mathfrak{P}_r(0,m)ts = t^m s$ and hence

$$\mathfrak{P}_r(1,0)ts\phi_r \cdots \phi_1 x = \mathfrak{P}_r(0,x)ts\phi_r \cdots \phi_1 x = t^x s\phi_r \cdots \phi_1 x = O_r ts\phi_r \cdots \phi_1 x$$

to give $O_r = \mathfrak{P}_r(1,0)$.

(c) Since

$$(\mathfrak{P}_{r'}(0,k))O_r t = O_r^k t = t_k$$

we see that the compound

$$(\mathfrak{P}_{r'}(0,\cdot))(\mathfrak{P}_r(1,0))$$

(for natural number arguments) gives the appropriate function. □

9.3 We require

$$F_s \mathsf{i} r x = ((ack_0^{i_0} \circ \cdots \circ ack_s^{i_s})f)^{r+1} x = (\mathfrak{P}\mathsf{i} ack f)^{r+1} x$$

where the second equality follows by Lemma 9.3.

The specs of the functions $F_0 = Ack_f, F_1, F_2, F_3$ are given in Exercise 7.32 and that for F_3 is repeated to the right. The specs for F_2, F_1, F_0 can be obtained by setting $l = 0$, $k = 0$, $j = 0$ in turn.

$(\perp)$	$F_3(0,0,0,0,0,x)$	$=$	$f(x)$
$(*)$	$F_3(l,k,j,i,r',x)$	$=$	$F_3(l,k,j,i,0,y)$
	where y	$=$	$F_3(l,k,j,i,r,x)$
(0)	$F_3(l,k,j,i',0,x)$	$=$	$F_3(l,k,j,i,x,x)$
(1)	$F_3(l,k,j',0,0,x)$	$=$	$F_3(l,k,j,x,0,x)$
(2)	$F_3(l,k',0,0,0,x)$	$=$	$F_3(l,k,x,0,0,x)$
(3)	$F_3(l',0,0,0,0,x)$	$=$	$F_3(l,x,0,0,0,x)$

The spec for a general s is shown to the right. In line $(*)$ we may choose a, b as we see fit. Thus $a = 0$, $b = r$ is the usual choice, but any $0 \leq a, b \leq r$ with $b+a = r$ will do. The bulk of the spec is lines (t) for $0 \leq t < s$. In line (t') the i' occurs in position t' of the multi-index followed by a 0 in position t. This 0 is replaced by x and i' decreased to i.

$(\perp)$	$F_3 00x$	$= fx$
$(*)$	$F_3 \mathsf{i} r' x$	$= F_s \mathsf{i} a y$
	where y	$= F_s \mathsf{i} b x$
		with $b + a = r$
(0)	$F_3 \mathsf{i}' 0x$	$= F_s \mathsf{i} x x$
(1)	$F_3(\mathsf{i}, i', 0)0x$	$= F_s(\mathsf{i}, i, x)0x$
$\vdots$		
(t')	$F_3(\mathsf{i}, i', 0, 0)0x$	$= F_s(\mathsf{i}, i, x, 0)0x$
$\vdots$		
(s)	$F_3(i', 0, 0)0x$	$= F_s(i, x, 0)0x$

This spec indicates that F_s is s''-recursive over f. It is a nested recursion of the s'' variables $i_s, \ldots, i_0, r$ with x as a parameter.

Let $G_s : \mathbb{M}_s \longrightarrow \mathbb{N}''$ be given by

$$G_s \mathsf{i} = F_s \mathsf{i} 0 = \mathfrak{P}\mathsf{i} ack f \quad \text{so that} \quad F_s \mathsf{i} r x = (G_s \mathsf{i})^{r+1} x$$

for all $\mathsf{i} \in \mathbb{M}_s$ and $r, x \in \mathbb{N}$. If we can specify G_s then

$$F_s\mathsf{i}0x = G_s\mathsf{i}x \qquad F_s\mathsf{i}r'x = G_s\mathsf{i}(F_s\mathsf{i}rx)$$

produces F_s. We can not produce a spec of G_s by setting $r = 0$ in the spec above. The Péter function does produce a spec for G_s, but this is not first order. We return to this in the next Solution. □

9.4 (a) We require

$$G_s\mathsf{i}x = (jmp_0^{i_0} \circ \cdots \circ jmp_s^{i_s})fx = \mathfrak{P}\mathsf{i}jmpfx$$

for $jmp = rob$ and $jmp = brw$. Extending the specs of Solution 7.33 leads to the spec shown right. In clause (00) the upper alternative gives $jmp = rob$ and the lower one gives $jmp = brw$. After the first three clauses ($\perp$, 00, 0+) we have $0 \le t < s$ to give s more clauses.

$(\perp)$	G_s0x	$= fx$
(00)	$G_s\mathsf{i}'0$	$= \begin{cases} G_s\mathsf{i}1 \\ 2 \end{cases}$
$(0+)$	$G_s(\mathsf{i}, \imath)x'$	$= G_s(\mathsf{i}, 1)y$ where $y = G_s(\mathsf{i}, \imath')x$
(1)	$G_s(\mathsf{i}, \imath', 0)x$	$= G_s(\mathsf{i}, \imath, x)x$
$\vdots$		
(t')	$G_s(\mathsf{i}, \imath', 0, 0)x$	$= G_s(\mathsf{i}, \imath, x, 0)x$
$\vdots$		
(s)	$G_s(\imath', 0, 0)x$	$= G_s(\imath, x, 0)x$

(b) For an arbitrary jmp (including ack) the spec of $\mathfrak{P}$ is shown to the right (where $0 \le t < s$ is the range of the index). Here the previous clauses (00, 0+) have been replaced by (step), but this, of course, is not first order. □

(base)	G_s0	$= f$
(step)	$G_s\mathsf{i}'$	$= jmp(G_s\mathsf{i})$
(leap$_1$)	$G_s(\mathsf{i}, \imath', 0)x$	$= G_s(\mathsf{i}, \imath, x)x$
$\vdots$		
(leap$_{t'}$)	$G_s(\mathsf{i}, \imath', 0, 0)x$	$= G_s(\mathsf{i}, \imath, x, 0)x$
$\vdots$		
(leap$_s$)	$G_s(\imath', 0, 0)x$	$= G_s(\imath, x, 0)x$

I.2 The long iterator

9.5 (a) For $g, h : \mathbb{N}'$

$$Gf = g \circ f \qquad Hf = f \circ h \qquad Lfg \circ f \circ h$$

(for $f : \mathbb{N}'$) give the appropriate operators.

(b) Given $h : \mathbb{N}'$ let $e_\bullet = \mathrm{Laur}\,Such\mathsf{zero}$ so that

$$e_0x = 0 \qquad e_{\alpha'}x = 1 + e_\alpha(hx) \qquad e_\mu x = e_{\mu[x]}x$$

hold for the usual α, μ, x. Given $g, l : \mathbb{N}$, let $l_\bullet = \mathrm{Laur}\,Sucghl$ as suggested. We show

$$\langle\alpha\rangle \quad (\forall x : \mathbb{N})[l_\alpha x = (g^{e_\alpha x} \circ l \circ h^{e_\alpha x})x]$$

by induction on α.

Since $l_0 = l$, the base case, $\langle\alpha\rangle$, is trivial.

For the recursion step, $\langle\alpha\rangle \rightarrow \langle\alpha'\rangle$, with $y = hx$ and $z = e_{\alpha'}x = 1 + e_\alpha y$, using the spec of $l_\bullet$, the induction hypothesis, and the spec of $e_\bullet$ we have

$$l_{\alpha'}x = (g \circ l_\alpha \circ h)x = g(l_\alpha y) = g((g^{e_\alpha y} \circ l \circ h^{e_\alpha y})y) = (g^z \circ l \circ h^z)x = (g^{e_{\alpha'}x} \circ l \circ h^{e_{\alpha'}x})x$$

as required.

The recursion leap to a limit ordinal is immediate. □

9.6 From Solution 8.11 we know that

$$\{0\} = \mathsf{zero} \qquad \{\alpha'\} = Suc \circ \{\alpha\} \qquad \{\mu\}x = \{\mu[x]\}x$$

(for the usual α, μ, x) is a recursive spec of the family $(\{\alpha\} \mid \alpha \in \mathbb{O})$ of functions. In other words, $\{\cdot\}$ is just the Slow-Growing hierarchy. □

9.7 We have

$$g_\bullet = \mathrm{Slow}\, g = \mathrm{Laur}\, g\, id_{\mathbb{N}}\mathsf{zero}$$

and hence, by Exercise 9.5,

$$g_\alpha x = (g^y \circ \mathsf{zero} \circ id_{\mathbb{N}}^y)x = g^y 0 \quad \text{where } y = e_\alpha x$$

for a certain family $e_\bullet$ of functions. From that solution we have

$$e_\bullet = \mathrm{Laur}\, Suc\, id_{\mathbb{N}}\mathsf{zero}$$

i.e. $e_\bullet$ is the Slow-Growing hierarchy. From Exercise 9.6 we have $e_\alpha = \{\alpha\}$ and so e_α is an exponential polynomial function, as required. □

9.8 (a) The spec to the right will do. Here, as usual, α is an arbitrary ordinal and μ is a limit ordinal. Also, $r, x \in \mathbb{N}$. The choice of a and b can be fixed as is suitable. You should compare this spec with that of Solution 9.3.

(base0)	$F00$	$= f$
(base+)	$F\alpha r'$	$= F\alpha a \circ F\alpha b$
	where $b + a = r$	
(step)	$F\alpha' 0x$	$= F\alpha xx$
(leap)	$F\mu 0x$	$= F\mu[x]0x$

Using part (b) (to be done next) and Theorem 9.6 we have

$$F(\mathrm{indi})rx = (\mathbb{G}(\mathrm{indi})ack f)^{r+1}x = (\mathfrak{P}iack f)^{r+1}x = F_s\mathsf{i}rx$$

for each $\mathsf{i} \in \mathbb{M}_s$, as required.

(b) We wish to show

$$F\alpha rx = (\mathbb{G}\alpha ack f)^{r+1}x$$

for all $\alpha \in \mathbb{O}$ and $r, x \in \mathbb{N}$.

First of all note that (base0, base+) give

$$F\alpha rx = (F\alpha 0)^{r+1}x$$

by induction on r. Thus (step) gives

$$F\alpha' 0x = F\alpha xx = (F\alpha 0)^{x+1}x = ack(F\alpha 0)x$$

for all $\alpha \in \mathbb{O}$, $x \in \mathbb{N}$. Hence we have

$$\text{(base)} \quad F00 = f \qquad \text{(step)} \quad F\alpha'0 = ack(F\alpha 0) \qquad \text{(leap)} \quad F\mu 0x = F\mu[x]0x$$

for the usual α, μ and $x \in \mathbb{N}$. Comparing this with the spec of $\mathbb{G}$ we obtain

$$(\forall \alpha : \mathbb{O})[F\alpha 0 = \mathbb{G}\alpha ack f]$$

by a routine induction over α.

Thus

$$F\alpha r x = (F\alpha 0)^{r+1}x = (\mathbb{G}\alpha ack f)^{r+1}x$$

as required.

(c) We have

$$F\omega^\omega 0x = F\omega^x 0x = F_x(1,0,\ldots,0)0x = ack_x f x = Pet^x ack f x = O_1 Pet ack f x$$

so that

$$F\omega^\omega 0 = O_1 O_0 ack f = E_1 ack f$$

where $E_1 = O_1 O_0$, as in Section 9.1.

An equality $F\omega^{\omega^\omega}0 = E_2 ack f$ can be shown by a rather long direct calculation, but it is easier to develop some general calculation techniques and then obtain $F\epsilon[r]0 = E_r ack f$ rather quickly for each $r < \omega$. This is done in Section 9.4. ☐

9.9 (a) For arbitrary $f : \mathbb{N}'$ and $jmp : \mathbb{N}''$ consider the spec

$$\text{(base)} \quad G0 = f \qquad \text{(step)} \quad G\alpha' = jmp(G\alpha) \qquad \text{(leap)} \quad G\mu = G \circ \mu[\cdot]$$

for the usual, α, μ.

(b) We have $G\alpha = \mathbb{G}\alpha jmp f$ by a routine induction over α.

(c) It can be shown that that $G\epsilon[r] = E_r jmp f$ for each $r < \omega$. This is best done using certain composition properties of $\mathbb{G}$ to be developed in Section 9.4. ☐

I.3 Limit creation and lifting

9.10 The operators named by the terms to the right will do. Here $\mathsf{I} = \mathsf{I}_\mathcal{N}$ and the prefix and body of each term have been displayed for ease of comparison. ☐

(exp)	$\lambda y : \mathcal{N}'$	$. \mathsf{I}(\lambda x : \mathcal{N} . y^2 x)$
(ack)	$\lambda y : \mathcal{N}', x : \mathcal{N}, u : \mathcal{N}$	$. \mathsf{I}yx(\mathsf{S}u)$
(rob)	$\lambda y : \mathcal{N}', x : \mathcal{N}$	$. \mathsf{I}y(y\ulcorner 1 \urcorner)$
(brw)	$\lambda y : \mathcal{N}'$	$. \mathsf{I}y\ulcorner 2 \urcorner$
(pol)	$\lambda y : \mathcal{N}', x : \mathcal{N}, u : \mathcal{N}$	$. y^2 x$

9.11 (a) The term

$$\delta = \lambda p : \mathcal{N} \to \mathcal{N}', u : \mathcal{N} . puu$$

will do.

(b) For each $r < \omega$ let

$$\delta(r) = \lambda p : (\mathcal{N} \to \mathcal{N}^{(r+1)}) . \lambda v_r : \mathcal{N}^{(r)}, \ldots, v_1 : \mathcal{N}' . \lambda u : \mathcal{N}^{(r)} . puv_r \cdots v_1 u$$

so that $\delta(0) = \delta$. We show that $\delta_r \rhd\!\!\!\rhd \delta(r)$ by induction on r. It suffices to give the induction step.

Since $\vdash \delta : \mathcal{L}(\mathcal{N}')$ we have $\vdash \delta_r : \mathcal{L}(\mathcal{N}^{(r+1)})$ for each r. In particular, with

$$\uparrow_r = \uparrow_{\mathcal{N}^{(r+1)}} = \lambda l : \mathcal{L}(\mathcal{N}^{(r+1)}), q : (\mathcal{N} \to \mathcal{N}^{(r+2)}), x : \mathcal{N}^{(r+1)} . l(\lambda u : \mathcal{N} . qux)$$

we have

$$\begin{array}{rl} \delta_{r'} & = \ \uparrow_r \delta_r \\ & \rhd\!\!\!\rhd \ \lambda q : (\mathcal{N} \to \mathcal{N}^{(r+2)}), x : \mathcal{N}^{(r+1)} . \delta_r(\lambda u : \mathcal{N} . qux) \\ & \rhd\!\!\!\rhd \ \lambda q : (\mathcal{N} \to \mathcal{N}^{(r+2)}), x : \mathcal{N}^{(r+1)} . \delta(r)(\lambda u : \mathcal{N} . qux) \\ & \rhd\!\!\!\rhd \ \lambda q : (\mathcal{N} \to \mathcal{N}^{(r+2)}), x : \mathcal{N}^{(r+1)} . \\ & \quad \lambda v_r : \mathcal{N}^{(r)}, \ldots, v_1 : \mathcal{N}' . \lambda u : \mathcal{N} . (\lambda u : \mathcal{N} . qux)uv_r \cdots v_1 u \\ & \rhd\!\!\!\rhd \ \lambda q : (\mathcal{N} \to \mathcal{N}^{(r+2)}), x : \mathcal{N}^{(r+1)} . \\ & \quad \lambda v_r : \mathcal{N}^{(r)}, \ldots, v_1 : \mathcal{N}' . \lambda u : \mathcal{N} . quxv_r \cdots v_1 u \qquad = \ \delta(r') \end{array}$$

to give the required result (up to alphabetic variance). □

I.4 Parameterized ordinal iterators

9.12 (a) The first three rules are

$$\text{(i)} \quad F^\alpha \circ F^\mu = F^{\mu+\alpha} \qquad \text{(ii)} \quad (F^\mu)^m = F^{\mu m} \qquad \text{(iii)} \quad (F^{\omega^\alpha})^\omega = F^{\omega^{\alpha'}}$$

where $\mu \gg \alpha$ in (i) and $\mu = \omega^\alpha$ in (ii). Here the left and right terms have been swapped round. The fourth rule is a higher order property

$$\text{(iv)} \qquad (\mathbb{G}_{\mathbb{S}}\omega L)^\alpha = \mathbb{G}_{\mathbb{S}}\omega^\alpha L$$

i.e. it shows how to calculate the αth power of an operator.

(b) From (iii) we have $(F^\omega)^\omega = F^{\omega^2}$ and $(F^{\omega^\omega})^\omega = F^{\omega^{\omega+1}}$ but $(F^\omega)^{\omega^\omega}$ is not so immediate. We have

$$(F^\omega)^{\omega^\omega} = ((\mathbb{G}_{\mathbb{S}}\omega^\omega L) \circ (\mathbb{G}_{\mathbb{S}}\omega L))F = (\mathbb{G}_{\mathbb{S}}\omega^{1+\omega} L)F = F^{\omega^{1+\omega}}$$

where the crucial second step is dealt with later in Exercise 9.16. Note that $1+\omega \neq \omega$, for these are *notations* for ordinals, not the ordinals themselves. □

9.13 (a) In fact

$$f^\alpha = \text{Laur}\, f id_{\mathbb{N}} id_{\mathbb{N}} = \mathbb{G}\alpha F id_{\mathbb{N}}$$

where $Fg = f \circ g$ for each $g : \mathbb{N}'$.

(b) The identity

$$(f^\omega)^\omega = f^{\omega^2}$$

does *not* hold in general. To see this let $m = f^\omega 2 = f^2 2$ so that

$$\begin{array}{lllll} (f^\omega)^\omega 2 = (f^\omega)^2 2 & = (f^\omega \circ f^\omega)2 & = (f^\omega \circ f^\omega)2 & = f^\omega m & = f^m m \\ (f^{\omega^2})2 = f^{\omega 2}2 & = f^{\omega+2}2 & = (f^2 \circ f^\omega)2 & = f^2(f^\omega 2) & = f^2 m \end{array}$$

which highlights the difference.

You should worry about this for a while. (If you are not convinced, you should calculate the two values $(S^\omega)^\omega 2$ and $S^{\omega^2}2$ explicitly. Both are small, but different.) With ordinal iteration of functions there are many pitfalls which are not so harmful at higher type levels. This is one of the many reasons why we start the iteration gadgets $\mathbb{G}_\bullet$ at one type level up from where you first think we should. □

9.14 We look at the four properties in turn.

(i) For a fixed limit ordinal μ we verify

$$\mu \gg \alpha \Rightarrow \mathbb{G}_\mathbb{S}(\mu+\alpha)Lt = \mathbb{G}_\mathbb{S}\alpha Lt \circ \mathbb{G}_\mathbb{S}\mu Lt$$

by induction on α with $\mathbb{S}, \mu, L, t$ fixed throughout.

The base case, $\alpha = 0$, is immediate.

For the induction step, $\alpha \mapsto \alpha'$, if $\mu \gg \alpha'$ then $\mu \gg \alpha$ so that

$$\mathbb{G}_\mathbb{S}(\mu+\alpha')Lt = t \circ \mathbb{G}_\mathbb{S}(\mu+\alpha)Lt = t \circ \mathbb{G}_\mathbb{S}\alpha Lt \circ \mathbb{G}_\mathbb{S}\mu Lt = \mathbb{G}_\mathbb{S}\alpha' Lt \circ \mathbb{G}_\mathbb{S}\mu Lt$$

as required. Here the first step uses the spec of ordinal addition, the second step uses the spec of $\mathbb{G}_\mathbb{S}$, the third uses the induction hypothesis, and the fourth uses the associativity of function composition and the spec of $\mathbb{G}_\mathbb{S}$.

For the induction leap to a limit ordinal $\alpha = \nu$, we note that if $\mu \gg \nu$ then both

$$(\mu+\nu)[x] = \mu + \nu[x] \qquad \mu \gg \nu[x]$$

hold for all $x \in \mathbb{N}$. For a given s let

$$s^+ = \mathbb{G}_\mathbb{S}\mu Lts \qquad p = \mathbb{G}_\mathbb{S}(\mu+\nu)[\cdot]Lts$$

so that, for each $x \in \mathbb{N}$,

$$px = \mathbb{G}_\mathbb{S}(\mu+\nu)[x]Lts = \mathbb{G}_\mathbb{S}(\mu+\nu[x])Lts = (\mathbb{G}_\mathbb{S}\nu[x]Lt \circ \mathbb{G}_\mathbb{S}\mu Lt)s = \mathbb{G}_\mathbb{S}\nu[x]Lts^+$$

using the above observations and the induction hypothesis. Thus, using the spec of $\mathbb{G}_\mathbb{S}$ twice, we have

$$\mathbb{G}_\mathbb{S}(\mu+\nu)Lts = Lp = \mathbb{G}_\mathbb{S}\nu Lts^+ = (\mathbb{G}_\mathbb{S}\nu[x]Lt \circ \mathbb{G}_\mathbb{S}\mu Lt)s$$

as required.

(ii) When $\alpha = 0$ we have $\mu = 1$ and the result is known. Otherwise, when $\alpha > 0$, we know that μ is a limit ordinal with $\mu \gg \mu m$ for all m. Then

$$\begin{array}{rcll} \mathbb{G}_\mathbb{S}(\mu m')Lts & = & \mathbb{G}_\mathbb{S}(\mu+\mu m)Lts & \\ & = & (\mathbb{G}_\mathbb{S}(\mu m)Lt \circ \mathbb{G}_\mathbb{S}\mu Lt)s & \\ & = & (\mathbb{G}_\mathbb{S}\mu Lt)^m(\mathbb{G}_\mathbb{S}\mu Lts) & = \quad (\mathbb{G}_\mathbb{S}\mu Lt)^{m'}s \end{array}$$

as required. Here the first step uses some ordinal arithmetic, the second uses property (i), the third uses the induction hypothesis, and the fourth uses a property of natural number iteration.

(iii) Let $t^+ = \mathbb{G}_{\mathbb{S}}\omega^\alpha Lt$ and observe that

$$\mathbb{G}_{\mathbb{S}}\omega Lts = Lq \quad \text{where } qx = \mathbb{G}_{\mathbb{S}}xLt^+s = (t^+)^x s$$

for each $x \in \mathbb{N}$. Similarly

$$\mathbb{G}_{\mathbb{S}}\omega^{\alpha+1}Lts = Lp$$

where

$$px = \mathbb{G}_{\mathbb{S}}(\omega^{\alpha+1})[x]Lts = \mathbb{G}_{\mathbb{S}}(\omega^\alpha x)Lts = (\mathbb{G}_{\mathbb{S}}\omega\alpha Lt)^x s = (t^+)^x s = qx$$

to give $q = p$. Here the third step uses property (ii). With this we have

$$\mathbb{G}_{\mathbb{S}}\omega^{\alpha+1}Lts = Lp = Lq = \mathbb{G}_{\mathbb{S}}\omega Lt^+s = (\mathbb{G}_{\mathbb{S}}\omega L \circ \mathbb{G}_{\mathbb{S}}\omega^\alpha L)ts$$

as required.

(iv) We verify this by induction on α.

The base case, $\alpha = 0$, is a simple calculation.

For the induction step, $\alpha \mapsto \alpha'$, observe that with $\Omega = \mathbb{G}_{\mathbb{S}}\omega L$ the spec of $\mathbb{G}_{\mathbb{S}'}$ gives

$$\mathbb{G}_{\mathbb{S}'}\alpha' L'\Omega = \Omega \circ \mathbb{G}_{\mathbb{S}'}\alpha L'$$

and hence

$$\mathbb{G}_{\mathbb{S}}\omega^{\alpha'} L = \Omega \circ \mathbb{G}_{\mathbb{S}}(\omega^\alpha)L = \Omega \circ (\mathbb{G}_{\mathbb{S}'}\alpha L')\Omega = (\mathbb{G}_{\mathbb{S}'}\alpha' L')\Omega$$

as required. Here the first step uses property (iii), the second uses the induction hypothesis, and the third uses the above observation.

For the leap to a limit ordinal $\alpha = \nu$ we have

$$\begin{array}{rclcrcl} \mathbb{G}_{\mathbb{S}}\omega^\nu Lts & = & Lp & \text{where} & px & = & \mathbb{G}_{\mathbb{S}}\omega^{\nu[x]}Lts \\ \mathbb{G}_{\mathbb{S}'}\nu L'\Omega t & = & L'q & \text{where} & qx & = & \mathbb{G}_{\mathbb{S}'}\nu[x]L'\Omega t \end{array}$$

(for the same Ω). The induction hypothesis gives $px = qxs$ and hence, by the lifting property, we have

$$\mathbb{G}_{\mathbb{S}}\omega^\nu Lts = Lp = L'qs = \mathbb{G}_{\mathbb{S}'}\nu L'\Omega ts$$

as required. □

9.15 (a) For each limit structure $\mathfrak{A} = (\mathbb{A}, a, A, \mathcal{A})$ the induced function is

$$\mathfrak{A}(\cdot) = \mathbb{G}_{\mathbb{A}}(\cdot)\mathcal{A}Aa$$

and every use of $\mathbb{G}_{\mathbb{A}}$ can be viewed in this way.

(b) Given a limit structure $\mathfrak{A}$ we define

$$\mathcal{A}_l : (\mathbb{M}_l \longrightarrow \mathbb{A}) \longrightarrow \mathbb{A}$$

for $l = 0, 1, 2$ by

$$\begin{array}{lcll} \mathcal{A}_0 p_0 & = & \mathcal{A} p_0 & \text{for } p_0 : \mathbb{M}_0 \longrightarrow \mathbb{A} \\ \mathcal{A}_1 p_1 & = & \mathcal{A} p_1(\cdot, 0) & \text{for } p_1 : \mathbb{M}_1 \longrightarrow \mathbb{A} \\ \mathcal{A}_2 p_2 & = & \mathcal{A} p_2(\cdot, 0, 0) & \text{for } p_2 : \mathbb{M}_2 \longrightarrow \mathbb{A} \end{array}$$

to produce a 3-structure with the same induced function. Clearly, not every 3-structure arises in this way.

(c) To fix notation let

$$\mathfrak{C} = (\mathbb{C}, c, C, \mathcal{C}) \quad \mathfrak{B} = (\mathbb{B}, b, B, \mathcal{B}) \quad \mathfrak{A} = (\mathbb{A}, c, C, \mathcal{C})$$

and let $\psi : \mathbb{C} \longrightarrow \mathbb{B}$ and $\phi : \mathbb{B} \longrightarrow \mathbb{A}$ be the given morphisms. Let $\theta = \phi \circ \psi$, so we must show that θ is a morphism. Only the preservation of limit creators is not immediate so we concentrate on that.

We are given

$$\psi \circ \mathcal{C} = \mathcal{B} \bullet \psi \qquad \phi \circ \mathcal{B} = \mathcal{A} \bullet \phi$$

and we must check that

$$\theta \circ \mathcal{C} = \mathcal{A} \bullet \theta$$

i.e. we are given

$$(\psi \circ \mathcal{C})r = \mathcal{B}(\psi \circ r) \qquad (\phi \circ \mathcal{B})q = \mathcal{A}(\phi \circ q)$$

for arbitrary $r : \mathbb{N} \longrightarrow \mathbb{C}$ and $q : \mathbb{N} \longrightarrow \mathbb{B}$ and we must check that

$$(\theta \circ \mathcal{C})r = \mathcal{A}(\theta \circ r)$$

holds. But we have

$$\begin{aligned} (\theta \circ \mathcal{C})r &= (\phi \circ \psi \circ \mathcal{C})r \\ &= \phi((\psi \circ \mathcal{C})r) = \phi(\mathcal{B}(\psi \circ r)) = (\phi \circ \mathcal{B})(\psi \circ = \mathcal{A}(\phi \circ \psi \circ r) \\ &= \mathcal{A}(\theta \circ r) \end{aligned}$$

to give the required result. □

9.16 By Theorem 9.18 we have $\mathbb{G}\alpha = \mathbb{G}_{\mathbb{S}}\alpha L$ where $\mathbb{S} = \mathbb{N}'$ and $L = \Delta$, and then

$$\begin{aligned} \mathbb{G}_{\mathbb{S}}\omega^3 LF &= (\mathbb{G}_{\mathbb{S}'}3L')(\mathbb{G}_{\mathbb{S}}\omega L)F = && O_0^3 F \\ \mathbb{G}_{\mathbb{S}}\omega^{\omega 2} LF &= (\mathbb{G}_{\mathbb{S}'}\omega 2L')(\mathbb{G}_{\mathbb{S}}\omega L)F = && ((\mathbb{G}_{\mathbb{S}'}\omega L')(\mathbb{G}_{\mathbb{S}}\omega L))^2 F = (O_1 O_0)^2 F \\ \mathbb{G}_{\mathbb{S}}\omega^{\omega^2} LF &= (\mathbb{G}_{\mathbb{S}'}\omega^2 L')(\mathbb{G}_{\mathbb{S}}\omega L)F = && (\mathbb{G}_{\mathbb{S}''}2L'')(\mathbb{G}_{\mathbb{S}'}\omega L')(\mathbb{G}_{\mathbb{S}}\omega L)F = \quad O_1^2 O_0 F \\ \mathbb{G}_{\mathbb{S}}\omega^{\omega^\omega} LF &= (\mathbb{G}_{\mathbb{S}'}\omega^\omega L')(\mathbb{G}_{\mathbb{S}}\omega L)F = && (\mathbb{G}_{\mathbb{S}''}\omega L'')(\mathbb{G}_{\mathbb{S}'}\omega L')(\mathbb{G}_{\mathbb{S}}\omega L)F = O_2 O_1 O_0 F \end{aligned}$$

using Lemma 9.17 and Theorem 9.16. Since

$$\alpha = \omega^{\omega^\omega} + \omega^{\omega^2} + \omega^{\omega 2} + \omega^3 + 7$$

is a meshing sum, the first composition property gives

$$\begin{array}{rcl} \mathbb{G}\alpha LF & = & \mathbb{G}_S 7LF \circ \mathbb{G}_S\omega^3 LF \circ \mathbb{G}_S\omega^{\omega 2} LF \circ \mathbb{G}_S\omega^{\omega^\omega} LF \\ & = & F^7 \circ O_0^3 F \circ (O_1 O_0)^2 F \circ (O_2 O_1 O_0) F \end{array}$$

as the final result. □

9.17 (a) With $\Omega = \mathbb{G}_{\mathbb{S}}\omega L$ we have

$$\mathbb{G}_{\mathbb{S}}\omega^{\alpha}L \circ \mathbb{G}_{\mathbb{S}}\omega^{\mu}L = (\mathbb{G}_{\mathbb{S}'}\alpha L')\Omega \circ (\mathbb{G}_{\mathbb{S}'}\mu L)\Omega = \mathbb{G}_{\mathbb{S}'}(\mu+\alpha)L'\Omega = \mathbb{G}_{\mathbb{S}}\omega^{\mu+\alpha}L$$

using the composition properties (iv,i,iv) of Lemma 9.17 in that order.

(b) With $s^{+} = t^{m}s$ we have

$$(\mathbb{G}_{\mathbb{S}}\omega Lt \circ \mathbb{G}_{\mathbb{S}}mLt)s = \mathbb{G}_{\mathbb{S}}\omega Lts^{+} = Lp$$

where

$$pu = \mathbb{G}_{\mathbb{S}}\omega[u]Lts^{+} = \mathbb{G}_{\mathbb{S}}uLts^{+} = t^{u}s^{+} = t^{u}(t^{m}s) = t^{m+u}s = \mathbb{G}_{\mathbb{S}}(m+u)Lts$$

so that $pu = \mathbb{G}_{\mathbb{S}}(m+\omega)[u]Lts$ for each $u \in \mathbb{N}$. Thus

$$\mathbb{G}_{\mathbb{S}}\omega Lt \circ \mathbb{G}_{\mathbb{S}}\mu Lt = \mathbb{G}_{\mathbb{S}}(m+\omega)Lt$$

as required.

(c) With $\Omega = \mathbb{G}_{\mathbb{S}}\omega L$ we have

$$\mathbb{G}_{\mathbb{S}}\omega^{\omega}L \circ \mathbb{G}_{\mathbb{S}}\omega L = (\mathbb{G}_{\mathbb{S}'}\omega L')\Omega \circ (\mathbb{G}_{\mathbb{S}'}1L')\Omega = \mathbb{G}_{\mathbb{S}'}(1+\omega)L'\Omega = \mathbb{G}_{\mathbb{S}}\omega^{1+\omega}L$$

using part (b) at the second step. The third step uses the appropriate generalization of the composition property (iv). □

9.18 We make use of the composition properties as stated in Lemma 9.17 as well as some notational generalizations. We also make use of Exercise 9.17.

(i) Using the fourth composition property we have

$$QP = (\mathbb{G}_{\mathbb{S}'}\omega L')(\mathbb{G}_{\mathbb{S}}\omega L) = \mathbb{G}_{\mathbb{S}}\omega^{\omega}L \qquad RQ = \mathbb{G}_{\mathbb{S}'}\omega^{\omega}L'$$

using the same argument twice. Thus

$$RQP = (\mathbb{G}_{\mathbb{S}'}\omega^{\omega}L')(\mathbb{G}_{\mathbb{S}}\omega L) = \mathbb{G}_{\mathbb{S}}\omega^{\omega^{\omega}}L$$

using the same property.

(ii) We have

$$R^{2} = (\mathbb{G}_{\mathbb{S}''}\omega L'') \circ (\mathbb{G}_{\mathbb{S}''}\omega L'') = \mathbb{G}_{\mathbb{S}''}\omega^{2}L''$$

by the third composition property. Hence

$$\begin{aligned} R^{2}Q &= (\mathbb{G}_{\mathbb{S}''}\omega^{2}L'')(\mathbb{G}_{\mathbb{S}'}\omega L') &= \mathbb{G}_{\mathbb{S}'}\omega^{\omega^{2}}L' \\ R^{2}QP &= (\mathbb{G}_{\mathbb{S}'}\omega^{\omega^{2}}L')(\mathbb{G}_{\mathbb{S}}\omega L) &= \mathbb{G}_{\mathbb{S}}\omega^{\omega^{\omega^{2}}}L \end{aligned}$$

by the fourth composition property.

(iii) Using Exercise 9.17(a)

$$(RQ)^{2} = (\mathbb{G}_{\mathbb{S}'}\omega^{\omega}L') \circ (\mathbb{G}_{\mathbb{S}'}\omega^{\omega}L') = \mathbb{G}_{\mathbb{S}'}\omega^{\omega+\omega}L' = \mathbb{G}_{\mathbb{S}'}\omega^{\omega 2}L'$$

and hence

$$(RQ)^{2}P = (\mathbb{G}_{\mathbb{S}'}\omega^{\omega 2}L')(\mathbb{G}_{\mathbb{S}}\omega L) = \mathbb{G}_{\mathbb{S}}\omega^{\omega^{\omega 2}}L$$

by the fourth composition property.

(iv) Using Exercise 9.17(a) we have

$$(RQP)^2 = (\mathbb{G}_{\mathbb{S}}\omega^{\omega^\omega}L) \circ (\mathbb{G}_{\mathbb{S}}\omega^{\omega^\omega}L) = \mathbb{G}_{\mathbb{S}}\omega^{\omega^\omega 2}L$$

for the required result. Notice how this differs from (iii).

(v) We have

$$(RQ)(QP) = (\mathbb{G}_{\mathbb{S}'}\omega^\omega L')(\mathbb{G}_{\mathbb{S}}\omega^\omega L) = \mathbb{G}_{\mathbb{S}}(\omega^\omega)^{\omega^\omega}L$$

using a notational extension of the fourth composition property. We may set

$$\omega \times \omega^\omega = \omega^{1+\omega}$$

where $1+\omega$ is as in 9.12(b). This gives

$$(RQ)(QP) = \mathbb{G}_{\mathbb{S}}\omega^{\omega^{1+\omega}}L$$

as the result.

We can also obtain this by a different method. Thus

$$\begin{aligned} (RQ)(QP) &= (\mathbb{G}_{\mathbb{S}'}\omega^\omega L')((\mathbb{G}_{\mathbb{S}'}\omega L')(\mathbb{G}_{\mathbb{S}}\omega L)) \\ &= ((\mathbb{G}_{\mathbb{S}'}\omega^\omega L') \circ (\mathbb{G}_{\mathbb{S}'}\omega L'))(\mathbb{G}_{\mathbb{S}}\omega L)) \\ &= (\mathbb{G}_{\mathbb{S}'}\omega^{1+\omega}L')(\mathbb{G}_{\mathbb{S}}\omega L) \qquad = \mathbb{G}_{\mathbb{S}}\omega^{\omega^{1+\omega}}L \end{aligned}$$

where the third step is a use of Exercise 9.17(c).

(vi) A notational extension of the fourth composition property gives

$$RQ^2 = (\mathbb{G}_{\mathbb{S}''}\omega L'')(\mathbb{G}_{\mathbb{S}'}\omega^2 L') = \mathbb{G}_{\mathbb{S}'}(\omega^2)^\omega L'$$

which is not too helpful. However, $Q^2 = (\mathbb{G}_{\mathbb{S}''}2L'')(\mathbb{G}_{\mathbb{S}'}\omega^2 L')$ so that

$$RQ^2 = ((\mathbb{G}_{\mathbb{S}''}\omega L'') \circ (\mathbb{G}_{\mathbb{S}''}2L''))(\mathbb{G}_{\mathbb{S}'}\omega L') = (\mathbb{G}_{\mathbb{S}''}(2+\omega)L'')(\mathbb{G}_{\mathbb{S}'}\omega L') = \mathbb{G}_{\mathbb{S}'}\omega^{2+\omega}L'$$

using Exercise 9.16(b) and a notational extension of the fourth composition property. Thus

$$RQ^2P = (\mathbb{G}_{\mathbb{S}'}\omega^{2+\omega}L')(\mathbb{G}_{\mathbb{S}}\omega L) = \mathbb{G}_{\mathbb{S}}\omega^{\omega^{2+\omega}}L$$

for the result.

(vii) Mimicking (vi) we have

$$RQP^2 = ((\mathbb{G}_{\mathbb{S}'}\omega^\omega L') \circ (\mathbb{G}_{\mathbb{S}'}2L'))(\mathbb{G}_{\mathbb{S}}\omega L) = (\mathbb{G}_{\mathbb{S}'}(2+\omega^\omega)L')(\mathbb{G}_{\mathbb{S}}\omega L) = \mathbb{G}_{\mathbb{S}}\omega^{2+\omega^\omega}L$$

where $\omega^{2+\omega^\omega}$ is a notation for a non-standard fundamental sequence for ω^ω.

(viii) Since $R : \mathbb{S}^{iv}$ and $(QP) : \mathbb{S}''$, this compound is meaningless.

We have expressed each compound in the form $\mathbb{G}_{\mathbb{S}}\omega^\alpha L$ with

(i) ω^ω (ii) ω^{ω^2} (iii) $\omega^{\omega 2}$ (iv) $\omega^\omega 2$ (v) $\omega^{1+\omega}$ (vi) $\omega^{2+\omega}$ (vii) $2+\omega^\omega$

as the exponents. Making a standard ordinal comparison we have

$$\text{(i)} < \text{(iv)} < \text{(iii)} < \text{(ii)}$$

and it seems reasonable that

$$\omega < 1+\omega < 2+\omega < \omega+\omega = \omega 2 \qquad \omega^\omega < 2+\omega^\omega < \omega^\omega + \omega^\omega = \omega^\omega 2 < \omega^{\omega 2}$$

in terms of strength. Thus we obtain

$$\text{(i)} < \left\{ \begin{matrix} \text{(v)} & < & \text{(vi)} \\ \text{(vii)} & < & \text{(iv)} \end{matrix} \right\} < \text{(iii)} < \text{(ii)}$$

as two chains of comparisons. It is plausible to suggest $2 + \omega^\omega < \omega^{1+\omega}$ to give (vii) < (iv) and hence we obtain

$$\text{(i)} < \text{(vii)} < \text{(v)} < \left\{ \begin{matrix} \text{(vi)} \\ \text{(iv)} \end{matrix} \right\} < \text{(iii)} < \text{(ii)}$$

with the comparison between

$$\text{(iv)} \;\; \omega^{\omega^{\omega 2}} \qquad \text{(vi)} \;\; \omega^{\omega^{2+\omega}}$$

harder to resolve (if it can be at all). Thus we have

$$RQP < RQP^2 < (RQ)(QP) < \left\{ \begin{matrix} (RQP)^2 \\ RQ^2P \end{matrix} \right\} < (RQ)^2P < R^2QP$$

as the comparisons of strength.

It is clear from these calculations that there is an arithmetic of ordinal notations which is similar to, but not the same as, the arithmetic of ordinals. It is beyond the scope of this book to investigate this arithmetic. □

I.5 How to name ordinal iterates

9.19 For each type σ and ordinal α we have a context $\Sigma = l : \mathcal{L}(\sigma), y : \sigma', x : \sigma$ and body term A_σ. We show

$$\langle\alpha\rangle \quad \text{For each type } \sigma,\ \Sigma \vdash A_\sigma : \sigma$$

by induction on the canonical notation for α. With this the required derivation $\vdash \alpha_\sigma : \mathcal{L}(\sigma) \to \sigma''$ follows by three Introductions. We use this consequence in some of the induction steps.

Following the construction of the notation, there are two base cases $\alpha = 0$ and $\alpha = \omega$, and three induction steps $\alpha \mapsto \alpha'$, $\alpha \mapsto \omega^\alpha$, $(\zeta, \eta) \mapsto \zeta + \eta$ which we look at in turn.

(0) Since the body is just x, the required derivation is a Projection.

(ω) The shape of the derivation is shown to the right.

$$\dfrac{\Sigma\vdash l:\mathcal{L}(\sigma) \qquad \dfrac{\dfrac{\dfrac{\dfrac{\Sigma,u:\mathcal{N}\vdash \mathsf{I}_\sigma : - \quad \Sigma,u:\mathcal{N}\vdash y:\sigma'}{\bullet} \quad \Sigma,u:\mathcal{N}\vdash x:\sigma}{\bullet}\quad \Sigma,u:\mathcal{N}\vdash u:\mathcal{N}}{\Sigma,u:\mathcal{N}\vdash \mathsf{I}_\sigma yxu:\sigma}}{\Sigma\vdash(-):\mathcal{N}\to\sigma}}{\Sigma\vdash A_\sigma:\sigma}$$

($\alpha \mapsto \alpha'$) The full derivation is shown right. The induction hypothesis is used at the top left. This is followed by a Weakening, and abnormality near the bottom right is intended.

$$\dfrac{\Sigma\vdash y:\sigma' \qquad \dfrac{\dfrac{\dfrac{\dfrac{\begin{array}{c}\vdots\\ \vdash \alpha_\sigma:\mathcal{L}(\sigma)\to\sigma''\end{array}}{\Sigma\vdash \alpha_\sigma:\mathcal{L}(\sigma)\to\sigma''}\quad \Sigma\vdash l:\mathcal{L}(\sigma)}{\bullet}\quad \Sigma\vdash y:\sigma'}{\bullet}\quad \Sigma\vdash x:\sigma}{\Sigma\vdash \alpha_\sigma lyx:\sigma}}{\Sigma\vdash A'_\sigma:\sigma}$$

($\alpha \mapsto \omega^\alpha$) The appropriate derivation is shown below. To save space some predicates are not given in full.

$$\dfrac{\dfrac{\dfrac{\dfrac{\dfrac{\begin{array}{c}\vdots\\ \vdash \alpha_{\sigma'}:\mathcal{L}(\sigma')\to\sigma'''\end{array}}{\Sigma\vdash \alpha_{\sigma'}:\mathcal{L}(\sigma')\to\sigma'''}\quad \dfrac{\dfrac{\begin{array}{c}\vdots\\ \vdash\uparrow_\sigma: -\end{array}}{\Sigma\vdash\uparrow_\sigma: -}\quad \Sigma\vdash l: -}{\Sigma\vdash l':\mathcal{L}(\sigma')}}{\Sigma\vdash(\alpha_{\sigma'}l'):\sigma'''}\quad \dfrac{\dfrac{\begin{array}{c}\vdots\\ \vdash\omega_\sigma: -\end{array}}{\Sigma\vdash\omega_\sigma: -}\quad \Sigma\vdash l: -}{\Sigma\vdash(\omega_\sigma l):\sigma''}}{\Sigma\vdash(\alpha_{\sigma'}l')(\omega_\sigma l):\sigma''}\quad \Sigma\vdash y:\sigma'}{\bullet}\quad \Sigma\vdash x:\sigma}{\Sigma\vdash A_\sigma:\sigma}$$

The induction hypothesis is used top left with a variation of the parameter σ.

($(\zeta,\eta)\mapsto\zeta+\eta$) Here

$$\dfrac{\dfrac{\dfrac{\dfrac{\begin{array}{c}\vdots\\ \vdash \eta_\sigma:\mathcal{L}(\sigma)\to\sigma''\end{array}}{\Sigma\vdash \eta_\sigma:\mathcal{L}(\sigma)\to\sigma''}\quad \Sigma\vdash l:\mathcal{L}(\sigma)}{\bullet}\quad \Sigma\vdash y:\sigma'}{\bullet}\qquad \dfrac{\dfrac{\dfrac{\begin{array}{c}\vdots\\ \vdash \zeta_\sigma:\mathcal{L}(\sigma)\to\sigma''\end{array}\quad \Sigma\vdash l:\mathcal{L}(\sigma)}{\bullet}\quad \Sigma\vdash y:\sigma'}{\bullet}\quad \Sigma\vdash x:\sigma}{\bullet}}{\Sigma\vdash A_\sigma:\sigma}$$

is the required derivation. There are two uses of the induction hypothesis. □

9.20 We proceed by induction on r with variation of σ.

Since $\epsilon[0]_\sigma = \omega_\sigma$, the base case, $r = 0$, is immediate.

For the induction step, $r \mapsto r'$, we have

$$\begin{array}{rcl} \epsilon[r']_\sigma & = & (\omega^{\epsilon[r]})_\sigma \\ & = & \boldsymbol{\lambda}_\sigma l, y, x\,.\,(\epsilon[r]_{\sigma'} l')(\omega_\sigma l)yx \\ & \rhd\!\!\!\rhd & \boldsymbol{\lambda}_\sigma l, y, x\,.\,\big((\boldsymbol{\lambda}_{\sigma'} l, y, x\,.\,E[\sigma', r, l]yx)l'\big)(\omega_\sigma l)yx \\ & \rhd\!\!\!\rhd & \boldsymbol{\lambda}_\sigma l, y, x\,.\,E[\sigma', r, l']\omega[\sigma, l]yx \qquad = \quad \boldsymbol{\lambda}_\sigma l, y, x\,.\,E[\sigma, r', l]yx \end{array}$$

as required. Here the third step uses the induction hypothesis, the fourth uses a careful β-reduction, and the fifth uses the construction of $E[\cdot,\cdot,\cdot]$. □

9.21 This is a rather tedious induction over the ordinal α viewed in Cantor normal form. There are no iterators in 0_σ, and there is one iterator in ω_σ. The use of (Step) or (Mesh) does not introduce any new iterators, but (Mesh) does increase the number of occurrences. Each use of (Exp) (to form $(\omega^\alpha)_\sigma$) requires a new iterator in $\alpha_{\sigma'}$. □

I.6 THE GODS

9.22 Let us consider the case $B_s = F_s$. The case $B_s = G_s$ is similar (and simpler).

For the case $f = Suc$, from Exercise 9.3 and Theorem 9.6 we have

$$F_s \mathsf{i} r = ((ack_0^{i_0} \circ \cdots \circ ack_s^{i_s})f)^{r+1} = (\mathfrak{P}\mathsf{i} ack f)^{r+1} = (\mathbb{G}\alpha f)^{r+1} = S_\alpha^{r+1}$$

where $\alpha = \mathrm{ind}\,\mathsf{i}$ for $\mathsf{i} \in \mathbb{M}_s$, $r \in \mathbb{N}$. Here, however, α is one of the arguments to the function, not a fixed index.

For each $\mathsf{i} \in \mathbb{M}_s$ we have $\alpha = \mathrm{ind}\,\mathsf{i} < \omega^{s+1}$ so that F_s is eventually dominated by $S_{\omega^{s+1}}$. With $\alpha = \omega^s x = \omega^{s+1}[x]$ and i with $\mathrm{ind}\,\mathsf{i} = \alpha$ we have

$$F_s \mathsf{i} 0 x = S_\alpha x = S_{\omega^{s+1}[x]} x = S_{\omega^{s+1}} x$$

so that $S_{\omega^{s+1}}$ is a particular section of F_s. Thus, in terms of complexity, F_s and $S_{\omega^{s+1}}$ are equivalent. □

9.23 (a) Proceeding by induction we show that

$$f_{\omega\alpha} = ack^\alpha f$$

holds for all $\alpha < \omega^\omega$.

The base case, $\alpha = 0$, is immediate.

For the induction step, $\alpha \mapsto \alpha + 1$, note first that the induction hypothesis gives

$$f_{\omega\alpha + r} = (ack^\alpha f)^{r+1}$$

(by a simple induction on r). This is because $|\omega\alpha + r| = \omega\alpha$. Thus, since we know $\omega(\alpha+1) = \omega\alpha + \omega$ and this is a meshing sum, we have

$$\omega(\alpha+1)[x] = \omega\alpha + x$$

so that

$$f_{\omega(\alpha+1)}x = f_{\omega\alpha+x}x = (ack^{\alpha}f)^{x+1}x = ack(ack^{\alpha}f)x = ack^{\alpha+1}fx$$

as required.

For the induction leap to a limit ordinal $\mu < \omega^{\omega}$ first note that

$$(\omega\mu)[x] = \omega\mu[x]$$

holds. (This follows by considering the decomposition of μ. You should do the calculation, for the identity is not true for some larger ordinals μ.) Thus

$$f_{\omega\mu}x = f_{(\omega\mu)[x]}x = f_{\omega\mu[x]}x = ack^{\mu[x]}fx = ack^{\mu}fx$$

as required.

(b) This tight relationship breaks down. The root cause of this is the identity $\omega\omega^{\omega} = \omega^{\omega}$. Calculate

$$f_{\omega\omega^{\omega}}x \qquad ack^{\omega^{\omega}}fx$$

for $x = 3$. □

POSTVIEW

The nine development chapters have taken you from the beginnings of propositional calculus to the classification of number theoretic functions. The pivotal idea is that of an applied λ-calculus, a simple kind of type system. Within these bounds a few topics have been looked at in some depth, but others have been merely touched on, and some have not even been mentioned. In these final pages I will make some comments on the available literature. This should help you understand more fully the material developed here, and fill in some of the missing details. I will also give a few pointers to what topics you might want to look at next.

1. DERIVATION SYSTEMS

This chapter introduces you to the judgemental style of derivation system for a small part of propositional logic. There is much more that could be done here, even within this restricted area. (There is easily enough for a whole book.) What is missing?

- Other connectives
- Derivation systems in the style of Gentzen
- Extension to classical logic
- Semantic aspects

Each of these topics can be studied independently of the others (but, of course, can not be fully understood in such isolation).

Unfortunately I know of no acceptable account of this material which does justice to the breadth and depth of this topic. Most accounts concentrate on just one or two systems, and rarely consider the interactions between different systems. The best general reference I can offer you is [22]. This is now a bit dated, but still worth reading.

There are, of course, other texts which discuss propositional calculus from a type theoretic perspective. The two works [13] and [26] form a nice contrast to the approach here.

All this is at the propositional level. Nowhere in this book do we look at any quantificational structures. This is a serious omission which, in the grand scheme of things, must be rectified.

The monograph [22] considers quantification rules. The book [11] contains the most comprehensive account of proof theory at this level. It is not easy going, but you should be aware of this material.

2. COMPUTATION MECHANISMS

This chapter is concerned with material that is much better documented. Almost anything you want to know about these aspects of combinator and λ-calculi can be found in [3]. The book [14] is a nice introduction to this material.

As Exercise 2.1 suggests, much of this topic, especially those aspects not covered here, is concerned with Term Rewriting Systems. A comprehensive account of this algorithmic idea is given in [17].

3. The Typed Combinator Calculus

This Chapter introduces you to the most primitive type system **C**. In itself this system is not very important, but it does provide a nice entry into more sophisticated systems. Any knowledge of combinators will help here, but you should think more in terms of the next chapter.

4. The Typed λ-calculus

The system $\boldsymbol{\lambda}$ is the first 'non-trivial' type system. It is found at the heart of almost every other type system. There are several texts that will help here. Of course [3] is useful. The article [5] by the same author is specifically about typed systems. The book [14] is well worth looking at. These references are not solely concerned with $\boldsymbol{\lambda}$, but the book [13] is. Furthermore, that book looks at things in a slightly different way, so you should delve into it. You will also find the survey paper [10] helpful.

5. Substitution Algorithms

Substitution is not often treated in such detail as here. More often than not a version of the algorithm on page 83 is used, and then terms are treated 'up to alphabetic variants'. For instance, you might like to look at the accounts in [14] and [13].

The algorithm of Section 5.3 is based on the paper [24]. This kind of substitution is now used in many other calculi. It is often called 'subtitution in context' or 'explicit substitution', but I prefer 'suspended substitution'. There is a plethora of λ-calculi which incorporate this kind of algorithm. These have names like $\lambda_\sigma, \lambda_\nu, \lambda_*, \ldots$ and not all have equal value.

6. Applied λ-calculi

This is the central chapter of the whole book. Chapters 1–5 are there merely to introduce this one. Chapters 7–9 provide a substantial example of the general material of this chapter.

There is much that is new here. The idea of treating a derivation as an object of interest in its own right is not usually developed in this depth. Doing so leads to the need for the arboreal code of a derivation, and the use of mutations (i.e. a substitution mechanism for arboreal code).

The whole approach in this book has been strongly influenced by the style of a Pure Type System. This is a comparativly recent notion, and is described in [4] and [5]. The idea grew out out of earlier work by de Bruijn who developed methods of formalizing mathematics in a mechanistic fashion. The book [20] discusses all aspects of this work.

Nowhere in this book do I describe in any detail the way we obtain a semantics of an applied λ-calculus. This, of course, is a serious omission. A discussion of these matters is given in [7].

That book shows how λ-calculi are related to some quite unexpected areas of mathematics; areas not usually associated with logic. The semantic approach to higher order systems has been studied quite intensively over the past 20 years or so. The best introduction to these ideas is [19]. The book [2] has a nice mix of semantic and syntactic aspects of λ-calculi, including some of the higher systems not covered here.

7. Multi-recursive Arithmetic

The subject matter of this chapter is a part of what is often called Recursive Function Theory. However, it is that part which is not often given the attention it deserves. The only two books that I know which consider this material in a serious way are [21] and [23]. The first of these should be read by anyone interested in this topic. It is a bit old fashioned, and can be hard going, but it contains a wealth of information.

8. Ordinals and Ordinal Notations

Ordinals (and 'ordinal notations') as set theoretic objects are well documented. However, ordinal notations as syntactic objects have received very little attention.

9. Higher Order Recursion

This final chapter is about the system called Gödel's T. To be precise, it is about the term calculus of Gödel's T, for the full system includes some equational reasoning. It has been known for many years that this system (with or without the equational reasoning) characterizes the class of what is often called the provably recursive functions, but the finer details are not often written down. An account of this is given in pages 98–108 of [23]. This outlines the connection between recursive complexity and ordinal height. A standard account of Gödel's T occurs in [11] from page 442 onwards. The paragraph beginning at the bottom of page 455 is particularly interesting. Neither of these accounts makes explicit use of the type structure and the λ-calculus.

The mechanism for naming ordinals in $\boldsymbol{\lambda G}$ is not often explained. The account here was greatly influenced by [16]. This paper is well worth reading as a general survey of the whole subject matter of this book.

And finally

As I mentioned in the Preview this book was inspired by [12]. By now you should have started reading that book. If you haven't do so, then start immediately.

BIBILIOGRAPHY

[1] S. Abramsky, D. Gabbay, and T. Maibaum, editors. *Handbook of Logic in Computer Science,* vol. 2. Oxford University Press, 1992. (Volumes 1, 3, and 4 not cited).

[2] A. Asperti and G. Longo. *Categories, Types, and Structure.* MIT Press, 1991.

[3] H. P. Barendregt. *The Lambda Calculus. Its syntax and semantics.* North Holland, 1981.

[4] H. P. Barendregt. Introduction to generalized type systems. *Journal of Functional Programming,* 1, 1991. (pp. 125–154).

[5] H. P. Barendregt. Lambda calculi with types. In *Handbook of Logic in Computer Science.* Oxford University Press, 1992. (pp. 117 – 309 of [1]).

[6] S. R. Buss, editor. *Handbook of Proof Theory.* North Holland, 1998.

[7] R. L. Crole. *Categories for Types.* Cambridge University Press, 1993.

[8] H. B. Enderton. *Elements of Set Theory.* Academic Press, 1977.

[9] M. Fairtlough and S. S. Wainer. Hierarchies of provably recursive functions. In *Handbook of Proof Theory.* North Holland, 1998. (pp. 149–207 of [6]).

[10] J. H. Gallier. Constructive logics. part 1: a tutorial on proof systems and typed λ-calculi. *Theoretical Computer Science,* 110, 1993. (pp. 247–339).

[11] J-Y. Girard. *Proof Theory and Logical Complexity,* vol. 1. Bibliopolis, 1987.

[12] J-Y. Girard, Y. Lafont, and P. Taylor. *Proofs and Types.* Cambridge University Press, 1996.

[13] J. R. Hindley. *Basic Simple Type Theory.* Cambridge University Press, 1997.

[14] J. R. Hindley and J. P. Seldin. *Introduction to Combinators and λ-calculus.* Cambridge University Press, 1986.

[15] G. Huet, editor. *Logical Foundations of Functional Programming.* Addison-Wesley, 1990.

[16] G. Huet. A uniform approach to types theory. In *Logical Foundations of Functional Programming.* Addison-Wesley, 1990. (pp. 337–397 of [15]).

[17] J. W. Klop. Term rewriting systems. In *Handbook of Logic in Computer Science.* Oxford University Press, 1992. (pp. 1–116 of [1]).

[18] K. Kuratowski and A. Mostowski. *Set Theory.* North Holland, 1968.

[19] C. McLarty. *Elementary Categories, Elementary Toposes.* Oxford University Press, 1995.

[20] R. P. Nederpelt, J. H. Geuvers, and R. C. de Vrijer. *Selected Papers on Automath.* North Holland, 1994.

[21] R. Péter. *Recursion Functions.* Academic Press, 1967.

[22] D. Prawitz. *Natural Deduction. A Proof-Theoretical Study.* Almqvist and Wiksell, 1965.

[23] H. E. Rose. *Subrecursion: Functions and Hierarchies.* Oxford University Press, 1984.

[24] A. Stoughton. Substitution revisited. *Theoretical Computer Science*, 59, 1988. (pp. 317–325).

[25] G. Takeuti and W. M. Zaring. *Introduction to Axiomatic Set Theory.* Springer, 1971.

[26] A. S. Troelstra and H. Schwichtenberg. *Basic Proof Theory.* Cambridge University Press, 1996.

Commonly Used Symbols

Some symbols used in this book have a unique meaning. Most of these are listed at the beginning of the Index. Other symbols are most commonly used to indicate a particular kind of gadget, and these are listed below. However, on occasions some of these symbols are used in a different way.

Types or formulas

The lower case greek letters

$$\alpha, \beta, \gamma, \delta, \epsilon, \zeta, \eta, \theta, \kappa, \lambda, \mu, \nu, \xi, \pi, \rho, \sigma, \tau, \phi, \chi, \psi, \omega$$

usually indicate types. Some of these (for instance β, γ, δ) often indicate a type of special shape. The later upper case roman letters

$$W, X, Y, Z$$

are used for type variables (primitive types). The symbols λ, ϵ also have a quite different meaning. The first, λ, is used as a binding symbol, and the second, ϵ, is used to indicate type erasure.

Greek letters are also used for ordinals.

Terms

Lower case roman letters, especially

$$p, q, r, s, t$$

usually indicate terms. Sometimes these are decorated with a subscript or superscript. The later lower case roman letters

$$w, x, y, z$$

are used as identifiers.

Context or hypothesis lists

These are indicated by certain upper case Greek letters such as

$$\Gamma, \Delta, \Lambda, \Xi, \Pi, \Sigma$$

and occasionally others.

Derivations

When viewed as trees or arboreal code these are indicated by

$$\nabla, P, Q, R, S, T, X, Y, Z$$

where ∇ is thought of as the 'generic' gadget.

Computations

When viewed as trees or arboreal code these are indicated by

$$\square, \mathsf{p}, \mathsf{q}, \mathsf{r}, \mathsf{s}$$

where $\square$ is thought of as the 'generic' gadget.

Combinators

Sans serif upper case letters

$$\mathsf{A}, \mathsf{B}, \mathsf{C}, \mathsf{D}, \mathsf{E}, \mathsf{F}, \mathsf{G}, \ldots$$

usually indicate combinators (in C or in $\boldsymbol{\lambda}$). Many of these have a fixed shape. However, the two letters H, N refer to particular derivation systems.

Substitution gadgets

Mutations are indicated by upper case gothic letters and the corresponding replacements by lower case gothic letters.

A	B	C	D	I	J	K	P	R	S	T	X	Y	Z
$\mathfrak{A}$	$\mathfrak{B}$	$\mathfrak{C}$	$\mathfrak{D}$	$\mathfrak{I}$	$\mathfrak{J}$	$\mathfrak{K}$	$\mathfrak{P}$	$\mathfrak{R}$	$\mathfrak{S}$	$\mathfrak{T}$	$\mathfrak{X}$	$\mathfrak{y}$	$\mathfrak{Z}$
$\mathfrak{a}$	$\mathfrak{b}$	$\mathfrak{c}$	$\mathfrak{d}$	$\mathfrak{i}$	$\mathfrak{j}$	$\mathfrak{k}$		$\mathfrak{r}$	$\mathfrak{s}$	$\mathfrak{t}$	$\mathfrak{x}$	$\mathfrak{y}$	$\mathfrak{z}$

For convenience the equivalent upper case roman letter is given. The symbols $\mathfrak{I}, \mathfrak{i}$ are the identity gadgets.

Concrete sets

As always $\mathbb{N}$ is the natural numbers. Other blackboard bold letters

$$\mathbb{R}, \mathbb{S}, \mathbb{T}, \ldots$$

indicate arbitrary sets.

Index